Additive Manufacturing

The book bridges the gap between theoretical knowledge and practical application, encouraging the integration of Industry 4.0 technologies, such as Artificial Intelligence (AI) and Machine Learning (ML), across diverse applications.

Features:

- Discusses the fundamentals of Additive Manufacturing (AM), materials selection, and future industrial applications.
- Highlights the integration of AM with advanced technologies such as Artificial Intelligence (AI) and Machine Learning (ML), as well as the importance of Industry 4.0 elements in smart manufacturing.
- Delves into the role of Reverse Engineering (RE) in healthcare for the development of customised medical devices and implants, as well as for future bioprinting applications.
- Explores the transformative role of AM in space exploration, spacecraft systems, and mission payloads.
- Covers concrete 3D printing for sustainable construction and discusses emerging 4D and 5D printing technologies that leverage smart materials, multi-axis control, and digital intelligence to enable adaptive, high-performance structures.

The book is primarily intended for polytechnic students, undergraduates, graduate students, medical students, and academic researchers in manufacturing engineering, mechanical engineering, industrial engineering, production engineering, aerospace engineering, automotive engineering, and biomedical engineering.

Additive Manufacturing

New Trends and Applications

Edited by
Mallikarjuna B., Deepak D. Patil, and
Santosh Kumar Mishra

CRC Press
Taylor & Francis Group
Boca Raton London New York

CRC Press is an imprint of the
Taylor & Francis Group, an **informa** business

Designed cover image: © Shutterstock Images

First edition published 2027
by CRC Press
2385 NW Executive Center Drive, Suite 320, Boca Raton FL 33431

and by CRC Press
4 Park Square, Milton Park, Abingdon, Oxon, OX14 4RN

CRC Press is an imprint of Taylor & Francis Group, LLC

ISBN: 978-1-032-96809-4 (hbk)
ISBN: 978-1-032-96810-0 (pbk)
ISBN: 978-1-003-59077-4 (ebk)

DOI: 10.1201/9781003590774

Typeset in Sabon
by Apex CoVantage, LLC

Contents

Contents

About the Editors

Mallikarjuna B. is currently working as Assistant Professor in the Department of Production Engineering, National Institute of Technology Tiruchirappalli (NITT), Tamil Nadu, India. Before he joined NITT, he served as Assistant Professor from August 2019 to October 2022 in the Mechanical Engineering department at Amrita Vishwa Vidyapeetham, Coimbatore, India. He received a Ph.D. in 2020 from the National Institute of Technology Karnataka, Surathkal, Mangaluru. During his doctoral research in the Additive manufacturing laboratory at NITK, he has published several peer-reviewed journals, book chapters, full papers, and abstract presentations. His team won first prize with a cash prize of Rs. 40,000 for the design and development of a concrete 3D printer and recently received a best paper award in an international conference at SRM University. He has filed an Indian patent on a 3D-printed soldering tool. He has conducted SERB-sponsored Karyashala and Vritika workshops. So far, he has presented and participated in 50 conferences, workshops, and webinars. He has given more than 20 invited lectures in various parts of the country. He has also worked as an application engineer at Cycloid System, Bengaluru. He has rich experience in designing and developing material extrusion-based Fused Deposition Modelling (FDM) printers and rapid manufacturing of polymers, including ABS, TPU, and PLA.

Deepak D. Patil is Assistant Professor in the School of Mechanical and Materials Engineering at IIT, Mandi. Earlier, he was Assistant Professor at the Department of Production Engineering at the National Institute of Technology (NIT) Tiruchirappalli, India. Prior to joining NIT Trichy, he was Postdoctoral Fellow at the Indian Institute of Science (IISc), Bangalore. He received his Ph.D. from the Indian Institute of Technology (IIT) Delhi, where his doctoral research on micro-nano manufacturing for biomedical applications was awarded a Distinction. He is a Gold Medalist at the Master's level from NIT Calicut and was the First Rank holder in his Bachelor's programme. Dr. Patil's research specialisation lies at the intersection of micro-nano manufacturing, precision engineering, and functional biomaterials. His current work focuses on fabricating bioinspired and mechanobactericidal surfaces, developing microfluidic devices, and advanced surface modifications for biomedical and sensory applications. An active contributor to the field, Dr. Patil has authored several high-impact journal publications in journals like the *Journal of Colloid and Interface Science*, *ACS Macro Letters*, and *Surfaces and Interfaces*. He holds published and filed Indian patents related to superhydrophobic surfaces and 3D-printed scaffolds. He has contributed book chapters to esteemed publishers, including CRC Press (on Focused Ion Beam

Machining). He serves as a reviewer for international journals such as *RSC Nanoscale* and is a life member of professional bodies, including the *Japan Society for Precision Engineering (JSPE)* and the *Indian Welding Society (IWS)*.

Santosh Kumar Mishra is currently serving as an Assistant Professor in the Department of Production Engineering at NIT, Trichy. He received his Ph.D. degree in Fluid Power Control from IIT Dhanbad in 2018. He received his B.Tech in Production Engineering from NIT Trichy in 2009. He has more than 10 years of experience in the design and development of fluid power systems. His current research interests include oil hydraulics, hydrostatic transmission systems, fault diagnosis, robotics, solar power systems with nano-materials, and additive manufacturing. He has more than 60 research publications comprising articles in both national and international journals, international and national conferences, and book chapters in esteemed internationally published books. He has also published nine patents. He is reviewer of many international journals. He has also given many invited talks at universities and workshops, and has been a resource person for faculty development programmes.

Contributors

Vedant Agarwal
Department of Mechanical Engineering
Indian Institute of Technology (IIT) Bombay
Mumbai, Maharashtra, India

Anupam Agrawal
Department of Materials
University of Manchester
Manchester, M13 9PL, UK

Dhanunjay Kumar Ammisetti
Department of Mechanical Engineering
Lakireddy Bali Reddy College of Engineering
Vijayawada, Andhra Pradesh, India.

Bhargavi Rani Anne
Department of Metallurgical and Materials Engineering
National Institute of Technology Raipur
Amanaka, Raipur, Chhattisgarh, India

Mallikarjuna B.
Department of Production Engineering
National Institute of Technology Tiruchirappalli
Tamil Nadu, India

Biswajoy Bhattacharjee
Department of Production Engineering
National Institute of Technology Agartala
India

Mikael Collan
LUT Business School
Yliopistonkatu
Lappeenranta, Finland

Vikas Diwakar
Department of Mechanical Engineering
Indian Institute of Technology (BHU)
Varanasi, India

Kartikeya Dixit
Department of Mechanical and Manufacturing Engineering
University of Calgary
Calgary, Alberta, Canada

Ashutosh Kumar Dubey
Department of Ceramic Engineering
Indian Institute of Technology (BHU)
Varanasi, Uttar Pradesh, India

Suresh Goka
Department of Mechanical Engineering
National Institute of Technology
Warangal, Telangana, India

Gopal Gote
Department of Mechanical Engineering
Indian Institute of Technology (IIT) Bombay
Mumbai, Maharashtra, India

Ri Han
Department of Mechanical Engineering
National University of Singapore
Singapore

Anupama Rajendra Hiremath
Manipal Institute of Technology
Manipal Academy of Higher Education
Manipal, Udupi, India

Shivashankarayya Hiremath
Manipal Institute of Technology
Manipal Academy of Higher Education
Manipal, Udupi, India

Monaf Hodhod
Department of Mechanical Engineering
College of Engineering, King Faisal University, Al-Ahsa, Saudi Arabia

Arunkumar Jayakumar
Department of Electrical and Electronics Engineering
St. Peter's Institute of Higher Education and Research,
Avadi, Tamil Nadu, India

Pushkar Kamble
Department of Mechanical Engineering
Birla Institute of Technology and Science
Pilani—Goa, India

T Karthik
Department of Computer Applications
National Institute of Technology Tiruchirappalli
Tamil Nadu, India

K. P. Karunakaran
Department of Mechanical Engineering
Indian Institute of Technology (IIT) Bombay
Mumbai, Maharashtra, India

Brahmansh Kaushik
Department of Mechanical Engineering
Northwestern University, Evanston, IL, USA

Keekyoung Kim
Department of Biomedical Engineering
University of Calgary
Calgary, Alberta, Canada

Prathamesh Rajan Korgaonkar
Department of Mechanical Engineering
Northwestern University, Evanston, IL, USA

Deepak Kumar
Department of Mechanical Engineering
National Institute of Technology Jamshedpur
Jharkhand, India

Hitendra Kumar
Department of Biosciences and Biomedical Engineering
Indian Institute of Technology Indore
Indore, Madhya Pradesh, India

Rakesh Kumar
Advanced Manufacturing Technology Lab
Department of Mechanical Engineering
Indian Institute of Technology Ropar
Rupnagar, Punjab, India

S Anand Kumar
Additive Manufacturing Research Laboratory
Indian Institute of Technology Jammu
Jammu, Jammu and Kashmir, India

Santosh Kumar
Department of Mechanical Engineering
Indian Institute of Technology (BHU)
Varanasi, Uttar Pradesh, India

Paras Mandal
Department of Mechanical Engineering
Indian Institute of Technology (IIT) Bombay
Mumbai, Maharashtra, India

Avinash Kumar Mehta
Department of Mechanical Engineering
Indian Institute of Technology (IIT) Bombay
Mumbai, Maharashtra, India

Santosh Kumar Mishra
Department of Production Engineering
National Institute of Technology Tiruchirappalli
Tamil Nadu, India

Syed Quadir Moinuddin
Department of Mechanical Engineering
College of Engineering,
King Faisal University
Al-Ahsa, Saudi Arabia

Yash Gopal Mittal
Department of Mechanical Engineering
Indian Institute of Technology (IIT) Bombay
Mumbai, Maharashtra, India

Shubhra Kamal Nandi
Department of Metallurgical and Materials Engineering
Indian Institute of Technology Kharagpur
West Bengal, Kharagpur, India

Akshay Pathania
Additive Manufacturing Research Laboratory
Indian Institute of Technology Jammu
Jammu, Jammu and Kashmir, India

Yogesh Patil
Department of Operations and Supply Chain Management
Indian Institute of Management (IIM)
Mumbai, Maharashtra, India

Madhukrishna Priyadarsini
Department of Computer Science and Engineering,
National Institute of Technology
Raipur, India

Vinay Raj
Department of Computer Applications
National Institute of Technology Tiruchirappalli

Mohammad Abdul Razack
Department of Mechanical Engineering
School of Technology
GITAM University, Andhra Pradesh, India

B Sharath Chander Reddy
Technology and Research
Rockwell Automation, Cleveland, USA

Mumtaz Rizwee
Department of Mechanical Engineering
National Institute of Technology Jamshedpur
Jharkhand, India

Prakash Kumar Sahu
Department of Production Engineering
National Institute of Technology Agartala
India

Shadi Samemaleki
LUT Business School
Yliopistonkatu
Lappeenranta, Finland

Jyrki Savolainen
LUT Business School
Yliopistonkatu
Lappeenranta, Finland

Shajahan Shaik
Department of Chemistry and Green-nano Materials Research Center
Kyungpook National University
Buk-gu, Daegu, South Korea.

Peijie Shangguan
Department of Mechanical Engineering
National University of Singapore
Singapore

Rahul Shukla
Department of Mechanical Engineering
Institute of Engineering & Technology
Bundelkhand University
Jhansi, Uttar Pradesh, India

Swee Leong Sing
Department of Mechanical Engineering
National University of Singapore
Singapore

Kunal Singh
Nokha AI Labs
Bikaner, Rajasthan, India

Niraj Sinha
Department of Mechanical Engineering
Indian Institute of Technology Kanpur
Kanpur, Uttar Pradesh, India

VVSS Srikanth
School of Engineering Sciences and Technology
University of Hyderabad
Gachibowli, Hyderabad, Telangana, India

Vadavalasa Harsha Vardhan
Department of Mechanical Engineering
Gandhi Institute for Technology
Bhubaneswar, Odisha, India

Meghanshu Vashista
Department of Mechanical Engineering
Indian Institute of Technology (BHU)
Varanasi, India

Govind Kumar Verma
Department of Mechanical Engineering
Indian Institute of Technology (BHU)
Varanasi, Uttar Pradesh, India

Dixita Arvind Yadav
Department of Mechanical Engineering
Indian Institute of Technology (IIT) Bombay
Mumbai, Maharashtra, India

Haoran Yin
Department of Mechanical Engineering
National University of Singapore
Singapore

Mohd Zaheer Khan Yusufzai
Department of Mechanical Engineering
Indian Institute of Technology (BHU)
Varanasi, India

Chapter 1

Introduction to Additive Manufacturing, Artificial Intelligence, and Machine Learning

Vinay Raj, T Karthik, B Sharath Chander Reddy, and Mallikarjuna B

1.1 HISTORY OF ADDITIVE MANUFACTURING

In the early 1980s, 3D printing, or rapid prototyping, emerged in Japan. Hideo Kodama invented 3D printing machines and filed a patent (JP S56–144478) in 1981 [1]. Hideo Kodama of the Nagoya Municipal Industrial Research Institute developed two innovative techniques for fabricating three-dimensional plastic prototypes. It was a thermoset polymer under UV exposure to build the 3D prototypes. He also published the work based on automated fabrication of 3D components in 1981 [2]. In a pivotal advancement for Additive Manufacturing (AM), Chuck Hull invented Stereolithography (SLA) and filed its patent (1984) [3]. His critical role in developing the Standard Tessellation Language (STL) file format made it the universal standard for 3D printing. Then Chuck Hull commercialised SLA technology by founding 3D Systems in 1986, launching the world's first commercial 3D printer, the SLA-1, in 1987 [4]. Now, numerous vat photopolymerisation techniques are available, such as Digital Light Processing (DLP), liquid crystal display/Masked SLA (LCD/SLA), continuous liquid interface production (CLIP), and two-photon polymerisation (2PP/TPP). All these techniques operate on the principle of light-curing a liquid resin; they differ significantly in their approach from point-by-point laser tracing (SLA) to full-layer projection (DLP, LCD) and even volumetric curing (CLIP) or nanoscale fabrication (2PP) [5, 6]. At that time, AM was called rapid prototyping because it was used solely for prototyping. Then, S. Scott Crump developed Fused Deposition Modelling (FDM) based on material extrusion in 1988. This patent was with Stratasys until 2009. Later, the technique flourished worldwide. Currently, FDM is one of the most commonly available techniques, costing around ₹10,000. Later, the companies have developed different FDM machines for printing various materials, including MakerBot (USA), 3D Systems (USA, Fabbster (Germany), Envision Tec (Germany), Aleph Objects (USA), Stratasys (USA-Israel), Wipro 3D (India), and Mingda (China). FDM techniques include Fused Filament Fabrication (FFF), FDM, Robocasting, and Fused Granular Fabrication (FGP). It can process a wide variety of polymers, elastomers, composites, and, recently, metals [7].

Dr. Carl Deckard developed the selective laser sintering technique in the mid-1980s under the guidance of Dr. Joe Beaman at the University of Texas at Austin [8]. This results in the formation of Desktop Manufacturing (DTM) Corp to design and develop Selective Laser Sintering (SLS) machines. The patent was filed in 1979 by R. F. Housholder but never became commercialised [9]. Subsequently, powder bed fusion (PBF) technologies emerged for the fabrication of metallic materials, alloys, and composites. In PBF processes, a thermal energy source, such as a laser or an electron beam, selectively melts or fuses specific regions of a powder bed, enabling the layer-by-layer construction of complex components. The unfused powder supports the part during the build, enabling the creation of highly complex geometries that would be impossible with traditional manufacturing methods. The technologies

DOI: 10.1201/9781003590774-1

include direct metal laser sintering (DMLS), Selective Laser Melting (SLM), multi-jet fusion (MJF), and Electron Beam Melting (EBM) [10]. The origins of modern metal AM can be traced to laser cladding technology developed in the 1970s–1980s. This foundational work later evolved into Directed Energy Deposition (DED) processes, giving rise to techniques such as Laser Engineered Net Shaping (LENS) (Inception around 1995 at Sandia National Laboratories, USA), Direct Metal Deposition (DMD) by the POM Group, and Electron Beam Freeform Fabrication. DED techniques focus thermal energy (laser/electron beam/plasma arc) to melt material as it is being deposited. The state of the material should be metal powder or wire, and it can process a wide variety of materials such as titanium alloys (Ti6Al4V, Ti48Al2Cr2Nb, etc.), nickel-based superalloys, stainless steel, cobalt chromium, aluminium, and metal matrix composites (TiC or TiB2 into a TiAl matrix) [11]. More recently, Wire Arc Additive Manufacturing (WAAM) has gained prominence by leveraging existing welding processes (MIG, TIG, Plasma) with wire feedstock for large-scale component fabrication. Further, the integration of DED with CNC machining has advanced the field through Hybrid Manufacturing, which enables additive construction and subsequent precision machining of a part within a single automated setup [11].

Another technology, Material Jetting, emerged in the 1990s, adapting inkjet printing principles to deposit UV-cured photopolymer droplets [12]. This process enables the creation of high-resolution, multi-material, and full-colour prototypes, making it ideal for detailed visual models, medical simulators, and casting patterns. Leading technologies in this category include Stratasys' PolyJet, 3D Systems' MultiJet Printing (MJP), and XJet's NanoParticle Jetting [13]. Binder Jetting, pioneered at Massachusetts Institute of Technology (MIT) in the early 1990s, employs a print head to selectively deposit a liquid binding agent onto a polymer, sand, or metal powder bed [14]. This process enables rapid production of large-scale models and sand-casting moulds, with growing industrial relevance for the creation of end-use metal parts through sintering. Prominent companies advancing this technology include ExOne (now part of Desktop Metal), HP with its Metal Jet technology, Voxeljet, and Digital Metal [15]. ASTM's last AM category is Sheet Lamination, also developed in the 1990s, which builds parts by bonding layers of material such as paper, metal, or polymer using adhesives or ultrasonic welding. Excess material is then trimmed away using a laser or blade [16]. Notable techniques include Ultrasonic Additive Manufacturing (UAM) and Laminated Object Manufacturing (LOM). Beyond these established methods, the AM landscape continues to evolve as novel technologies emerge [17]. **Figure 1.1** shows the the evolution of 3D printing technology.

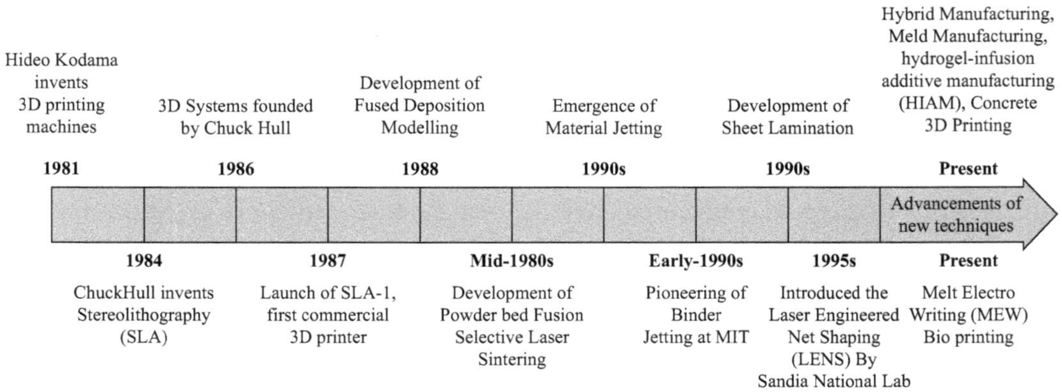

Figure 1.1 The evolution of 3D printing technology.

1.2 INTRODUCTION TO ADDITIVE MANUFACTURING

AM, also known as 3D printing (3DP), is a process of fabricating three-dimensional objects. Earlier, geographically, AM was called rapid prototyping, DTM, solid freeform fabrication, digital manufacturing, etc. Hence, in 2009, the ASTM F42 committee established definitions and classifications for AM. The definition of AM, or 3DP, is that it is a process of joining materials to make parts from 3D model data, usually layer by layer, as opposed to subtractive and formative manufacturing methodologies.

Also, ASTM classified AM into seven categories: vat photopolymerisation, material extrusion, PBF, DED, binder jetting, material jetting, and sheet lamination, as shown in **Figure 1.2** [18–21]. However, advancements in current 3D printing techniques are evolving, such as concrete 3D printing, melt electrowriting (MEW), bioprinting, melt manufacturing, hybrid manufacturing, and WAAM [22, 23].

1.2.1 VAT Photopolymerisation

It is a process of curing liquid photopolymer resin in a vat using UV light. SLA and DLP are popular technologies that provide high surface polish and detail for jewellery, dentistry, and micro-parts, but are limited to photosensitive materials [24].

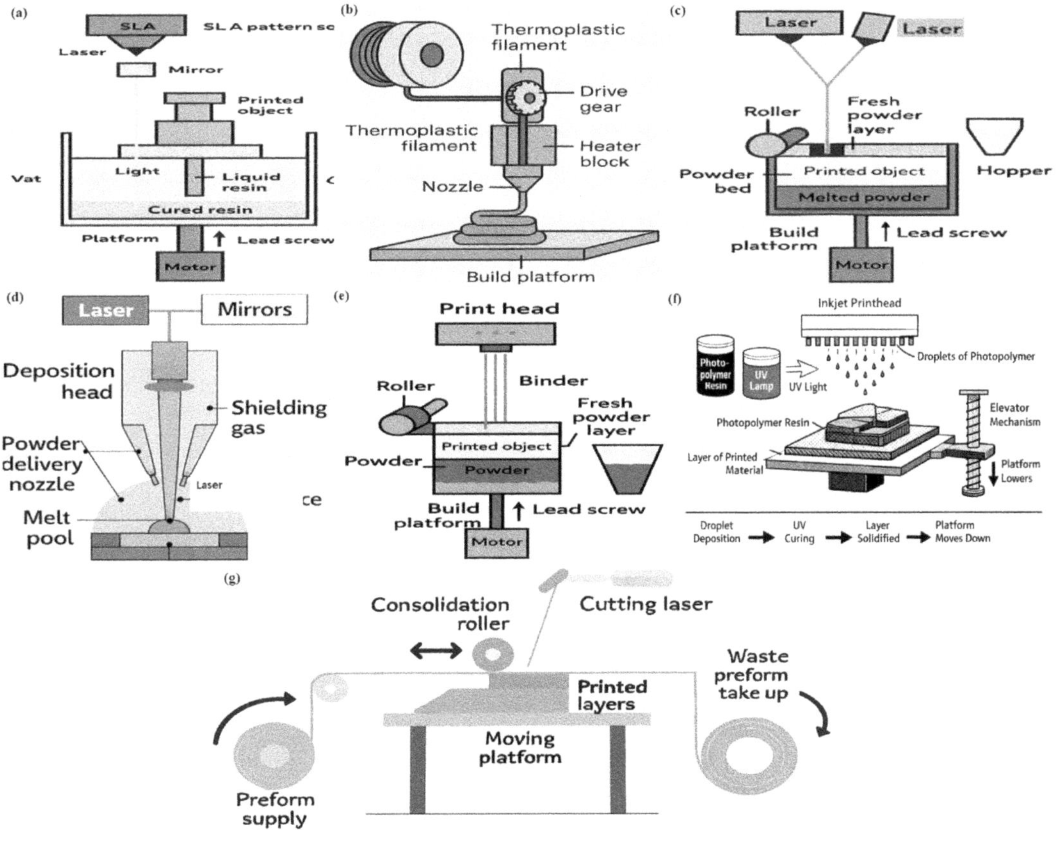

Figure 1.2 ASTM classifications of seven additive manufacturing techniques.

1.2.2 Material Extrusion

The most accessible AM approach, as shown by FDM, extruded, involves heating and extruding thermoplastic filaments via a nozzle. It is commonly used for prototyping because of its low cost and ease of use, although its anisotropy and lower resolution limit its use [25].

1.2.3 Powder Bed Fusion (PBF)

It uses a laser or electron beam to selectively melt powder layers. Polymer subtypes include SLS, whereas metals use SLM or EBM. PBF excels at making dense, functional parts but has high costs and powder management issues [26].

1.2.4 Directed Energy Deposition (DED)

It uses directed thermal energy, such as a laser or electron beam, to melt and fuse materials as they are deposited. Commonly used for fixing high-value components or adding functionality to existing ones, it allows multi-material deposition, though the resulting surfaces may be rough and require machining [27].

1.2.5 Binder Jetting

Selectively deposits a liquid binding agent onto a powder bed to connect particles. This is perfect for sand moulds and full-colour prototypes. Metals, ceramics, and polymers are all examples of materials. The advantages include rapid speed and the absence of support structures; nevertheless, post-processing, such as sintering, is often necessary for strength [28].

1.2.6 Material Jetting

Similar to inkjet printing, droplets of photopolymer or wax are deposited and cured using UV light. High resolution and multi-material capabilities make it ideal for intricate models and medical applications, albeit materials may be fragile [29].

1.2.7 Sheet Lamination

It is the process of bonding sheets of material (paper, metal, or plastic) together using adhesives or ultrasonic welding and then cutting them to form. It is cost-effective for big pieces, but trimming generates waste. Although less common today, this method has found niche applications in rapid prototyping and the fabrication of composite parts [30].

1.2.8 Hybrid Manufacturing

Hybrid manufacturing in 3D printing is an integrated process that combines additive and subtractive techniques within a single automated workstation [31]. The component is first built up layer-by-layer through material deposition until its near-net shape is complete. Following this, the same machine utilises precision machining tools to finish the part, achieving the required dimensional accuracy and surface quality. This technique offers significant advantages over standalone 3D printing methods by eliminating the need to transfer a part between separate machines, thereby reducing both production time and overall costs. A prominent category within this field employs DED for the additive phase, which is then followed by computerised milling. Industry-leading systems that utilise this method include the Lasertec series by DMG Mori and the Integrex i-AM series by Mazak. Furthermore, another hybrid approach integrates material extrusion, such as FDM, with CNC milling to finish critical features. Commercial examples of this technology are available from companies like Desktop Metal and CEAD.

1.3 OTHER ADDITIVE MANUFACTURING TECHNIQUES

These techniques often fall outside the standard ASTM classifications or are hybrid processes that combine AM with other principles. **Figure 1.3** shows the detailed breakdown of the other emerging techniques.

1.3.1 Melt Electrospinning Writing (MEW)

A high-voltage electric field draws a jet of molten polymer from a nozzle. Unlike regular FDM, the electrostatic drawing allows for the creation of incredibly thin fibres (down to the micron scale) with precise placement [32]. A computer-controlled stage collects these fibres to build a structure layer by layer. The key feature is MEW's ultra-fine resolution (sub-micron to single-digit microns) and the ability to create highly porous, delicate 3D scaffolds that mimic the natural extracellular matrix. Primary Application is creating scaffolds for cell growth in areas like nerve regeneration, blood vessel development, and skin grafts.

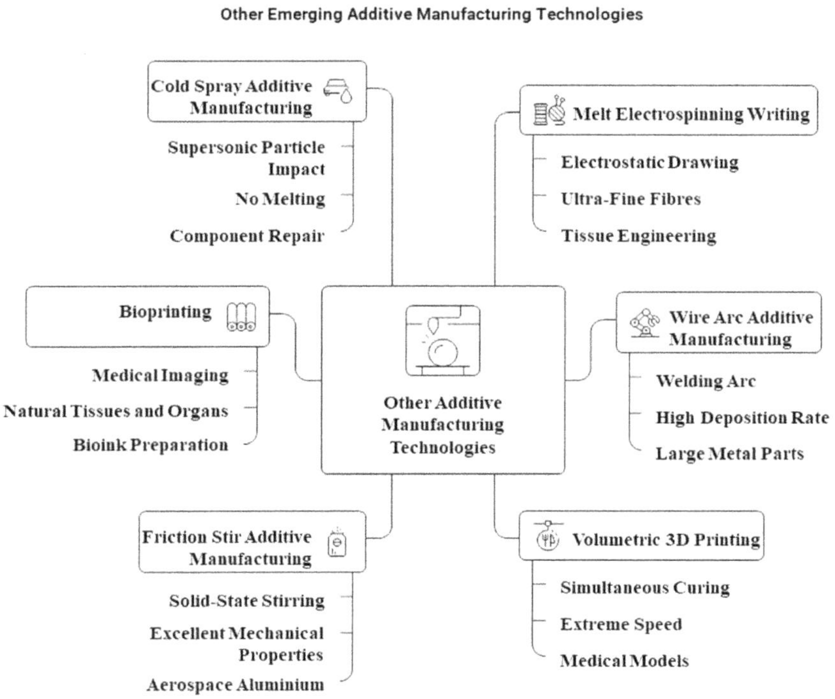

Figure 1.3 Other emerging additive manufacturing technologies.

1.3.2 Wire Arc Additive Manufacturing (WAAM)

In the WAAM process, a welding power supply that generates an electric arc is used to melt metal wire feedstock and deposit material layer-by-layer (e.g., MIG, TIG, or Plasma Arc). It is a form of DED. A robotic arm melts the wire and deposits it layer by layer onto a substrate. It has high deposition rates (several kgs per hour) compared to powder-based metal AM. It is excellent for building very large, near-net-shape metal parts (meters in size). The primary application is repairing and rebuilding ship propellers, aircraft landing gear components, and creating large prototypes for the aerospace and maritime industries. It's cost-effective for low-volume, large parts [33].

1.3.3 Volumetric 3D Printing

This is a radical departure from layer-by-layer printing. The object is created all at once inside a rotating volume of transparent photopolymer resin. Multiple lasers from different angles are projected into the resin vial. The lasers are precisely calibrated so that they only intersect at specific points within the volume, curing the entire 3D shape simultaneously in a matter of seconds. It has excellent speed (prints in tens of seconds) and geometric freedom without the need for support structures. Its primary focus is on research and development stages but holds promise for medical applications (printing soft tissue-like structures) and rapid prototyping of complex, delicate objects [34].

1.3.4 Friction Stir Additive Manufacturing (FSAM)

It is a hybrid solid-state process. A rotating, non-consumable tool with a specially designed pin and shoulder is plunged into multiple layers of metal plates or feedstock. Friction and pressure plasticise the metal without melting it, stirring it together to form a solid, bonded block. This block can then be machined into a final part. FSAM is a solid-state process, which means no melting, resulting in excellent mechanical properties, fine-grained microstructures, and no residual stresses associated with melting and solidification. It is used in manufacturing large, high-strength aluminium components for the aerospace industry (e.g., wing spars, ribs), where traditional forging would be too expensive or wasteful [35].

1.3.5 Cold Spray Additive Manufacturing (CSAM)

It is a solid-state process. Metal powder particles are accelerated to supersonic speeds (via a heated gas) and aimed at a substrate. Upon impact, the particles undergo plastic deformation and bond mechanically and metallurgically, building up a shape. CSAM builds parts without melting, resulting in very low thermal stress, the ability to mix dissimilar metals, and the repair of temperature-sensitive components. It's used for repairing high-value components (e.g., helicopter gearboxes, aircraft skins), adding features to existing parts, and manufacturing corrosion-resistant coatings [36]. **Table 1.1** presents a comparison of emerging AM technologies.

1.4 ADDITIVE MANUFACTURING GENERIC PROCESS CHAIN FOR THE BUILDING OF COMPONENTS

AM follows the generic component-building process chain. As we know, the AM process starts with the CAD model of the product or part. The conversion of a digital CAD model into a physical object through AM follows a structured sequence of eight steps [38], as shown in **Figure 1.4**.

Table 1.1 Comparison of Emerging AM Technologies [37]

Technology	*Acronym*	*Principle*	*Key Advantage*	*Primary Application*
Melt Electrospinning Writing	MEW	Electrostatic drawing of molten polymer	Ultra-fine fibres, high porosity	Tissue engineering scaffolds
Wire Arc Additive Manufacturing	WAAM	Welding arc melts the wire feedstock	Very high deposition rate, large parts	Large-scale metal industrial parts
Volumetric 3D Printing	-	Simultaneous curing in a resin volume	Extreme speed, no supports	Medical models, rapid prototyping
Friction Stir Additive Manufacturing	FSAM	Solid-state stirring and bonding	Excellent mechanical properties	High-strength aerospace aluminium
Electron Beam Additive Manufacturing	EBAM	Electron beam melts wire in a vacuum	High purity, high deposition	Large titanium aerospace structures
Cold Spray Additive Manufacturing	CSAM	Supersonic particle impact bonding	No melting, repair capability	Component repair, dissimilar metals

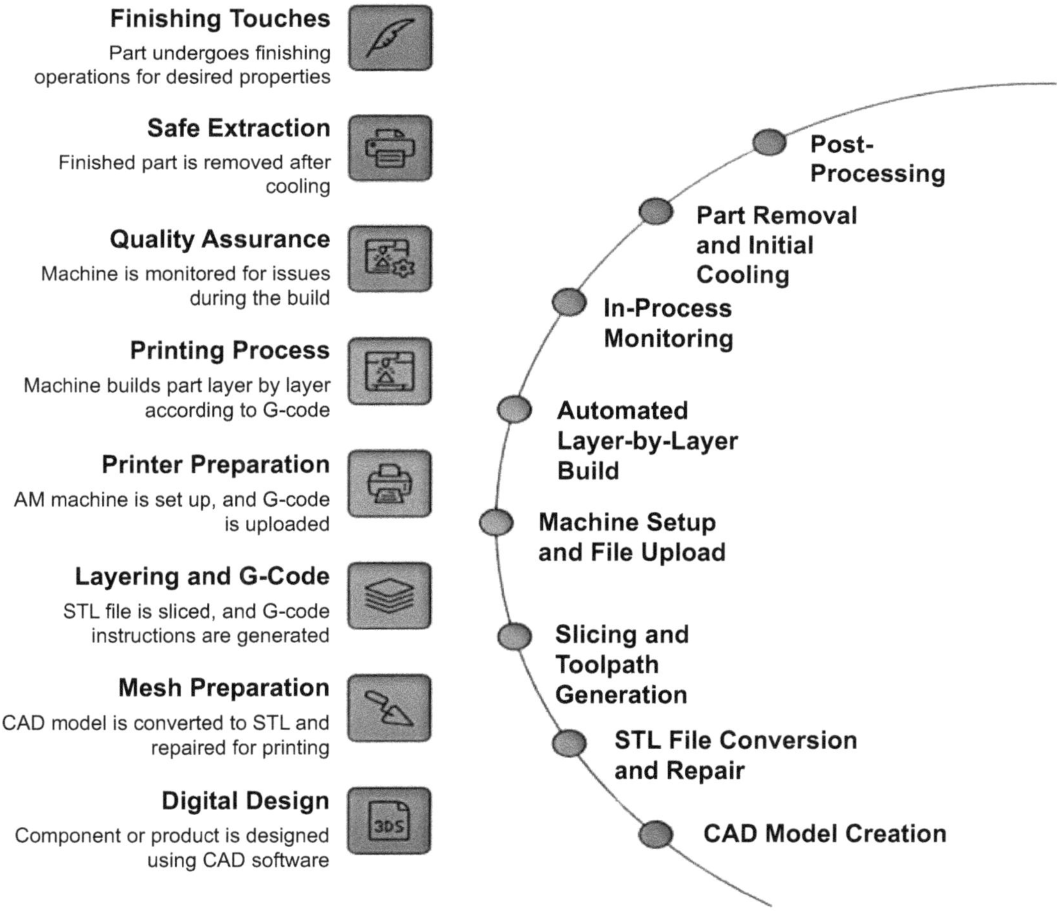

Figure 1.4 AM process chain illustrates the transition from CAD to physical parts.

1.4.1 CAD Model Creation

The process begins with designing the component or product using Computer-Aided Design (CAD) software (e.g., CATIA, SolidWorks). The output is a native file format specific to that software (e.g., CAT Part for CATIA). However, this file must be converted to the STL file format. Then it is used in the slicing software for tool path generation.

1.4.2 STL File Conversion and Repair

The CAD model is exported into a universal format understood by AM systems, typically the STL file. This file approximates the model's surfaces using a triangular mesh. This STL file is then checked and repaired using specialised software to correct errors such as missing facets, holes, or overlapping geometry that could disrupt the printing process. In addition, recently, slicing software has become capable of accepting the advanced file formats, including AMF, 3MF, and OBJ.

1.4.3 Slicing and Toolpath Generation

The repaired STL file is imported into a "slicer" software. Here, the 3D model is digitally sliced into thin horizontal layers. The user defines critical process parameters (e.g., layer height, print speed, build orientation), and the software generates the machine instructions, known as G-code, which dictate the toolpath for the machine's movements.

1.4.4 Machine Setup and File Upload

The physical AM machine is prepared, which includes loading the build platform and the feedstock material (filament, powder, or resin). The G-code file is then transferred to the AM machine's controller.

1.4.5 Automated Layer-by-Layer Build

The machine executes G-code instructions, building the component layer by layer using the specific AM process (e.g., material extrusion, vat photopolymerisation, PBF).

1.4.6 In-Process Monitoring

Throughout the build cycle, the machine is monitored for any issues such as material jams, power loss, or platform misalignment to ensure a successful print.

1.4.7 Part Removal and Initial Cooling

After deposition is complete, the build chamber is often allowed to cool to a safe handling temperature. The finished part is then carefully removed from the build platform using appropriate tools while adhering to safety protocols.

1.4.8 Post-Processing

The "printed" part often requires finishing operations. These can include removal of support structures, sanding or polishing to improve surface finish, machining for dimensional accuracy, and thermal treatments (e.g., stress relief, annealing) to achieve the desired mechanical properties.

1.5 MATERIALS AND PROCESS PARAMETERS

In AM, the material and the process parameters are inextricably linked. Optimising these parameters is essential for achieving the desired density, mechanical properties, dimensional accuracy, and surface finish for a specific application. This optimisation is often performed using Design of Experiments (DoE) to establish a reliable "process window" for each material [39]. So far, materials processed using various AM techniques spanning polymers, metals, ceramics, composites, and emerging smart or biomaterials are shown in **Figure 1.5** [40, 41, 42].

The effectiveness of each AM technology is critically dependent on a specific set of process parameters that directly control the quality, properties, and integrity of the final part. **Figure 1.6** shows the process parameters of seven AM techniques. In FDM/FFF, which processes thermoplastics such as PLA, ABS, and Nylon, the outcome is primarily determined by thermal and mechanical settings. The nozzle temperature must be optimised to melt the filament for smooth extrusion, while the build plate temperature ensures proper adhesion and prevents warping. The layer height dictates the trade-off between surface roughness and print speed, and the print speed influences layer bonding and dimensional accuracy [43]. For example, FDM with ABS, setting the nozzle temperature to 230°C, the build plate to 110°C, the layer height to 0.1 mm, and the print speed to 50 mm/s, promotes better filament fusion, reducing inter-layer voids. This leads to tensile strength of about 45 MPa, approximately 6% elongation, and a Shore D hardness of about 80, where higher temperatures and controlled cooling increase strength and fatigue resistance [44, 45]. In SLA/DLP and other vat polymerisation techniques using photopolymer resins, the curing process is key. Layer height determines the Z-axis resolution, while laser power (in SLA) and exposure time (in DLP) control the degree of polymerisation and the dimensional accuracy of each cured layer. The orientation of the part during printing is crucial to minimise suction forces, support structures, and the "stair-stepping" effect on curved surfaces. SLS for polymers like Nylon (PA 11/12) and TPU relies on precisely fusing powder particles. The laser power and scan speed work in tandem to deliver the optimal energy density to melt the powder without degradation. The bed temperature is maintained just below the melting point to facilitate sintering and reduce thermal distortion, while the layer thickness affects both the resolution and the build time. For metal PBF techniques like SLM/DMLS, used for alloys such as Ti-6Al-4V and Inconel, the parameter interplay is complex. Laser power, scan speed, and hatch spacing collectively define the volumetric energy density, which controls melt pool stability, porosity, and mechanical properties. The scan strategy (the pattern the laser follows) is vital for managing residual stresses and controlling the microstructure of the solidified metal. For instance, SLM of Ti-6Al-4V alloy demonstrates that carefully optimised parameters, such as laser power (200 W), scan speed (1200 mm/s), hatch spacing (0.1 mm), and layer thickness (30 μm), yield a fine martensitic microstructure with very low porosity. This results in tensile strength around 1100 MPa, elongation near 12%, and hardness of about 370 HV. The use of specific scan strategies manages residual stress and, combined with post-process heat treatment, enhances ductility and toughness [46, 47]. EBM, which also processes conductive metals like Ti-6Al-4V in a vacuum, uses a different energy source. Here, the beam current and scan speed define the energy input. A key differentiator is the requirement for a high preheat temperature of the powder bed, which sinters the surrounding material, reduces residual stresses, and enables the construction of crack-prone materials. The focus offset adjusts the beam's spot size to control the energy density. EBM of Ti-6Al-4V uses a beam current of 20 mA, a scan speed of 4000 mm/s, and a powder bed preheat at 700°C to generate a coarser but highly dense microstructure with over 99.5% density. This yields tensile strength near 990 MPa, elongation about 15%, and hardness around 350 HV, with the high preheat helping reduce residual stresses and cracking [48–51]. In DED processes such as LENS, where material is

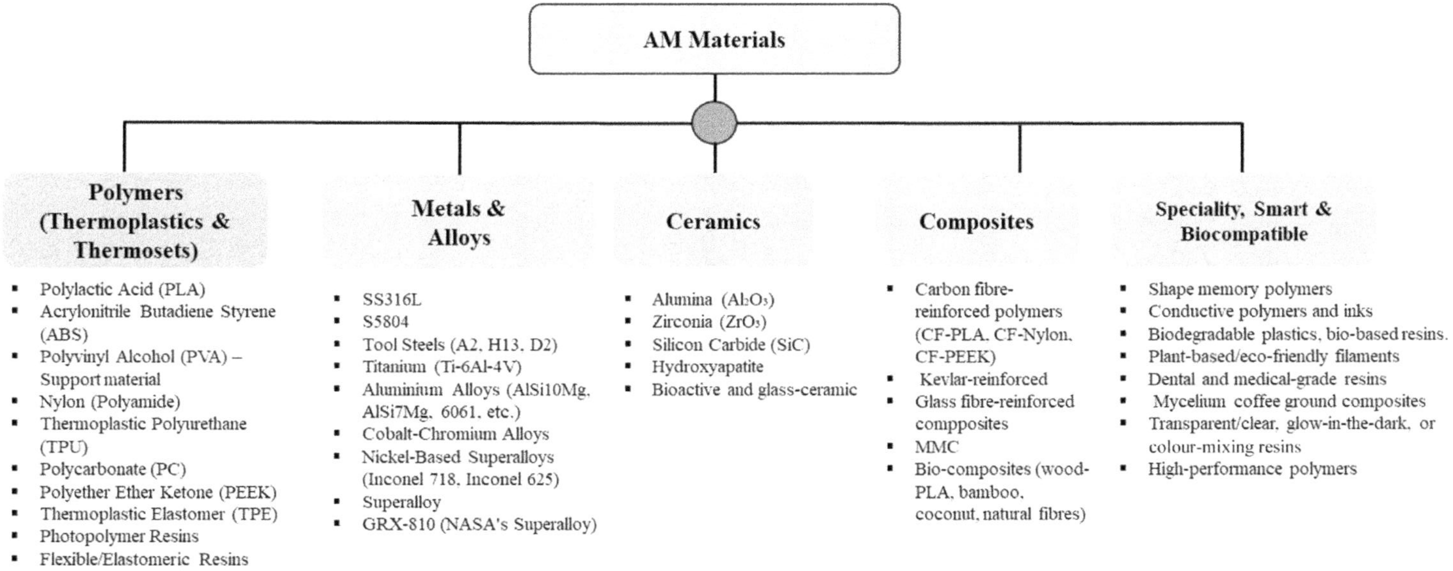

Figure 1.5 These materials have been thoroughly characterised for mechanical, thermal, chemical, and biocompatibility requirements, making them widely suitable for prototyping, tooling, medical, automotive, aerospace, and consumer applications.

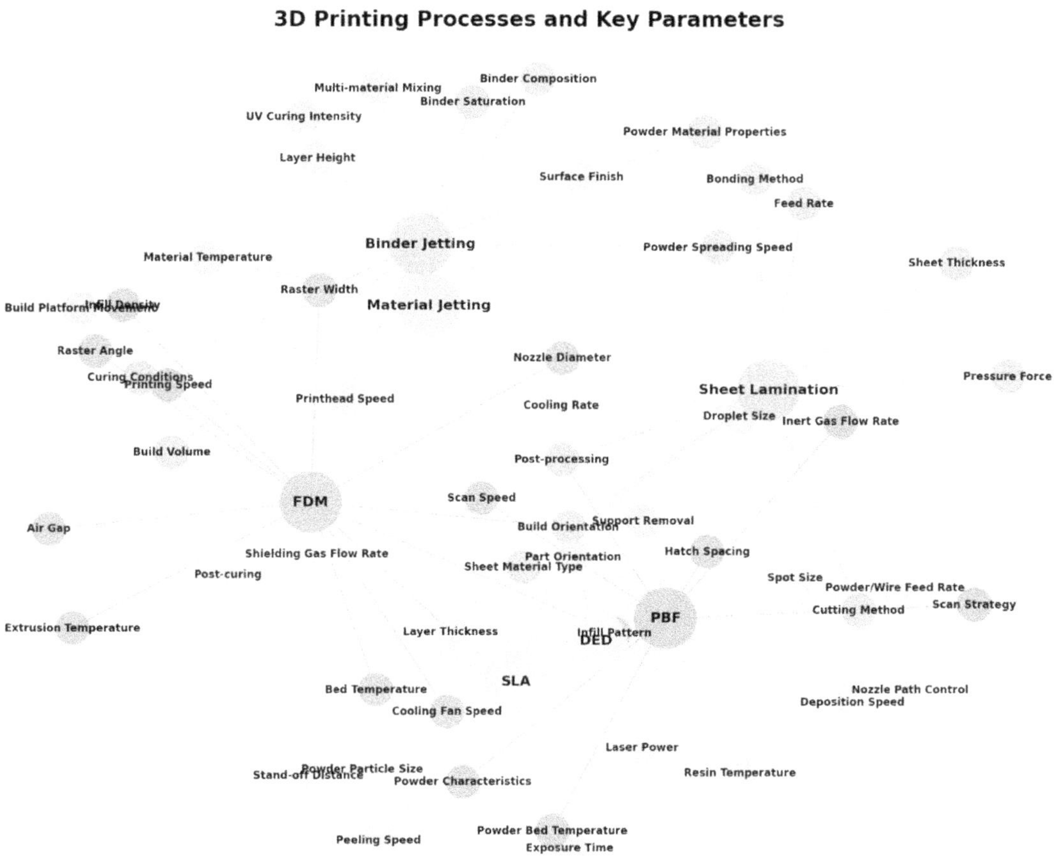

Figure 1.6 Seven additive manufacturing processes and their operating parameters.

simultaneously fed and melted, the main parameters are the laser or arc power, the material feed rate (powder or wire), and the deposition head traverse speed. The ratio of these parameters determines the geometry of the deposited track, the dilution with the substrate, and the resulting microstructure. For example, the study quantitatively predicts tensile residual stresses of approximately 300 MPa near the top surface and compressive residual stresses of about –150 MPa near the substrate in γ titanium aluminide thin-wall structures. Laser power variations from 300 W to 600 W resulted in stress increases of up to 25%, whereas increasing the scan speed from 5 mm/s to 15 mm/s reduced residual stresses by nearly 30%. Powder feed rate adjustments also influenced stress magnitude by approximately 10–15%. These quantitative insights enable precise control of residual stress through process optimisation [52, 53]. Finally, Binder Jetting, used for sand, metals, and ceramics, involves a two-step process. During printing, binder saturation and layer thickness control the green part's strength and resolution. However, the most critical parameter is often the post-sintering temperature, which determines the final density, shrinkage, and mechanical properties of the consolidated part. In binder jetting of stainless-steel parts, processed with 100% binder saturation, 50 μm layer thickness, and sintered at 1300°C for 4 hours, achieves partial densification with roughly 5% residual porosity, resulting in tensile strength near 450 MPa, elongation around 8%, and hardness near 180 HV. The sintering temperature and duration are pivotal for achieving adequate density and mechanical strength [54, 55].

1.6 APPLICATIONS OF ADDITIVE MANUFACTURING

Since its inception in 1988, the applications of AM have grown exponentially. Begun as a niche technology for rapid prototyping, it has evolved into a vital production method for fully functional parts across numerous sectors. Today, AM is widely used in aerospace, automotive, biomedical, space exploration, apparel and fashion, thermal and nuclear power plants, food production, construction, architecture, electronics, sports, footwear, and energy systems. Additionally, it is a game changer in tissue engineering and regenerative medicine, enabling the creation of custom implants, scaffolds, and organs [56–60].

The technology continues to expand into emerging fields such as microfluidics, dental and orthodontic appliances, defence and military components, the jewellery industry, educational models, robotics, and even the production of sustainable materials. Its unmatched design freedom, ability to consolidate parts, facilitate mass customisation, and produce complex geometries have made AM indispensable for innovation and accelerated product development across a broad spectrum of industries [61–65]. **Figure 1.7** shows the applications of additive manufacturing/3D printing.

1.7 WHY ADDITIVE MANUFACTURING?

AM offers a paradigm shift in product design and manufacturing, enabling significant improvements in speed, cost, and functionality. The core benefits are outlined next [66].

1.7.1 Rapid Product Development through Time Compression

AM fundamentally streamlines the path from concept to final product by eliminating traditional tooling, such as moulds and jigs. This facilitates on-demand, low-cost rapid prototyping, allowing functional prototypes to be produced in hours or days rather than weeks. Consequently, the product design cycle is dramatically accelerated; for instance, injection moulds can become operational within two to three months, compared to four to six months required by conventional processes. This compression of timelines provides a critical competitive advantage by bringing products to market with unprecedented speed. For example, Making AM do part consolidation of aircraft duct redesign [67].

1.7.2 Design Freedom and Part Consolidation

A principal advantage of AM is its ability to create geometrically complex shapes, such as lattice structures, at little to no extra cost. This enables a "design-for-function" approach, liberating engineers from traditional "design-for-manufacturability" constraints. This freedom allows for the consolidation of multiple components into a single, integrated part. This simplification reduces assembly requirements, minimises material waste, cuts overall weight, and enhances product reliability. Adapting AM to do part consolidation of aircraft duct redesign is shown in **Figure 1.8 (a)** [68, 69].

1.7.3 High-Performance Light Weighting

AM is intrinsically linked to advanced design techniques such as topology optimisation, which algorithmically distributes material only where it is structurally needed. This enables the fabrication of high-strength, lightweight components that are often impossible to produce with other methods. This can lead to weight reductions of 20–50%, which is crucial for

Figure 1.7 Applications of additive manufacturing (AM) across different industries.

applications in aerospace and automotive industries, resulting in improved fuel efficiency and lower emissions [71].

1.7.4 Advanced Thermal Management and Material Gradients

AM provides unique capabilities for embedding conformal cooling channels directly into tools, such as injection moulds. These channels follow the mould surface contour, enabling

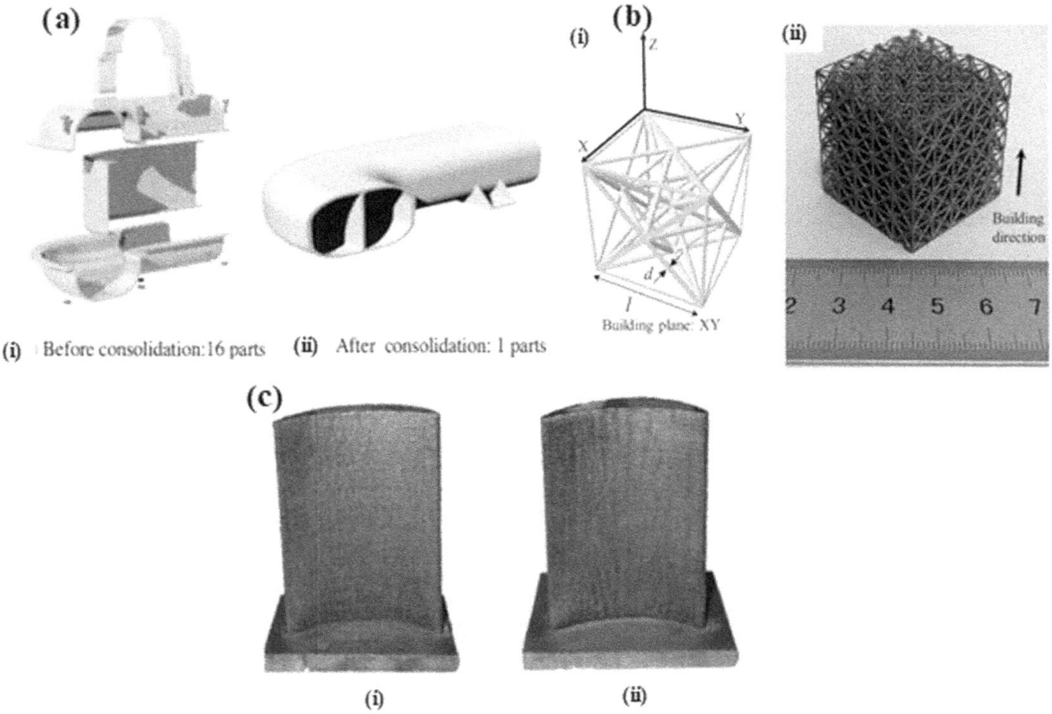

Figure 1.8 (a) Adapting AM to do part consolidation of aircraft duct redesign before (i) and after consolidation (ii), (b) Proposed unit cell type of SC-BCC-FCC (i) and EBM-built lattice structure (ii), (c) Turbine blades fabricated using the laser direct deposition process (i) the original as deposited blade, (ii) the standard restored blade (Reproduced from the Permission of Publisher and Copyright Clearance Center) [68–70].

more uniform heat dissipation than straight-drilled channels. This can reduce cooling times by up to 50–70%, significantly increasing production throughput. Furthermore, technologies such as DED enable the creation of Functionally Graded Materials (FGMs), in which the composition and properties of a component can be gradually varied within a single build to meet specific local requirements [72].

1.7.5 Efficient Repair and Refurbishment

AM, particularly DED processes, serves as a powerful tool for repairing and refurbishing high-value components, such as turbine blades or forging dies. By adding material precisely to worn or damaged areas, AM can extend the service life of expensive tooling and parts, avoiding the cost and lead time of full replacements. This capability is also valuable for correcting machining errors and for efficiently implementing last-minute design changes [70].

1.8 WHAT ARE THE EXISTING CHALLENGES OF ADDITIVE MANUFACTURING AND WHY ARTIFICIAL INTELLIGENCE IS REQUIRED?

Layer thickness (affecting resolution and build time), build orientation (affecting strength owing to anisotropy), infill density (balancing weight and durability), and process-specific elements like laser power in PBF or extrusion temperature in FDM are all important considerations in determining AM quality. AM relies heavily on materials science. Polymers like PLA and ABS dominate extrusion-based techniques because they are easy to use, although metals such as titanium alloys (Ti-6Al-4V) and stainless steel are preferred in PBF for aerospace applications owing to their high strength-to-weight ratios [67]. Emerging materials include biocompatible resins for medical implants and smart materials that react to stimuli. AM's benefits go beyond creative freedom: it shortens lead times from weeks to hours, reduces material waste (by up to 90% compared with subtractive processes), and enables on-demand manufacturing, which supports sustainability objectives. However, obstacles remain, including high starting prices, post-processing requirements (e.g., heat treatment for stress relief), and quality assurance concerns, such as porosity or dimensional errors.

AM part quality is determined not only by material choice but also by a wide range of process parameters [36]. Key variables such as layer thickness directly influence surface resolution and build time, while build orientation governs anisotropy, affecting mechanical strength and fatigue performance. Similarly, infill density represents a trade-off between structural durability and lightweight design [37, 38]. Process-specific parameters, including laser power and scanning speed in powder bed fusion (PBF) or extrusion temperature and nozzle velocity in fused deposition modelling (FDM), must be carefully calibrated to balance melt pool stability, bonding, and dimensional accuracy.

In the ever-changing world of contemporary production, where efficiency and creativity are critical, AM stands out as a game-changer [73]. Today, industries ranging from aerospace to healthcare depend on it to manufacture anything from lightweight aircraft components to personalised medical implants, reducing waste by up to 90% and significantly shortening production times. However, despite its promise, AM is not without challenges: inconsistent part quality, the difficulty of optimising process settings, and the sheer volume of data generated during builds all make achieving consistent outcomes difficult [74, 75].

Selecting these parameters is a complex optimisation task. Traditional approaches such as trial-and-error testing, DOE, statistical modelling (e.g., response surface methodology), and numerical modelling have been widely employed to tune settings. However, given the nonlinear interactions among parameters, such methods can be time-consuming and costly [76]. Improper parameter selection leads to defects, including porosity, delamination, warping, incomplete fusion, and residual stresses, which not only degrade mechanical properties but also increase post-processing needs.

To address these challenges, Machine Learning (ML) and Artificial Intelligence (AI) are increasingly integrated into AM workflows. Supervised ML models can predict tensile strength, surface roughness, or porosity based on process inputs, enabling data-driven parameter optimisation. Deep learning approaches, particularly Convolutional Neural Networks (CNNs), are used for in-situ defect detection by analysing thermal images, melt pool monitoring, or acoustic signals in real time. Reinforcement Learning (RL) and Bayesian optimisation methods further enable adaptive process control, in which parameters are dynamically adjusted to ensure quality consistency [77, 78]. Thus, the fusion of materials science, process parameter optimisation, and AI-driven monitoring is paving the way for reliable, defect-minimised, and sustainable AM.

In the context of Industry 4.0, AM works with digital twins, IoT sensors, and simulation software to improve predictive maintenance and optimisation. For example, finite element analysis (FEA) mimics thermal distortions to help optimise designs before construction. As of 2025, advances such as hybrid manufacturing (combining AM with CNC) and large-scale AM (e.g., for building) are broadening its use. Understanding these basics is critical as AM advances, especially with the incorporation of AI and ML, which help manage variability and improve process predictability [79].

AI and its powerful subset, ML, have been very successful at solving many engineering problems. These technologies are changing AM by addressing those specific pain areas [80]. AI mimics human decision-making, but ML systems sift through massive amounts of information to identify patterns, forecast outcomes, and enhance processes without requiring ongoing human intervention. Imagine a system that analyses thermal pictures in real time during a print, recognises abnormalities such as porosity or fractures, and changes laser power on the fly. This is the kind of closed-loop control AI provides, with accuracy reaching over 95% in certain circumstances. This integration is more than theoretical; new research shows that ML models can optimise PBF parameters, minimising trial-and-error iterations and material costs. At its heart, AM begins with a digital design created using CAD software, which is then cut into layers and materialised using processes such as FDM for plastics and SLM for metal. However, the magic comes when AI is involved in the design and execution process [81, 82]. Generative design techniques driven by ML examine thousands of iterations based on limitations such as weight or strength, generating optimised topologies that conventional engineers may overlook, for example, 40% lighter components for drones or UAV. In metal AM, where faults such as thermal distortions hamper production, Physics-Informed Neural Networks (PINNs) combine physical variables with data-driven insights to produce accurate simulations, speeding up development while accounting for real-world constraints [83–86]. A comprehensive network of research trends in AM with AI and ML, along with their related topics, is presented in **Figure 1.9.**

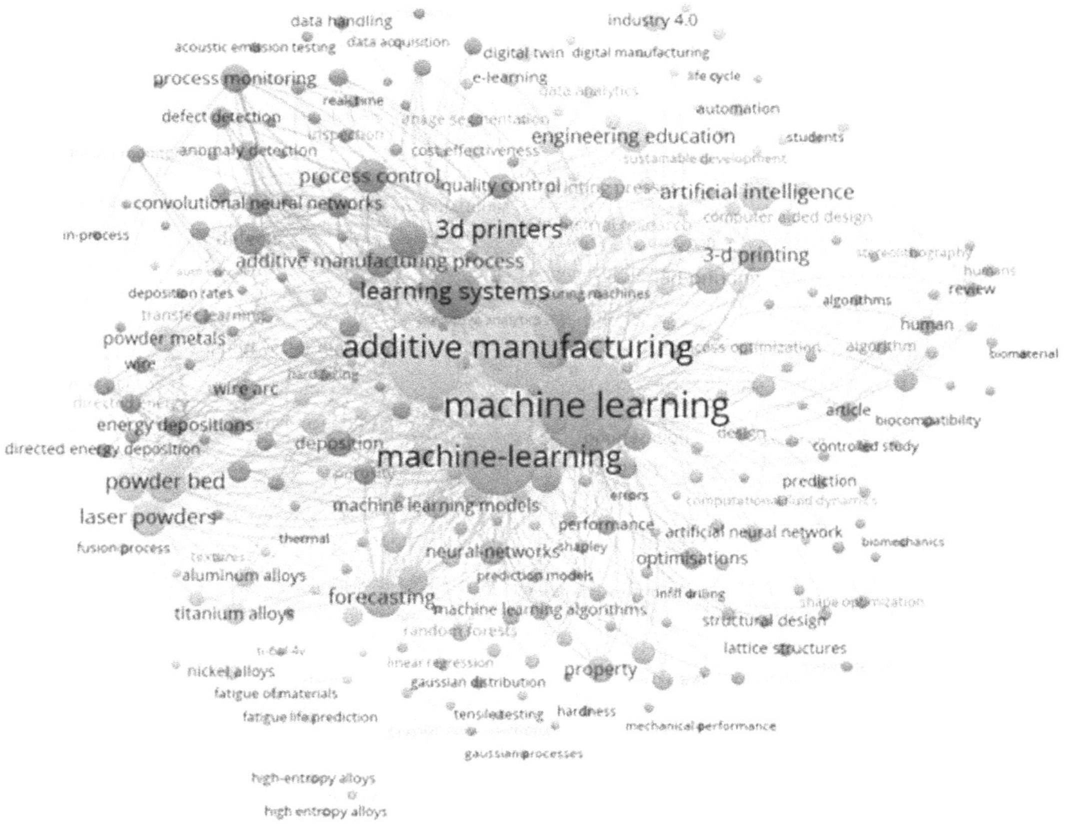

Figure 1.9 Comprehensive network of additive manufacturing with AI and ML, along with their related topics.

The convergence of AM with AI/ML is especially crucial in the transition to Industry 4.0, where smart factories need predictive maintenance and sustainable practices [87, 88]. For example, ML may predict equipment failures by monitoring printer sensor data, reducing downtime, and even optimising energy consumption in large-scale buildings, aligning with global environmental objectives. However, challenges remain, including data scarcity, high computational demands, and the "black box" nature of certain ML models, which may undermine confidence, particularly in regulated industries such as aerospace. Still, ongoing research works find solutions for these problems, like hybrid models using AI alongside classical simulations or proposing standardised datasets to improve models' transferability [89, 90]. In the future, this combination could allow for self-correcting and creative systems from custom-made prostheses in health care to stronger supply chains for the automotive industry. This chapter provides a comprehensive overview of AM and Artificial Intelligence/ Machine Learning (AI/ML), exploring their foundational principles, practical integration through case studies, and the emerging challenges and opportunities. By highlighting current breakthroughs, this chapter equips readers to understand and leverage the strategic potential of these converging technologies [91].

1.9 ARTIFICIAL INTELLIGENCE AND MACHINE LEARNING

1.9.1 Introduction to Artificial Intelligence (AI)

AI refers to the simulation of human intelligence processes by machines, particularly computer systems. These processes include learning (the acquisition of information and rules for using the information), reasoning (using rules to reach approximate or definite conclusions), and self-correction [92]. AI systems are designed to perform tasks that typically require human intelligence, such as visual perception, speech recognition, decision-making, and language translation. Unlike traditional programming, where explicit instructions are provided for every scenario, AI enables machines to mimic cognitive functions, adapt to new inputs, and improve over time through iterative processes.

The foundations of AI can be traced back to the 1950s, with seminal contributions from pioneers such as Alan Turing, who proposed the Turing Test in 1950 to evaluate a machine's ability to exhibit intelligent behaviour indistinguishable from a human, and John McCarthy, who coined the term "Artificial Intelligence" at the Dartmouth Conference in 1956. Early AI research focused on symbolic reasoning and rule-based systems, but progress was hampered by limited computational power. Recent surges in AI development have been propelled by exponential increases in processing power (e.g., GPUs), the availability of massive datasets from the internet and IoT devices, and breakthroughs in algorithms such as neural networks and deep learning [93].

AI is broadly classified into two types:

- **Narrow AI (Weak AI)**: This is task-specific and excels in performing a single function, such as voice assistants (e.g., Siri or Alexa) for natural language processing, recommendation systems (e.g., Netflix algorithms), or image recognition in autonomous vehicles. Narrow AI dominates current applications and lacks general cognitive abilities.
- **General AI (Strong AI)**: This speculative form aims for human-like adaptability across diverse tasks, with the ability to understand, learn, and apply knowledge in any intellectual endeavour. Achieving general AI remains a long-term goal, with ongoing debates about its feasibility and ethical implications.

Key AI techniques extend beyond ML to include knowledge representation (e.g., ontologies and semantic networks), search and optimisation (e.g., A* algorithm for pathfinding), and expert systems (rule-based decision-making) [94]. For instance, the A* search algorithm balances exploration and exploitation using a heuristic cost function:

$$f(n)=g(n)+h(n) \quad (1.1),$$

where g(n) is the exact cost from the start to node n, and h(n) is the heuristic estimate from n to the goal. This equation ensures optimal pathfinding in complex spaces, such as robotics navigation.

Another foundational technique is logic programming, exemplified by Prolog, which uses Horn clauses for inference:

$$\forall x(P(x) \leftarrow Q(x) \land R(x)) \quad (1.2)$$

This represents implications for automated reasoning in expert systems. As AI and ML represent distinct contexts, the differences between the two paradigms are presented in **Figure 1.10 (a)**. Also, there are different types of algorithms in ML, mainly supervised, unsupervised and RL. The characteristics of these three types are presented in **Figure 1.10 (b)**.

A comparative overview of AI and ML, based on their definitions, scopes, and learning paradigms, is presented in **Figure 1.10 (a)**. While AI is a vast field that aims to replicate human intelligence and cognitive processes such as perception, reasoning, and problem-solving, ML is a subset of AI that focuses on enabling systems to automatically learn patterns from data without explicit programming. While ML has developed more recently due to improvements in data availability and processing power, AI was first developed in the 1950s by pioneers like Alan Turing and John McCarthy. The figure illustrates how ML works across three learning paradigms: supervised, unsupervised, and RL. These paradigms form the foundation for most intelligent systems in use today. The three major learning paradigms in ML—supervised, unsupervised, and RL—are highlighted in **Figure 1.10 (b)** with respect to their data requirements, objectives and applications. Labelled data is used in supervised learning to train predictive models for regression or classification tasks, including defect detection. Using techniques such as K-Means or Principal Component Analysis (PCA) for anomaly identification and clustering, unsupervised learning uses unlabelled data to discover hidden patterns and groupings. RL, which is frequently used in robotics and autonomous systems, allows agents to learn optimal behaviours through interaction with their surroundings and the rewards or penalties they receive for their actions. When combined, these paradigms demonstrate how various ML techniques create intelligent, flexible solutions across a range of fields.

1.10 INTRODUCTION TO MACHINE LEARNING (ML)

ML, a subset of AI, focuses on developing algorithms that enable computers to learn from data and make predictions or decisions without being explicitly programmed for each task. ML algorithms identify patterns in historical data and use them to forecast future outcomes, gradually improving performance through experience. This paradigm shift from rule-based programming to data-driven inference has revolutionised fields such as healthcare, finance, and manufacturing [95, 96].

The historical roots of ML trace back to the 1950s alongside AI, with Arthur Samuel's checkers-playing programme in 1959 defining ML as "the field of study that gives computers

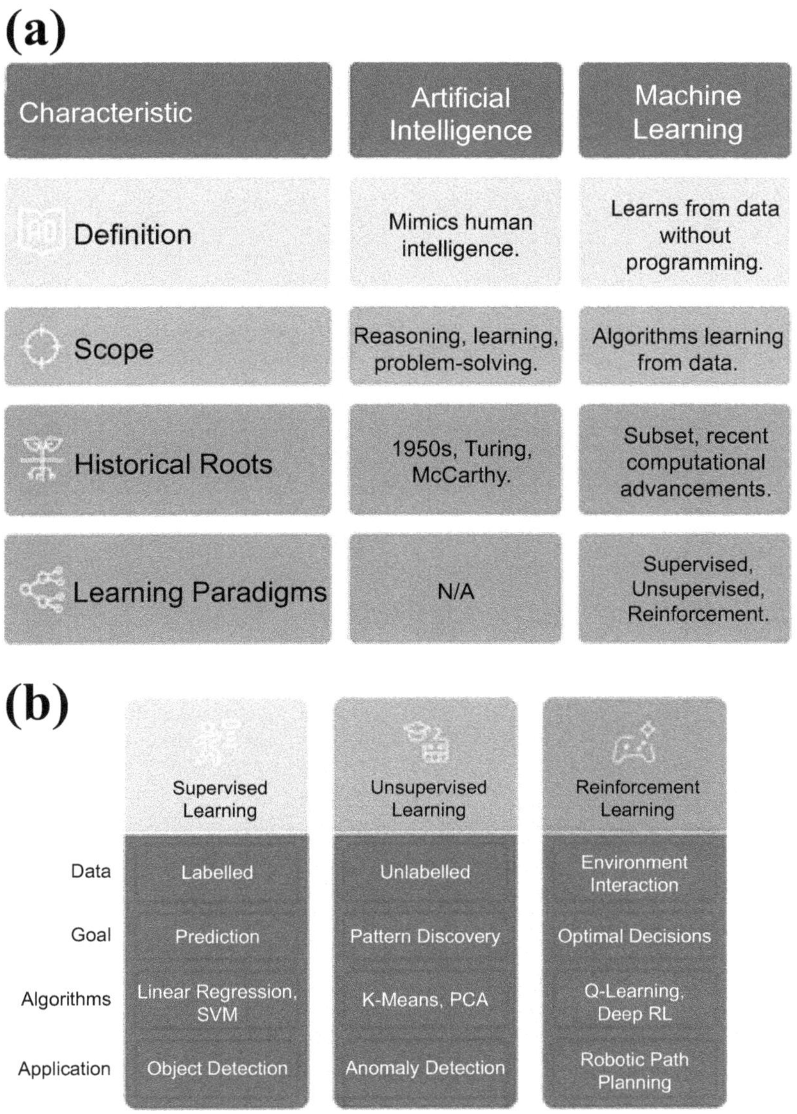

Figure 1.10 (a) Infographic comparing AI and machine learning characteristics, scope, and paradigms; (b) Comparison of machine learning types.

the ability to learn without being explicitly programmed." Modern ML has been accelerated by big data and computational advances, leading to widespread adoption.

ML operates on data using statistical models, and performance is measured by metrics such as accuracy, precision, and recall. Core to ML is the bias-variance trade-off, formalised as:

$$\text{Total Error} = \text{Bias}^2 + \text{Variance} + \text{Irreducible Error}$$

This equation guides model selection to balance underfitting (high bias) and overfitting (high variance).

1.10.1 Three Paradigms of ML

ML is categorised into three primary paradigms: Supervised Learning, Unsupervised Learning, and RL [97], as shown in **Figure 1.11** (**a, b, c**). Each paradigm addresses different data characteristics and objectives, with applications tailored to real-world problems, such as those in manufacturing.

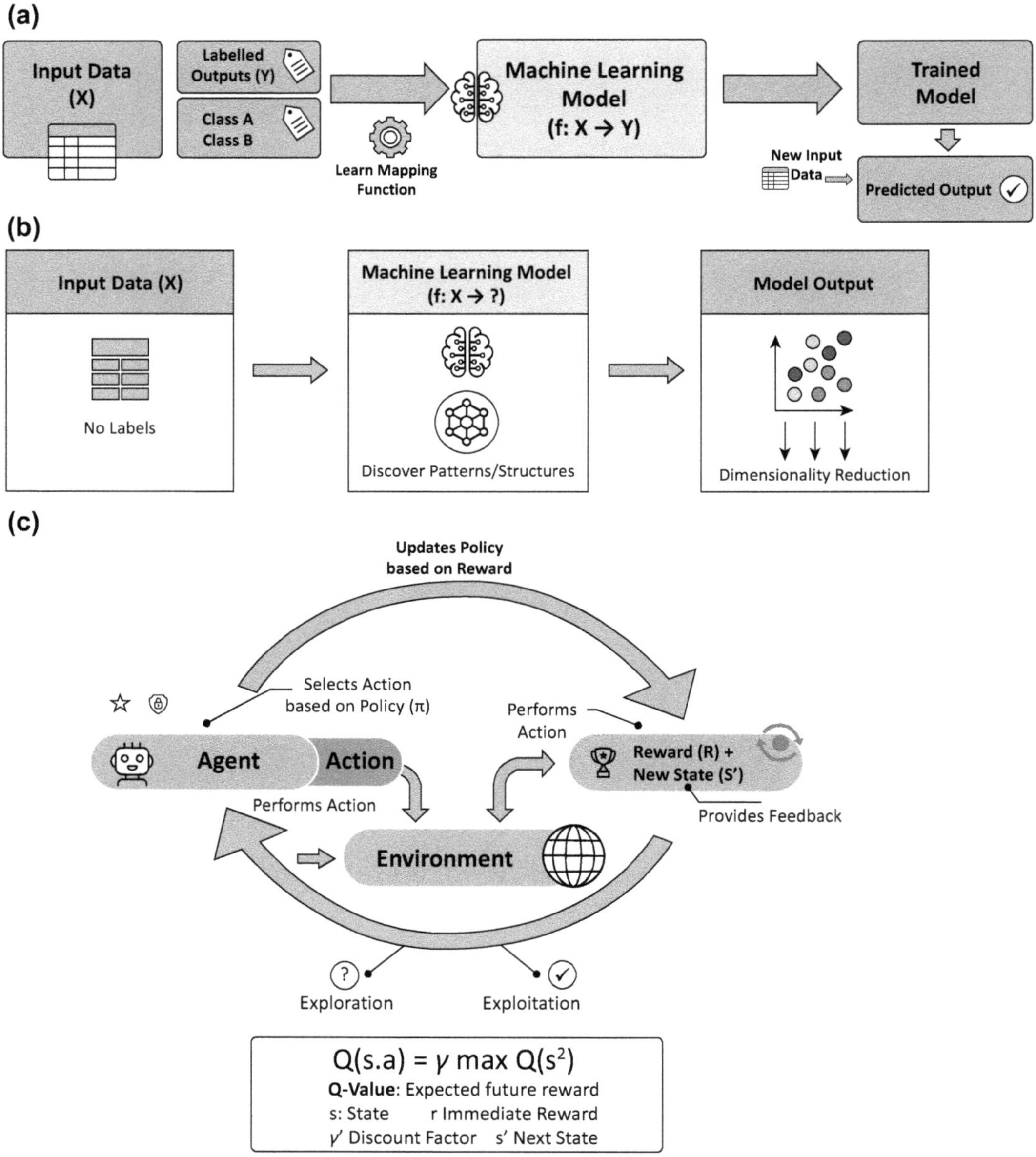

Figure 1.11 (a) Workflow of supervised learning algorithms, (b) Workflow of unsupervised learning algorithms, (c) Reinforcement learning workflow.

1.10.1.1 Supervised Learning

Supervised Learning is a fundamental approach in ML that relies on labelled datasets, where each training example consists of an input and its corresponding output (**Figure 1.11** (**a**)). The goal is to learn a mapping function from inputs to outputs that can generalise well to unseen data. By minimising prediction errors during training, supervised algorithms learn to capture relationships in the data and make accurate predictions on new instances. This paradigm is widely used in both regression and classification tasks, where continuous values are predicted and inputs are assigned to discrete categories, respectively. Its effectiveness depends on the quality and quantity of labelled data, making it highly applicable in domains such as finance, healthcare, and natural language processing [98].

It depicts (**Figure 1.11** (**a**)) a ML model that trains a mapping function f: X • Y, processing input data (X) with matching labelled outputs (Y) for classes A and B. A predicted output is subsequently generated by applying the trained model to new input data. Based on labelled data, the procedure embodies model learning, training, and prediction. The workflow of an unsupervised learning algorithm is shown in **Figure 1.11** (**b**). When input data (X) is fed into a ML model that discovers hidden patterns or structures, the model displays the data without labels. Pattern recognition and dimensionality reduction are two examples of model outputs that demonstrate how unsupervised learning finds structure in unlabelled datasets. The RL process is shown in **Figure 1.11** (**c**), where an agent interacts with its environment by acting and receiving feedback in the form of rewards and new states. To make better decisions in the future, the agent updates its policy based on the rewards it earns. The diagram shows the exploration and exploitation cycles to illustrate how the agent strikes a balance between acquiring new tactics and refining those already in place.

The goal is to learn a mapping function f: X→Y that minimises prediction error from inputs X to outputs Y. Training involves an objective function, such as Mean Squared Error (MSE) for regression:

$$\text{MSE} = \frac{1}{n}\sum\nolimits_{i=1}^{n}(y_i - \hat{y}_i)^2 \tag{1.3}$$

where yi is the true label and $\hat{y}_i$ is the predicted value. Common algorithms include Linear Regression for continuous predictions, which assumes a linear relationship:

$$y=\beta 0+\beta 1x+\epsilon \tag{1.4}$$

Here, β0 and β1 are estimated via least squares. For classification, Support Vector Machines (SVM) find the hyperplane that maximises the margin between classes:

$$\min_{\mathbf{w},b}\frac{1}{2}\parallel \mathbf{w}\parallel^2 + C\sum\nolimits_{i=1}^{n}\xi_i \tag{1.5}$$

subject to yi(w·xi+b)≥1−ξi, where C controls the trade-off between margin and misclassification.

1.10.1.2 Unsupervised Learning

Unsupervised Learning is a ML approach that works with unlabelled data, meaning the algorithm must identify hidden patterns, groupings, or structures without predefined outputs (**Figure 1.11** (**b**)). Unlike supervised learning, which learns from input-output pairs, unsupervised learning focuses on exploring the data distribution to uncover insights such as clusters, anomalies, and lower-dimensional representations. This makes it particularly valuable in situations where labelled data is scarce, costly, or impractical to obtain [99].

Key techniques include Clustering, such as K-Means, which partitions data into K clusters by minimising intra-cluster variance:

$$J = \sum_{i=1}^{K} \sum_{\mathbf{x} \in C_i} \| \mathbf{x} - \mu_i \|^2 \quad (1.6),$$

where C_i is the i-th cluster, and µi is its centroid.

Dimensionality reduction via PCA projects data onto principal components that capture maximum variance:

$$Z = XV \quad (1.7),$$

where V are the eigenvectors of the covariance matrix.

1.10.1.3 Reinforcement Learning

RL is a ML paradigm in which an agent learns to make decisions by interacting with an environment and receiving feedback in the form of rewards or penalties (**Figure 1.11** (**c**)). Unlike supervised learning, which relies on labelled input-output pairs, RL focuses on trial-and-error learning, where the agent explores different actions to maximise long-term cumulative reward. The learning process is guided by a policy that defines the agent's strategy for choosing actions in various states and is refined through feedback using value functions or models of the environment. This approach enables agents to adapt dynamically, making RL particularly powerful for tasks such as robotics, autonomous driving, game playing, and resource optimisation, where sequential decision-making and adaptation to uncertainty are critical [100].

RL can be understood as a continuous learning loop in which an Agent interacts with an Environment to improve its decision-making ability. The agent, represented conceptually by the learner or decision maker (e.g., a robot or a game player), follows a strategy known as a Policy (π) that determines which action to take in a given situation. Once the agent performs an Action, it directly influences the environment, such as a robot moving forward or a player pressing a button. The Environment then responds by providing feedback in two forms: a Reward (R), which can be positive or negative depending on the action's benefit, and a New State (S′), which represents the updated situation after the action. This feedback helps the agent evaluate the quality of its choices and adjust its policy accordingly.

This interaction creates a continuous feedback loop: the agent acts, the environment responds, and the agent learns from the resulting reward and state to improve future actions. A crucial aspect of RL lies in balancing exploration and exploitation. Exploration involves trying out new actions to discover potentially better strategies, while exploitation focuses on leveraging the knowledge already gained to maximise rewards. A proper balance is essential; excessive exploration may waste time on suboptimal choices, whereas overexploitation may cause the agent to miss better long-term strategies.

The mathematical foundation of RL is captured by the Bellman Equation, which underpins Q-learning, one of the most widely used RL algorithms. In this framework, the Q-value function Q(s, a) represents the expected cumulative reward of taking action *a* in state *s*. It incorporates both the immediate reward *r* and the potential long-term rewards from future states, discounted by a factor γ (gamma), which determines the importance of future rewards relative to immediate ones. The equation can be expressed as:

$$Q(s,a) = r+\gamma \max Q(s') \quad (1.8)$$

Here, s' is the next state resulting from action a. By iteratively updating the Q-values in this way, the agent gradually refines its policy and improves its performance over time through experience.

The agent maximises cumulative reward over time via a Markov Decision Process (MDP), defined by states S, actions A, transition probabilities P, and reward function R.

The Bellman equation underpins value function estimation:

$$V^{\pi}(s)=\sum_{a\in A}\pi(a|s)\ \sum_{s',r}P(s',r|s,a)[R(s,a,s',r)+\gamma V^{\pi}(s')] \quad (1.9),$$

where π is the policy, and γ is the discount factor. Q-Learning, a model-free algorithm, updates action-value functions:

$$Q(s,a)\leftarrow Q(s,a)+\alpha[r+\gamma \max_{a'}Q(s',a')-Q(s,a)] \quad (1.10)$$

Deep RL extends this with neural networks, as in AlphaGo, which combines Monte Carlo Tree Search with deep Q-networks.

In manufacturing, AI/ML analyses large volumes of sensor data to detect failures, optimise settings, and enable autonomous systems. In AM, where operations produce terabytes of data (e.g., thermal pictures, layer scans), ML models may forecast part quality in real time, minimising iterations. AI/ML collaboration with AM offers smarter, more efficient production, shifting from reactive to predictive manufacturing.

1.11 KEY ML TECHNIQUES FOR ADDITIVE MANUFACTURING PARADIGMS

AM, commonly known as 3D printing, encounters various challenges, including process variability (e.g., inconsistencies in material properties or environmental conditions), defect prediction (e.g., identifying porosity, warping, or cracks), and optimisation (e.g., fine-tuning parameters like laser power and scan speed for superior build quality). ML techniques, specifically tailored for AM, utilise data from sensors, simulations, and in-situ monitoring to mitigate these issues, as depicted in **Figure 1.12**.

Key ML methods, including regression, classification, clustering, deep learning, RL, and hybrid models, are outlined in **Figure 1.12**, along with their applications in AM. Design, process planning, in-situ monitoring, quality assurance, and supply chain integration are the workflow steps to which these methodologies are mapped. Examples of how ML improves accuracy, automation, and optimisation across the AM process include melt-pool temperature prediction, defect detection, build-failure diagnosis, distortion minimisation, and AI-driven supply forecasting.

1.11.1 Regression Models

Regression models are supervised learning algorithms that forecast continuous outcomes, such as melt pool temperature or residual stresses in AM. They model non-linear dependencies between inputs (e.g., laser power P, scan speed v) and outputs, supporting real-time parameter adjustments [92].

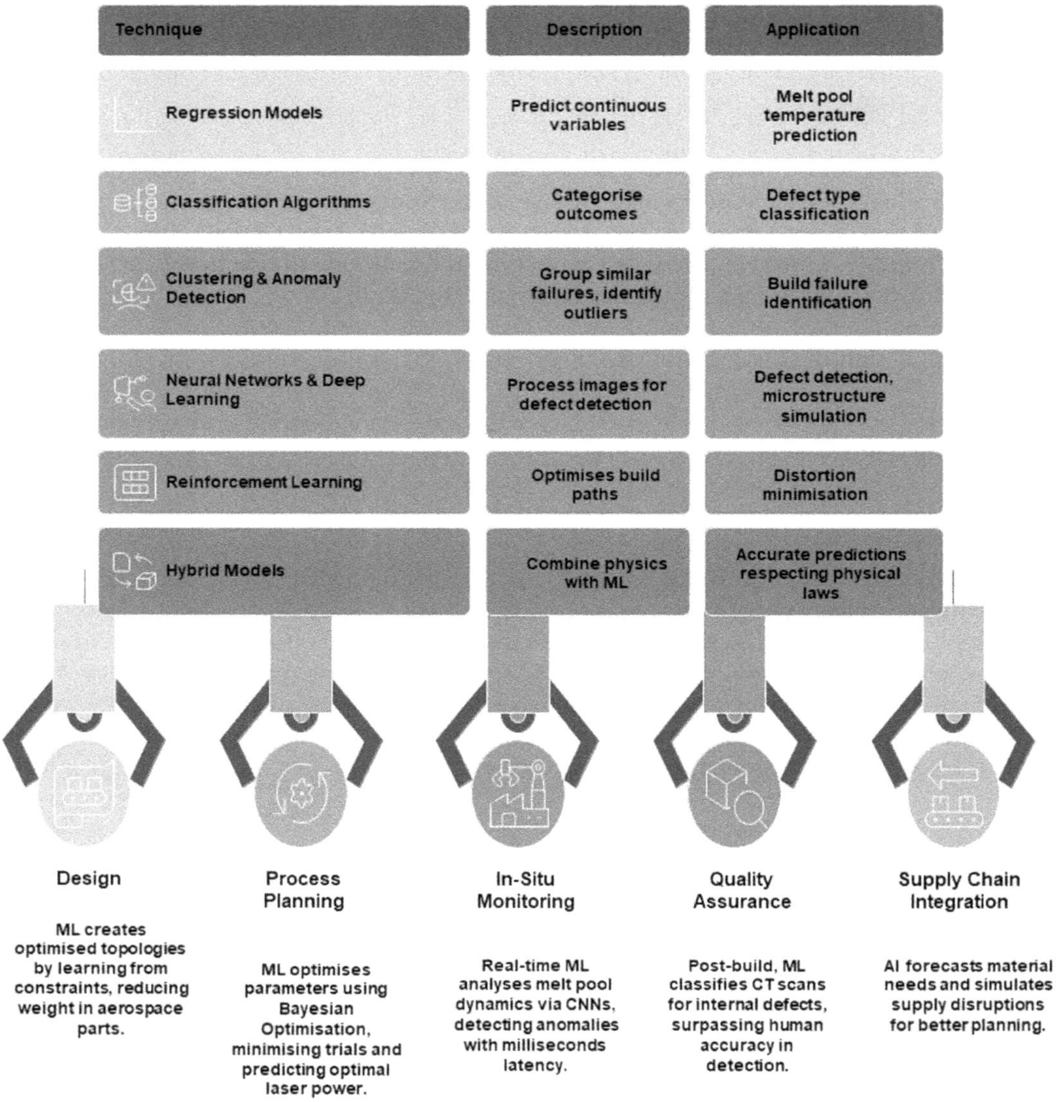

Figure 1.12 (a) Machine learning techniques for additive manufacturing, (b) Machine learning application in additive manufacturing.

1.11.1.1 Gaussian Process Regression (GPR)

Gaussian Process Regression (GPR) is a probabilistic, nonparametric method ideal for quantifying uncertainty in AM processes affected by noise (e.g., powder-bed inconsistencies). It treats the target function $f(\mathbf{x})$ as drawn from a Gaussian Process prior, yielding predictions with confidence bounds. For instance, GPR predicts residual stresses σ from P and v.

The core of GPR is the predictive distribution for a test point $\mathbf{x}_*$ given training data $\mathbf{X}, \mathbf{y}$:

$$f_* \mid \mathbf{X}, \mathbf{y}, \mathbf{x}_* \sim \mathcal{N}\left(\mu_*, \sigma_*^2\right) \tag{1.11}$$

- Mean Prediction (μ_*):

$$\mu_* = \mathbf{k}_*^T (\mathbf{K} + \sigma_n^2 \mathbf{I})^{-1} \mathbf{y} \tag{1.12}$$

  - $\mathbf{K}$ is the $n \times n$ kernel (covariance) matrix for training inputs $\mathbf{X}$, where n is the number of samples.
  - $\mathbf{k}_*$is the $n \times 1$ vector of covariances between $\mathbf{X}$ and $\mathbf{x}_*$.
  - σ_n^2 is the noise variance (diagonal addition to $\mathbf{K}$ for regularisation).
  - $\mathbf{y}$ are the observed targets (e.g., measured stresses).
  - Explanation: This is a weighted average of training outputs, weighted by similarities (via kernel) to the test point. The inverse matrix solves for optimal weights, akin to kernel ridge regression.
- Predictive Variance (σ_*^2):

$$\sigma_*^2 = k(\mathbf{x}_*, \mathbf{x}_*) - \mathbf{k}_*^T (\mathbf{K} + \sigma_n^2 \mathbf{I})^{-1} \mathbf{k}_* \tag{1.13}$$

  - $k(\mathbf{x}_*, \mathbf{x}_*)$ is the prior variance at $\mathbf{x}_*$.
  - Explanation: Measures remaining uncertainty after incorporating training data; high values indicate extrapolation regions (e.g., untested parameter combinations in AM).

  Common kernel: Squared Exponential (RBF):

$$k(\mathbf{x}_i, \mathbf{x}_j) = \sigma_f^2 \exp\left(-\frac{\| \mathbf{x}_i - \mathbf{x}_j \|^2}{2\ell^2}\right) \tag{1.14}$$

  - σ_f^2: Signal variance (amplitude).
  - ℓ: Length scale (smoothness).
- Step-by-Step: (1) Compute $\mathbf{K}$ from training data. (2) Invert (Cholesky decomposition for efficiency). (3) Plug into mean/variance for prediction. In AM, GPR achieves <5% error in PBF melt pool forecasting [101].

1.11.2 Classification Algorithms

These supervised methods assign categorical labels to data, such as classifying defects (e.g., porosity vs fractures) from sensor arrays (e.g., acoustic or thermal data) [102].

1.11.2.1 Random Forests (RF)

RF ensembles multiple decision trees, each trained on bootstrapped data subsets, for robust classification resilient to AM's noisy environments.

Trees split data using impurity measures; RF aggregates via voting.

- Gini Impurity (for a node):

$$G = 1 - \sum_{k=1}^{C} p_k^2 \tag{1.15}$$

  - C: Number of classes (e.g., 2 for defect/no-defect).
  - p_k: Proportion of class k in the node.

- Explanation: Measures node "purity" (0 = pure, 0.5 = max impurity for binary). Splits minimise weighted Gini: $G_{\text{parent}} - \frac{N_l}{N} G_l - \frac{N_r}{N} G_r$, where N_l, N_r are child sizes. Step-by-Step: (1) Bootstrap data. (2) Grow trees randomly selecting features. (3) Average predictions. RF excels in multi-fault detection (85–95% accuracy).

1.11.2.2 Support Vector Machines (SVMS)[103]

SVMs maximise margins in high-dimensional spaces, which are perfect for multi-sensor AM data.

1.11.2.3 Formula Setup and Explanation

Finds hyperplane $\mathbf{w} \cdot \mathbf{x} + b = 0$ maximising margin.

- Optimisation Problem:

$$\min_{\mathbf{w},b,\xi} \frac{1}{2} \| \mathbf{w} \|^2 + C \sum_{i=1}^{n} \xi_i \quad \text{s.t.} \quad y_i \left(\mathbf{w} \cdot \phi(\mathbf{x}_i) + b \right) \geq 1 - \xi_i, \xi_i \geq 0 \tag{1.16}$$

- $\phi(\mathbf{x}_i)$: Feature map (e.g., RBF kernel $k(\mathbf{x}_i, \mathbf{x}_j) = \exp\left(-\gamma \| \mathbf{x}_i - \mathbf{x}_j \|^2\right)$).
- C: Trade-off parameter (higher = softer margins).
- ξ_i: Slack for misclassifications.

1.11.3 Clustering and Anomaly Detection

These unsupervised methods group similar data points without requiring labelled data, identifying patterns in unlabelled sensor streams (e.g., grouping comparable build failures) or detecting outliers for early intervention. One effective technique is Density-Based Spatial Clustering of Applications with Noise (DBSCAN), which clusters data based on density and marks low-density points as anomalies. This is particularly useful in AM for spotting irregular thermal signatures that may indicate cracks.

1.11.3.1 Core Distance and Epsilon Neighbourhood

A point p is classified as a core point if it has at least MinPts neighbours within a radius ϵ:

$$N_\epsilon(p) = \{q \in D \mid \text{dist}(p,q) \leq \epsilon\} \quad \text{where} \quad |N_\epsilon(p)| \geq \text{MinPts} \tag{1.17}$$

- $N_\epsilon(p)$: The set of points q within distance ϵ from p in dataset D.
- $\text{dist}(p,q)$: Typically, the Euclidean distance is calculated as:

  $$\text{dist}(p,q) = \sqrt{\Sigma(p_i - q_i)^2} \tag{1.18},$$

  where p_i and q_i are the coordinates of points p and q.
- MinPts: The minimum number of points required to form a dense region.
- Explanation: Clusters are formed by connecting the neighbourhoods of core points. Points that do not belong to a cluster are considered noise. This approach enables DBSCAN to handle irregularly shaped clusters, making it suitable for identifying anomalies in AM sensor data.

- Step-by-Step: (1) Select an unvisited point p. (2) If p is a core point, expand the cluster by including all points in $N_\epsilon(p)$. (3) Repeat until no new points can be added. (4) Mark non-core, non-clustered points as noise.

In AM, DBSCAN analyses accelerometer data to cluster failure modes (e.g., delamination groups) with 88% purity, enabling proactive pauses in DED builds.

1.11.4 Deep Learning and Neural Networks

Deep Learning (DL) excels at processing complex, unstructured data such as images or time-series from in-situ monitoring, scaling effectively to the large datasets typical in AM. Two key techniques are CNNs and Generative Adversarial Networks (GANs) [104].

1.11.4.1 Convolutional Neural Networks (CNNS)

CNNs detect defects in photos (e.g., keyhole pores) by learning hierarchical features through convolutional operations, achieving high accuracy in AM applications [105, 106].

Convolution Operation

For an input image I and a kernel K:

$$(O)_{i,j} = (I^{*}K)_{i,j} = \sum_{m}\sum_{n} I_{i+m,j+n}K_{m,n} \tag{1.19}$$

- $(O)_{i,j}$: The output value at position (i,j) in the convolved feature map.
- $I_{i+m,j+n}$: The input image pixel values.
- $K_{m,n}$: The kernel weights.
- Explanation: The convolution slides the kernel over the input image, computing a weighted sum at each position to extract features like edges or textures. This is followed by pooling (e.g., max pooling):

$$P_{i,j} = \max_{m,n\in\text{window}} O_{i+m,j+n} \tag{1.20}$$

and activation functions such as sigmoid:

$$\sigma(z) = \frac{1}{1+e^{-z}} \tag{1.21}$$

or ReLU:

$$\max(0,z) \tag{1.22}$$

The network concludes with fully connected layers, and the loss for classification is cross-entropy:

$$-\Sigma y\log(\hat{y}) \tag{1.23}$$,

where y is the true label and $\hat{y}$ is the predicted probability.

- Step-by-Step: (1) Apply convolution to extract features. (2) Use pooling to reduce spatial dimensions. (3) Pass through activation and fully connected layers. (4) Optimise using backpropagation with cross-entropy loss. CNNs achieve >95% accuracy in detecting keyhole defects in laser scans.

1.11.4.2 Generative Adversarial Networks (GANS)

GANs generate synthetic microstructures to accelerate material design in AM, employing a generator G and a discriminator D in a competitive framework [107].

Key Equation: Min-Max Loss

$$\min_{G}\max_{D} V(D,G) = \mathbb{E}_{x\sim p_{\text{data}}}\left[\log D(x)\right] + \mathbb{E}_{z\sim p_z}\left[\log\left(1 - D\left(G(z)\right)\right)\right] \tag{1.24}$$

- $\mathbb{E}_{x\sim p_{\text{data}}}\left[\log D(x)\right]$: The expected log-likelihood of the discriminator correctly classifying real data x from the true data distribution p_{data}.
- $\mathbb{E}_{z\sim p_z}\left[\log\left(1 - D\left(G(z)\right)\right)\right]$: The expected log-likelihood of the discriminator rejecting fake data $G(z)$ generated from noise z with distribution p_z.
- Explanation: The generator G aims to minimise this loss by producing realistic samples, while the discriminator D maximises it by distinguishing real from fake. This adversarial process iteratively improves both models.
- Step-by-Step: (1) Train D to maximise V. on real and fake data. (2) Train G to minimise V by adjusting to fool D. (3) Repeat until convergence. In AM, GANs simulate grain structures for alloy optimisation, reducing the number of experimental trials by 70%.

1.11.5 Reinforcement Learning (RL)

RL is a paradigm for optimising sequential decision-making problems. In AM, particularly in DED, RL can plan optimal toolpaths or parameter adjustments step by step, minimise distortions, and improve overall build quality. RL trains an agent to maximise a cumulative reward signal, such as minimising residual stress or warpage in the final part [108].

1.11.5.1 Proximal Policy Optimisation (PPO)

PPO is a widely used policy-gradient method that balances exploration (trying new actions) with stability (avoiding drastic changes in strategy). It improves the agent's decision-making policy while ensuring updates remain within a safe and steady range.

Key Formula: Clipped Objective

$$L\_CLIP(\theta) = E\_t\left[min\left(r_{t(\theta)} * A_{t_{hat}}, clip\left(r_{t(\theta)}, 1-\epsilon, 1+\epsilon\right) * A\_t\_hat\right)\right] \tag{1.25}$$

- Policy Ratio ($r_t(\theta)$):

$$r_t(\theta) = \frac{\pi_\theta(a_t \mid s_t)}{\pi_{\text{old}}(a_t \mid s_t)} \tag{1.26}$$

  - Compares the probability of action a_t under the new policy π_θ with the old policy π_{old}.
  - Intuition: Measures how much the new strategy differs from the previous one.
- Advantage Function ($\hat{A}_t$): Quantifies how much better (or worse) an action was compared to the expected outcome, often estimated using Generalised Advantage Estimation (GAE):

$$\hat{A}_t = \delta_t + (\gamma\lambda)\delta_{t+1} + (\gamma\lambda)^2\delta_{t+2} + \cdots \tag{1.27}$$

where:

- δ_t: Difference between expected reward and actual reward (temporal difference error).
- γ: Discount factor ($0 < \gamma < 1$), reduces the weight of future rewards.
- λ: Balancing factor ($0 < \lambda < 1$), which trades off variance vs. bias.

- Clipping Function:

$$clip(r_{t(\theta)}, 1-\epsilon, 1+\epsilon \tag{1.28}$$

Ensures $r_t(\theta)$ stays within $[1-\epsilon, 1+\epsilon]$, preventing large and unstable updates.

Step-by-Step PPO Process

1. Collect Trajectories: The agent interacts with the AM process (e.g., DED path planning) and records states, actions, and rewards.
2. Compute Advantages: Evaluate how effective each action was using $\hat{A}_t$.
3. Update Policy: Optimise the clipped objective with stochastic gradient ascent, making controlled improvements.

1.11.5.2 Application in AM

Using PPO in multi-layer DED builds has been shown to reduce warpage by up to 40%, thereby improving accuracy and mechanical properties of fabricated parts. RL thus serves as a key enabler for adaptive, intelligent process control in AM.

1.11.6 Hybrid Models

Hybrid models integrate ML with physics-based simulations to produce interpretable, physically consistent predictions.

1.11.6.1 Physics-Informed Neural Networks (PINNS)

PINNs embed governing equations (e.g., heat transfer equations in AM) into the loss function to ensure predictions align with physical principles.

Key Equation: Augmented Loss

$$\begin{aligned} \mathcal{L} = \mathcal{L}_f + \mathcal{L}_u + \mathcal{L}_b = \frac{1}{N_f}\sum_{i=1}^{N_f} | \mathcal{N}\left[u\left(\mathbf{x}_i, t_i; \theta\right)\right] |^2 + \frac{1}{N_u}\sum_{i=1}^{N_u} | u\left(\mathbf{x}_i, t_i\right) - u_i |^2 \\ + \sum_k \frac{1}{N_b^k}\sum_{i=1}^{N_b^k} | u\left(\mathbf{x}_i^k, t_i^k\right) - g_k |^2 \end{aligned} \tag{1.29}$$

- $\mathcal{L}_f$: The physics loss, enforcing the residual of the governing PDE to zero (e.g., for the heat equation $\frac{\partial u}{\partial t} = \alpha \nabla^2 u$, $\mathcal{N}[u] = 0$ represents this constraint.
- $\mathcal{L}_u$: The data loss, measuring the difference between predicted $u(\mathbf{x}_i, t_i)$ and observed u_i values.
- $\mathcal{L}_b$: The boundary loss, ensuring boundary conditions g_k are met at points $\left(\mathbf{x}_i^k, t_i^k\right)$.
- N_f, N_u, N_b^k: Number of collocations, data, and boundary points, respectively.

- Explanation: The neural network $u(\mathbf{x},t;\theta)$ is trained to minimise this combined loss, embedding physical laws into the learning process.
- Step-by-Step: (1) Define the PDE and boundary conditions. (2) Sample collocation points for $\mathcal{L}_f$, data points for $\mathcal{L}_u$, and boundary points for $\mathcal{L}_b$. (3) Use automatic differentiation to compute PDE residuals. (4) Optimise with gradient descent. PINNs predict thermal fields in PBF with 10x faster computation than finite element methods, ensuring physical consistency [109].

1.12 INTEGRATION OF AI/ML IN ADDITIVE

Integrating AI and ML into AM spans the entire product lifecycle from conceptual design and process planning to in-process monitoring and post-process inspection and is rapidly shifting AM from manual tuning and offline testing towards data-driven, adaptive production (**Figure 1.12 (b)**). The following subsections summarise key use cases and give literature-backed examples.

1.12.1 Design and Generative Design

At the design stage, ML models support generative design and design-for-additive-manufacturing (DfAM) by automatically generating optimised topologies and lattice structures under manufacturing constraints. Autodesk's generative design tools, informed by ML, have demonstrated up to 40% weight reduction in aerospace parts while meeting structural requirements. Recent studies highlight how ML-enabled generative design tools and explainable AI methods enhance product design lifecycles, leading to lightweight, high-performance structures.

1.12.2 Process Planning and Parameter Optimisation

AM process planning involves selecting parameters such as layer thickness, hatch spacing, scan speed, extrusion temperature, and build orientation, which are often nonlinear and interdependent. Bayesian optimisation and surrogate modelling approaches significantly reduce the need for trial-and-error, especially in SLM, where optimised laser power/scan speed pairs prevent defects, such as balling and porosity. AI-augmented planning frameworks have been reported to accelerate metal AM process development and alloy qualification cycles through high-throughput testing combined with ML [110].

1.12.3 In-Situ Monitoring and Closed-Loop Control

In-situ monitoring leverages ML to interpret sensor data from thermal cameras, melt pool imaging, photodiodes, and acoustic signals. CNNs trained on thermal images detect anomalies within milliseconds, enabling closed-loop control of melt pool stability [111, 112]. Advanced frameworks, such as digital twins and long short-term memory (LSTM)-based neural operators, couple physics-based models with ML to achieve real-time adaptive control, thereby reducing porosity, delamination, and thermal deviations [113, 114].

1.12.4 Post-Processing and Quality Assurance

Beyond the build process, ML supports post-build inspection and predictive quality control. Deep learning applied to CT scans and X-ray images outperforms manual inspection in identifying internal voids, porosity clusters, and cracks. Transfer learning approaches extend defect recognition across different machines and materials, while multimodal ML models combine acoustic and thermal signals for robust fatigue life prediction [115–118]. These methods improve quality assurance and reduce post-processing costs, thereby increasing production efficiency.

1.12.5 Supply Chain and Lifecycle Integration

AI integration is not limited to the printing process. In the AM supply chain, ML forecasts material demand, predicts machine maintenance schedules, and simulates supply disruptions to ensure just-in-time production and reduce inventory costs [119–121]. AI-based lifecycle assessment (LCA) models also evaluate the sustainability impacts of AM processes, such as energy consumption and material waste, enabling eco-efficient design and manufacturing strategies [120].

1.13 LITERATURE REVIEW

The integration of AI and ML into AM has become a major research direction, producing measurable gains in process optimisation, material discovery, and quality assurance across AM technologies. Multiple comprehensive reviews synthesise this trend, reporting how supervised and deep learning models are applied to predict mechanical properties, reduce costs through parameter tuning, and detect defects in metal and polymer processes [122–127]. These surveys collectively emphasise that ML methods, ranging from regression ensembles [128] and explainable models to deep neural networks, are being deployed throughout the AM workflow: at the design stage (CAD and DfAM), during slicing and process planning, and in in-process monitoring and post-process inspection. The body of work shows consistent improvements in yield and part quality when data-driven feedback loops are integrated into production pipelines, while also documenting domain-specific successes such as polymer property prediction and metal melt-pool stabilisation.

Applied studies extend these survey findings by demonstrating concrete, domain-specific benefits and methodological advances. Research on responsive and functional composites illustrates how ML enables tailored material behaviour and even programmable/4D-printed functionality, expanding AM beyond structural parts into active-materials design. For metal AM, recent work on high-throughput characterisation and AI-assisted alloy discovery accelerates the materials qualification cycle and improves process stability, while studies on melt-pool monitoring and closed-loop control (including digital-twin frameworks with LSTM, neural operators, and surrogate models) show how coupling physics-based models with ML enables real-time adaptive control that reduces defects such as porosity and thermal excursions [129, 130]. Special issues and targeted reviews also highlight cross-cutting frameworks that bridge materials science, process monitoring, and automated design generation, signalling a shift from isolated ML experiments towards integrated, production-scale strategies. **Table 1.2** lists different areas of scope where AI/ML can be integrated to help solve current challenges.

Despite these advances, the literature repeatedly identifies critical barriers that must be overcome for widespread industrial adoption. Chief among these are limited availability and heterogeneity of high-quality datasets, differences in machines, material batches, sensor setups, and environmental conditions, which reduce model transferability and generalisation [146]. Real-time deployment of ML for closed-loop control faces computational and latency constraints, and many predictive models lack robust uncertainty quantification and

Table 1.2 Classification of Works in Additive Manufacturing by References

Area of Work	*Related Problems*	*References*
Additive Manufacturing Modification	Design optimisation; Historical manufacturing data analysis; Efficiency improvement; Lack of standardised datasets; High computational cost.	[131, 132]
AM Quality Control and Process Optimisation	Cost reduction; Data-driven decision-making; Quality control; Quality Assurance; Defect detection; Energy efficiency; Large-scale production.	[133, 134]
Predictive Multiscale Design in AM	Physical performance prediction; Process parameter optimisation; Deviation detection; Reliability improvement; Multi-level material interaction modelling; Model transferability across materials.	[135, 136]
AM Production Control	Smart production; Product quality improvement; Real-time error detection; Durable production; Real-time processing delays.	[78, 79]
Additive and Subtractive Manufacturing	AI-driven process automation; Intelligent decision-making; Hybrid production; Incompatible monitoring data; Cross-platform challenges.	[137, 138]
AM Lifecycle	Process planning; Testing and verification; Process design adaptation; Insufficient prediction accuracy; Lack of real-time feedback loops.	[139, 140]
Advanced Manufacturing Systems	Process adjustment, predictive maintenance, high data storage requirements, and cybersecurity risks.	[141–143]
AM Design and Process Selection	Production function integration; Process monitoring; Process selection; Lack of labelled datasets;	[144]
Efficiency Gains	Production time optimisation; Resource utilisation; High computational overhead of AI models; Limited generalisability of efficiency models; Integration of AI for dynamic resource allocation.	[77–79]
Scalability	Scaling AI models for large-scale AM; Handling high-dimensional data; Limited computational resources for real-time scalability; Cross-machine compatibility issues.	[145]
Sustainability	Material waste reduction; Energy-efficient AM processes; Lack of sustainability-focused datasets; AI model optimisation for eco-friendly processes.	[119–121]
Real-Time Benefits	Real-time process monitoring; Adaptive control systems; Low-latency AI inference; Data streaming challenges; Hardware limitations for real-time AI; Model drift in dynamic AM environments.	[18, 113, 142]

interpretability, which are required for operator trust and certification in regulated industries [147]. Surface quality and cross-material generalisation remain open problems, motivating research into hybrid physics-ML models, transfer learning, standardised benchmark datasets, human-AI teaming for curated data extraction, and edge/cloud architectures for low-latency inference.

Recent reviews emphasise that supervised learning, deep neural networks, and hybrid models are now applied across the AM workflow, from design to post-processing, enabling real-time decision-making and reducing trial-and-error [122–124].

At the design stage, DL models are widely used for generative design and optimisation. Neural networks trained on CAD and simulation data can propose lightweight structures and predict mechanical properties before fabrication, reducing design cycles and material waste [124, 127]. In polymer AM, DL has been shown to predict part performance such as strength and thermal stability, while in metal AM, DL models are applied to understand microstructure-property relationships and surface quality outcomes [129].

During printing and process control, RL provides a powerful framework for adaptive, real-time adjustment. RL agents learn to control laser power, scan speed, or extrusion rates based on sensor feedback, stabilising melt pools and minimising porosity [105, 110]. This "learning by doing" approach is particularly valuable in metal PBF and DED, where small process deviations can result in part rejection. Coupled with digital twin frameworks, RL has been shown to improve reliability and defect prevention in closed-loop AM systems [110, 129, 131].

For in-situ monitoring and defect detection, transformer-based architectures are emerging as a new direction. Originally developed for natural language processing, transformers excel at handling sequential and high-dimensional data streams. In AM, they are applied to continuous sensor inputs (thermal imaging, acoustic emissions, layer-wise cameras) to detect anomalies at high speed and low latency [107]. This capability allows real-time defect correction, reducing scrap rates and ensuring higher part quality.

A further innovation is transfer learning and spin-transfer methods, which aim to reuse knowledge across processes, materials, and machines. Because AM datasets are often limited and heterogeneous, transfer learning allows a model trained on one material system (e.g., polymer extrusion) to be adapted for another (e.g., metal laser sintering), improving generalisation with minimal additional data [100, 147]. Spin-transfer frameworks extend this idea by sharing insights across different AM platforms or organisations while protecting sensitive data, which is critical for industrial adoption [148, 149].

In post-processing and lifecycle optimisation, ML supports prediction of surface finish, fatigue life, and residual stress, guiding heat treatment and machining steps [123, 126]. At the production scale, AI also contributes to predictive maintenance, scheduling, and energy optimisation, helping AM transition from lab-scale innovation to factory-scale manufacturing [63, 81, 90, 141].

Addressing these gaps, standardising datasets, improving model transparency, and designing systems for real-time, uncertainty-aware control will be essential to move AM from experimental demonstrations to reliable, autonomous manufacturing ecosystems.

1.13.1 Case Study: Non-Destructive Fatigue Life Prediction for Additively Manufactured Metal Parts through a Multimodal Transfer Learning Framework with AI [117]

In this study, researchers tested a Multimodal Transfer Learning (MMTL) framework to predict the fatigue life of Ti-6Al-4V parts made using Laser Powder Bed Fusion (L-PBF). They used two sets of specimens produced on an EOS M290 machine, with laser power and scanning speed varied to create defects such as keyholes (KHs), lack of fusion (LOF), and gas-entrapped pores (GEPs). The first set of five parts was scanned using high-resolution X-ray computed tomography (XCT) to identify 1531 defects (68 KHs, 1308 LOFs, and 155 GEPs larger than 10 μm). Features like defect volume, surface area, axis lengths, shape ratios, and flatness were extracted and used to train a Hierarchical Graph Convolutional Network (HGCN) for defect classification. The second set of 13 parts was stress-relieved, XCT-scanned, machined into standard fatigue specimens, and tested under cyclic loading at 450, 500, and 600 MPa. Critical defects were identified and matched to XCT data with 81.25% accuracy, resulting in 63 key defects for fatigue prediction.

The researchers created graph representations of the defects and process parameters to help the AI model understand relationships. Integrating process parameters with defect features made it easier to differentiate defects and showed how energy input affects defect shapes. For defect classification, the HGCN outperformed other methods (HNN, GCN, standard neural networks), achieving high accuracy even with limited data.

For fatigue life prediction, MMTL was compared with other pre-trained models and a baseline neural network. Using metrics that handle wide variations in fatigue life, MMTL showed the best performance, reducing errors by 19–14% compared to other approaches. When applied to new specimens, the pre-trained HGCN could classify defects from 80% of the fatigue data and accurately predict the remaining 20%, with predicted fatigue lives very close to experimental results, even for challenging defects like LOF and KH.

The framework used a relatively small number of parameters (2532 in total) and had a total training time of 35.5 seconds, with predictions taking only milliseconds, making it suitable for real-time assessments. Overall, MMTL successfully combined defect information, process parameters, and physics-based understanding to accurately and efficiently predict fatigue life in L-PBF parts, even with limited data.

1.13.2 Case Study: Using Machine Learning to Predict Tensile Strength in 3D-Printed Concrete [118]

In this study, researchers aimed to predict the tensile strength of 3D-printed concrete, the material's ability to resist being pulled apart. 3D printing in construction can cause the concrete to behave differently in different directions (anisotropy), making it challenging to estimate strength using traditional methods. To address this, the researchers collected data from 25 previous studies, resulting in 49 concrete mix designs specifically tested for tensile strength. The inputs for prediction included material composition such as water, cement, fly ash, silica fume, aggregates, fibres (with properties like length, diameter, and strength), and additives that improve workability. Printing parameters like nozzle size and printing speed were also included, as they affect the final structure. Four supervised ML models were tested: GPR, Decision Tree Regression (DTR), Support Vector Machine Regression (SVMR), and XGBoost Regression (XGBR). Before training, the data was cleaned, normalised, and statistically analysed. Model performance was measured using standard metrics like MSE, Root Mean Squared Error (RMSE), Coefficient of Determination (R^2), and Mean Absolute Error (MAE).

The results showed that the SVM model with a linear kernel achieved the most accurate predictions, with an R^2 of 0.8454 and very low MAE (0.6108 MPa) and RMSE (0.8126 MPa). Sensitivity analysis indicated that the tensile strength of fibres in the concrete was the most important factor influencing overall tensile strength. Although DTR and XGBR were prone to overfitting, GPR performed moderately well, and SVM clearly provided the best results. The models were tested with six new concrete mixes, showing a very small mean prediction error of ±0.26 MPa. This confirms that the approach can reliably predict tensile strength for new mixes. By using this method, engineers can reduce the number of physical experiments needed, saving time and resources, and better optimise 3D-printed concrete mixtures for construction.

1.14 CHALLENGES AND LIMITATIONS

The integration of AI and ML into AM holds immense potential to revolutionise production processes, yet it faces myriad challenges and limitations that must be addressed to realise its

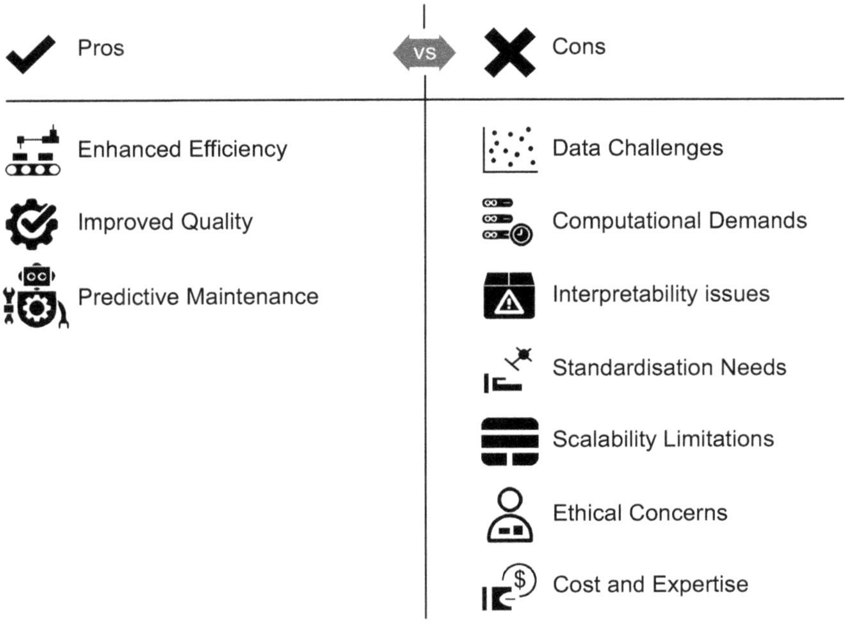

Figure 1.13 The pros and cons of Artificial Intelligence in additive manufacturing.

full potential [63, 85, 87]. These hurdles span technical, operational, ethical, and regulatory domains, necessitating innovative solutions and interdisciplinary collaboration, as shown in **Figure 1.13**.

1.15 PROS OF ADDITIVE MANUFACTURING (AM)

1.15.1 Enhanced Efficiency

AM improves efficiency because it can directly make parts from digital designs without extra moulds or tooling. This reduces steps, saves time, and reduces waste compared to traditional methods. For example, in wire arc AM, smart monitoring helps adjust process conditions to keep the machine running smoothly. Research shows this can shorten production lead times by as much as 30%, which is important when making many parts quickly [50, 81, 89, 124, 131, 141]. This efficiency makes AM suitable for industries like automotive and aerospace, where speed matters.

1.15.2 Improved Quality

AM improves part quality by checking for problems while the part is still being built. Sensors and cameras can detect cracks, porosity, and warping, allowing corrections before printing is finished. This is better than older inspection methods, which only find issues after production. In metal AM, real-time quality monitoring has been shown to improve surface finish and mechanical strength [38, 78, 79, 115, 129]. Some studies report accuracies of more than 95% in detecting defects. This reliability is important for high-performance parts in medical and aerospace uses.

1.15.3 Predictive Maintenance

AM machines can wear down over time, causing problems during production. To prevent sudden breakdowns, AM systems track vibrations, temperature, and usage. By studying this data, failures can be predicted in advance, allowing maintenance to be scheduled at the right time. This method has reduced unexpected downtime by up to 50% in some factories [76, 77]. For example, in laser-based printing, predictive maintenance extends machine lifespan. This approach improves reliability and lowers repair costs.

1.16 CONS OF ADDITIVE MANUFACTURING (AM)

1.16.1 Data Challenges

AM produces huge amounts of data about temperatures, material flow, and machine behaviour. However, this data is often incomplete or unreliable, leading to manufacturing errors. Inconsistent datasets can make it difficult to predict or control outcomes, especially in polymer AM, where materials exhibit different behaviour across batches [62, 126]. Without good data, part quality and repeatability suffer. Standardised methods are needed to improve data use and make AM results more consistent.

1.16.2 Computational Demands

AM processes, such as thermal flow or melt-pool simulations, require very powerful computers. These simulations help predict part behaviour but need expensive hardware. Running them on regular machines slows production and raises costs by two to three times [80, 146]. In many cases, companies cannot afford the required computing power. This creates a gap between what research can achieve in controlled environments and what industries can use in daily practice.

1.16.3 Interpretability Issues

Many AM systems act like "black boxes," making it unclear how process changes affect the final parts. For example, adjusting the laser speed or energy may change the quality, but the reasons are not always obvious. This reduces user trust and makes it harder to certify safety-critical parts [105, 147]. For industries like aerospace, where certification is strict, this lack of transparency is a big problem. Improving understanding of AM processes will make adoption easier and safer.

1.16.4 Standardisation Needs

Different AM systems use different data formats, machine settings, and metadata. For example, some use STL files, while others use AMF. This makes it hard to transfer designs and processes across machines from different vendors. The lack of shared standards limits collaboration and slows adoption [88, 111]. Standardisation efforts are ongoing, but until they are complete, interoperability will remain a challenge. Companies need open frameworks to make AM systems more compatible.

1.16.5 Scalability Limitations

Scaling AM from lab prototypes to large-scale production is not always smooth. Changes in machine calibration, material batches, or environmental noise can reduce part performance.

Models that work in controlled conditions often fail in industrial environments [109, 142, 144]. For example, in wire arc AM, scaling from small to large parts causes new problems. Addressing these scalability gaps is important for AM to fully support mass production.

1.16.6 Ethical Concerns

AM raises ethical issues, such as unfair access to designs, high energy use, and the risk of tampering with sensitive parts. In the defence and medical fields, a hacked file could produce dangerous defects. Data privacy is also important when sharing part designs across global supply chains [119, 143, 150]. To address these challenges, ethical guidelines and robust security measures are needed. Responsible use will help balance innovation with safety.

1.16.7 Cost and Expertise

Adopting AM requires expensive machines, advanced software, and trained workers. Many small companies cannot afford the investment or do not have staff with the right knowledge. This slows down adoption even when AM is useful [82, 104, 106]. Training programmes and open-access platforms can reduce the skills gap. With better support, AM could become more widely available and affordable, even for smaller businesses.

1.17 FUTURE DIRECTIONS

The future of AI/ML in AM is poised for revolutionary development, driven by creative approaches that overcome current limitations and open new possibilities, as shown in **Figure 1.13**.

1.17.1 Autonomous AM Systems

The future of autonomous AM systems, such as the Autonomous Manufacturing System for Phononic Crystals (AMSPnC), lies in making them more intelligent, efficient, and independent [145]. One promising direction is the use of reinforcement learning (RL), which enables the system to automatically adjust manufacturing settings in real time. For example,

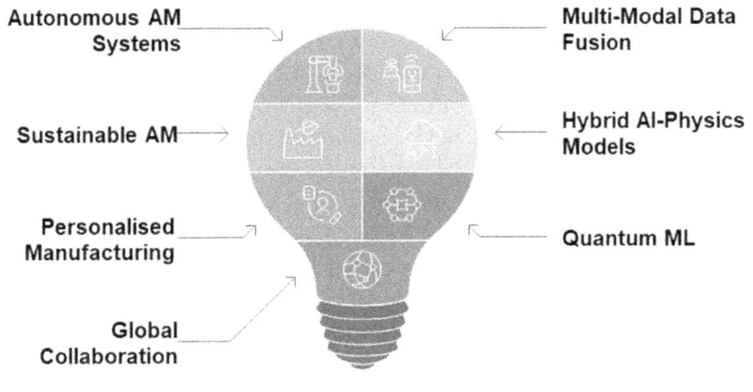

Figure 1.14 Future directions in AI/ML for additive manufacturing.

sensors that capture images, sound, and temperature can provide live feedback, helping the system detect and correct issues such as filament buckling or material variations during the printing process. By combining this sensor data with physics-based simulations (e.g., finite element models that predict wave behaviour), manufacturers can reduce the gap between predicted performance and real outcomes. This integration can support "zero-shot optimisation," meaning the system can successfully produce new, complex designs like phononic crystals or auxetic structures without lengthy trial-and-error testing. The addition of advanced sensors—such as cameras that watch filament flow or sensors that track environmental changes—will make these systems more reliable. With RL, they can learn to self-correct and continuously improve, even in large-scale production. Another important step will be the use of multi-objective optimisation, which enables the system to balance trade-offs among different goals, such as achieving a specific frequency range while minimising material waste [114].

Finally, moving beyond a single technology (such as FFF) to hybrid methods that combine extrusion, laser sintering, or other AM techniques will expand applications into areas such as soft robotics, dielectric metasurfaces, and new functional materials. This shift will reduce reliance on slow, manual trial-and-error approaches and enable faster, cheaper, and more accessible advanced material discovery.

1.17.2 Sustainable AM

Recent work has shown that AM, especially metal-based methods such as Laser Powder Bed Fusion (LPBF)/Selective Laser Melting (SLM), has substantial potential to advance sustainability but also faces important trade-offs. For example, the authors in [62] find that LPBF can significantly reduce material waste (thanks to near-net-shape manufacturing and the reuse/recycling of unused powder) and produce lightweight complex parts which, during their service life (e.g., in aerospace or automotive), yield fuel or energy savings that offset part of the higher energy consumption during printing. However, the study also notes that stages such as powder atomisation, auxiliary systems (chillers, post-processing support), and electricity use are major contributors to environmental impact, making the energy source, machine efficiency, and post-print processing sites potential areas for improvement. Similarly, the authors in [119] argue for a more systemic, design-level approach to sustainable AM: rather than optimising a single process or material, they suggest combining AM with conventional manufacturing, sustainable design methods, and life-cycle thinking from the outset. They emphasise that AM should not be treated as automatically "green" but that its environmental, economic, and social impacts should be evaluated holistically—including design for sustainability, material selection, energy use, and how AM integrates into broader manufacturing and supply chains.

There is also work aimed at forecasting and optimising energy demand in AM systems using AI/ML methods. For instance, the authors in [121] propose a cross-patch graph transformer model, guided by contrastive information fusion, to better predict energy demand across different operating conditions. Such prediction models enable planning and control strategies to reduce peak energy use, manage loads more effectively, and thus improve the overall environmental impact and cost efficiency of AM operations. Together, these findings suggest that sustainable AM requires combining process improvements (e.g., reducing waste, improving machine utilisation, optimising powder handling), design strategies (lightweight, functional integration, recycling), energy management (including renewable energy, forecasting, efficient auxiliary systems), and system-level integration (how AM fits in supply chains, lifecycle, and policies).

1.17.3 Hybrid AI-Physics Model

The next stage of autonomous AM systems, such as the Autonomous Manufacturing System for Phononic Crystals (AMSPnC), will be shaped by hybrid approaches that combine physics and ML. One example is the use of PINNs, which embed governing equations of physics (e.g., elastic wave propagation in phononic crystals) directly into the neural network's loss function. This allows the model to maintain physical consistency while achieving faster convergence and greater predictive accuracy than purely data-driven methods [150]. Such hybrid AI–physics models help reduce the gap between computational predictions, such as finite element simulations, and the actual behaviour of printed parts, addressing key issues like filament buckling and material variability during complex 3D metamaterial fabrication. In parallel, RL can be integrated with PINNs to enable self-optimising manufacturing environments. With the support of multi-modal sensors—including cameras, acoustic detectors, and thermal sensor systems can monitor the process in real time, predict defects, and automatically adjust printing parameters to prevent failures. This creates a closed-loop control mechanism that reduces reliance on traditional, labour-intensive design-of-experiments approaches. Moreover, tunable loss functions, enhanced with physics-based constraints, can be extended to multi-objective optimisation frameworks. Such frameworks balance trade-offs between key performance targets, for example, achieving desired bandpass frequencies or bandwidths while minimising material usage.

Beyond FFF, these methods can be applied to hybrid AM modalities, such as combining extrusion with laser sintering, thereby expanding applications in areas including soft robotics, dielectric metasurfaces, and structural metamaterials. This broader adoption will accelerate the low-cost discovery of functional materials and strengthen the foundation for scalable, autonomous manufacturing systems that require minimal human oversight [114, 150].

1.17.4 Personalised Manufacturing

Personalised manufacturing represents an emerging frontier in AM, where AI-driven customisation enables the production of components tailored to specific performance or user requirements. Drawing inspiration from healthcare applications where patient-specific implants and prostheses are designed using anatomical imaging data, AM systems can extend this concept to advanced metamaterials and structural components [60, 61]. For example, autonomous manufacturing systems for phononic crystals (AMSPnC) can integrate application-specific performance targets, such as acoustic bandpass frequencies or mechanical flexibility, into their cognition modules to generate bespoke designs optimised for end-use scenarios. The integration of RL with real-time sensor feedback, including in-situ machine vision for filament monitoring and environmental sensors for temperature control, allows these systems to iteratively refine part geometries during production, ensuring alignment with customised design objectives [114]. This approach reduces the reliance on trial-and-error methods while ensuring that manufactured parts achieve high precision and tailored performance.

Furthermore, multi-objective optimisation frameworks that balance trade-offs between competing requirements, such as strength versus weight in aerospace components or flexibility versus durability in biomedical devices, can be embedded into the learning algorithms. By incorporating physics-informed constraints, similar to those applied in PINNs, AM systems can respect fundamental material and structural laws while pursuing personalisation [121]. These capabilities position personalised manufacturing as a key enabler for scalable applications across fields such as aerospace, biomedical engineering, dielectric metasurfaces, and soft robotics, where bespoke performance criteria drive innovation and sustainability [120, 121].

1.17.5 Global Collaboration

Global collaboration in AM will play a major role in the future by connecting researchers, industries, and digital platforms worldwide. Since AM relies heavily on design files, material data, and machine parameters, sharing this information through cloud platforms and secure digital ecosystems can speed up innovation and reduce costs [63, 81, 90]. For example, a design created in one country can be instantly manufactured in another, while material performance data collected in one lab can help improve production elsewhere. Collaboration also enables the development of standards, best practices, and large shared datasets more quickly, helping AI and ML models become more accurate and widely applicable [107, 146]. In short, global collaboration will turn AM into a connected, knowledge-driven network where ideas, resources, and expertise are exchanged seamlessly to accelerate innovation and make manufacturing more efficient worldwide [142, 143].

1.17.6 Quantum ML

The future of autonomous AM systems, such as the Autonomous Manufacturing System for Phononic Crystals (AMSPnC), can advance significantly through the use of quantum machine learning (QML). QML methods offer a way to handle complex multi-objective optimisation and large-scale process simulations that are difficult for conventional computing [114]. Algorithms such as the variational quantum eigensolver (VQE) and the quantum approximate optimisation algorithm (QAOA) can efficiently explore high-dimensional design spaces, allowing optimisation of material composition, structural topology, and process conditions for advanced 3D metamaterials like phononic crystals [121]. For example, QML can improve topology optimisation by solving combinatorial problems that lead to better acoustic filtering or mechanical strength, while also addressing process challenges such as filament buckling or material variability [150]. Another promising direction is quantum federated learning (QFL) [151], which enables secure, collaborative model training across institutions. By using quantum properties such as entanglement and superposition, QFL can improve both privacy and computational efficiency when sharing and learning from open datasets of AM process parameters or material properties. Processing these datasets with quantum-enhanced models can enable more accurate prediction of thermal histories or melt pool dynamics at lower computational cost [152].

QML can also work with real-time data from sensors, such as cameras, microphones, and thermal detectors, to dynamically adjust manufacturing parameters through quantum-enhanced RL. This integration can support accurate defect prediction and prevention during production. In the near term, progress will rely on hybrid quantum-classical systems, where current Noisy Intermediate-Scale Quantum (NISQ) devices are used for optimisation tasks while classical computers handle data pre-processing, helping overcome hardware challenges such as quantum noise and limited coherence times [153]. Scaling these QML approaches to hybrid AM processes, for example, combining extrusion with laser sintering, could enable the cost-effective production of complex, high-performance materials. Applications may include aerospace structures, biomedical devices, dielectric metasurfaces, and soft robotic actuators, where both performance customisation and efficiency are critical.

1.17.7 Multi-Modal Data Fusion

Multi-Modal Data Fusion in AM involves combining information from multiple sources, such as cameras, temperature sensors, acoustic signals, and simulation outputs, into a single system. Right now, most monitoring methods rely on only one type of data, which can miss important defects or quality issues. In the future, by fusing all these data streams together, AI and ML models will be able to build a more complete picture of what is happening during printing. This approach can improve real-time defect detection, part quality prediction, and adaptive process control [18, 107, 142]. It also supports the development of digital twin frameworks, where virtual models and physical processes continuously update each other to improve reliability and scalability in AM [19, 105]. In simple terms, multi-modal fusion will give AM "multiple senses" like sight, hearing, and touch that work together to make production more accurate, efficient, and trustworthy.

1.18 CONCLUSION

In conclusion, combining AM with AI and ML is driving a major shift in how products are designed and manufactured. AI and ML give AM new strengths in design optimisation, process control, and quality checking, helping to reduce errors and speed up production. Case studies already show real benefits, such as fewer defects in metal printing and faster discovery of new materials. At the same time, challenges remain, including limited data, high computing needs, and the need for models that people can easily understand and trust. Looking ahead, using advanced methods like hybrid physics ML models, RL, and transfer learning can help make AM systems more autonomous, reliable, and sustainable. Global collaboration will also be key, especially for sharing standardised datasets and building ethical AI practices. By linking smart algorithms with AM machines, industries can unlock higher efficiency and lower costs at scale. Healthcare, aerospace, and automotive sectors are likely to be the first to benefit from personalised parts and lightweight structures. Sustainability will also improve, as AI-guided AM reduces waste and energy usage. Overall, the future of AM with AI and ML is not just about making parts faster, but about creating an interconnected, intelligent, and greener manufacturing ecosystem.

ETHICS STATEMENT

We confirm that in accordance with the Taylor & Francis AI Policy, AI tools were used to improve the readability and language of the manuscript. After using the tools, the authors reviewed and edited the content as needed and took full responsibility for the published article.

Nomenclature or Abbreviation

3DP	Three-Dimensional Printing
AI	Artificial Intelligence
AM	Additive Manufacturing
DL	Deep Learning
DLP	Digital Light Processing
CAD	Computer-Aided Design
CNN	Convolutional Neural Network

DTR	Decision Tree Regression
FDM	Fused Deposition Modelling
FEM	Finite Element Method
GPR	Gaussian Process Regression
MAE	Mean Absolute Error
ML	Machine Learning
MSE	Mean Squared Error
QML	Quantum Machine Learning
RL	Reinforcement Learning
RMSE	Root Mean Squared Error
SLA	Stereolithography Apparatus
SLS	Selective Laser Sintering
SVM	Support Vector Machine
XGBR	XGBoost Regression

REFERENCES

1. JP-S56-144478, 1981. JP Patent: S56-144478-3D figure production device, issued 10, November.
2. Kodama, H., 1981. A scheme for three-dimensional display by automatic fabrication of three-dimensional model. *Journal of IEICE*, 64.
3. U.S. Patent 4,575,330 apparatus for production of three-dimensional objects by Stereolithography.
4. Stereolithography/3D Printing/Additive Fabrication. Photopolymers. Savla Associates. Archived from the original on 14 February 2008. Retrieved 10 August 2017 & Hull, C., 2012. On stereolithography. *Virtual and Physical Prototyping*, 7(3), p.177.
5. RSILVERS, 2019 March 2. On the difference between DLP and LCD based SLA printers/Matter Replicator. Archived from the original on 2023-11-13. Retrieved 2025-09-17.
6. Ali, F., 2024 January 19. Difference between DLP and LCD-based SLA printers. Lets3dPrint. Retrieved 2024-09-02
7. Chua, C.K., Leong, K.F. and Lim, C.S., 2003. *Rapid Prototyping: Principles and Applications*. Singapore: World Scientific, p.124. ISBN 9789812381170
8. Deckard, C., 1989 September 5. Method and apparatus for producing parts by selective sintering. U.S. Patent 4,863,538, filed October 17, 1986.
9. Housholder, R., 1981 January 27. Molding process. U.S. Patent 4,247,508, filed December 3, 1979.
10. Mallikarjuna, B., Bhargav, P., Hiremath, S. et al., 2025. A review on the melt extrusion-based fused deposition modeling (FDM): Background, materials, process parameters and military applications. *International Journal on Interactive Design and Manufacturing*, 19, pp.651–665. https://doi.org/10.1007/s12008-023-01354-0
11. Mallikarjuna, B. and Reutzel, E.W., 2022. Reclamation of intermetallic titanium aluminide aero-engine components using directed energy deposition technology. *Manufacturing Review*, 9, p.27.
12. Webster, E., 2000. *Print Unchained: Fifty Years of Digital Printing, 1950–2000 and Beyond: A Saga of Invention and Enterprise*. West Dover, VT: DRA of Vermont, Inc., p.116. ISBN 0-9702617-0-5. OCLC 46611664
13. Gregory, P., ed., 1996. *Chemistry and Technology of Printing and Imaging Systems*. London: Blackie Academic & Professional. ISBN 978-94-011-0601-6. OCLC 669699931
14. 3D systems completes the acquisition of Z corp and vidar. 3D Systems. January 3, 2012. Retrieved September 24, 2025.
15. Bai, Y. and Williams, C.B., 2018. Binder jetting additive manufacturing with a particle-free metal ink as a binder precursor. *Materials & Design*, 147, pp.146–156.
16. How laminated object manufacturing works. THRE3D.com. Archived from the original on 20 February 2014. Retrieved September 3, 2025.
17. Rapid prototyping: LOM. http://www.efunda.com/processes/rapid_prototyping/lom.cfm. Retrieved September 8, 2025.

18. Fang, Q., Xiong, G., Zhou, M., Tamir, T.S., Yan, C.B., Wu, H., Shen, Z. and Wang, F.Y., 2022. Process monitoring, diagnosis and control of additive manufacturing. *IEEE Transactions on Automation Science and Engineering*, 21(1), pp.1041–1067.
19. Kumar, S., Gopi, T., Harikeerthana, N., Gupta, M.K., Gaur, V., Krolczyk, G.M. and Wu, C., 2023. Machine learning techniques in additive manufacturing: A state of the art review on design, processes and production control. *Journal of Intelligent Manufacturing*, 34(1), pp.21–55.
20. Tariq, U., Joy, R., Wu, S.H., Mahmood, M.A., Malik, A.W. and Liou, F., 2023. A state-of-the-art digital factory integrating digital twin for laser additive and subtractive manufacturing processes. *Rapid Prototyping Journal*, 29(10), pp.2061–2097.
21. Chen, K., Zhang, P., Yan, H., Chen, G., Sun, T., Lu, Q., Chen, Y. and Shi, H., 2024. A review of machine learning in additive manufacturing: Design and process. *The International Journal of Advanced Manufacturing Technology*, 135(3), pp.1051–1087.
22. He, F., Yuan, L., Mu, H., Ros, M., Ding, D., Pan, Z. and Li, H., 2023. Research and application of artificial intelligence techniques for wire arc additive manufacturing: A state-of-the-art review. *Robotics and Computer-Integrated Manufacturing*, 82, p.102525.
23. Pires, J.N., Azar, A.S., Nogueira, F., Zhu, C.Y., Branco, R. and Tankova, T., 2022. The role of robotics in additive manufacturing: Review of the AM processes and introduction of an intelligent system. *Industrial Robot: The International Journal of Robotics Research and Application*, 49(2), pp.311–331.
24. Maturi, M., Locatelli, E., de Leon, A.S., Franchini, M.C. and Molina, S.I., 2025. Sustainable approaches in vat photopolymerization: Advancements, limitations, and future opportunities. *Green Chemistry*, 27(29), pp.8710–8754.
25. Gupta, V., Bankapalli, N.K., Saxena, P., Bajpai, A. and Ruan, D., 2025. Additive manufacturing of fiber-reinforced polymer matrix composites through material extrusion: A comprehensive review on filament fabrication, printing, testing methods, applications, and challenges. *Advanced Engineering Materials*, p.2500676.
26. Dong, Y., Li, H., Wang, L., Chen, J., Sucala, V.I. and Jiang, J., 2025. Review on recycled materials for additive manufacturing. *International Journal of Precision Engineering and Manufacturing-Green Technology*, pp.1–32.
27. Nagalingam, A.P., Shamir, M., Tureyen, E.B., Sharman, A.R.C., Poyraz, O., Yasa, E. and Hughes, J., 2025. Recent progress in wire-arc and wire-laser directed energy deposition (DED) of titanium and aluminium alloys. *The International Journal of Advanced Manufacturing Technology*, 136(5), pp.2035–2073.
28. Yu, K., Zhang, Y., Cai, J., Zhang, B. and Qu, X., 2025. Research progress on densification behavior of binder jetting additive manufacturing. *Review of Materials Research*, p.100073.
29. Stampone, B., Giorleo, L. and Trotta, G., 2025. The behaviour of micro-injection moulding inserts produced with material jetting technology. *Composites Part B: Engineering*, 288, p.111913.
30. Choi, J., Sung, K., Hyun, J. and Shin, S., 2025. Sheet-laminated additive manufacturing of bacterial cellulose nanofiber-reinforced hydrogels. *Carbohydrate Polymers*, 349, p.122972.
31. Khakzad, M., Khademi, M., Perruci, G.F., Lu, H., Jack, D.A. and Minary-Jolandan, M., 2025. Hybrid manufacturing of ceramic-metal composites by vat polymerization 3D printing and pulse electroplating. *Journal of Manufacturing Processes*, 144, pp.157–169.
32. Wang, Y.H., Liu, T.T., Guo, Y.P., Zhu, S.J., Liao, Z.M., Song, J.M., Zhu, X.M., Liang, J.L., Nasser, M.I., Liu, N.B. and Chang, D.H., 2025. Integrating melt electrospinning writing and microfluidics to engineer a human cardiac microenvironment for high-fidelity drug screening. *Bioactive Materials*, 45, pp.551–566.
33. Meng, X. and Gardner, L., 2025. Hybrid construction featuring wire arc additive manufacturing: Review, concepts, challenges and opportunities. *Engineering Structures*, 326, p.119337.
34. Bernal, P.N., Florczak, S., Inacker, S., Kuang, X., Madrid-Wolff, J., Regehly, M., Hecht, S., Zhang, Y.S., Moser, C. and Levato, R., 2025. The road ahead in materials and technologies for volumetric 3D printing. *Nature Reviews Materials*, pp.1–16.
35. Nouranga, K.N., Prashanth, B.N. and Ramprabhu, T., 2025. A comprehensive review of solid-phase additive techniques: Insights into friction stir additive manufacturing (FSAM) and additive friction stir deposition (AFSD). *Results in Engineering*, p.106604.

36. Kafle, A., Lu, S., Silwal, R. and Zhu, W., 2025. A review on material dynamics in cold spray additive manufacturing: Bonding, stress, and structural evolution in metals. *Metals*, 15(2), p.187.
37. Park, S., Shou, W., Makatura, L., Matusik, W. and Fu, K.K., 2022. 3D printing of polymer composites: Materials, processes, and applications. *Matter*, 5(1), pp.43–76.
38. Seifi, S.H., Nellippallil, A.B., Tian, W., Yadollahi, A., Bian, L. and Prabhu, R.K., 2025. A robust design exploration approach for the integrated inverse design of additive manufacturing process chain. *International Journal on Interactive Design and Manufacturing (IJIDeM)*, 19(3), pp.2303–2315.
39. Kim, H., Lin, Y. and Tseng, T.L.B., 2018. A review on quality control in additive manufacturing. *Rapid Prototyping Journal*, 24(3), pp.645–669.
40. Mittal, R.K., Haleem, A. and Kumar, A., eds., 2022. *Advances in Additive Manufacturing: Artificial Intelligence, Nature-Inspired, and Biomanufacturing*. Elsevier.
41. Singh, S., Ramakrishna, S. and Berto, F., 2020. 3D printing of polymer composites: A short review. *Material Design & Processing Communications*, 2(2), p.e97.
42. Ranjan, R., Kumar, D., Kundu, M. and Moi, S.C., 2022. A critical review on Classification of materials used in 3D printing process. *Materials Today: Proceedings*, 61, pp.43–49.
43. Blanco, I., 2020. The use of composite materials in 3D printing. *Journal of Composites Science*, 4(2), p.42.
44. Nasrin, T., Pourkamali-Anaraki, F. and Peterson, A.M., 2024. Application of machine learning in polymer additive manufacturing: A review. *Journal of Polymer Science*, 62(12), pp.2639–2669.
45. Ramiah, K. and Pandian, P., 2023. Effect of process parameters on the strength of ABS based FDM prototypes: Novel machine learning based hybrid optimization technique. *Materials Research Express*, 10(2), p.025305.
46. Mallikarjuna, B., Bontha, S., Krishna, P. and Balla, V.K., 2021. Characterization and thermal analysis of laser metal deposited γ-TiAl thin walls. *Journal of Materials Research and Technology*, 15, 6231–6243.
47. Ni, C., Zhu, J., Zhang, B., An, K., Wang, Y., Liu, D., . . . and Liu, C., 2025. Recent advance in laser powder bed fusion of Ti–6Al–4V alloys: Microstructure, mechanical properties and machinability. *Virtual and Physical Prototyping*, 20(1), p.e2446952.
48. Yang, H., Rao, P., Simpson, T., Lu, Y., Witherell, P., Nassar, A.R., Reutzel, E. and Kumara, S., 2020. Six-sigma quality management of additive manufacturing. *Proceedings of the IEEE*, 109(4), pp.347–376.
49. Kim, R.G., Abisado, M., Villaverde, J. and Sampedro, G.A., 2023. A survey of image-based fault monitoring in additive manufacturing: Recent developments and future directions. *Sensors*, 23(15), p.6821.
50. Niu, J., Wei, M., Hao, M., Zhang, X., Zhang, B., Li, Q., . . . and Wang, J., 2025. Improved mechanical properties of high strength titanium alloy with bi-lamellar microstructure via in-situ heat treatment during additive manufacturing. *Materials Science and Engineering: A*, 922, p.147611.
51. Dzogbewu, T.C. and du Preez, W.B., 2021. Additive manufacturing of Ti-based intermetallic alloys: A review and conceptualization of a next-generation machine. *Materials*, 14(15), p.4317.
52. Balichakra, M., Bontha, S., Krishna, P. and Balla, V.K., 2019. Prediction and validation of residual stresses generated during laser metal deposition of γ titanium aluminide thin wall structures. *Materials Research Express*, 6(10), p.106550.
53. Mallikarjuna, B., Kumar, A., Jayanth, N. and Sathies, T., 2025. Residual stress in laser cladding material. In *Tailoring of Engineering Material Properties through Laser Cladding* (pp.197–227). CRC Press.
54. Esmati, K., Chakraborty, A., Pendurti, S., Natarajan, A. and Martin, É., 2024. Anisotropic sintering behavior of stainless steel 316L printed by binder jetting additive manufacturing. *Materials Today Communications*, 41, p.110528.
55. Do, T., Kwon, P. and Shin, C.S., 2017. Process development toward full-density stainless steel parts with binder jetting printing. *International Journal of Machine Tools and Manufacture*, 121, pp.50–60.

56. Mohanavel, V., Ali, K.A., Ranganathan, K., Jeffrey, J.A., Ravikumar, M.M. and Rajkumar, S.J.M.T.P., 2021. The roles and applications of additive manufacturing in the aerospace and automobile sector. *Materials Today: Proceedings*, 47, pp.405–409.
57. Mallakpour, S., Radfar, Z. and Hussain, C.M., 2024. Advanced application of additive manufacturing in the footwear industry: From customized insoles to fully 3D-printed shoes. In *Medical Additive Manufacturing* (pp.153–178). Elsevier.
58. Abdulrahman, K.O., Akinlabi, E.T. and Mahamood, R.M., 2019. Characteristics of laser metal deposited titanium aluminide. *Materials Research Express*, 6(4), p.046504.
59. Mallikarjuna, B., Govindaraju, M., Akshay, P., Anand Kumar, S. and Nagesha, B.K., 2025. Facile production of titanium aluminide intermetallic powder for powder metallurgy and metal additive manufacturing. *Progress in Additive Manufacturing*, 10(5), pp.3751–3760.
60. Mallikarjuna, B., Hareeswar, M.N.S. and Sharath, P.C., 2023. Applications of additive manufacturing in construction and building industries. In *Practical Implementations of Additive Manufacturing Technologies* (pp.251–272). Singapore: Springer Nature Singapore.
61. Ukwaththa, J., Herath, S. and Meddage, D.P.P., 2024. A review of machine learning (ML) and explainable artificial intelligence (XAI) methods in additive manufacturing (3D Printing). *Materials Today Communications*, 41, p.110294.
62. Liu, C., Tian, W. and Kan, C., 2022. When AI meets additive manufacturing: Challenges and emerging opportunities for human-centered products development. *Journal of Manufacturing Systems*, 64, pp.648–656.
63. Mizuta, K., Oishi, W., Sakata, K., Gokan, K. and Kakehi, K., 2025. Characterization of Ti-48Al-2Cr-2Nb fabricated by directed energy deposition. *Journal of Materials Research and Technology*, 37, pp.750–759.
64. Wang, J., Shao, C., Wang, Y., Sun, L. and Zhao, Y., 2020. Microfluidics for medical additive manufacturing. *Engineering*, 6(11), 1244–1257.
65. Boretti, A., 2025. Metal additive manufacturing for internal combustion engine components: A narrative review of applications, materials, processes, challenges, and future directions. *The International Journal of Advanced Manufacturing Technology*, 1–43.
66. Toyserkani, E., Sarker, D., Ibhadode, O.O., Liravi, F., Russo, P. and Taherkhani, K., 2021. *Metal Additive Manufacturing*. John Wiley & Sons.
67. Pham, D. and Dimov, S.S., 2012. *Rapid Manufacturing: The Technologies and Applications of Rapid Prototyping and Rapid Tooling*. Springer Science & Business Media.
68. Yang, S., Tang, Y. and Zhao, Y.F., 2015. A new part consolidation method to embrace the design freedom of additive manufacturing. *Journal of Manufacturing Processes*, 20, 444–449.
69. Yu, G., Li, X., Dai, L., Xiao, L. and Song, W., 2022. Compressive properties of imperfect Ti-6Al-4V lattice structure fabricated by electron beam powder bed fusion under static and dynamic loadings. *Additive Manufacturing*, 49, 102497.
70. Wilson, J.M., Piya, C., Shin, Y.C., Zhao, F. and Ramani, K., 2014. Remanufacturing of turbine blades by laser direct deposition with its energy and environmental impact analysis. *Journal of Cleaner Production*, 80, 170–178.
71. Boschetto, A., Bottini, L., Macera, L. and Vatanparast, S., 2023. Additive manufacturing for lightweighting satellite platform. *Applied Sciences*, 13(5), 2809. https://doi.org/10.3390/app13052809
72. Troxler, C.J., Freeman, T.B., Odukomaiya, A. and Boetcher, S.K.S., 2026. Experimental characterization of a triply periodic minimal surface PCM-to-air thermal energy storage device. *International Journal of Heat and Mass Transfer*, 255, p.127716. https://doi.org/10.1016/j.ijheatmasstransfer.2025.127716.
73. Kumar, A., Kumar, P., Sharma, N. and Srivastava, A.K., eds., 2024. *3D Printing Technologies: Digital Manufacturing, Artificial Intelligence, Industry 4.0*. Walter de Gruyter GmbH & Co KG.
74. Xiong, Y., Tang, Y., Zhou, Q., Ma, Y. and Rosen, D.W., 2022. Intelligent additive manufacturing and design: State of the art and future perspectives. *Additive Manufacturing*, 59, p.103139.
75. Plathottam, S.J., Rzonca, A., Lakhnori, R. and Iloeje, C.O., 2023. A review of artificial intelligence applications in manufacturing operations. *Journal of Advanced Manufacturing and Processing*, 5(3), p.e10159.

76. Liu, J., Ye, J., Silva Izquierdo, D., et al., 2023. A review of machine learning techniques for process and performance optimization in laser beam powder bed fusion additive manufacturing. *Journal of Intelligent Manufacturing*, 34, pp.3249–3275. https://doi.org/10.1007/s10845-022-02012-0
77. Baumann, F.W., Sekulla, A., Hassler, M., Himpel, B. and Pfeil, M., 2018. Trends of machine learning in additive manufacturing. *International Journal of Rapid Manufacturing*, 7(4), pp.310–336.
78. Sarkon, G.K., Safaei, B., Kenevisi, M.S., Arman, S. and Zeeshan, Q., 2022. State-of-the-art review of machine learning applications in additive manufacturing; from design to manufacturing and property control. *Archives of Computational Methods in Engineering*, 29(7), pp.5663–5721.
79. Jandyal, A., Chaturvedi, I., Wazir, I., Raina, A. and Haq, M.I.U., 2022. 3D printing–A review of processes, materials and applications in industry 4.0. *Sustainable Operations and Computers*, 3, pp.33–42.
80. Ciccone, F., Bacciaglia, A. and Ceruti, A., 2023. Optimization with artificial intelligence in additive manufacturing: A systematic review. *Journal of the Brazilian Society of Mechanical Sciences and Engineering*, 45(6), p.303.
81. Mattera, G., Nele, L. and Paolella, D., 2024. Monitoring and control the Wire Arc Additive Manufacturing process using artificial intelligence techniques: A review. *Journal of Intelligent Manufacturing*, 35(2), pp.467–497.
82. Banerjee, D.K., Kumar, A. and Sharma, K., 2025. Artificial intelligence on additive manufacturing. *International IT Journal of Research*, ISSN, pp.3007–6706.
83. Elambasseril, J. and Brandt, M., 2022. Artificial intelligence: Way forward to empower metal additive manufacturing product development–an overview. *Materials Today: Proceedings*, 58, pp.461–465.
84. Choi, W., Advincula, R.C., Wu, H.F. and Jiang, Y., 2023. Artificial intelligence and machine learning in the design and additive manufacturing of responsive composites. *MRS Communications*, 13(5), pp.714–724.
85. Jan, Z., Ahamed, F., Mayer, W., Patel, N., Grossmann, G., Stumptner, M. and Kuusk, A., 2023. Artificial intelligence for industry 4.0: Systematic review of applications, challenges, and opportunities. *Expert Systems with Applications*, 216, p.119456.
86. Mypati, O., Mukherjee, A., Mishra, D., Pal, S.K., Chakrabarti, P.P. and Pal, A., 2023. A critical review on applications of artificial intelligence in manufacturing. *Artificial Intelligence Review*, 56(Suppl 1), pp.661–768.
87. Shahrubudin, N., Lee, T.C. and Ramlan, R.J.P.M., 2019. An overview on 3D printing technology: Technological, materials, and applications. *Procedia Manufacturing*, 35, pp.1286–1296.
88. Iqbal, A., Zhao, G., Suhaimi, H., He, N., Hussain, G. and Zhao, W., 2020. Readiness of subtractive and additive manufacturing and their sustainable amalgamation from the perspective of Industry 4.0: A comprehensive review. *The International Journal of Advanced Manufacturing Technology*, 111(9), pp.2475–2498.
89. Hamrani, A., Agarwal, A., Allouhi, A. and McDaniel, D., 2024. Applying machine learning to wire arc additive manufacturing: A systematic data-driven literature review. *Journal of Intelligent Manufacturing*, 35(6), pp.2407–2439.
90. Milazzo, M. and Libonati, F., 2022. The synergistic role of additive manufacturing and artificial intelligence for the design of new advanced intelligent systems. *Advanced Intelligent Systems*, 4(6), p.2100278.
91. Westphal, E. and Seitz, H., 2024. Generative artificial intelligence: Analyzing its future applications in additive manufacturing. *Big Data and Cognitive Computing*, 8(7), p.74.
92. Charniak, E., 1985. *Introduction to Artificial Intelligence*. Pearson Education India.
93. Raj, V. and Phridviraj, M.S.B., 2022, December. A generative model based chatbot using recurrent neural networks. In *International Conference on Advanced Network Technologies and Intelligent Computing* (pp.379–392). Cham: Springer Nature Switzerland.
94. Raj, V. and Ravichandra, S., 2021, December. Enhanced service point approach for microservices based applications using machine learning techniques. In *International Conference on Advanced Informatics for Computing Research* (pp.78–90). Cham: Springer International Publishing.

95. Badillo, S., Banfai, B., Birzele, F., Davydov, I.I., Hutchinson, L., Kam-Thong, T., Siebourg-Polster, J., Steiert, B. and Zhang, J.D., 2020. An introduction to machine learning. *Clinical Pharmacology & Therapeutics*, 107(4), pp.871–885.
96. Simeone, O., 2018. A brief introduction to machine learning for engineers. *Foundations and Trends® in Signal Processing*, 12(3–4), pp.200–431.
97. Raj, V., Bhattacharjee, S. and Nekkala, S., 2025, March. Predicting if SQL Injection Query can get access to Database using machine learning techniques. In *2025 6th International Conference on Recent Advances in Information Technology (RAIT)* (pp.1–6). IEEE.
98. Singh, A., Thakur, N. and Sharma, A., 2016, March. A review of supervised machine learning algorithms. In *2016 3rd International Conference on Computing for Sustainable Global Development (INDIACom)* (pp.1310–1315). IEEE.
99. Peng, H., Wang, J., Pérez-Jiménez, M.J. and Riscos-Núñez, A., 2015. An unsupervised learning algorithm for membrane computing. *Information Sciences*, 304, pp.80–91.
100. Shakya, A.K., Pillai, G. and Chakrabarty, S., 2023. Reinforcement learning algorithms: A brief survey. *Expert Systems with Applications*, 231, p.120495.
101. Inayathullah, S. and Buddala, R., 2025. Review of machine learning applications in additive manufacturing. *Results in Engineering*, 25, p.103676.
102. Kaushik, N., Kumar, H. and Raj, V., 2024, December. Workload prediction and its impact on QoS of microservices architecture based applications. In *2024 International Conference on Progressive Innovations in Intelligent Systems and Data Science (ICPIDS)* (pp.22–27). IEEE.
103. Babu, S.S., Mourad, A.H.I., Harib, K.H. and Vijayavenkataraman, S., 2023. Recent developments in the application of machine-learning towards accelerated predictive multiscale design and additive manufacturing. *Virtual and Physical Prototyping*, 18(1), p.e2141653.
104. Rojek, I., Kopowski, J., Lewandowski, J. and Mikołajewski, D., 2024. Use of machine learning to improve additive manufacturing processes. *Applied Sciences*, 14(15), p.6730.
105. Trovato, M., Belluomo, L., Bici, M., Prist, M., Campana, F. and Cicconi, P., 2025. Machine learning in design for additive manufacturing: A state-of-the-art discussion for a support tool in product design lifecycle. *The International Journal of Advanced Manufacturing Technology*, pp.1–24.
106. Kaur, R., Kaur, M. and Singh, M., 2025. Artificial intelligence in additive manufacturing (AI-in-AM): A scientometric analysis. *International Journal of Information Technology & Decision Making*, pp.1–43.
107. Jin, Z., Zhang, Z., Demir, K. and Gu, G.X., 2020. Machine learning for advanced additive manufacturing. *Matter*, 3(5), pp.1541–1556.
108. Morales, E.F. and Zaragoza, J.H., 2012. An introduction to reinforcement learning. In *Decision Theory Models for Applications in Artificial Intelligence: Concepts and Solutions* (pp.63–80). IGI Global Scientific Publishing.
109. Wang, C., Tan, X.P., Tor, S.B. and Lim, C.S., 2020. Machine learning in additive manufacturing: State-of-the-art and perspectives. *Additive Manufacturing*, 36, p.101538.
110. Adapa, V.S.K., Kalidindi, S.R. and Saldana, C.J., 2025. Rapid development of metal additive manufacturing using artificial intelligence/machine learning and high-throughput material testing. *Annual Review of Materials Research*, 55.
111. Ben Amor, S., Elloumi, N., Eltaief, A., Louhichi, B., Alrasheedi, N.H. and Seibi, A., 2024. Digital twin implementation in additive manufacturing: A comprehensive review. *Processes*, 12(6), p.1062.
112. Lin, Y., Lu, G., Li, T., Liu, C., Zhang, H. and Peng, S., 2025. Rapid life cycle assessment for metal additive manufactured product: A multi-feature based predictive approach. *Journal of Cleaner Production*, 521, p.146221.
113. Cheepu, M., 2023. Machine learning approach for the prediction of defect characteristics in wire arc additive manufacturing. *Transactions of the Indian Institute of Metals*, 76(2), pp.447–455.
114. Zhang, Z., George, A., Ferdous Alam, M., Eubel, C., Prasad Vallabh, C.K., Shtein, M., Barton, K. and Hoelzle, D.J., 2024. An additive manufacturing testbed to evaluate machine learning-based autonomous manufacturing. *Journal of Manufacturing Science and Engineering*, 146(3), p.031008.

115. Batu, T., Lemu, H.G. and Shimels, H., 2023. Application of artificial intelligence for surface roughness prediction of additively manufactured components. *Materials*, 16(18), p.6266.
116. Pereira, A.G., Barbosa, G.F., Godinho Filho, M., Shiki, S.B. and da Silva, A.L., 2025. Quality control in extrusion-based additive manufacturing: A review of machine learning approaches. *IEEE Transactions on Cybernetics*, 55(6), pp.2522–2534.
117. Li, A., Poudel, A., Shao, S., Shamsaei, N. and Liu, J., 2025. Nondestructive fatigue life prediction for additively manufactured metal parts through a multimodal transfer learning framework. *IISE Transactions*, 57(11), pp.1344–1359.
118. Ali, A., Riaz, R.D., Malik, U.J., Abbas, S.B., Usman, M., Shah, M.U., Kim, I.H., Hanif, A., Faizan, M., 2023. Machine learning-based predictive model for tensile and flexural strength of 3D-printed concrete. *Materials (Basel)*, 16(11), p.4149. https://doi.org/10.3390/ma16114149.
119. Graziosi, S., Faludi, J., Stanković, T., Borgianni, Y., Meisel, N., Hallstedt, S.I. and Rosen, D.W., 2024. A vision for sustainable additive manufacturing. *Nature Sustainability*, 7(6), pp.698–705.
120. Rahmani, R., Bashiri, B., Lopes, S.I., Hussain, A., Maurya, H.S. and Vilu, R., 2025. Sustainable additive manufacturing: An overview on life cycle impacts and cost efficiency of laser powder bed fusion. *Journal of Manufacturing and Materials Processing*, 9(1), p.18.
121. Wang, K., Lin, H., Fang, N., Xu, J., Zhang, S., Tan, J., Qin, J. and Liang, X., 2025. Cross-patch graph transformer enforced by contrastive information fusion for energy demand forecasting towards sustainable additive manufacturing. *Journal of Industrial Information Integration*, 45, p.100795.
122. Equbal, A., Equbal, A., Khan, Z.A. and Badruddin, I.A., 2025. Machine learning in additive manufacturing: A comprehensive insight. *International Journal of Lightweight Materials & Manufacture*, 8(2).
123. Soori, M., Jough, F.K.G., Dastres, R. and Arezoo, B., 2025. Additive manufacturing modification by artificial intelligence, machine learning, and deep learning: A review. *Additive Manufacturing Frontiers*, p.200198.
124. Rojek, I., Mikołajewski, D., Kempiński, M., Galas, K. and Piszcz, A., 2025. Emerging applications of machine learning in 3D printing. *Applied Sciences*, 15(4), p.1781.
125. Jiang, J., 2023. A survey of machine learning in additive manufacturing technologies. *International Journal of Computer Integrated Manufacturing*, 36(9), pp.1258–1280.
126. Ng, W.L., Goh, G.L., Goh, G.D., Ten, J.S.J. and Yeong, W.Y., 2024. Progress and opportunities for machine learning in materials and processes of additive manufacturing. *Advanced Materials*, 36(34), p.2310006.
127. Xiao, S., Li, J., Wang, Z., Chen, Y. and Tofighi, S., 2024. Advancing additive manufacturing through machine learning techniques: A state-of-the-art review. *Future Internet*, 16(11), p.419.
128. Raj, V., Prakash, B.K., Kumar, A. and Garg, P., 2024, December. Optimize the time a Mercedes-Benz spends on the test bench using stacking ensemble learning. In *2024 International Conference on Progressive Innovations in Intelligent Systems and Data Science (ICPIDS)* (pp.445–450). IEEE.
129. Ladani, L.J., 2021. Applications of artificial intelligence and machine learning in metal additive manufacturing. *Journal of Physics: Materials*, 4(4), p.042009.
130. Patel, S., Nathani, A., Poozesh, A., Xu, S., Kazempoor, P. and Ghamarian, I., 2024. Combining neural networks and genetic algorithms to understand composition–microstructure–property relationships in additively manufactured metals. *Journal of Manufacturing and Materials Processing*, 8(6), p.269.
131. Grierson, D., Rennie, A.E. and Quayle, S.D., 2021. Machine learning for additive manufacturing. *Encyclopedia*, 1(3), pp.576–588.
132. Saimon, A.I., Yangue, E., Yue, X., Kong, Z. and Liu, C., 2025. Advancing additive manufacturing through deep learning: A comprehensive review of current progress and future challenges. *IISE Transactions*, pp.1–24.
133. Zhou, H.R., Yang, H., Li, H.Q., Ma, Y.C., Yu, S., Shi, J., Cheng, J.C., Gao, P., Yu, B., Miao, Z.Q. and Wei, Y.P., 2024. Advancements in machine learning for material design and process optimization in the field of additive manufacturing. *China Foundry*, 21(2), pp.101–115.

134. Charalampous, P., Kostavelis, I. and Tzovaras, D., 2020. Non-destructive quality control methods in additive manufacturing: A survey. *Rapid Prototyping Journal*, 26(4), pp.777–790.
135. Jiang, J., Xiong, Y., Zhang, Z. and Rosen, D.W., 2022. Machine learning integrated design for additive manufacturing. *Journal of Intelligent Manufacturing*, 33(4), pp.1073–1086.
136. Xames, M.D., Torsha, F.K. and Sarwar, F., 2023. A systematic literature review on recent trends of machine learning applications in additive manufacturing. *Journal of Intelligent Manufacturing*, 34(6), pp.2529–2555.
137. Habeeb, H.A., Wahab, D.A., Azman, A.H. and Alkahari, M.R., 2023. Design optimization method based on artificial intelligence (hybrid method) for repair and restoration using additive manufacturing technology. *Metals*, 13(3), p.490.
138. Doran, E., Bommer, S. and Badiru, A., 2022. Integration of human factors, cognitive ergonomics, and artificial intelligence in the human-machine interface for additive manufacturing. *International Journal of Mechatronics and Manufacturing Systems*, 15(4), pp.310–330.
139. Razvi, S.S., Feng, S., Narayanan, A., Lee, Y.T.T. and Witherell, P., 2019, August. A review of machine learning applications in additive manufacturing. In *International Design Engineering Technical Conferences and Computers and Information in Engineering Conference* (Vol. 59179, p.V001T02A040). American Society of Mechanical Engineers.
140. Raza, A., Deen, K.M., Jaafreh, R., Hamad, K., Haider, A. and Haider, W., 2022. Incorporation of machine learning in additive manufacturing: A review. *The International Journal of Advanced Manufacturing Technology*, 122(3), pp.1143–1166.
141. Wang, Y., Zheng, P., Peng, T., Yang, H. and Zou, J., 2020. Smart additive manufacturing: Current artificial intelligence-enabled methods and future perspectives. *Science China Technological Sciences*, 63(9), pp.1600–1611.
142. Sani, A.R., Zolfagharian, A. and Kouzani, A.Z., 2024. Artificial intelligence-augmented additive manufacturing: Insights on closed-loop 3D printing. *Advanced Intelligent Systems*, 6(10), p.2400102.
143. Pancholi, S., Gupta, M.K., Bartoszuk, M., Vashishtha, G., Ross, N.S., Korkmaz, M.E., Krolczyk, G.M. and Petru, J., 2025. Transforming additive manufacturing with artificial intelligence: A review of current and future trends. *Archives of Computational Methods in Engineering*, pp.1–32.
144. Yadav, A., Srivastava, M., Jain, P.K. and Rathee, S., 2025. Machine learning in solid state additive manufacturing: State-of-the-art and future perspectives. *International Journal on Interactive Design and Manufacturing (IJIDeM)*, 19(4), pp.2317–2336.
145. Fan, H., Liu, C., Bian, S., Ma, C., Huang, J., Liu, X., Doyle, M., Lu, T., Chow, E., Chen, L. and Fuh, J.Y.H., 2025. New era towards autonomous additive manufacturing: A review of recent trends and future perspectives. *International Journal of Extreme Manufacturing*, 7, p.3.
146. Yang, J., Chen, Y., Huang, W. and Li, Y., 2017, September. Survey on artificial intelligence for additive manufacturing. In *2017 23rd International Conference on Automation and Computing (ICAC)* (pp.1–6). IEEE.
147. Safdar, M., Xie, J., Mircea, A. and Zhao, Y.F., 2025. Human–artificial intelligence teaming for scientific information extraction from data-driven additive manufacturing literature using large language models. *Journal of Computing and Information Science in Engineering*, 25(7), p.074501.
148. Chang, L.K., Chen, P.C., Tsai, M.C., Lee, R.M., Huang, J.C., Chang, T.W. and Horng, M.H., 2025. Enhancing prediction of magnetic properties in additive manufacturing products through a 3D convolutional vision transformer model. *The International Journal of Advanced Manufacturing Technology*, pp.1–17.
149. Tripathi, S., Ansari, A.A., Singh, M., Dash, M., Kumar, P., Singh, H., Panda, B., Nukavarapu, S., Camci-Unal, G., Li, B. and Jain, P.K., 2025. Transforming surgical planning and procedures through the synergistic use of additive manufacturing, advanced materials and artificial intelligence: Challenges and opportunities. *Materials Horizons*, 12(19), pp.7814–7864.
150. Narwane, V.S., Gunasekaran, A., Gardas, B.B. and Sirisomboonsuk, P., 2025. Quantum machine learning a new frontier in smart manufacturing: A systematic literature review from period 1995 to 2021. *International Journal of Computer Integrated Manufacturing*, 38(1), pp.116–135.
151. Chandu, S.V., Varri, U.S. and Raj, V., 2025. Federated learning in detecting fake news: A survey. *Procedia Computer Science*, 260, pp.457–467.

152. Eider, G.P., Meritxell, G.O., Jon, L. and Naroa, M., 2025. Quantum surrogate modelling to solve phase field simulation problems in metal solidification. *The Journal of Supercomputing*, 81(5), p.656.
153. Okokpujie, I.P. and Tartibu, L.K., 2025. Study of the economic viability of internet of things (IoTs) in additive and advanced manufacturing: A comprehensive review. *Progress in Additive Manufacturing*, 10(5), pp.3175–3194.

Chapter 2

Role of Additive Manufacturing in Industry 4.0

Madhukrishna Priyadarsini, Mallikarjuna B, and Santosh Kumar Mishra

2.1 INTRODUCTION TO ADDITIVE MANUFACTURING

Additive Manufacturing (AM) [1], often referred to as 3D printing [2], is a process that constructs complex components layer by layer directly from digital CAD models, in contrast to traditional subtractive methods. AM technologies are commonly grouped into seven categories: vat photopolymerisation, material extrusion, powder bed fusion, binder jetting, material jetting, directed energy deposition, and sheet lamination [3]. AM has revolutionised prototyping and end-use part production by significantly reducing lead times and enabling unprecedented levels of customisation. By building components layer by layer, it promotes highly efficient material use and facilitates innovative, complex designs that are often impossible to achieve with traditional manufacturing methods [4, 5]. This technology has found diverse applications across critical industries, including aerospace, where it produces lightweight, high-strength parts; automotive, for custom tools and components; biomedical, for patient-specific implants and prosthetics; and even construction and fashion, where it enables unique structural and aesthetic possibilities. AM has a profound impact on the existing manufacturing [6]. Distributed manufacturing models decentralise production by leveraging localised on-demand fabrication to place manufacturing centres closer to consumers, thereby streamlining supply chains. Then the precision of AM significantly reduces material waste, while its decentralised nature shortens supply chains, thereby diminishing logistical footprints and lowering inventory carrying costs. Lastly, the creation of entirely new business models based on digital product distribution and cloud manufacturing services [3, 7]. In addition, the AM market demonstrates remarkable growth potential, with projections indicating it will reach US$314.0 million by 2030, expanding at a compound annual growth rate (CAGR) of 26.3% [5]. However, realising this potential requires overcoming significant technical challenges, particularly in parameter optimisation. Successful AM production requires careful calibration of multiple process variables, including layer height, print speed, build orientation, and energy source settings. This optimisation process typically involves considerable experimentation, often resulting in material waste, extended production times, and increased costs during the initial phases [3, 8–10].

However, AM perfectly aligns with Industry 4.0's vision of smart [11], interconnected production environments. Its digital workflow seamlessly integrates design with fabrication, enabling fully automated manufacturing systems and supporting the development of intelligent factories. The capability of technology to produce complex geometries on demand [12] facilitates more agile supply chains and distributed manufacturing models, thereby significantly reducing transportation needs and associated carbon emissions. Furthermore, AM contributes to improved product lifecycle management through digital inventories and localised production, while simultaneously reducing material waste and energy consumption compared to conventional manufacturing approaches. These advantages position AM as the

DOI: 10.1201/9781003590774-2

cornerstone technology in the transition to more sustainable [13, 14], efficient, and intelligent industrial systems through Industry 4.0 [15]. The main advantages of AM include mass customisation, reduced material wastage, lightweight design, and rapid prototyping, making it indispensable in the aerospace, healthcare, automotive, and energy sectors [10].

Recent advances in artificial intelligence (AI), machine learning (ML), and the Industrial Internet of Things (IIoT) are further expanding AM's capabilities by enabling adaptive, data-driven manufacturing environments [16, 17]. The integration of real-time sensor networks and digital twins enables predictive maintenance, process monitoring, and closed-loop control, thereby improving the reliability and repeatability of builds [18]. Blockchain and cloud-based platforms are being explored to secure design files, manage intellectual property, and improve traceability in distributed manufacturing networks. Meanwhile, edge and quantum computing paradigms promise to accelerate process optimisation by reducing computational latency and improving multi-objective decision-making in high-fidelity simulations [19]. These technological synergies transform AM from a standalone fabrication technique into a core pillar of smart, autonomous and sustainable industrial ecosystems, marking a decisive step towards Industry 5.0, where human-machine collaboration and environmental intelligence converge.

2.2 INDUSTRIAL REVOLUTIONS

It is well known that four industrial revolutions have occurred since 1760, marking significant transformations in manufacturing and technological advancement. Each revolution plays a significant role in advancing product manufacturing. Industry 1.0 (1760–1840) production shifted from manual craftsmanship to machine-based methods through the introduction of mechanisation, steam engines, and waterpower, marking the start of industrialised manufacturing. Industry 2.0 (late 19th–early 20th century) manufacturing entered an era of mass production made possible by widespread electrification, the development of assembly lines, and the structured division of labour, leading to significant gains in efficiency and production scale. Then, Industry 3.0 (1970s–2000s) manufacturing sector advanced further with the integration of computers, electronics, and early robotics, ushering in automated production processes and setting the stage for digital manufacturing. Modern factories in Industry 4.0 (2011-present) are being transformed by the convergence of smart technologies, creating interconnected, intelligent, and autonomous manufacturing environments. **Figure 2.1** illustrates the evolution of industrial revolutions.

2.3 INDUSTRY 4.0

Industry 4.0 [20], also known as the fourth industrial revolution [21], refers to the integration of advanced digital technologies into manufacturing systems. Key technologies driving this transformation include the Internet of Things (IoT) [22], the Industrial Internet of Things (IIoT) [23], AM, ML, AI, digital twin simulations [24], big data analytics [25], virtual and augmented reality, as well as advanced automation and robotics. The adoption of these Industry 4.0 technologies continues to grow rapidly across manufacturing sectors. Industry 4.0 in manufacturing is characterised by enhanced inter-connectivity, intelligent systems [26], and widespread automation. The concept gained prominence after its introduction at the Hannover Messe industrial fair in Germany in 2011. Industry 4.0 represents the convergence of cyber-physical systems, IoT, cloud computing, and real-time data analytics, leading to the development of smart factories. Within this framework, AM, commonly referred to as 3D printing [27], serves as a cornerstone of this industrial evolution, enabling unprecedented

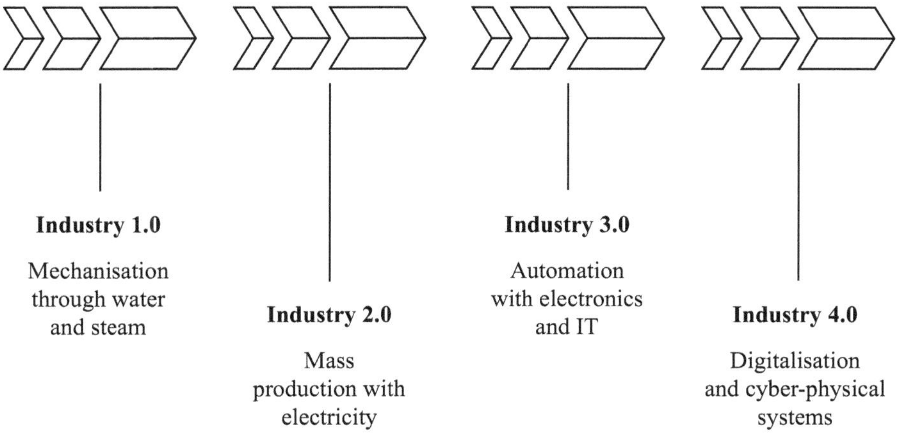

Figure 2.1 Illustration of four industrial revolutions.

flexibility and innovation in production processes. **Figure 2.2** shows the market revenue, forecast, and growth rate for Industry 4.0. The market revenue of Industry 4.0 was USD 135.32 billion and is expected to grow to USD 634.94 billion, with a compound annual growth rate of 18.74% (2023–2032). This concludes that the Industry 4.0 market is strong, changes the manufacturing industry and then influences any country's GDP [5, 28].

2.3.1 Relevance of Industry 4.0 in AM

It is well established that the fabrication of a component through AM constitutes a highly structured process comprising eight distinct stages, each characterised by specific requirements and critical process parameters that collectively determine the quality and performance of the final product. These stages typically include Computer-Aided Design (CAD) modelling, conversion to an STL file, digital slicing, identification of process parameters, machine setup, part fabrication, removal from the build platform, and post-processing [3, 29]. Crucially, each step is not isolated but is part of a complex digital thread where data integrity and flow are paramount. The journey begins with the CAD model, which contains the full geometric data of the part. This data is then approximated and simplified into an STL file, a process that can introduce geometric errors. Sending this file converts it into a series of thin, printable layers (slicing), generating critical toolpath data (G-code). This G-code must then be integrated with specific material and machine parameters (e.g., laser power, temperature, speed) to create a build file [29]. This file is loaded onto the AM machine, initiating the physical build, during which the machine's internal computer interprets the instructions. Modern systems with in-situ monitoring generate vast amounts of real-time sensor data (thermal, visual, acoustic) to validate the process.

Finally, data guides post-processing, such as specifying support-removal locations or heat-treatment cycles based on recorded build parameters. A failure in data fidelity or flow at any step from a corrupted STL file to incorrect parameters or misinterpreted sensor feedback can lead to catastrophic build failure, underscoring that AM is as much a data management challenge as a manufacturing one [30]. The integration of Industry 4.0 technologies fundamentally enhances AM by enabling smarter, more efficient, and interconnected production systems for all these steps. Through IoT, real-time monitoring of 3D printers enables quality control and predictive maintenance, while AI and ML optimise process parameters, such as

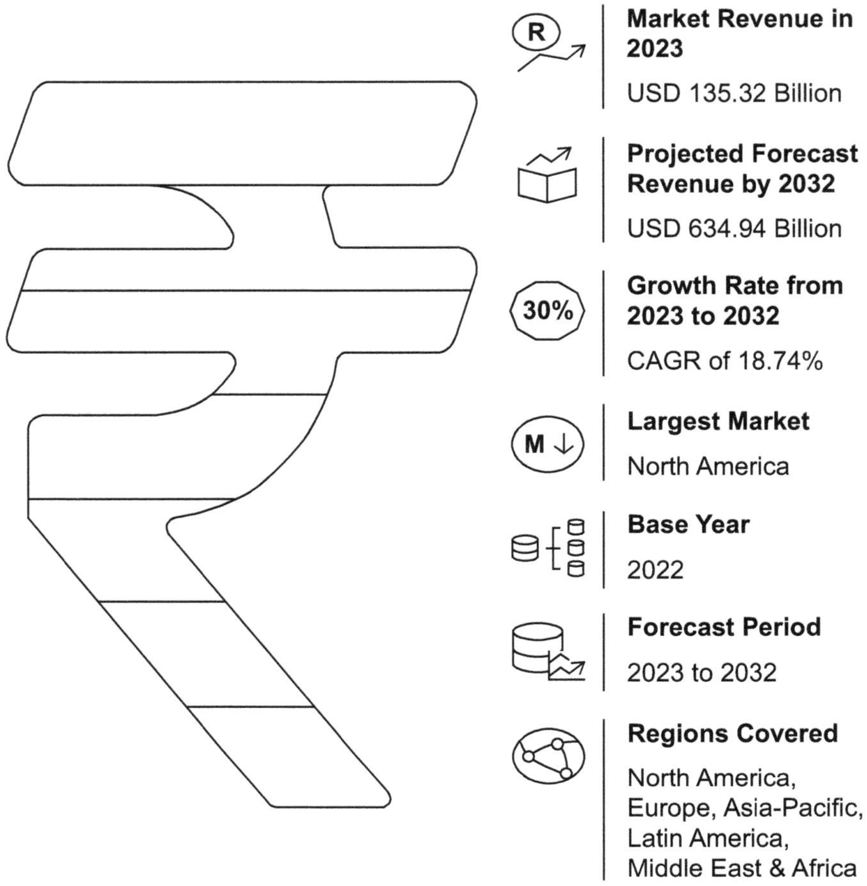

Figure 2.2 Global market overview.

layer adhesion and energy use. Digital twins virtually simulate printing processes, reducing material waste and iteration time. Cloud computing facilitates decentralised production, allowing global collaboration and on-demand manufacturing. Big data analytics drives continuous improvement in design and material usage. Together, these advancements align AM with Industry 4.0 goals: agile, customisable, and sustainable manufacturing, transforming supply chains, and enabling mass customisation in aerospace, healthcare, and automotive industries [31]. **Figure 2.3** shows how the inefficient AM can be converted into a more flexible, efficient, and agile manufacturing process.

2.3.2 Integration of AM with Industry 4.0 Technologies

The true potential of AM is realised when it is integrated with broader Industry 4.0 technologies that enable intelligent, connected, and adaptive production. **Figure 2.4** illustrates the connection of AM machinery with IIoT sensors, monitoring screens, gateways, and data processing components such as AI, ML, and cloud systems within an Industry 4.0 environment. The following subsections describe the key technologies that involve the integration of AM-IIoT, along with the visual representation in **Figure 2.5**.

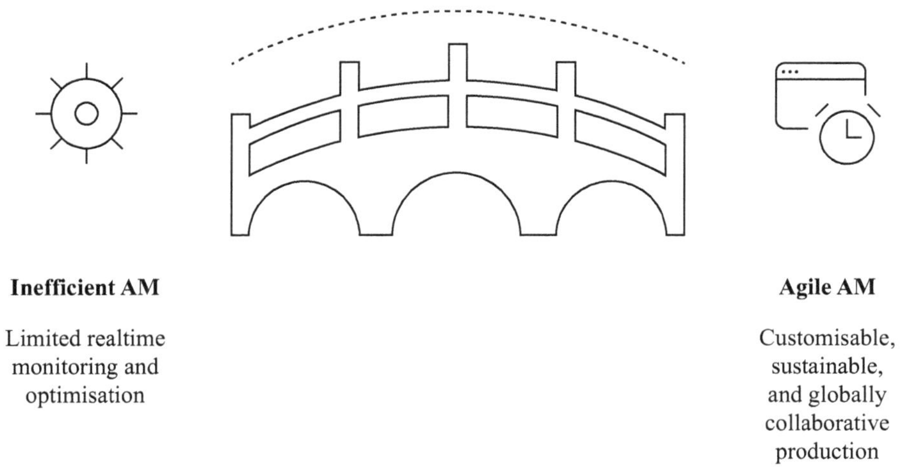

Figure 2.3 Industry 4.0 technologies transform additive manufacturing into agile, customisable, and sustainable production.

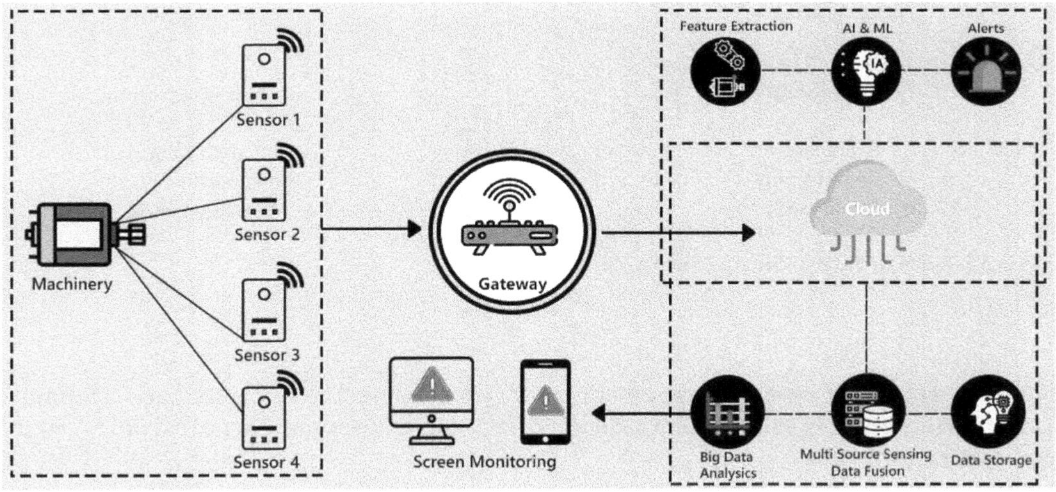

Figure 2.4 IIoT sensors and key processes involved in Industry 4.0.

2.3.2.1 Cyber-Physical Systems

Cyber-Physical Systems (CPS) embed computation, sensing, and control into AM equipment, creating a tight integration between physical processes and digital intelligence. In a CPS-enabled AM setup, sensors continuously capture melt pool dimensions, layer uniformity, and chamber conditions, while controllers adjust process parameters in real time. This fusion enables adaptive manufacturing, where deviations in temperature or scan paths are corrected instantly. For example, EOS's EOSTATE monitoring suite operates as a CPS, linking optical sensors and control units for dynamic adjustments. Similarly, Siemens MindSphere integrates AM machines into CPS networks, making them responsive nodes within larger digital factories. By doing so, CPS transforms AM from isolated printers into autonomous, networked

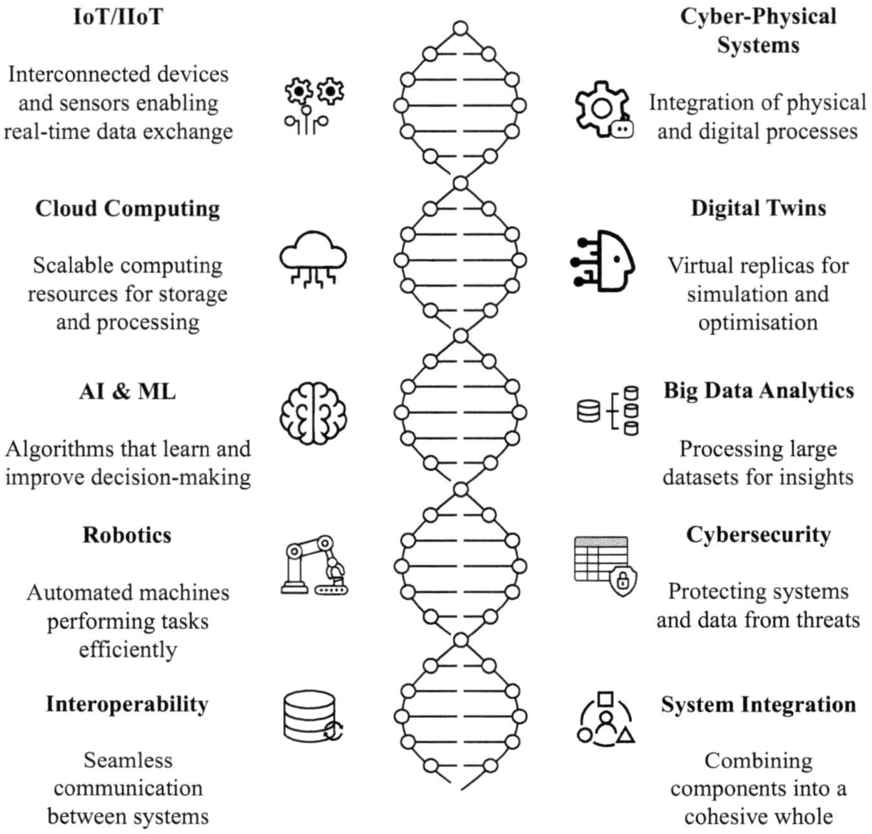

Figure 2.5 Industry 4.0 foundation technologies.

entities that can self-monitor and self-correct. By embedding AM machines into CPS frameworks, manufacturers gain resilience, adaptiveness, and self-optimisation capabilities central to smart factories [32].

2.3.2.2 Internet of Things (IoT), Industrial Internet of Things (IIoT)

IoT refers to interconnected devices that share data through networks, while IIoT extends this concept to industrial environment. Both IoT and IIoT provide the backbone for connectivity in AM. Machines embedded with thermal, acoustic, and vibration sensors share data over IIoT gateways, enabling distributed coordination. This connectivity enables cross-site synchronisation, such as aligning production schedules across AM facilities in different countries. HP's Multi Jet Fusion printers [33] use IoT sensors and cloud dashboards for fleet management, allowing engineers to monitor utilisation and uptime across locations. In aerospace, MTU (Motoren- und Turbinen-Union) Aero Engines [34] leverages IIoT-enabled EOS systems to connect in-situ monitoring with enterprise MES (Manufacturing Execution Systems), ensuring quality consistency across distributed plants. By embedding AM into IIoT networks, industries achieve remote visibility, data-driven decision-making, and scalable global manufacturing ecosystems [34].

2.3.2.3 Digital Twins

A digital twin is a dynamic virtual replica of a physical system. In AM, it mirrors the build process by integrating real-time sensor data with predictive simulations, supporting proactive quality control, predictive maintenance, and lifecycle management of manufactured parts. By linking sensor streams with predictive simulations, digital twins [35] enhance quality control by continuously comparing expected and actual outcomes. For example, if melt pool dimensions deviate from digital predictions, the system can proactively adjust laser power or scan strategy. This allows manufacturers to move from reactive inspections to proactive quality assurance, significantly reducing scrap rates. In AM-IIoT ecosystems, digital twins [36] also support lifecycle management, enabling parts to be monitored even during field deployment. The diagram shown in **Figure 2.6** is the integration of the digital twin implementation of AM and IIoT.

2.3.2.4 Artificial Intelligence

AI applies computational intelligence to analyse, optimise, and automate decision-making. AI transforms AM by enabling intelligent optimisation and anomaly detection across every build. By analysing thermal images, acoustic data, and laser signals, AI algorithms can optimise scan strategies, predict failure points, and recommend corrective actions. In healthcare, Materialise AI software personalises implant designs by integrating patient data with AM constraints, ensuring better fit and reliability. Siemens NX with AI-driven topology optimisation generates lightweight aerospace parts optimised for AM. Furthermore, AI-driven exposure pattern monitoring in EOS systems helps detect irregularities mid-build. With AI, AM shifts from static, pre-programmed workflows to adaptive, learning-driven manufacturing pipelines capable of delivering customised, defect-free parts at scale [37]. In addition, AI is increasingly being deployed in AM to accurately predict and validate mechanical properties, such as tensile strength, compressive strength, and fatigue life. By analysing the complex relationships between process parameters (e.g., laser power, scan speed, hatch spacing) and the resulting material microstructure, AI models can forecast a part's performance before it is even built. The implementation of AI yields significant improvements, such as increased prediction accuracy, improved defect detection, and fatigue life forecasting [37].

2.3.2.5 Cybersecurity

The integration of AM and IIoT raises cybersecurity concerns, as design files, process parameters, and monitoring data are transmitted across networks. Without robust protection, sensitive intellectual property or build logs could be stolen, altered, or sabotaged. To counter this, manufacturers employ end-to-end encryption, secure device authentication, and blockchain-based traceability [38]. For example, HP's Digital Production Suite enforces mutual Transport Layer Security (TLS) encryption to ensure only authorised users can access printers or logs. Research projects with Lockheed Martin and the Defence Advanced Research Projects Agency (DARPA) have explored embedding digital watermarks into AM files to detect tampering. Cybersecurity is thus not just an IT layer but a core enabler of trust, ensuring that certified AM parts remain authentic and tamper-proof throughout their lifecycle.

2.3.2.6 Cloud Computing

Cloud platforms provide scalable, centralised infrastructure for AM data storage, simulation, and coordination. By hosting digital inventories and process logs in the cloud, companies achieve distributed manufacturing across global networks. Siemens Additive Manufacturing

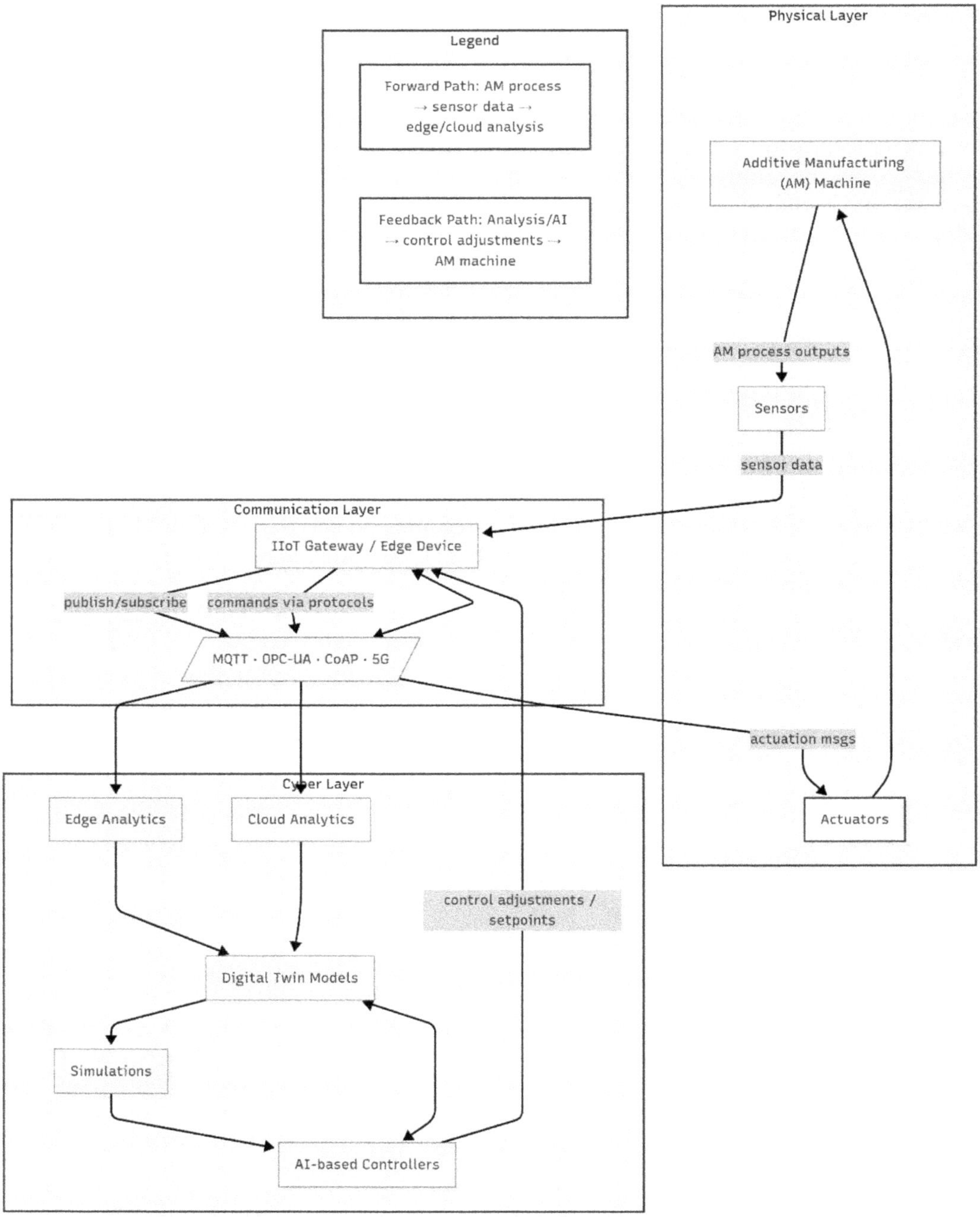

Figure 2.6 Digital twin implementation in AM-IIoT environment.

Network (AMN) connects designers, suppliers, and customers through a cloud-based marketplace for AM parts, reducing lead times and enhancing supply chain transparency. HP 3D Centre cloud dashboards enable global job scheduling and status monitoring across printer fleets. Cloud systems also host computationally intensive tasks, such as digital twin simulations and fleet-wide defect analytics, enabling real-time optimisation and collaboration.

However, hybrid approaches combining cloud and edge are increasingly adopted to balance scalability with low-latency control [5].

2.3.2.7 Machine Learning (ML)

ML, a subset of AI, uses algorithms trained on past data to recognise patterns and improve predictions without explicit programming. ML enables AM-IIoT systems [39] to learn from historical data and continuously improve build outcomes. Supervised learning algorithms such as SVMs, Random Forests, and regression models classify defects and predict part properties from process parameters. Unsupervised methods like clustering and PCA help identify hidden patterns in sensor data for anomaly detection. Deep learning approaches (CNNs, RNNs, autoencoders) analyse thermal images, acoustic signals, and time-series data for accurate in-situ monitoring. Reinforcement learning is increasingly applied for adaptive control, dynamically tuning laser power and scan strategies. Hybrid frameworks that combine ML with physics-based models further improve interpretability and reliability. Collectively, these algorithms enable real-time decision-making, predictive quality assurance, and scalable AM production, advancing the integration of AM into Industry 4.0 ecosystems. For example, MTU Aero Engines and EOS apply ML models to reduce reliance on costly CT scans, using melt pool data to qualify parts. Research at ETH Zurich applies ML to predict microstructural outcomes in laser powder bed fusion (LPBF), guiding laser settings dynamically. By combining ML with AI-driven optimisation, AM systems evolve into self-learning factories where every build enhances the predictive power of the next [34].

2.3.2.8 Big Data

AM processes generate massive data streams, ranging from terabytes of melt-pool images to petabytes of lifecycle-monitoring logs. Big data platforms enable industries to collect, integrate, and analyse this information for fleet-wide insights. Aerospace companies like GE aggregate melt-pool information, powder batch histories, and part inspection data into centralised big-data repositories via Predix. This allows statistical modelling of defect probabilities across entire fleets of builds. Big data analytics also enables supply chain optimisation by linking material usage, machine performance, and customer demand. Without big data, AM-IIoT integration would remain siloed; with it, manufacturers unlock predictive quality assurance, real-time traceability, and continuous improvement at scale [25].

2.3.2.9 Robotics and Automation

Robotics and automation integrate robotic arms, conveyors, and automated handling systems with AM machines. They support material handling, powder removal, support separation, and finishing tasks, enabling lights-out manufacturing and large-scale industrialisation of AM. Siemens has demonstrated fully automated AM factories in which robots connect AM machines to machining centres and inspection stations [40]. Daimler Trucks uses robotics to automate post-processing of AM spare parts, reducing turnaround times. When linked with IIoT-enabled MES, robotic systems execute builds around the clock, enabling lights-out manufacturing. This integration transforms AM from a prototyping tool into a scalable industrial production solution [32].

The diagram in **Figure 2.7** illustrates Industry 4.0 implementation for AM using a layered architecture. On the shop floor, AM machines, in-situ sensors, and robotics generate process data within cyber-physical systems. The connectivity and edge layer processes this data in real time via IIoT gateways and edge analytics, enabling low-latency defect detection. The cloud

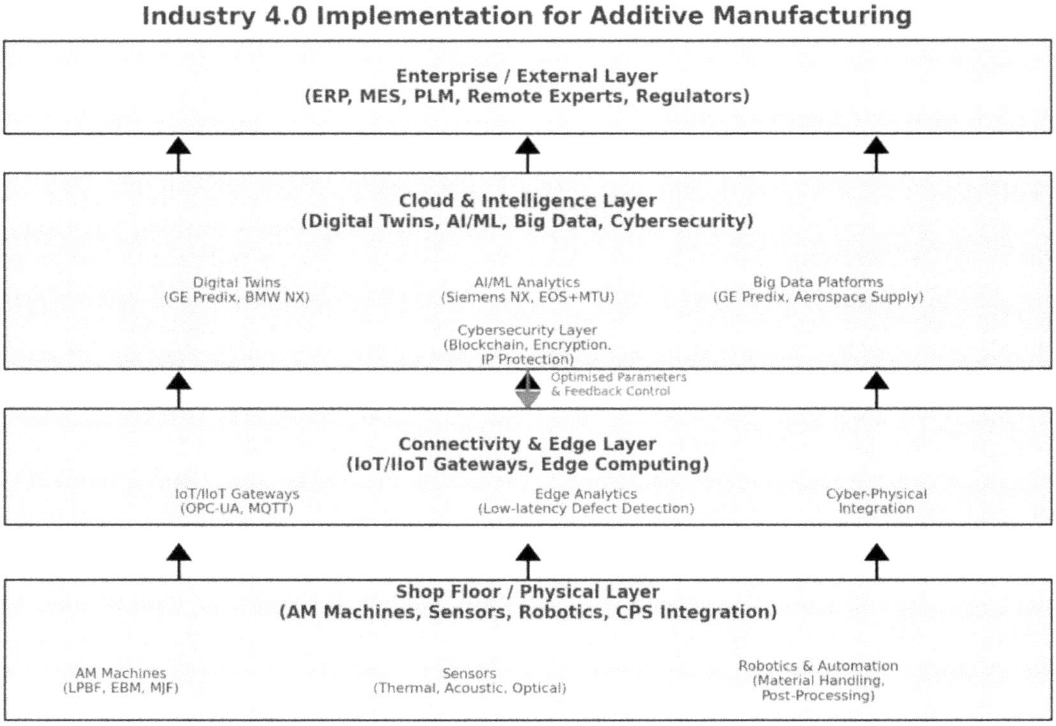

Figure 2.7 Industry 4.0 implementation for Additive Manufacturing showing layered architecture.

and intelligence layer hosts digital twins, AI/ML models, big data platforms, and cybersecurity, transforming raw signals into optimised feedback and certification-ready logs. Finally, the enterprise layer integrates ERP, MES, PLM systems, and external regulators, ensuring global coordination, traceability, and compliance [22].

2.4 APPLICATIONS OF AM-IIOT INTEGRATION

The fusion of AM and the IIoT enables next-generation applications that go beyond prototyping, supporting intelligent, adaptive, and sustainable manufacturing systems. By embedding sensors, connectivity, and analytics into AM workflows, industries achieve real-time decision-making, predictive quality control, and distributed production capabilities. The application areas of AM, IIoT, and Industry 4.0 integration are shown in **Figure 2.8**. Major applications are discussed in the following subsections:

2.4.1 Smart Production Lines and On-Demand Manufacturing

Traditional production lines are often rigid and capital-intensive. The AM-IIoT convergence transforms them into flexible, adaptive systems capable of producing parts on demand. IIoT-enabled AM machines are integrated into smart production cells, where real-time monitoring ensures minimal downtime. On-demand manufacturing enables companies to fabricate spare parts or customised components instantly, reducing reliance on warehouses. In addition, Digital

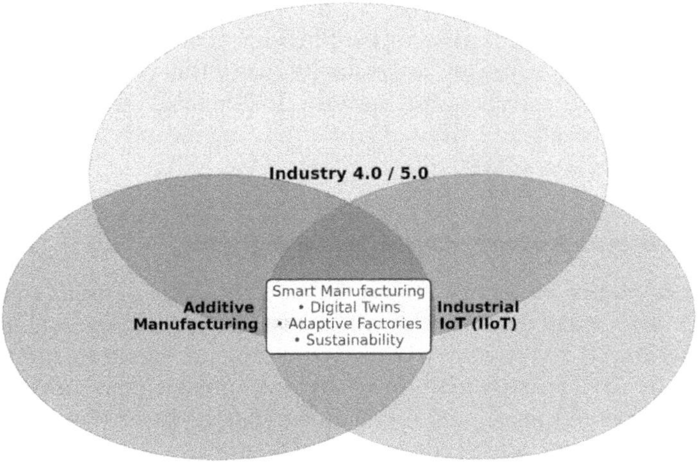

Figure 2.8 Application areas of 3 key technologies: AM, IIoT, and Industry 4.0 integration.

twins simulate line efficiency, enabling predictive scheduling and reducing bottlenecks. The real-life example is as follows: Siemens' IIoT-enabled AM facilities in Germany demonstrate automated production scheduling using real-time machine health data [41]. On an analytical note: if P(t) is the production capacity at time t and d(t) is the real-time demand, then on-demand production aims to minimise:

$$\min \int_0^T \left|P(t) - d(t)\right| dt \qquad (2.1),$$

which ensures supply matches demand with minimal surplus.

2.4.2 Supply Chain and Digital Inventory Management

The integration of AM and IIoT supports a shift from physical stockpiles to digital inventories. CAD/STL files are stored in the cloud and printed locally on IIoT-enabled AM systems. IoT sensors track raw material usage, machine readiness, and logistics status. Blockchain frameworks secure design IP and validate the authenticity of produced parts. For instance, the U.S. Navy uses IIoT-enabled AM on ships for digital spare parts, reducing supply chain risks. Let Id be the digital inventory size (number of digital designs), and Ip be the physical inventory. The integration reduces:

$$I_p \ll I_d \text{ while ensuring Availability}(I_d) \geq \text{Availability}(I_p) \qquad (2.2),$$

achieving higher availability at lower cost.

2.4.3 Healthcare and Biomedical Implants

Healthcare is profoundly transformed by the synergy of AM and the Industrial Internet of Things (IIoT), enabling the creation of patient-specific, IoT-enabled medical devices. This integration spans three key areas: (1) AM facilitates the production of customised prosthetics, orthopaedic implants, and dental crowns; (2) IIoT enhances these devices with embedded biosensors to monitor healing, infection, or load-bearing conditions in real time; and (3) Telemedicine integration allows surgeons to remotely access this data via secure IoT channels for continuous patient management. For instance, IoT-enabled 3D-printed prosthetic limbs provide live feedback on gait and stress distribution, significantly improving rehabilitation outcomes and personalised care [41].

2.4.4 Aerospace and Automotive Lightweighting

Aerospace and automotive industries require lightweight, high-strength components, which are ideal for AM. IIoT integration further enhances safety, reliability, and lifecycle monitoring. In this application, (i) AM produces lattice structures and topology-optimised designs for weight reduction, (ii) IIoT Sensors monitor in-flight or in-vehicle performance (temperature, stress, vibration) of printed parts, and (iii) digital twins predict fatigue failure and recommend proactive replacements, which comes under life cycle tracking of the devices. Example: GE Aviation's 3D-printed fuel nozzles [10] are IoT monitored for wear and performance during operation, improving safety.

2.4.5 Energy and Sustainability Applications

AM-IIoT integration supports sustainable energy systems by optimising resource use and enabling localised, eco-friendly production. (i) The energy sector applications cover AM of turbine blades, heat exchangers, and renewable energy components; (ii) IIOT's role is to monitor performance (temperature, stress, vibration) of AM parts in real-world energy plants; and (iii) sustainability is applied by the digital twins, which predict material wear, ensuring longer component lifetimes and reduced waste. Example: IIoT-enabled AM of wind turbine spare parts in remote areas reduces downtime and transportation emissions. This section demonstrates how AM provides structural and material innovation, while IIoT delivers intelligence, connectivity, and adaptability across industrial domains.

2.5 ARCHITECTURES AND FRAMEWORKS

In this section, the architecture and framework for AM and the Industrial Internet of Things (IIoT) are described in detail.

2.5.1 Reference Architectures for AM-IIoT

Reference architectures serve as blueprints for unifying AM and IIoT platforms. They define layers, communication flows, and decision-making pathways. The Industrial Internet Reference Architecture (IIRA) [42] and RAMI 4.0 [43] can be extended to support AM-IIoT workflows, ensuring data consistency and standardisation. A five-layer reference architecture diagram with AM equipment, sensors, edge and cloud layers, and applications is depicted in **Figure 2.9**. The perception layer shows the sensors embedded in AM machines to collect and process environmental data. The network layer consists of communication protocols (5G, OPC-UA, MQTT, CoAP) to transmit data. The edge layer does the pre-processing with low-latency AI models for defect detection. The cloud layer is used for centralised analytics, digital twins, and inventory management. Finally, the application layer supports decision-making, scheduling, and user dashboards.

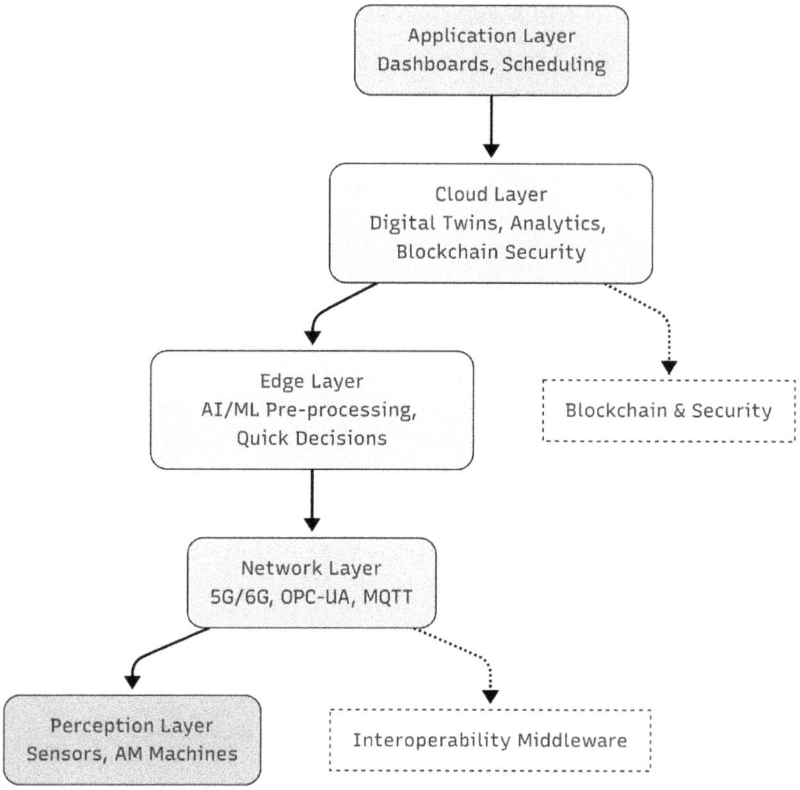

Figure 2.9 Integrated framework for AM-IIoT convergence.

2.5.2 Data Flow Models and Interoperability

The effectiveness of AM-IIoT depends on the smooth flow of data between machines, sensors, and analytics platforms. The upstream flow processes environmental data (melt pool temperature, vibration, humidity) and forwards it to the edge nodes and then to the cloud analytics. The downstream flow optimises process parameters, further collects digital twin insights, and forwards them to actuator-level feedback for AM machines. In cases of interoperability needs, vendors use proprietary formats, so middleware is required for standardised communication.

2.5.3 Cybersecurity and Trust Issues

AM-IIoT increases exposure to cyber threats, particularly in design file theft, process sabotage, and counterfeit parts. There can be three types of theft scenarios: (i) hackers can alter laser power and weaken aerospace parts, (ii) IP theft is possible from intercepted STL/CAD files, and (iii) malicious code injection in machine firmware. As countermeasures to this, blockchain can be used for immutable design provenance and process logs, TLS and PKI encryptions can be used for secure transmission, and continuous authentication across nodes can be provided by zero trust.

2.5.4 AI/ML Integration for Adaptive Manufacturing

AI/ML integration can transform AM-IIoT systems into self-learning, adaptive ecosystems in terms of (i) defect detection: CNNs can analyse thermal camera images for porosity and cracks, (ii) predictive maintenance: ML models can predict AM machine failures using vibration/acoustic signals, and (iii) generative design: AI algorithms create optimised geometries for AM, validated by IIoT feedback.

2.6 INDUSTRY CASE STUDY: METAL AM AND IIOT INTEGRATION BY GE ADDITIVE

A generic pipeline for integrating AM-IIoT into industries is shown in **Figure 2.10**, and a detailed explanation of the specific case studies for GE Additive (Concept Laser/ArCAM) and MTU Aero Engines is presented next.

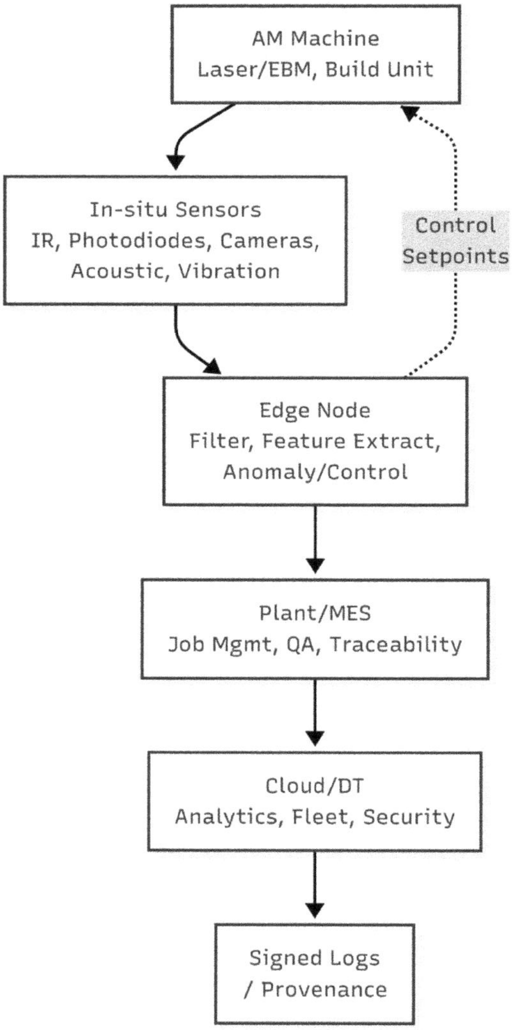

Figure 2.10 Generic AM-IIoT pipeline.

2.6.1 GE Additive (Concept Laser/Arcam)

The detailed working principle of GE Additive (Concept Laser/Arcam) in line with AM and IIoT integration is shown in **Figure 2.11**, and the explanation of the figure in terms of layers and data flow is as follows:

2.6.1.1 Machine Layer (AM Systems)

This is the foundation on which the physical AM processes in Concept Laser LPBF and Arcam EBM systems operate. The powder bed chamber, or deposition chamber, holds the raw material, while laser or e-beam optics control the energy source for selective melting. Subsystems like recoaters spread powder layers uniformly, and embedded sensors, such as the QM Multipool 3D and QM Coating, continuously monitor the melt pool and powder layer quality. Acoustic and vibration sensors provide additional signals to detect cracks or delamination events. This layer ensures that the AM process [44] is fully instrumented, generating the raw thermal, optical, and acoustic data that fuels the digital thread.

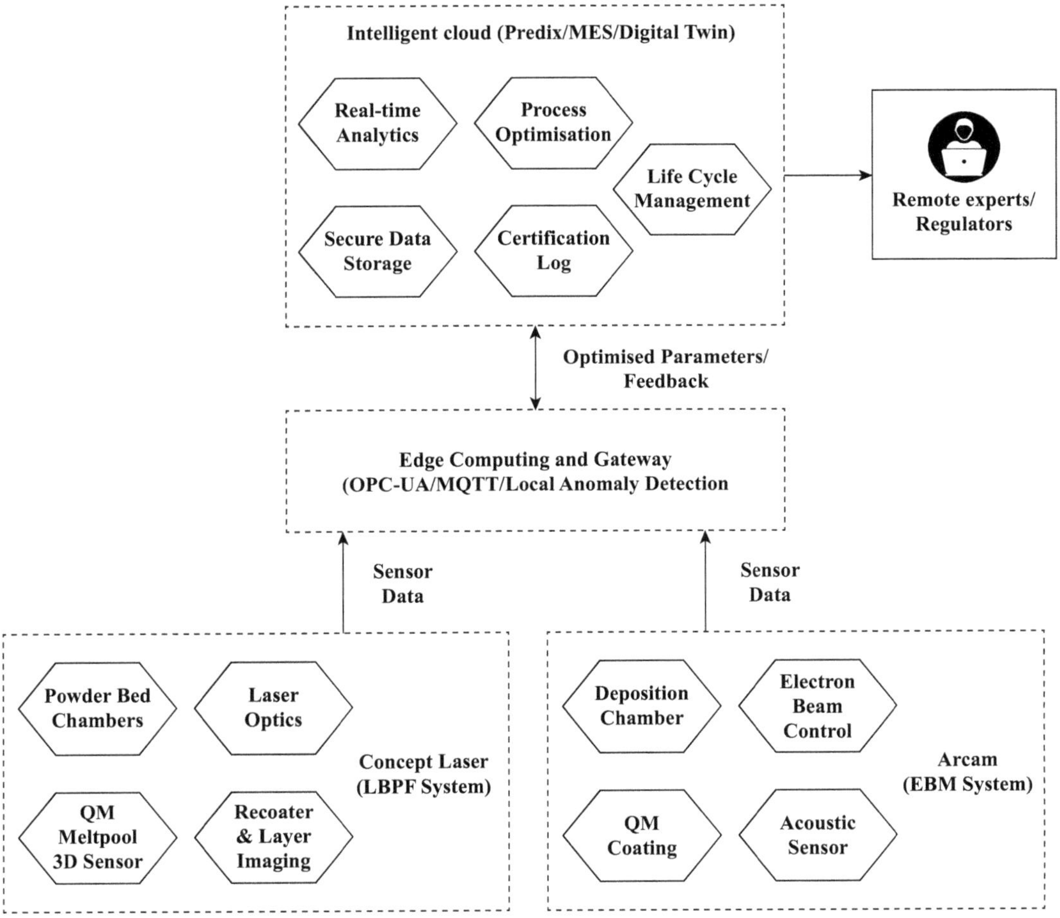

Figure 2.11 AM-IIoT integration with examples of Concept Laser LPBF, Arcam EBM systems and their sub-components.

2.6.1.2 Edge and Gateway Layer

Here, real-time data filtering and first-level analytics are performed at the machine level. Since melt pool imaging or acoustic sensing can generate gigabytes of data per build, the edge node extracts meaningful features, such as melt pool size, thermal gradients, or layer irregularities, before transmitting. Lightweight anomaly detection algorithms run locally and can raise alarms or pause the build when deviations exceed thresholds. The gateway implements industrial communication protocols such as OPC-UA and MQTT, converting heterogeneous sensor data into standardised formats. This reduces data overload, ensures secure transfer, and supports low-latency control actions, which are essential for process stability.

2.6.1.3 Digital Twin Layer

This is the intelligence hub of the GE Additive ecosystem, built on platforms such as Predix, MES, or PLM systems. All edge-filtered logs and features are uploaded to the cloud for deep analytics, historical comparison, and optimisation. A digital twin of the build process simulates expected outcomes, and deviations trigger adjustments in machine parameters. The cloud also acts as a secure repository, storing certification-ready logs in compliance with aerospace (FAA) and medical (FDA) regulations. In addition, it supports process optimisation, predictive maintenance, and lifecycle management, linking design, build, inspection, and field use into a unified digital thread.

2.6.1.4 Remote Experts and Regulators Layer

This layer ensures human oversight and compliance across distributed manufacturing sites. Remote engineers and service providers can log into cloud dashboards to review real-time builds, troubleshoot anomalies, and fine-tune process parameters. Regulators or certification authorities can access immutable build records, ensuring that each part meets stringent quality requirements for safety-critical sectors such as aerospace and healthcare. This setup enables global collaboration and validation, allowing subject matter experts to contribute without being physically present on site. It also accelerates certification cycles by providing regulators with traceable, tamper-proof data for audits.

2.6.1.5 Data Flow among Layers

Upward flows carry sensor data from the machines to the edge, then onward to the cloud for deeper analytics. Downward flows send optimised parameters and corrective actions back to the machines, closing the loop in real-time. Side flows enable experts and regulators to intervene remotely or validate process compliance. This integrated flow guarantees traceability, adaptiveness, and trust, ensuring that every manufactured part is optimised, monitored, and certified as part of a continuous digital lifecycle.

GE Additive integrated its Concept Laser (Germany) and Arcam (Sweden) systems with the Predix Industrial IoT platform to create a connected AM ecosystem. This integration enabled real-time monitoring, predictive maintenance, and digital twin-driven optimisation of build parameters. Through data-driven feedback loops, GE achieved a success rate of up to 25–30%. Through data-driven feedback loops, GE achieved up to a 20% improvement in build quality consistency and faster certification cycles for aerospace and medical components. These results illustrate how linking AM with IIoT and AI analytics can enhance both productivity and reliability in large-scale industrial production.

2.6.2 MTU Aero Engines: AM-IIoT Integration for Aerospace Components

MTU Aero Engines AG, a leading German aerospace company, has been at the forefront of adopting AM for critical aero engine components. With strict aerospace certification requirements, MTU focused on integrating AM with Industrial IoT (IIoT), CPS, and MES to ensure quality assurance, traceability, and scalability. The details of the AM-IIoT integration process are explained as follows:

2.6.2.1 AM Systems and Monitoring

MTU employs EOS LPBF machines for producing turbine parts, including borescope bosses for the Airbus A320neo engine. Each machine is equipped with EOSTATE Melt Pool Monitoring and Optical Tomography (OT). EOSTATE Melt Pool monitors melt pool intensity and stability using photodiodes, whereas Optical Tomography captures high-resolution images of each layer to detect defects such as lack of fusion, porosity, or recoating errors. These systems act as cyber-physical extensions, streaming data in real time to edge analytics modules.

2.6.2.2 IIoT and Edge Integration

The processed sensor data is transmitted via IIoT gateways using protocols such as OPC-UA and MQTT. Edge analytics preprocesses melt pool signals and tomography images to detect anomalies locally with low latency. This reduces reliance on time-consuming post-build inspections and allows in-situ defect detection. Critical events are flagged immediately, enabling corrective actions during the build rather than after completion.

2.6.2.3 Cloud, AI/ML, Big Data Integration

MTU integrates the monitoring process into MES and PLM systems using big data platforms. ML models (trained on historical builds) predict porosity or cracking likelihood from melt-pool signatures. AI algorithms optimise laser-scanning strategies and energy inputs to reduce material waste. Big data analytics links powder batch history, process signatures, and inspection results to ensure consistency across the entire fleet.

2.6.2.4 Quality Assurance and Certification

Instead of relying solely on costly X-ray CT scans, MTU combines in-situ monitoring logs, ML defect predictions, and selected Non-Destructive Testing (NDT) validation for certification. Digital build logs are stored in MES, ensuring traceability of every layer and batch of material. This hybrid certification process reduces inspection time and cost while still meeting the requirements of the EASA/FAA aerospace.

2.6.2.5 Robotics and Automation

Robotics is employed in MTU for powder removal, support separation, and part handling, further integrating AM into automated production lines. This reduces manual intervention and ensures consistent post-processing quality. MTU Aero Engines implemented EOS LPBF machines equipped with EOSTATE Melt Pool Monitoring and Optical Tomography, integrated into its MES and big data analytics frameworks. The use of ML models for defect prediction and in-situ monitoring reduced post-build inspection time by 40% and material waste by 15%. Furthermore, integrating IIoT gateways and edge analytics helped achieve an approximate 25% reduction in total production downtime while maintaining the aerospace quality and traceability required by EASA/FAA certification standards. This case study demonstrates how the strategic integration of AM with IIoT, in situ monitoring, and MES systems can deliver certified aerospace components with reduced inspection costs, improved scalability, and increased trust, proving that AM-IIoT ecosystems are viable for critical safety applications.

2.7 DISCUSSIONS

2.7.1 Integration with Legacy Systems and Additive Manufacturing Practices

One of the major hurdles in adopting AM-IIoT solutions is the coexistence of new digital workflows with traditional manufacturing infrastructure. Most factories already operate with legacy systems such as ERP (Enterprise Resource Planning), MES (Manufacturing Execution Systems), SCADA (Supervisory Control and Data Acquisition), and CNC-based production lines. These systems were not originally designed to handle the high-volume, high-frequency, and multimodal data streams generated by AM monitoring sensors (melt pool imaging, acoustic emissions, layer scans). Ensuring interoperability requires middleware and communication standards such as OPC-UA, MTConnect, or API-based gateways to bridge the gap. From the AM practices perspective, traditional quality assurance relies heavily on ex-situ Non-Destructive Testing (NDT), while AM-IIoT introduces in-situ monitoring and digital certification. This often leads to duplicated workflows where a part is both monitored digitally and physically inspected. This can increase costs unless a harmonised validation protocol is developed. Additionally, legacy supply chain practices are typically based on physical inventory and batch production, whereas AM-IIoT pushes towards digital inventories, on-demand production, and distributed manufacturing. Integrating these approaches requires both technical adaptation (standard data models, interoperability layers, cloud connectors) and organisational change (new certification workflows, operator training, cultural acceptance).

Ultimately, successful integration will depend on hybrid frameworks that allow legacy machines and processes to coexist with AM-IIoT-enabled production lines, gradually phasing in advanced digital practices without disrupting established operations. This means embracing interoperability standards, modular IoT gateways, and digital twin frameworks that can overlay legacy environments, enabling a smoother transition to fully adaptive, connected manufacturing ecosystems.

2.7.2 Standardisation of Materials and Processes

A key challenge in advancing AM-IIoT adoption lies in the lack of universal standards for materials, processes, and quality metrics. Unlike conventional manufacturing, where material grades and process parameters are well-defined (e.g., ASTM steel grades, ISO machining tolerances), AM often deals with customised powders, resins, and process settings that vary by machine vendor, material supplier, or application. This variability can cause significant challenges in data interoperability, quality assurance, and certification when IoT-based monitoring is introduced.

On the materials side, differences in powder size distribution, flowability, porosity, and alloy composition can lead to inconsistent build outcomes across systems. Without standardised specifications for feedstock preparation, storage, and reuse, IoT-enabled data comparisons between builds or across sites become unreliable. On the process side, parameters like laser power, scan speed, hatch spacing, layer thickness, and recoating strategies differ not only between technologies (FDM, SLS, EBM) but also across vendors. This creates a challenge for digital thread traceability, as the same "material" or "process" might mean different things in practice.

Efforts by organisations such as ASTM International (Committee F42 on AM Technologies), ISO/TC 261, and SAE are addressing this gap by developing global standards for material qualification, process calibration, and in-situ monitoring protocols. For example, standards are emerging for powder characterisation (ASTM F3049), process control (ISO/ASTM 52900 series), and digital data traceability. When linked to IIoT platforms, these standards will enable cross-machine comparisons, cloud-based certification, and secure sharing of digital build records. Ultimately, standardisation ensures that AM-IIoT can scale beyond isolated pilots to become a mainstream manufacturing solution. It allows industries such as aerospace, automotive, and biomedical to rely on consistent material properties, predictable process outputs, and trusted digital records, thereby reducing costs, simplifying certification, and enabling globally distributed, yet interoperable, AM supply chains.

2.7.3 Quality Assurance and Certification for Additively Manufactured Components

Quality assurance in AM traditionally relies on ex-situ Non-Destructive Testing (NDT) methods such as X-ray CT, ultrasonic inspection, and tensile testing of coupons. However, as AM scales, these methods become time-consuming and costly, creating bottlenecks in production. With the integration of IIoT, in-situ monitoring systems, such as melt-pool sensors, layer imaging (QM Coating), and acoustic/vibration analysis, provide continuous data streams that can serve as digital evidence of build quality. This enables a shift towards process-based certification, where the integrity of the part is validated not only by post-build tests but also by real-time digital monitoring logs.

For regulatory compliance, industries like aerospace (FAA, EASA) and biomedical (FDA) require traceable, tamper-proof process data. IIoT platforms, combined with digital twins and blockchain, create immutable certification logs that link each component to its material source, process parameters, and monitor data. Standardisation bodies such as ASTM F42 and ISO/ASTM 52900 are working to align in-situ data with certification protocols, so that digital monitoring can reduce or even replace redundant destructive testing. Ultimately, this integration enhances trust, efficiency, and scalability, allowing AM components to meet the same rigorous safety and performance standards as conventionally manufactured parts, while leveraging the speed and adaptability of IIoT-enabled certification workflows.

2.8 CHALLENGES AND RESEARCH OPPORTUNITIES

2.8.1 Interoperability, Latency, and Scalability Issues

AM-IIoT systems bring together diverse printers, sensors, and platforms, but interoperability remains a hurdle due to proprietary data formats and inconsistent semantics. Real-time AM monitoring demands low latency, yet congested networks and cloud-heavy analytics often exceed control loop budgets. Scalability is another challenge, as hundreds of AM machines streaming video/thermal feeds generate bursty traffic that overwhelms middleware and storage. Moving computation to the edge, adopting standards such as OPC-UA/MTConnect, and using adaptive streaming can help address these issues. Research is needed on open AM ontologies, latency-aware scheduling, and telemetry compression to ensure stable control under large-scale deployments.

2.8.2 Data Security and IP Protection in AM-IIoT

The convergence of AM and IIoT exposes designs and processes to cyber threats, including CAD file theft, toolpath tampering, and counterfeit parts. Attackers may alter laser power or material feeds, creating hidden defects that compromise safety. Traditional encryption and TLS secure communication, while blockchain-based traceability and design watermarking protect intellectual property. However, these add computational overhead, risking increased process latency. Research opportunities lie in robust watermarking methods, post-quantum cryptography, secure multiparty analytics, and lightweight anomaly detection that link sensor traces with toolpaths to identify sabotage without disrupting print performance.

2.8.3 Lack of Standards and Protocols

Although frameworks like RAMI 4.0 [43], IIRA [42], and ISO/ASTM 529xx provide partial guidelines, AM-IIoT still lacks robust standards for real-time sensor data exchange, melt-pool monitoring semantics, and provenance tracking. This lack complicates certification for safety-critical sectors such as aerospace and healthcare, where regulators require uniform evidence of design-to-part traceability. Proprietary slicers and monitoring systems hinder cross-vendor interoperability, while cyber assurance norms for AM-IIoT ecosystems are immature. Future research should develop an open telemetry schema, provenance graph models linking design to toolpath to process to part, and NIST-like conformance testbeds to validate interoperability and security across platforms.

2.8.4 Research Gaps: Quantum AM, Blockchain-Enabled Supply Chain, Generative Design with IoT Feedback

Emerging opportunities lie in quantum computing, blockchain, and generative design. Quantum-inspired solvers can accelerate AM design optimisation and enable new defect detection with quantum sensors, while post-quantum cryptography ensures long-term security. Blockchain can enable trusted digital supply chains, but challenges remain in balancing scalability, privacy, and storage across hybrid on- and off-chain models. Generative design, when combined with IoT feedback loops, can evolve geometries based on in situ monitoring and real-world performance data, closing the loop between design, manufacturing, and lifecycle use. Future work should focus on multi-objective optimisation frameworks, scalable blockchain protocols, and lifecycle-aware generative design pipelines.

2.9 FUTURE DIRECTIONS

The convergence of AM and IIoT is moving towards a future defined by autonomous, adaptive, and sustainable factories under Industry 5.0. Next-generation directions include the following:

1. The adoption of digital marketplaces and distributed manufacturing networks, where secure cloud repositories and blockchain-enabled supply chains enable local, on-demand production with global traceability.
2. AI-driven generative design, coupled with IoT feedback from both in-situ monitoring and field performance, will create closed-loop innovation cycles, continuously improving part design and functionality.
3. Equally important is the integration of human-centric approaches, where operators, engineers, and AI collaborate through immersive interfaces and augmented reality, ensuring not just efficiency and resilience but also sustainability, inclusivity, and societal benefit.

2.10 CONCLUSION

The integration of AM with the Industrial Internet of Things (IIoT) is transforming traditional production into a connected, intelligent, and adaptive ecosystem. By combining AM's design freedom and material efficiency with IIoT sensing, connectivity, and analytics, industries can achieve real-time optimisation, predictive quality assurance, and secure digital supply chains. The chapter has outlined fundamentals, convergence frameworks, applications across sectors, and challenges, including interoperability, cybersecurity, and standardisation, while highlighting emerging opportunities such as digital twins, blockchain, generative design, and quantum-enabled AM. The presented case studies highlight the application of AM-IIoT integration in Industry 4.0, which can be extended to other applications. In addition, challenges and opportunities are discussed along with quality assurance and certification components. Ultimately, this convergence not only advances the vision of Industry 4.0 but also lays the foundation for Industry 5.0, where manufacturing will be human-centric, sustainable, and resilient.

ETHICS STATEMENT

We confirm that in accordance with the Taylor and Francis AI Policy, AI tools such as ChatGPT, Napkin AI, and Grammarly were used to improve the readability and language of the manuscript. After using the tools, the authors reviewed and edited the content as needed and took full responsibility for the published article.

REFERENCES

1. I. Gibson, D. Rosen, B. Stucker. 2015. *Additive Manufacturing Technologies: 3D Printing, Rapid Prototyping, and Direct Digital Manufacturing*. 2nd ed. New York: Springer.
2. B. Berman. 2012. 3-D printing: The new industrial revolution. *Business Horizons* 55:155–162.
3. B. Mallikarjuna, P. Bhargav, S. Hiremath, et al. 2025. A review on the melt extrusion-based fused deposition modeling (FDM): Background, materials, process parameters and military applications. *International Journal on Interactive Design and Manufacturing* 19:651–665.
4. M. Balichakra, S. Bontha, P. Krishna, V.K. Balla. 2019. Prediction and validation of residual stresses generated during laser metal deposition of γ titanium aluminide thin wall structures. *Materials Research Express* 6:106550.
5. Precedence research. Industry 4.0 Market Will Grow at CAGR of 18.74% By 2032. downloaded on 23-09-2025.
6. S. Ford, M. Despeisse. 2016. Additive manufacturing and sustainability: An exploratory study of the advantages and challenges. *Journal of Cleaner Production* 137:1573–1587.
7. C.P. Paul, A.N. Jinoop. 2022. Additive manufacturing: Bringing a paradigm shift. In *New Horizons in Metallurgy, Materials and Manufacturing*. Springer Nature 8:35–50.
8. N. Jayanth, P. Senthil, B. Mallikarjuna. 2022. Experimental investigation on the application of FDM 3D printed conductive ABS-CB composite in EMI shielding. *Radiation Physics and Chemistry* 198:110263.
9. B. Mallikarjuna, E.W. Reutzel. 2022. Reclamation of intermetallic titanium aluminide aero-engine components using directed energy deposition technology. *Manufacturing Review* 9:27.
10. S. Hiremath, J.F. Dsouza, D.S. Chiniwar, V. HM, B. Mallikarjuna. 2024. Exploring the impact of epoxy coated 3D-Printed polymers on surface roughness and mechanical behavior: An experimental and numerical study. *Results in Engineering* 23:102779.
11. S. Dehghan, S. Sattarpanah Karganroudi, S. Echchakoui, N. Barka. 2025. The integration of additive manufacturing into Industry 4.0 and Industry 5.0: A bibliometric analysis (trends, opportunities, and challenges). *Machines* 13:62.

12. A.J. Vayre, F. Vignat, F. Villeneuve. 2012. Designing for additive manufacturing. *Procedia CIRP* 3:632–637.
13. M. Fathi, K. Daneshmand-Mehr, R. Harrison. 2022. Industrial cyber-physical systems and digital twins in Industry 4.0. *IEEE Transactions on Industrial Informatics* 18:1383–1395.
14. J. Lee, H.A. Kao, S. Yang. 2014. Service innovation and smart analytics for Industry 4.0 and big data environment. *Procedia CIRP* 16:3–8.
15. J. Ni, J. Lee, H.R.L. Fernandes. 2020. Machine learning approaches for intelligent additive manufacturing. *Procedia Manufacturing* 48:126–134.
16. H. Li, J. Tang, S. Hu. 2023. Industrial Internet of Things-enabled smart additive manufacturing systems: Architecture, challenges, and opportunities. *IEEE Internet of Things Journal* 10:4051–4063.
17. F. Ding, Y. Liu, Z. Chen. 2023. Digital twin-driven process optimization for metal additive manufacturing. *IEEE Transactions on Automation Science and Engineering* 20:176–188.
18. M. Kumar, T. Singh, A. Kaur. Blockchain-enabled secure data framework for distributed additive manufacturing networks. *IEEE Access* 11:98250–98262.
19. P.P. Rai, A.K. Gupta, R. Bandyopadhyay. 2024. Convergence of AI, IoT, and blockchain for autonomous and sustainable additive manufacturing. *IEEE Transactions on Industrial Electronics* 71: 9067–9078.
20. H. Lasi, P. Fettke, H.G. Kemper, T. Feld, M. Hoffmann. 2014. Industry 4.0. *Business & Information Systems Engineering* 6:239–242.
21. C. Hopkinson, R. Hague, P. Dickens. 2006. *Rapid Manufacturing: An Industrial Revolution for the Digital Age*. Chichester, UK: Wiley.
22. A. Kumar, R. Shukla, P. Dwivedi, R. Dixit 2022. The role of IoT in additive manufacturing: Trends, opportunities, and challenges. *Materials Today: Proceedings* 49:2211–2216.
23. L. Lu, L. Wang, Y. Li, S. Gao. 2019. Industrial Internet of Things: Architecture and key technologies. *IEEE Internet of Things Journal* 6:9449–9465.
24. R. Xu, Y. Chen, Y. Guo. 2020. Digital twin-driven additive manufacturing: Frame-work and applications. *IEEE Access* 8:94410–94423.
25. D. Mourtzis, E. Vlachou, N. Milas. 2016. Industrial big data as a result of IoT adoption in manufacturing. *Procedia CIRP* 55:290–295.
26. A. Nassehi, S.T. Newman, R. Duflou. 2020. A review of the role of additive manufacturing in the digital transformation of manufacturing. *International Journal of Advanced Manufacturing Technology* 111:1–14.
27. M. Attaran. 2017. The rise of 3-D printing: The advantages of additive manufacturing over traditional manufacturing. *Business Horizons* 60:677–688.
28. C.K. Chua, K.F. Leong, C.S. Lim. 2010. Rapid prototyping: Principles and applications, 3rd Edition. *World Scientific*. https://doi.org/10.1142/6665
29. L.J. Ladani. 2021. Applications of artificial intelligence and machine learning in metal additive manufacturing. *Journal of Physics: Materials* 4:042009.
30. B. Mallikarjuna, K. Jayachristiyan. 2013. Innovative modeling and rapid prototyping of turbocharger impeller. In *Proceedings of the International Conference on Advanced Materials, Manufacturing and Management in Thermal Sciences* (vol. 2, pp. 1426–1432). Innovative Modeling and Rapid Prototyping of Turbocharger Impeller, IJERT.
31. Y. Liu, S. Xu, J. Li, J. Tang. 2019. Industrial IoT for 3D printing: Design, challenges, and applications. *IEEE Communications Standards Magazine* 3:42–47.
32. V. Karkaria, et al. 2024. Towards a digital twin framework in additive manufacturing: Machine learning and bayesian optimization for time series process optimization. *arXiv preprint arXiv:2402.17718*.
33. S. Mondal, D. Gwynn, A. Ray, A. Basak. 2020. Investigation of melt pool geometry control in additive manufacturing using hybrid modeling. *Metals* 10:683.
34. J. Bamberg, K. Dusel, W. Satzger. 2014. Overview of additive manufacturing activities at MTU aero engines. *Review of Progress in Quantitative Nondestructive Evaluation*a 34:156–163.
35. R. Jayaraman et al. 2021. The role of digital twins in additive manufacturing. *Journal of Manufacturing Systems* 59:498–507.
36. M.M. Ahsan, Y. Liu, S. Raman, Z. Siddique. 2024. Digital twins in additive manufacturing: A systematic review. *arXiv preprint arXiv:2409.00877*.

37. X. Zhang, Y. Guo, L. Wang. 2023. Intelligent additive manufacturing: Integrating artificial intelligence and big data for predictive process control. *IEEE Transactions on Industrial Informatics* 19:2176–2188.
38. S. Tao, W. Qi, Z. Li. 2021. Blockchain-enabled secure additive manufacturing framework for Industry 4.0. *IEEE Transactions on Industrial Informatics* 17:7650–7659.
39. W. He, L. Zhu, C. Liu, H. Jiang. 2025. Metal additive manufacturing and molten pool dynamic characterization monitoring: Advances in machine learning for directed energy deposition. *Metals* 15:106.
40. P. Becker, C. Eichmann, A. Roennau, R. Dillmann. 2020. Automation of post-processing in additive manufacturing with industrial robots. In *International Conference on Automation Science and Engineering (CASE)*. Automation of Post-Processing in Additive Manufacturing with Industrial Robots, IEEE Conference Publication.
41. A. Bibb. 2012. Medical modelling: The application of 3D printing in surgery and medicine. *Medical Engineering & Physics* 34:1441–1453.
42. R.S. Duflou, et al. 2012. Environmental assessment of additive manufacturing processes: Methodology and case studies. *CIRP Annals* 61:761–777.
43. RAMI 4.0. https://ec.europa.eu/futurium/en/system/files/ged/a2-schweichhart-reference_architectural_model_industrie_4.0_rami_4.0.pdf (accessed August 2025) 36. The Industrial Internet Reference Architecture IIRA. https://www.iiconsortium.org/iira/ (accessed August 2025).
44. Grand View Research. Market report analysis. Industry 4.0 Market (2023–2030) (accessed September 2025).

Chapter 3

Role of Additive Manufacturing in Reverse Engineering for the Healthcare Sector

Govind Kumar Verma, Santosh Kumar, and Ashutosh Kumar Dubey

3.1 INTRODUCTION

Recreating human body parts, particularly broken bone replacements, with artificial components raises several issues due to varying requirements. Anatomical accuracy is related to custom fit because every patient's anatomy is unique; hence, a standard artificial implant cannot ensure compatibility with the human body. The prosthetics may not always provide a good, optimal fit; sometimes, they may lead to discomfort and even compromise functionality, creating inevitable complications. Therefore, creating potential implants based on patient data and their design, such as body shape, prosthetics, surface texture, restoration of shape, and anatomical body, is very important preoperatively [1]. The creation of highly accurate implants conforms precisely to their body anatomy. It involves selecting manufacturing technology, which depends on several factors such as the compatibility of the processing material, shape complexity, production volume, dimensional accuracy, and surface finish [2]. However, the manufacturing process sometimes needs to address a few issues, such as the unavailability of digital data for the physical models of body parts, which may have been designed in the pre-digital era, legal restrictions, etc. [3]. To counteract these manufacturing issues in creating customised body parts, Reverse Engineering (RE) is a proven approach for analysing the parts and acquiring their design information [4, 5]. Unlike forward engineering, RE begins with part manufacturing aided by systematic, logical designs; it has been applied to a wide range of applications, such as defence, automotive, aerospace, medical, sports equipment, jewellery, and software [6]. For physical parts, the RE digitalises the physical products by acquiring a 3D surface model and converting it into a computer-aided design (CAD) model. During RE, the digitalisation process can be achieved by utilising 3D scanners, which capture the part's physical attributes, such as shape, size, and, in a few cases, colour. The use of 3D scanners is beneficial compared with traditional approaches for obtaining oral impressions, such as plaster models, which have several drawbacks [7]. 3D scanning technology has made significant breakthroughs in orthodontics.

The RE is considered a revolutionary approach and can facilitate deeper awareness of the human body even in the absence of design information. It utilises 3D medical imaging to scan internal body parts and 3D scanners to scan external body parts. Data collected from 3D medical imaging can be applied to the manufacturing of implants [8], and the research trend towards 3D scanning applications in the medical field is growing rapidly in the present era [9]. The RE is useful in healthcare for various medical applications, including implants, patient size, shape, surface area, body-part data, and digital models for virtual and holographic applications. The implant manufacturing process can be accelerated by using 3D scanners and post-processing software. The RE, together with Additive Manufacturing (AM)/3D printing, has several benefits. The AM can produce parts with intricate shapes at low cost in small lots, thereby amplifying the benefits of RE [10]. The integration of AM and

DOI: 10.1201/9781003590774-3

RE has initiated a transformative shift in the medical field, transitioning from traditional mass production to the creation of patient-specific constructs, and the medical application of RE is shown in **Figure 3.1**.

Hence, the AM technology became instrumental in reverse engineering, e-manufacturing, product design, rapid tooling, and development, and numerous healthcare domains such as medicinal education and training, design and implementation of medical devices, design of customised implants, tissue engineering, orthotics, prosthetics, mechanical bone replicas, forensics, and has the ability to resolve the issues of dentistry [11]. In addition, this approach has the potential to start up a new market to support humanity. However, apart from the several benefits pointed out in the aforementioned discussion for the healthcare sector, they have some challenges that hinder its widespread adoption, which are as follows: (i) the utilisation of scanning and imaging is a financial burden to the patient [12, 13], (ii) design and customisation needs high financial assistance to use high-end design and AM technology [14], (iii) post process, sterilisation and surgeries increases the overall cost of the medical implants [15]. Apart from these challenges, there are some issues like (i) data acquisition and accuracy based on the high-quality scan data, data processing, and segmentation, (ii) material changing options are limited based on biocompatibility with surrounding tissues, (iii) in design and manufacturing, customised parts concerning the minimisation of time and cost of production are ongoing problems [16]. Therefore, addressing these challenges requires continuous interdisciplinary collaboration among engineers, biologists, materials scientists, clinicians, and regulatory bodies to make incremental progress in the field and ensure the safe and effective application of reverse engineering and AM in the healthcare sector. The three essential steps involved in fabricating a medical model using RE and AM are schematically illustrated in **Figure 3.2.**

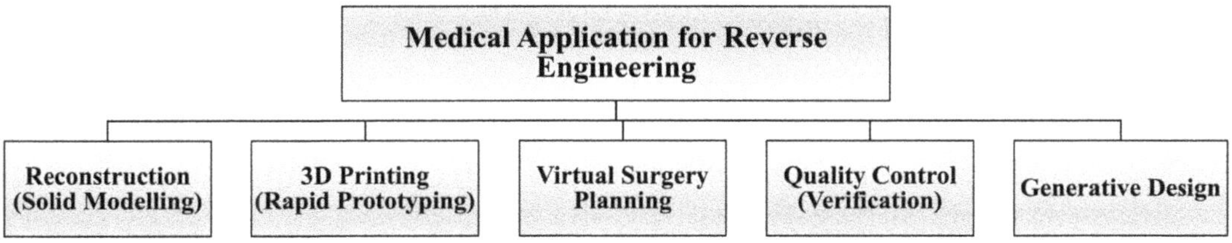

Figure 3.1 A summary of the healthcare sector for reverse engineering.

Source: Adapted from [1] under the Creative Commons Attribution 4.0 (CC BY 4.0) License.

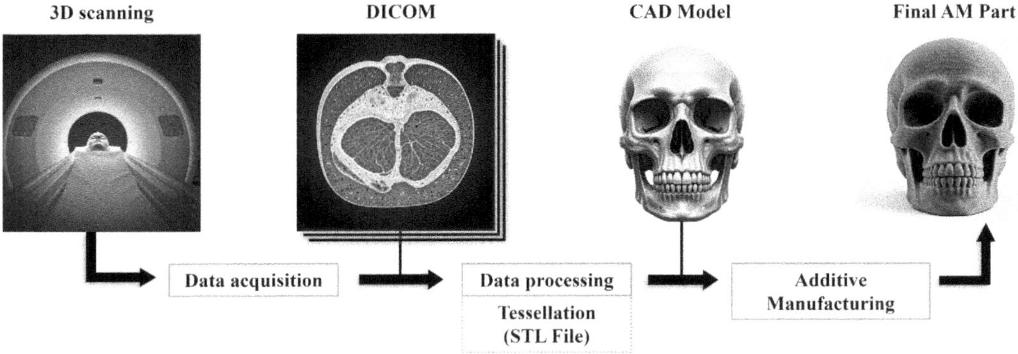

Figure 3.2 Graphical diagram from reverse engineering to additive manufacturing for healthcare.

Source: Adapted from [17] under the CC BY 4.0 License.

3.2 REVERSE ENGINEERING

Reverse engineering refers to the process of analysing existing products to extract their design and functional information. Unlike forward engineering, which begins with a planned, logical design to develop a product, RE works in the opposite direction—starting with the final product to understand its construction. This technique finds applications across various industries, including healthcare, aerospace, defence systems, sports equipment, toys, and jewellery. In the process of reinventing a product, RE prioritises analysis and evaluation over creativity and originality, which are typically emphasised during the initial design phase. **Figure 3.3** highlights the fundamental differences between RE and forward engineering design processes as they apply to manufacturing workflows [1].

RE is seeing increased adoption in product development and spare parts creation, particularly when CAD data isn't available. This approach enables the extraction of design data from existing prototypes, facilitates iterative redesign improvements, and ensures quality assurance by detecting deviations from intended specifications. In the context of spare parts, RE proves indispensable for reproducing original equipment manufacturer (OEM) components, modifying or repairing elements without access to original design records, and generating digital models when CAD files are unavailable. Essentially, RE involves systematically analysing existing product components, their geometry, and how they function together, either before or during the replication process. Such methods enable the reproduction, restoration, or adaptation of products even in the absence of original technical drawings [18]. In medical applications, the convergence of RE with 3D scanning has shown effectiveness in verifying the dimensional accuracy of surgical guides, including those applied in dental implantology. This integration significantly enhances precision and strengthens quality control within healthcare practices. Moreover, RE has made it possible to create patient-specific equipment like implants and prosthetics, leading to better surgical outcomes. Customised implants, in particular, minimise micromovement at fracture sites, which promotes bone healing and lowers the risk of complications [19].

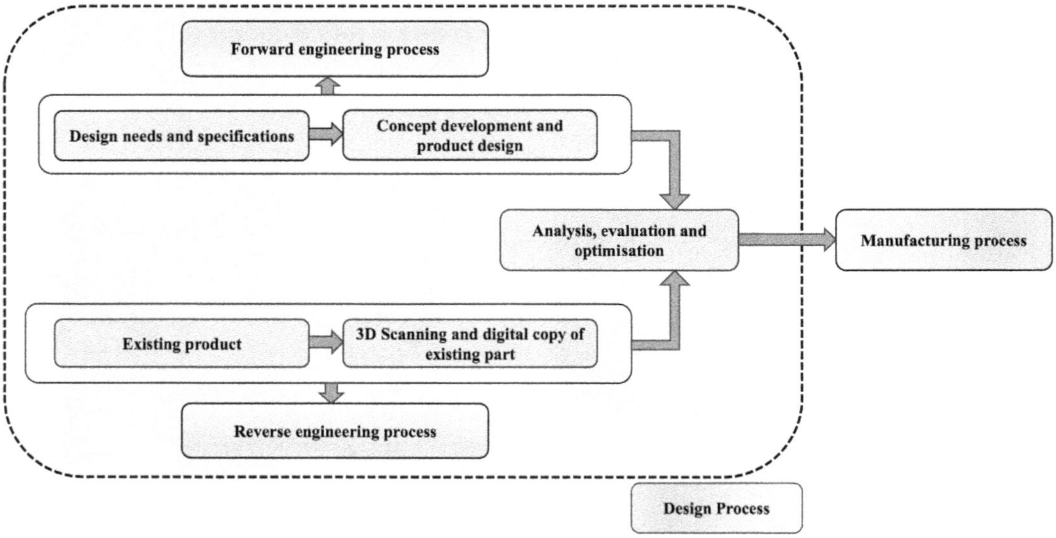

Figure 3.3 Comparison of forward and reverse engineering methodologies in the design process.

Source: Adapted from [1] under the CC BY 4.0 License.

3.2.1 Definition of Reverse Engineering

The term "Reverse Engineering" has multiple definitions based on context. Aqeel et al. [20] define RE as the process of analysing an object or existing system, whether hardware or software, to identify its components, comprehend their interrelationships, and assess its functionality, with the objective of redesigning or reproducing it without access to the original design specifications. Wang [19] describes RE as an approach that encompasses the measurement, analysis, and assessment of an object to replicate it or derive insights regarding a historical design or event. It functions as a technology for reinvention, providing a framework for reconstruction and replication, and can be seen as the practical science of maintaining the original design intent of a component. Cooper [21] characterises RE as a technique employed to reproduce components in the absence of original technical blueprints. The procedure, when integrated with AM technology, entails transforming a physical object into a digital model, which can subsequently be employed to create either exact reproductions or altered iterations of the original component.

3.2.2 Significance of RE in Developing Products

Manufacturers are progressively acknowledging RE as a vital element for sustaining competitiveness. Although it is often applied to detect and correct design flaws, its broader potential across industries is increasingly being realised. The incorporation of CAD tools has further enhanced RE's role in modern product development. Several key factors contribute to the growing adoption of RE in the product development process [6].

3.2.2.1 Addressing Design Information Gaps

RE serves as an essential tool for recovering precise design data from both outdated and newly developed components. Its importance is especially evident in Computer-Aided Manufacturing (CAM), where it ensures accuracy and efficiency in production processes, particularly when original design records are missing or insufficient.

3.2.2.2 Investigating Product Failures

RE is often used to investigate the root causes of product failure and extend product lifespan. For instance, in orthopaedic implants such as hip or knee replacements, premature loosening, excessive wear, or fracture may occur. By analysing stress concentrations, geometric inconsistencies, and material degradation in the failed implant, RE enables the redesign of improved, patient-specific models with better biomechanical compatibility and extended durability.

3.2.2.3 Enhancing Manufacturing Efficiency

In addition to design analysis, RE also contributes to the optimisation of manufacturing processes. It helps identify cost-effective and time-efficient production methods while maintaining product quality, thereby supporting the implementation of lean manufacturing practices.

3.2.2.4 Promoting Competitive Analysis

RE empowers manufacturers to analyse competing products, providing insights into their design strengths and weaknesses. This strategic evaluation fosters responsive innovation and contributes to a competitive marketplace that drives continuous improvement.

3.2.2.5 *Driving Innovation*

A major contribution of RE lies in its ability to drive innovation. By leveraging existing designs, engineers can develop products with improved performance, additional functionality, greater compactness, and greater adaptability—without having to start from scratch. Consequently, RE not only expedites the conversion of ideas into practical solutions but also promotes innovation across organisational levels.

3.2.3 Fundamental Process for Reverse Engineering

The RE process involves capturing the physical dimensions of an existing object and digitally reconstructing it into a three-dimensional model. Traditionally, this was a manual task, relying on tools such as tape measures and thickness gauges to record dimensions, which were then hand-sketched and later converted into CAD models. This conventional method was time-consuming, expensive, and prone to human error, especially when dealing with complex geometries that are difficult to measure manually [6]. However, modern advancements such as Computed Tomography (CT), Magnetic Resonance Imaging (MRI), and 3D scanning have significantly streamlined the process, enabling rapid, accurate capture of millions of 3D data points, even from intricate or irregular surfaces. The gathered data is typically converted into a polygonal mesh or STL file, which serves as the basis for generating the final 3D CAD model. Physical measurements can also be acquired using two-dimensional scanning technologies such as Coordinate Measuring Machines (CMMs), laser scanners, structured-light digitisers, industrial CT scanning, or photogrammetry (which extracts geometric data from overlapping photographs). While this data initially exists as a point cloud, it lacks topological information and therefore requires further processing into usable formats like triangular meshes, NURBS surfaces, or comprehensive CAD models [1]. The RE process generally comprises three main phases: (1) Digitisation, (2) Reconstruction, and (3) Manufacturing, as depicted in **Figure 3.4**.

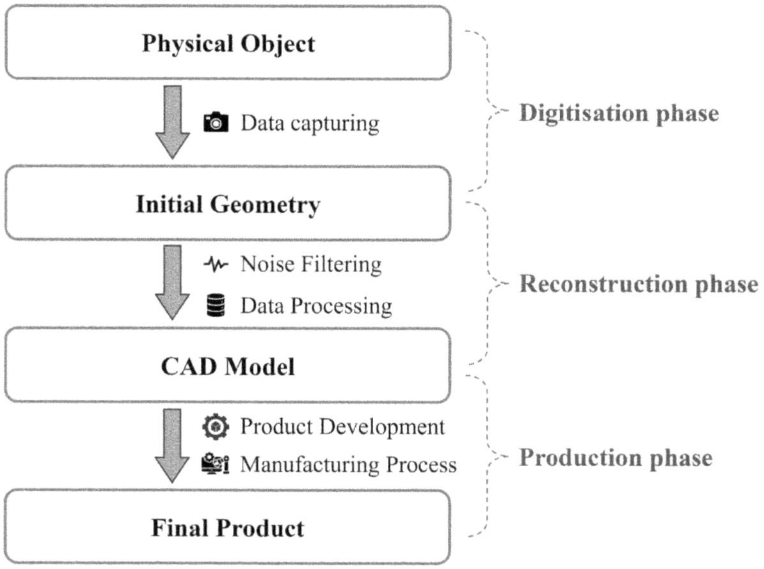

Figure 3.4 Processes of reverse engineering.

Source: Adapted from [22] under the CC BY 3.0 License.

3.2.3.1 Digitisation—Data Capturing, Imaging, and Visualisation

The digitisation step is fundamental to digital reverse engineering (RE), which converts real-world objects into accurate three-dimensional digital models for industrial and medical use. This essential preliminary phase entails a methodical approach to selecting an optimal scanning technique, setting up the object, and recording complex geometric features, such as steps, slots, pockets, and holes, to ensure accuracy in the resulting model. **Figure 3.5** illustrates that this phase incorporates advanced technology and strategic planning to lay the foundation for subsequent RE processes, enabling applications ranging from component redesign to personalised medical treatments.

In industrial environments, scanning technologies are categorised into contact and non-contact methods, each designed for particular specifications. Contact methods, including human measurements with vernier callipers or automatic CMMs, engage directly with the object's surface. Although economical, manual techniques are prone to human error, whereas CMMs improve precision by using touch probes to collect data and dynamically adjusting scan density to adapt to intricate surface characteristics. Non-contact methods such as laser triangulation, structured light projection, and digital photogrammetry provide high-precision capture of surface geometry without physical interaction. Laser triangulation maps the surface by analysing reflections of projected laser beams, while structured light techniques project predefined patterns to generate dense, accurate point clouds. Photogrammetry reconstructs 3D models from multiple overlapping images, offering versatility across different surface textures and environmental conditions. These techniques yield detailed point cloud data that faithfully represent the object's external geometry, facilitating seamless integration with CAD and CNC manufacturing systems.

Conversely, medical imaging focuses on visualising internal anatomical structures rather than external surface features. Techniques such as MRI, CT, positron emission tomography (PET),

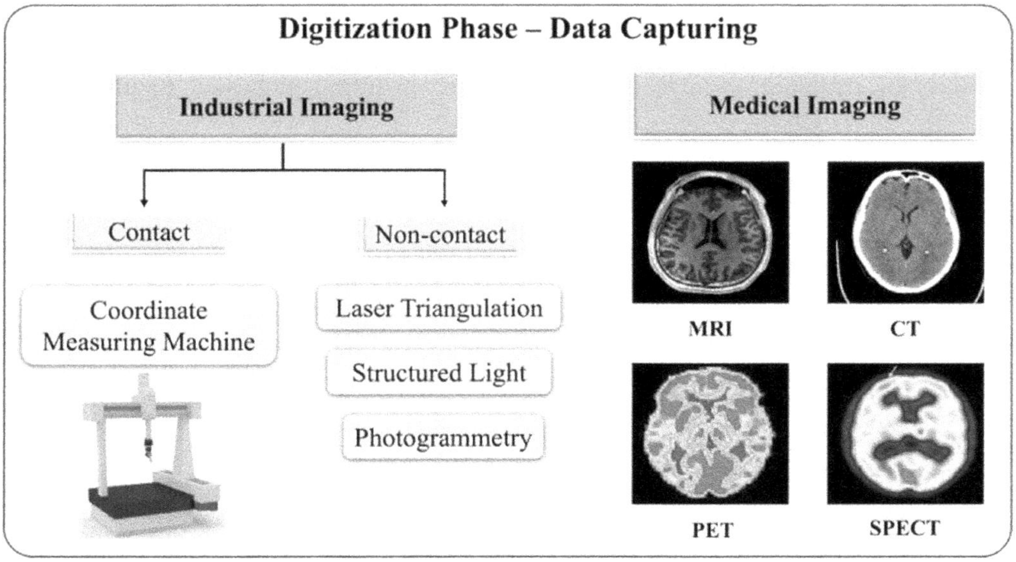

Figure 3.5 Data capturing methods illustrating contact and non-contact industrial imaging, as well as medical imaging techniques for reverse engineering and 3D model creation.

Source: Adapted from [23] under the CC BY 4.0 License.

and single-photon emission computed tomography (SPECT) generate three-dimensional volumetric data by stacking two-dimensional slices, typically with a slice resolution of 0.5–1 mm. These diagnostic modalities require skilled operators to ensure accurate data acquisition and interpretation. Unlike industrial methods that primarily capture external geometry, medical imaging captures internal anatomical details, which are then reconstructed into surface models for applications such as surgical planning and custom implant design.

The digitisation step ensures the production of high-fidelity digital models by aligning scanning technology with the object's attributes and intended use. **Figure 3.5** depicts the organised workflow, from item preparation to data acquisition, emphasising the unique yet synergistic methodologies of industrial and medical scanning. This step lays a solid foundation for reverse engineering, facilitating a smooth transition to subsequent stages in industrial design, manufacturing, and biomedical engineering.

3.2.3.2 Reconstruction—Image Processing

In both advanced manufacturing and medical visualisation, the generation and processing of point cloud data from industrial and medical scans are essential, as illustrated in **Figure 3.6.** This data enables the creation of precise 3D models customised for a wide range of applications, from structural design to biomedical engineering.

Industrial 3D scanners generate point clouds that include x, y, z coordinates along with metadata such as scanner type and resolution. However, these raw datasets often contain sampling errors, noise, and uneven point densities, making pre-processing essential to ensure accuracy. The workflow typically begins with importing the point cloud into specialised software, followed by noise filtering and point reduction techniques, such as voxel grid down-sampling, to reduce file size while maintaining geometric fidelity. For complex geometries, multiple scans are aligned and merged using algorithms such as Iterative Closest Point (ICP), and gap-filling algorithms are applied to address missing data. Advanced filtering and processing steps then convert the cleaned point cloud into usable 3D formats, such as polygonal meshes or CAD-compatible models. Careful planning of the scanning environment, including optimal lighting and object positioning, further minimises errors

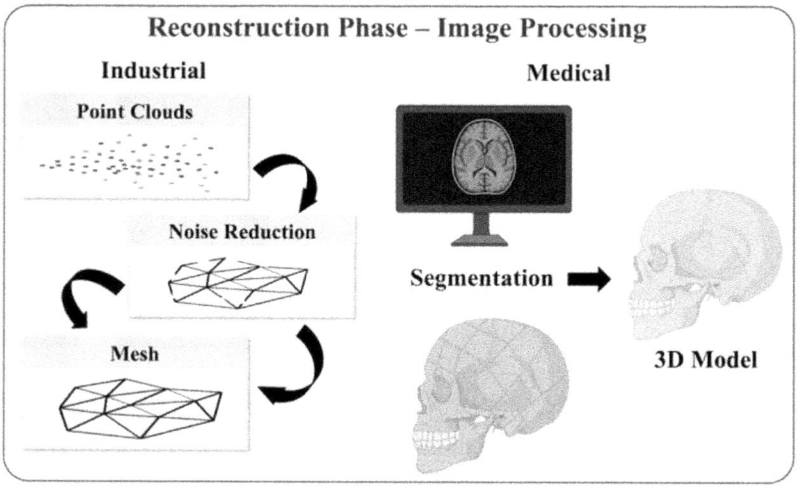

Figure 3.6 Image processing workflows illustrating industrial point cloud processing and medical image segmentation, followed by data conversion for 3D modelling.

and ensures the generation of a clean, accurate point cloud suitable for reverse engineering or AM.

Medical imaging, mainly used for diagnostics, produces Digital Imaging and Communications in Medicine (DICOM) files from CT or MRI scans, containing volumetric data represented in Hounsfield Units. As DICOM files are not directly compatible with CAD systems, they must first be imported into segmentation software. In this step, specific grayscale thresholds are applied to isolate targeted anatomical structures, resulting in highly accurate 3D organic surface models. The segmented data is then exported in formats such as STL or NURBS, facilitating integration into industrial design workflows or 3D printing applications. In contrast to industrial point clouds, medical imaging datasets typically yield continuous, gap-free surfaces due to their inherently high-resolution volumetric nature. This quality makes them ideal for applications such as patient-specific implant design, surgical simulation, and biomedical device development. Ultimately, both industrial and medical reverse engineering workflows aim to transform raw scan data into structured, functional 3D models. While industrial workflows emphasise noise reduction, alignment, and geometric optimisation, medical workflows prioritise anatomical accuracy and format interoperability, bridging diagnostic imaging with advanced manufacturing and biomedical engineering applications.

3.2.3.3 *Manufacturing—Product Development*

This stage represents a critical phase in the RE process, as it directly aligns with project objectives like design validation, quality control, or biomedical adaptation. The procedure involves importing the original CAD model and overlaying it with the scanned point cloud data of the physical component to create a digital prototype. Specialised RE software then analyses these datasets, identifying deviations between the intended design and the manufactured part or reconstructed geometry. **Figure 3.7** illustrates that this process typically includes aligning the point cloud with the CAD model, conducting comparative evaluations, and refining the digital prototype to meet defined specifications. In industrial applications, this ensures dimensional precision and design integrity. In the medical domain, the same methodology

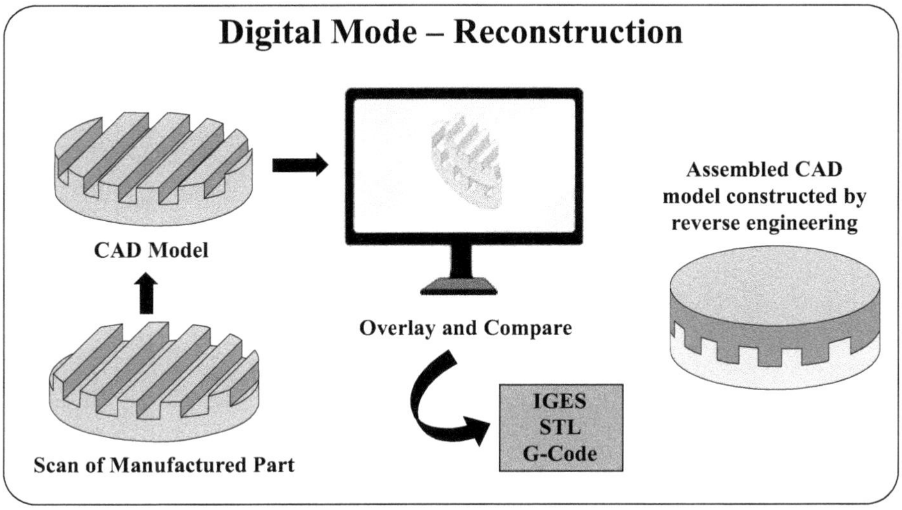

Figure 3.7 Digital model development phase showing CAD model overlay, point cloud comparison, and export of personalised designs in standard file formats.

is used to scan anatomical structures, such as hip joints, for the development of patient-specific implants, enabling highly accurate, personalised medical solutions. By leveraging each patient's anatomical data, highly customised joint replacements can be designed, significantly enhancing fit and performance, key aspects of personalised medicine. The final step involves exporting the validated digital model in commonly used proprietary formats like . IGES, .STL, .DXF, .OBJ, or .GCODE facilitates further manufacturing or surgical planning.

3.2.4 Reverse Engineering in Biomedical Applications

When RE is applied in biomedical contexts, it is called medical reverse engineering (MRE) and advances healthcare applications and research. MRE involves utilising patient data or biomedical objects to create 3D models of anatomical structures, enabling the development of personalised medical products and research tools [24]. Regardless of the technique employed, the primary objective of RE is to capture the 3D geometry of a patient's anatomy. Accurate processing of this data is essential for representing patient-specific anatomical features, which is crucial for the design and manufacture of biomedical solutions. This process aligns with the generic RE workflow outlined in Section 4.2.3, which also includes an important stage, as Phase IV: Biomedical application development and research [24], as presented in **Figure 3.8.**

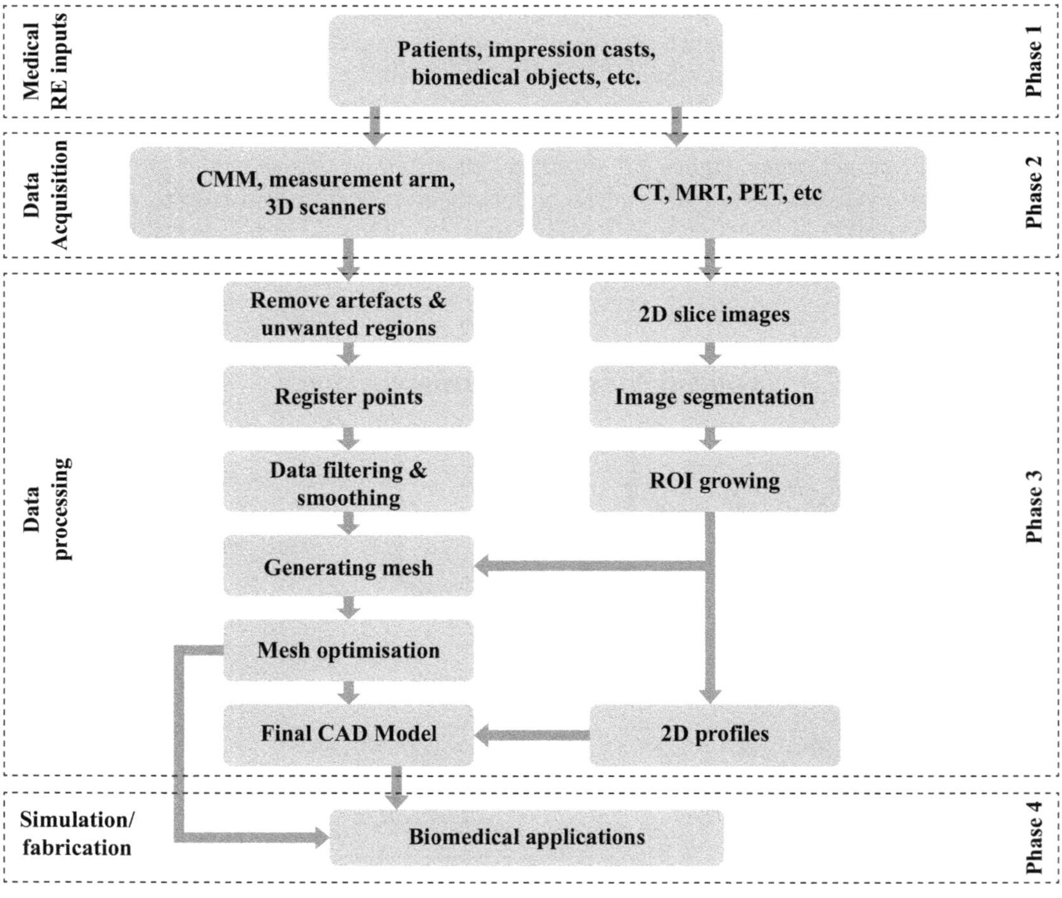

Figure 3.8 RE process: detailing steps from the diagnosis of anatomical defects through the creation of CAD models to the fabrication of organs for medical use.

Design methodologies are integrated throughout the MRE pipeline, which defines core principles and highlights distinctions from industrial RE practices.

Figure 3.9 illustrates the implementation of the RE process used to create patient-specific cranial implants for cranioplasty. In this context, RE deconstructs anatomical and defect-specific information to generate a precise digital model, which serves as the basis for a customised medical device. This strategy is particularly valuable in addressing complex cranial defects where conventional solutions are inadequate [25]. The process begins with detailed data acquisition. A clinical image of the patient's scalp reveals clear signs of atrophy and scarring due to prior failed cranioplasties, an important reference for assessing the site. A subsequent CT scan provides the volumetric dataset needed to reconstruct cranial anatomy, with the defective region precisely highlighted to enable quantitative assessment of the missing structure.

The CT data is then processed to produce a 3D reconstruction of the skull, marking a fundamental step in transforming raw imaging into a usable geometric model. Subsequent refinement includes surface mapping and mesh overlay. These techniques enable accurate representation of the bone defect and assessment of soft-tissue aspects, such as scalp shrinkage and contour, to ensure that the final implant is both functionally compatible and aesthetically aligned with the surrounding tissues. The reconstruction phase transitions into the design stage, where the cranial structure is virtually restored to its intended anatomical form. This reverse design step digitally restores the missing section of the skull. Using the restored geometry, a patient-specific porous titanium implant is developed. The porosity is a deliberate biomedical design feature intended to support osseointegration and tissue integration. The finalised digital model of the implant is then depicted, showcasing its refined contours and personalised features derived from the patient's anatomical data. This is followed by the virtual placement of the implant onto the reconstructed 3D skull. This step is critical for confirming anatomical fit and alignment, ensuring the implant precisely conforms to the patient's

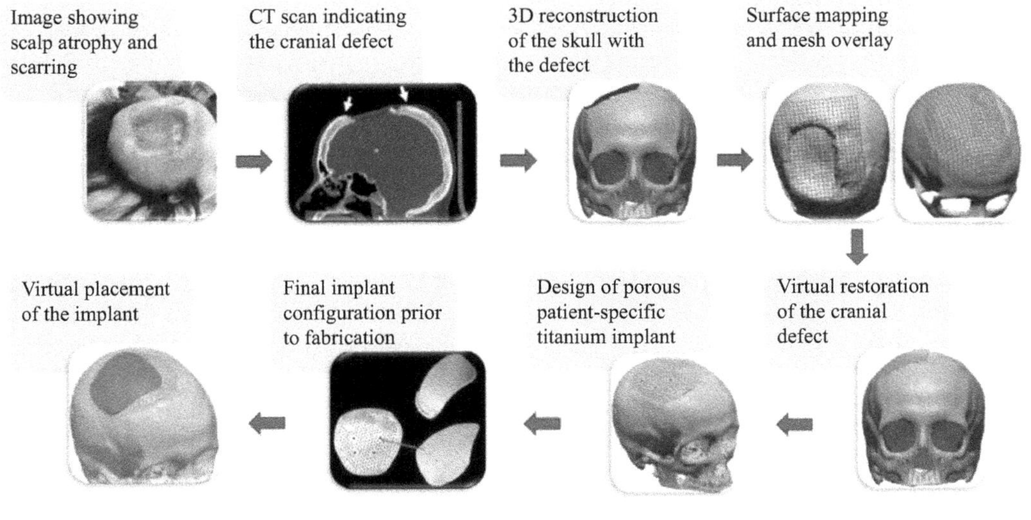

Figure 3.9 Reverse engineering workflow for patient-specific cranioplasty: this process involves capturing a clinical image of scalp atrophy and a CT scan of the cranial defect, followed by 3D skull reconstruction. Surface mapping assesses skin shrinkage and scalp geometry, enabling virtual defect restoration and porous patient-specific titanium implant design. The final implant configuration is prepared, and its anatomical fit is verified through virtual placement.

Source: Adapted from [25] under the CC BY 4.0 License.

anatomy. Such digital validation minimises intraoperative adjustments and enhances surgical accuracy and patient safety. Overall, this RE process transforms raw biomedical imaging into highly customised, anatomically precise implants, underscoring its potential for superior patient-specific solutions in healthcare [25, 26].

3.3 AM INTEGRATION WITH RE FOR HEALTHCARE

The anatomical CAD model, generated during the concluding phase of RE, is then utilised for simulation or further manufacturing. However, conventional fabrication methods, such as manual sculpting, moulding, or subtractive machining, struggle to accurately replicate intricate biological structures. These traditional approaches yield compromised anatomical fidelity, time-consuming fabrication, and limited patient-specific customisation, especially for complex geometries or porous structures vital for precise surgical planning and implantation. The inability to perfectly match individual anatomy often results in suboptimal fits, extended surgical times, and increased complication risks. AM has emerged as a transformative solution by building structures layer by layer from CAD designs. AM inherently overcomes these limitations, enabling the creation of highly intricate, custom-fit, patient-specific anatomical models, surgical guides, and implants with unparalleled precision and efficiency.

3.3.1 AM Methods for Medical Applications

AM processes can be categorised by their material-deposition techniques, energy sources, or operational principles. However, a more standardised classification, developed by the International Organisation for Standardisation (ISO) and the American Society for Testing and Materials (ASTM), delineates AM techniques into seven distinct categories [27], as shown in **Figure 3.10**. The selection of particular AM techniques in the healthcare sector is predominantly determined by factors such as material type, required precision, and the complexity of the medical application. This systematic categorisation improves understanding and facilitates the selection of the most appropriate AM technique for diverse biomedical needs.

AM processes are ideal for printing biocompatible polymers, while others perform better with ceramics or metals. The careful selection of an AM process directly impacts the final mechanical properties, surface finish, and crucial biological response of the fabricated medical device or anatomical model, as demonstrated in **Table 3.1**.

3.3.2 Importance of AM in RE for Part Manufacturing

Understanding and manipulating the geometry of biomedical objects and patient data are crucial for developing biomedical engineering applications. Medical Reverse Engineering (MRE) reconstructs 3D anatomical models for medical product design and manufacturing, as well as research and development. Since the 1990s, RE has been significant in developing personalised implants, surgical tools, and medical training models, utilising patient data [24].

The different RE applications in **Figure 3.11** show how important AM is for addressing the intricacies of human anatomy. AM enables the fabrication of complex, patient-specific structures directly from digital models, something that is often not possible with traditional manufacturing methods. The graphic shows 3D head models generated from RE data to create helmets that fit perfectly, as well as broken skull shapes that have been reconstructed to create tailored cranioplasty implants. These implants are designed to integrate seamlessly with RE and Rapid Prototyping, a vital AM technology. More instances show how important

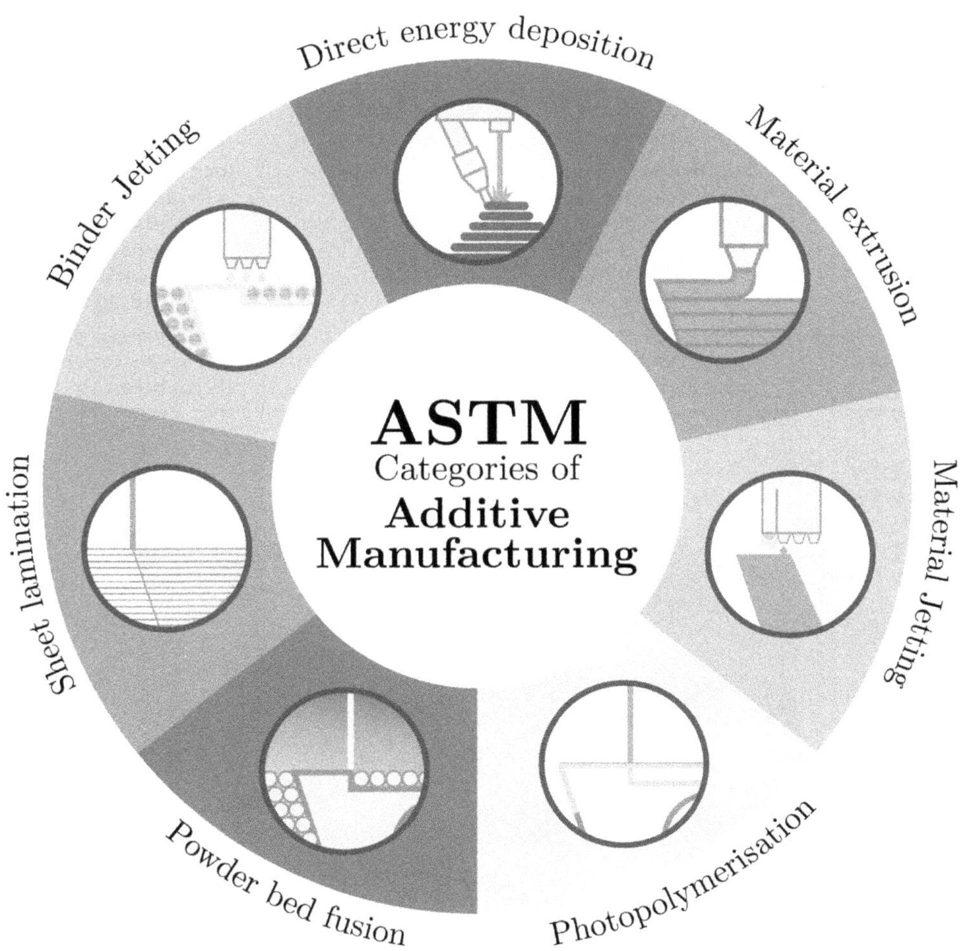

Figure 3.10 ASTM categories of additive manufacturing.

Source: Reproduced with permission from [27], Copyright © 2024 Elsevier.

Table 3.1 ASTM Category of AM Processes Most Commonly Used in the Healthcare Sector [9, 28]

S.No.	*AM Category*	*AM Technique*	*Materials*	*Medical Applications*
1	Vat Photopolymerisation	Stereolithography (SLA); Digital light processing (DLP)	Photopolymers, Ceramics, Metal resin	Dental implants and models, Bone implants, Hearing aids, Tools for biomedical use
2	Material Jetting	Poly-jet printing (PJ); Multi-jet modelling (MJM)	Polymers, Plastics	Dental implant guides and casts, medical models
3	Binder Jetting	3D printing (3DP)	Polymer, Ceramic, Metal	Knee and Hip joint replacements, Dental prosthetics, Bone scaffolds
4	Material extrusion	Fused deposition modelling (FDM); Direct ink writing (DIW)	Thermoplastic Polymer (ABS, nylon, PEEK, PLA, PVA), metal, Ceramic, Bioink	Rapid prototyping exoskeleton, Antibiotic delivery systems, Scaffolds, Medical devices, Porous structures

(Continued)

Table 3.1 (*Continued*) ASTM Category of AM Processes Most Commonly Used in the Healthcare Sector [9, 28]

S.No.	*AM Category*	*AM Technique*	*Materials*	*Medical Applications*
5	Sheet lamination	Laminated object Manufacturing (LOM)	Plastic, Paper, Sheet metals	Orthopaedic modelling of bone surfaces
6	Powder bed fusion	Selective laser Sintering (SLS); Selective laser melting (SLM); Electron beam melting (EBM)	Polymer, Metal, Ceramic	Joint implants, Scaffolds (Tissue Engineering), Craniofacial implants, Vertebral body replacement, Cervical, Porous dental implants, Hip and Knee implants, Press-fit
7	Direct Energy Deposition	Laser metal deposition (LMD); Wire-arc additive manufacturing (WAAM)	Metals, titanium, cobalt chrome	Repair existing parts and build a very big (large) parts

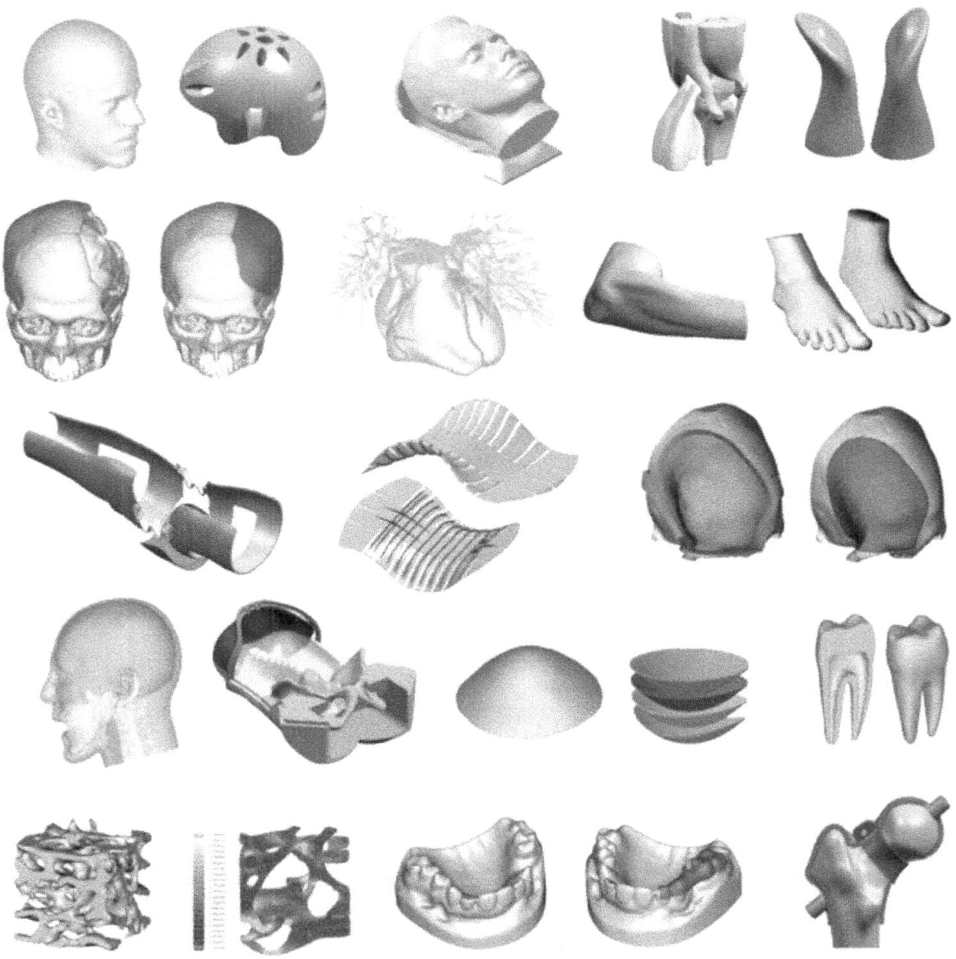

Figure 3.11 Typical RE applications in various healthcare sectors, including customised implants; dental prosthetics and their digital modelling; surgical tool development; surgical education and simulation; contributions to vision science and optometry; orthopaedic solutions; ergonomic product design; orthotic devices; and applications in tissue engineering.

Source: Reproduced from [24], © 2010 University of Greenwich.

AM is: 3D knee models made from CT or MRI data support biomechanical studies and make it easier to produce implants with complex internal structures that work better. AM technology also enables the production of patient-specific orthoses for the elbow and foot that fit the body's contours very precisely and at a low cost. AM's ability to fabricate highly realistic and anatomically accurate replicas has enabled applications such as next-generation contact lenses and surgical training models for complex procedures. Anatomical structures are naturally quite complicated, from the ear on the outside to the organs and bones on the inside. AM's accuracy and flexibility are essential. AM is an important link in providing personalised biomedical solutions by turning RE data into physical models and medical devices.

3.3.3 Sector-Wise Contribution of AM and RE for Healthcare

AM plays a significant role in RE, particularly in medical applications. It addresses a broad spectrum of challenges in medical education and training, device and implant development, tissue engineering, prosthetics, orthotics, mechanical bone replication, forensics, and dental solutions. AM enhances these areas through improved accuracy, speed, customisation, visualisation, and modelling, while also supporting clinical decision-making. As summarised in **Table 3.2**, AM is shaping a transformative market for healthcare innovation through RE. **Figure 3.12** further illustrates this impact, showing that Engineering and Materials dominate the research landscape with a 56% share, followed by Medicine, Dentistry, and Biotechnology at 17%, Basic and Computer Sciences at 21%, and other fields contributing 6%. This distribution underscores the strong influence of engineering and materials, driven by AM's capability for tool-less, rapid production of complex geometries [11].

3.4 CONCLUSIONS

The integration of AM and RE has undeniably revolutionised the healthcare sector, offering unprecedented capabilities for personalised medicine. As detailed throughout this chapter, the inherent complexity and uniqueness of human anatomy, which traditionally posed insurmountable manufacturing challenges, are now effectively addressed by AM's ability to create intricate, custom-fit structures directly from RE-generated digital designs. This powerful synergy enables the precise fabrication of patient-specific implants, anatomical models for surgical planning and training, and specialised medical devices, thereby enhancing accuracy, reducing surgical time, and improving patient outcomes. Despite these significant advancements and the demonstrated potential across diverse medical applications, from craniomaxillofacial reconstruction and orthopaedics to tissue engineering and forensics, challenges persist. These include the high initial costs associated with advanced scanning and AM technologies, complexities in data processing and material selection for biocompatibility, and the ongoing need to minimise production time and cost for customised parts. Overcoming these hurdles necessitates sustained interdisciplinary collaboration among engineers, material scientists, clinicians, and regulatory bodies. By continuously refining workflows, developing new biocompatible materials, and innovating AM processes, the combined power of AM and RE is poised to unlock further transformative solutions, fundamentally advancing healthcare and improving human well-being.

ACKNOWLEDGEMENTS

The authors sincerely thank Ms. Aditi Harsha, School of Biomedical Engineering, Indian Institute of Technology (BHU), Varanasi, for her assistance in figure preparation. They also

Table 3.2 Major Healthcare Sectors in Which AM and RE Integration Is Implemented [11]

S.No.	*Healthcare Sector*	*Purpose*	*Key Benefits*	*Ref.*
1	Surgical planning	• The primary objective is to improve surgical planning with AM. • These models help surgeons and doctors plan surgeries visually. • Assessing the patient's bone structure before surgery reduces time, expense, and risk.	• This technology predicted the problem cause and provided diagnostic quality during operation. • Complex anomalies and procedures are best understood by AM models. • In heart, spine, maxillofacial, and craniofacial procedures, these models aid in deformity or anatomical abnormality surgery.	[29, 30]
2	Education and training in medicine	• The main objective is to present internal and external human anatomical structures more effectively using this technology. • These multicoloured models are utilised in medical education for teaching and research purposes.	• AM models are used for better illustration in schools and museums. • Medical students and young doctors utilise these models to grasp surgical procedures and problems without inflicting patient distress.	[31, 32]
3	Design and development of medical devices and instrumentation	• This technology is meant to help with the design and development of medical instruments and devices.	• Manufacturing healthcare devices and instruments, AM is employed to design, develop, and build medical equipment or devices. • It includes hearing aids, dental tools, and surgical instruments.	[33]
4	Custom implant design	• This technology aims to manufacture customised fittings and implants. • An intricate shape can be built in a short period.	• The use of RE and AM technology enables the creation of customised implants that suit patients comfortably and affordably. • AM creates precise implants for patients, greatly benefiting them more than standard-sized implants. • Customised implant manufacture diminishes risk and reduces surgical duration.	[30, 34]
5	Scaffolds and tissue engineering	• The main objective is to create implants with unique geometrical properties, like scaffolds for tissue restoration. • It supersedes traditional scaffold manufacturing techniques.	• The scaffold is a supportive framework that offers assistance and guidance to impaired patient bone or damaged developing tissue. • AM techniques such as FDM, SLS, and 3D printing can yield controlled porosity structures employing biomaterials for tissue engineering and scaffolding. • AM technique extended the capacity to fabricate intricate geometric products with greater precision.	[35]

Table 3.2 (Continued) Major Healthcare Sectors in Which AM and RE Integration Is Implemented [11]

S.No.	*Healthcare Sector*	*Purpose*	*Key Benefits*	*Ref.*
6	Prosthetics and orthotics	• This method is advantageous in the field of prosthetics and orthotics, beginning with specific patient anatomy.	• In this framework, accurate patient alignment is crucial for biomechanical geometry development, comfort, and stability. • AM creates personalised prostheses that fit patients accurately and affordably, similar to dental crowns.	[29, 31]
7	Mechanical bone replicas	• How the AM method is used to make mechanical bone models. • This technique effortlessly mimics material variations.	• AM can make a composite structure that has properties akin to bone. • These bones may impart strength under numerous conditions. • This is also helpful to replicate the stresses, fractures, and other changes in bones that happen over time.	[36]
8	Forensics	• AM offers a more effective tool for investigating crimes, like murder cases, where the crime scene is recreated for the inquiry.	• These models collect evidence for criminal investigations and construct scaffolds for investigators to answer questions. • The accurate creation of events and scenes, often facilitated by this, helps in the swift resolution of cases.	[32]

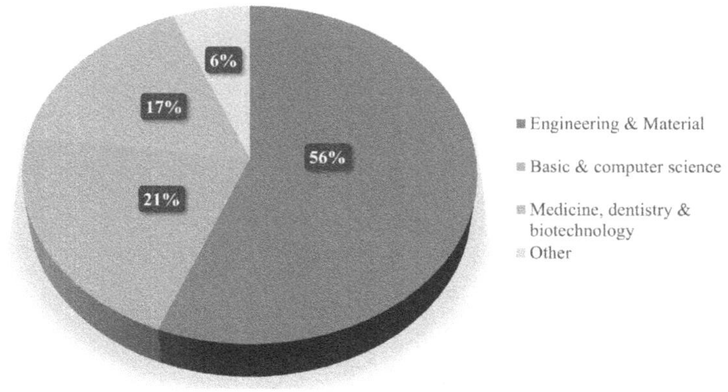

Figure 3.12 Area-wise contribution of AM in the healthcare sector.

gratefully acknowledge the support of Mr. Rajat Jain, Department of Mechanical Engineering, National Institute of Technology, Warangal, and Dr. Venumurali Jagannati, Department of Mechanical Engineering, Annamacharya Institute of Technology and Sciences, Tirupati.

REFERENCES

[1] Y. Wakjira, N.S. Kurukkal and H.G. Lemu, *Reverse engineering in medical application: Literature review, proof of concept and future perspectives*, Nature Scientific Reports 14 (2024), p. 23621.

[2] J. López and C. Vila, *An approach to reverse engineering methodology for part reconstruction with additive manufacturing*, IOP Conference Series: Materials Science and Engineering 1193 (2021), p. 012047.

[3] R.J. Urbanic, *A design and inspection based methodology for form-function reverse engineering of mechanical components*, The International Journal of Advanced Manufacturing Technology 81 (2015), pp. 1539–1562.

[4] F. Buonamici, M. Carfagni, R. Furferi, L. Governi, A. Lapini and Y. Volpe, *Reverse engineering of mechanical parts: A template-based approach*, Journal of Computational Design and Engineering 5 (2018), pp. 145–159.

[5] W. Wang, *Reverse Engineering*, CRC Press, 2010.

[6] Z. Geng and B. Bidanda, *Review of reverse engineering systems—current state of the art*, Virtual and Physical Prototyping 12 (2017), pp. 161–172.

[7] E. Taneva, B. Kusnoto and C.A. Evans, 3D scanning, imaging, and printing in orthodontics, in *Issues in Contemporary Orthodontics*, InTech, 2015.

[8] E. Huotilainen, R. Jaanimets, J. Valášek, P. Marcián, M. Salmi, J. Tuomi et al., *Inaccuracies in additive manufactured medical skull models caused by the DICOM to STL conversion process*, Journal of Cranio-Maxillofacial Surgery 42 (2014), pp. e259–e265.

[9] A. Haleem and M. Javaid, *3D scanning applications in medical field: A literature-based review*, Clin Epidemiol Glob Health 7 (2019), pp. 199–210.

[10] H. Bikas, A.K. Lianos and P. Stavropoulos, *A design framework for additive manufacturing*, The International Journal of Advanced Manufacturing Technology 103 (2019), pp. 3769–3783.

[11] M. Javaid and A. Haleem, *Additive manufacturing applications in medical cases: A literature based review*, Alexandria Journal of Medicine 54 (2018), pp. 411–422.

[12] S. Sedghi, M. Sanderson and P. Clough, *Medical image resources used by health care professionals*, Aslib Proceedings 63 (2011), pp. 570–585.
[13] J.M. Friend and B.A. Thayer, *Brain Imaging and Political Behavior*, pp. 231–255. Emerald Group Publishing Limited, 2011.
[14] D. Eyers and K. Dotchev, *Technology review for mass customisation using rapid manufacturing*, Assembly Automation 30 (2010), pp. 39–46.
[15] R. Bibb, D. Eggbeer, P. Evans, A. Bocca and A. Sugar, *Rapid manufacture of custom-fitting surgical guides*, Rapid Prototyping Journal 15 (2009), pp. 346–354.
[16] S. Lei, M.C. Frank, D.D. Anderson and T.D. Brown, *A method to represent heterogeneous materials for rapid prototyping: The Matryoshka approach*, Rapid Prototyping Journal 20 (2014), pp. 390–402.
[17] M. van Eijnatten, J. Koivisto, K. Karhu, T. Forouzanfar and J. Wolff, *The impact of manual threshold selection in medical additive manufacturing*, International Journal of Computer Assisted Radiology and Surgery 12 (2017), pp. 607–615.
[18] F. Buonamici, M. Carfagni, R. Furferi, L. Governi, A. Lapini and Y. Volpe, *Reverse engineering of mechanical parts: A template-based approach*, Journal of Computational Design and Engineering 5 (2018), pp. 145–159.
[19] W. Wang, *Reverse Engineering*, CRC Press, 2010.
[20] A. Bin Aqeel, M.I. Aziz, U.K. uz Zaman and N. Aafaq, Reverse engineering, in *Handbook of Manufacturing Systems and Design*, CRC Press, Boca Raton, 2023, pp. 283–316.
[21] K. Cooper, *Rapid Prototyping Technology*, CRC Press, 2001.
[22] S. Rooppakhun, N. Chantarapanich and K. Sitthiseripratip, Advanced medical imaging and reverse engineering technologies in craniometric study, in *Forensic Medicine—From Old Problems to New Challenges*, InTech, 2011.
[23] N. Negied and A. SeragEldin, *Automatic detection of Alzheimer disease from 3D MRI images using deep CNNs*, International Journal of Advanced Computer Science and Applications 13 (2022).
[24] L.C. Hieu, J. Vander Sloten, L.T. Hung, L. Khanh, S. Soe, N. Zlatov, L.T. Phuoc and P.D. Trung, Medical reverse engineering applications and methods, in *Proceedings of International Conference on Innovations, Recent Trends and Challenges in Mechatronics, Mechanical Engineering and New High-Tech Products Development (MECAHITECH'10)*, 2010, pp. 186–196. INCD-MTM, Bucharest, Romania. https://gala.gre.ac.uk/id/eprint/11735
[25] A. Marzola, F. Buonamici, R. Furferi, L. Governi, L. Genitori and F. Mussa, *Additive manufacturing and reverse engineering in cranioplasty: A personalized approach to minimize skin flap complications*, Applied Sciences 11 (2021), p. 4926.
[26] F. Calignano, M. Galati, L. Iuliano and P. Minetola, *Design of additively manufactured structures for biomedical applications: A review of the additive manufacturing processes applied to the biomedical sector*, Journal of Healthcare Engineering 2019 (2019), pp. 1–6.
[27] J.L. Amaya-Rivas, B.S. Perero, C.G. Helguero, J.L. Hurel, J.M. Peralta, F.A. Flores et al., *Future trends of additive manufacturing in medical applications: An overview*, Heliyon 10 (2024), p. e26641.
[28] M. Salmi, *Additive manufacturing processes in medical applications*, Materials 14 (2021), p. 191.
[29] J. Faber, P.M. Berto and M. Quaresma, *Rapid prototyping as a tool for diagnosis and treatment planning for maxillary canine impaction*, American Journal of Orthodontics and Dentofacial Orthopedics 129 (2006), pp. 583–589.
[30] H. Zenha, L. Azevedo, L. Rios, A. Pinto, M. Luz Barroso, C. Cunha et al., *The application of 3-D biomodeling technology in complex mandibular reconstruction—experience of 47 clinical cases*, European Journal of Plastic Surgery 34 (2011), pp. 257–265.
[31] K. Mori, T. Yamamoto, K. Oyama and Y. Nakao, *Modification of three-dimensional prototype temporal bone model for training in skull-base surgery*, Neurosurgical Review 32 (2009), pp. 233–239.
[32] Q. Liu, M.C. Leu and S.M. Schmitt, *Rapid prototyping in dentistry: Technology and application*, The International Journal of Advanced Manufacturing Technology 29 (2006), pp. 317–335.

[33] R. van Noort, *The future of dental devices is digital*, Dental Materials 28 (2012), pp. 3–12.
[34] W.-Y. Yeong, C.-K. Chua, K.-F. Leong and M. Chandrasekaran, *Rapid prototyping in tissue engineering: Challenges and potential*, Trends in Biotechnology 22 (2004), pp. 643–652.
[35] D.W. Hutmacher, M. Sittinger and M. V Risbud, *Scaffold-based tissue engineering: Rationale for computer-aided design and solid free-form fabrication systems*, Trends in Biotechnology 22 (2004), pp. 354–362.
[36] S.M. Peltola, F.P.W. Melchels, D.W. Grijpma and M. Kellomäki, *A review of rapid prototyping techniques for tissue engineering purposes*, Annals of Medicine 40 (2008), pp. 268–280.

Chapter 4

The Contemporary Prospect of Machine Learning in 3D Printing of Medical Devices and Implants

Ri Han, Haoran Yin, and Swee Leong Sing

4.1 INTRODUCTION

3D printing has revolutionised the medical field by enabling the production of highly customised medical devices and implants with intricate designs and complex geometries. This advanced technology provides unique advantages in developing patient-specific solutions for applications such as tissue scaffolds, prosthetics, and orthopaedic implants. However, ensuring optimal design, efficient production, and quality control throughout the 3D printing process for clinical applications remains a challenge. Therefore, Artificial Intelligence (AI) and Machine Learning (ML) are powerful tools for refining and optimising various aspects of 3D printing to achieve greater precision and customisation in medical device manufacturing (Mamo et al. 2023).

In the field of biomedical applications, Fused Deposition Modelling (FDM), Stereolithography (SLA), Selective Laser Sintering (SLS), and Laser Powder Bed Fusion (LPBF) can be categorised as the most prominent and widely used additive manufacturing technologies. FDM is a well-known extrusion-based technique due to its affordability and ease of use, which make it suitable for medical prototypes, prosthetics, and anatomical models. SLA can deliver high-resolution printing and smooth surface finishes, making it ideal for dental prosthetics, surgical models, and tissue scaffolds. SLS is a powder-bed fusion technique for fabricating complex, porous structures without the need for support materials, and it is particularly valuable for bone scaffolds and patient-specific prosthetics. LPBF, another leading powder-bed fusion technology, can produce strong and high-density medical implants in orthopaedic and dental applications. Beyond these methods, other 3D printing technologies, such as Digital Light Processing (DLP) and Inkjet Bioprinting, are also being explored for medical applications with various biomaterials. Each of these 3D printing methods plays a vital role in advancing medical device fabrication by offering customisation, precision, and design flexibility to meet the growing demands of personalised healthcare and regenerative medicine.

Machine learning is significantly enhancing 3D printing technology for medical devices and implants by optimising design, automating manufacturing processes, and driving innovation through various techniques (Inayathullah and Buddala 2025). For example, supervised learning can be used in diagnostic imaging analysis, predictive maintenance of 3D printers, and the development of smart implants that monitor patient health in real-time by training models on labelled datasets. Unsupervised learning can help identify patterns from unlabelled data, which is important for optimising design and analysing potential failure modes in 3D-printed implants to improve their performance, functionality, and durability. In addition, medical device design can be enhanced by semi-supervised learning because it combines labelled and unlabelled data when expert annotations are scarce. This method enables optimisation of material composition and structural parameters while reducing reliance on expensive labelled datasets. The application of reinforcement learning (RL), in which

DOI: 10.1201/9781003590774-4

AI agents learn through trial and error, is becoming increasingly common in the control of 3D printing processes. Dynamic parameter adjustment through RL will precisely control printing speed, temperature, and material flow to improve implant fabrication accuracy, structural integrity and biocompatibility. Overall, machine learning is revolutionising 3D printing in biomedical engineering by enabling advanced customisation, quality control, and design optimisation for medical implants. These advancements pave the way for more effective, patient-specific treatments and improved clinical outcomes.

The integration of machine learning into 3D printing for medical devices and implants is driving unprecedented advancements, and this chapter will explore AI and ML applications across all stages of the 3D printing process, including pre-printing, printing, and post-printing. In the pre-printing stage, ML can provide intelligent material design by predicting mechanical properties, optimising composite structures, and accelerating the discovery of novel biomaterials. For example, generative models, convolutional neural networks (CNNs), and Bayesian optimisation have already been studied to refine biocompatible polymers and multi-material printing strategies. During the printing process, ML-based real-time monitoring techniques, such as acoustic and optical imaging, can enhance defect detection and quality control, ensuring greater consistency in implant fabrication. In the post-printing phase, ML-powered vision systems and predictive modelling will facilitate automated quality inspection, surface optimisation, and post-processing adjustments to improve implant durability and osseointegration. As AI and ML continue to evolve, they will unlock new possibilities in the design and production of next-generation medical devices and implants, further advancing precision medicine and patient-specific treatments.

4.2 3D PRINTING TECHNOLOGY APPLIED FOR MEDICAL DEVICES AND IMPLANTS

Fused Deposition Modelling (FDM) is a widely used 3D printing technology known for its simplicity, affordability, and capability to fabricate complex geometries. This technology was invented by Scott Crump in 1988, who subsequently co-founded Stratasys Ltd. in 1989 to commercialise the method (Singh and Jonnalagadda 2020). In 1992, Stratasys introduced its first commercial 3D printer, making FDM one of the earliest additive manufacturing technologies accessible to the public. As an extrusion-based 3D printing technology that builds objects layer by layer, it operates by heating thermoplastic filament and extruding it through a nozzle along a programmed path defined by G-code to build objects with customised designs. The advantages of FDM include its affordability, accessibility, and compatibility with various materials. It is suitable for creating patient-specific devices and prototypes. However, this technology also has some limitations, such as lower resolution and rougher surface finishes compared to other 3D printing methods. Some post-processing techniques have been developed to address the rough surfaces of FDM 3D-printed samples, but chemical solutions may also raise toxicity concerns for subsequent cell culture and animal studies if they are not thoroughly removed. Apart from these challenges, the versatility and material diversity of FDM continue to make it a valuable tool in biomedical engineering.

Stereolithography (SLA) was invented by Chuck Hull in the 1980s, marking a breakthrough in additive manufacturing. Hull co-founded 3D Systems and launched the SLA-1, the world's first commercially available 3D printing system, in 1987 (Savini and Savini 2015). As one of the most precise 3D printing technologies, SLA uses an ultraviolet laser to cure liquid photopolymer resin layer by layer into solid objects. Its ability to achieve exceptional resolution and fabricate complex geometries has made SLA a cornerstone technology in the production of medical devices and implants. However, several limitations and challenges still hinder its widespread adoption. From an engineering perspective, the major limitation of SLA is its inability to print with multiple resins in a single structure. Compared to FDM and other extrusion-based 3D printing technologies, SLA also faces challenges such as higher costs, slower production speeds, complex post-processing, and limited scalability.

Selective Laser Sintering (SLS) is a powder bed fusion 3D printing technology that was first developed in the 1980s by Dr. Carl Deckard and Dr. Joe Beaman at the University of Texas at Austin (Han et al. 2022). It involves a high-powered laser (typically a CO_2 laser) to selectively fuse powder particles layer by layer to create a solid 3D structure from a digital design. During this fabrication, a thin layer of powder is evenly distributed across the build platform. The laser then scans and sinters targeted regions while leaving the surrounding powder loose. Once a layer is processed, the build platform is lowered slightly, and a new layer of powder is deposited, repeating the process. Unlike SLA or FDM, SLS does not require support structures, as the surrounding, unsintered powder acts as a natural support for building complex geometries and porous structures ideal for medical implants, bone scaffolds, and customised prosthetics. While SLS offers significant advantages in biomedical applications, it also presents several process-related challenges that limit its scalability and clinical adoption. One major limitation is the high processing temperature and energy consumption, as the powder bed must be maintained slightly below the material's melting point and requires substantial laser power. This makes SLS more energy-intensive than other additive manufacturing techniques, such as SLA and FDM.

Laser powder bed fusion (LPBF) is an advanced technique used to produce highly dense, mechanically robust, and patient-specific medical implants and devices. The technology originated in the 1990s through a research collaboration between the Fraunhofer Institute for Laser Technology (ILT) in Germany and industry partners, which led to its official patent in 1996 (Shanthar et al. 2023). After that, LPBF has evolved into one of the most widely used metal 3D printing methods in biomedical, aerospace, and automotive applications. Similar to SLS, a thin layer of metal powder (typically 15–45 μm thick) is evenly spread across the build platform within an inert gas environment (e.g., argon or nitrogen) to prevent oxidation. A high-powered laser (e.g., ytterbium fibre laser or CO_2 laser) then selectively scans the cross-section of the digital 3D model to fully melt the powder particles and form a solid layer. Unlike SLS, which only sinters powder particles together, LPBF fully melts the powder to fabricate samples with mechanical properties comparable to those of cast or forged metals. In addition, LPBF technology can reduce material waste and minimise the need for extensive machining to improve manufacturing efficiency. However, LPBF also faces some challenges in medical applications. High processing temperatures can induce residual stress and cause part distortion, requiring post-processing treatments such as heat treatment or hot isostatic pressing (HIP) to enhance mechanical stability (**Figure 4.1**).

4.3 MACHINE LEARNING IN MEDICAL 3D PRINTING

Supervised learning is a fundamental machine learning approach in which a model is trained on labelled datasets to predict or classify outputs from given inputs. By minimising the error between predicted and actual outputs, the model learns patterns in the data, typically using optimisation algorithms such as gradient descent. This approach is primarily applied in classification tasks (e.g., diagnosing medical conditions) and regression tasks (e.g., predicting treatment outcomes). A key advantage of supervised learning is its ability to achieve high accuracy when sufficient labelled data and computational resources are available. It has been widely applied in the development of medical devices and implants. Advances in deep learning have further enhanced its capabilities by enabling breakthroughs in tasks such as diagnostic

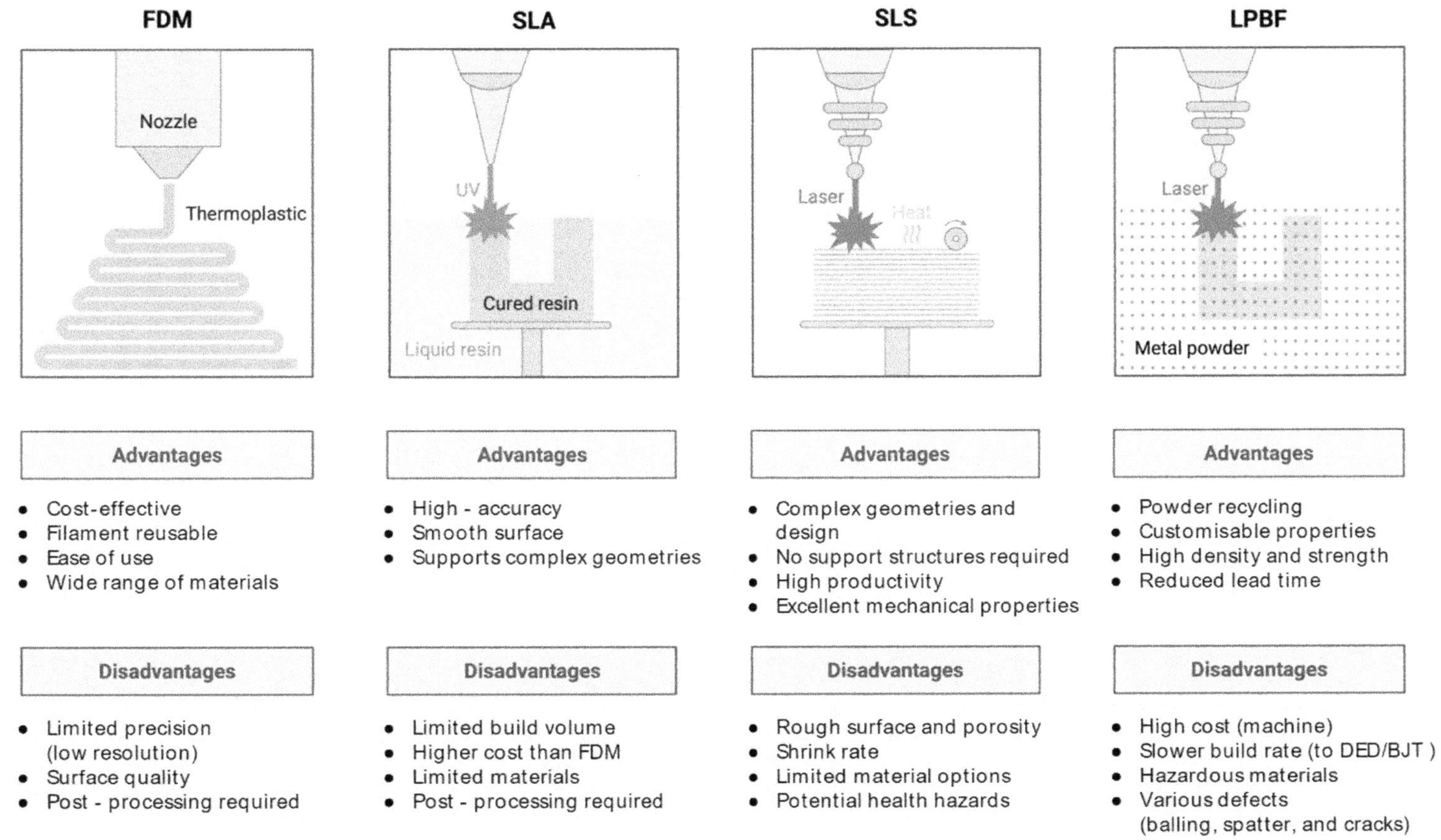

Figure 4.1 Typical 3D printing technologies and their advantages and disadvantages.

imaging, predictive maintenance of medical equipment, and personalised treatment planning. In summary, supervised learning remains a cornerstone of artificial intelligence, driving innovation in the development and optimisation of medical devices and implants.

Unsupervised learning is a machine learning approach in which models are trained on unlabelled datasets to identify hidden patterns, structures, or relationships within the data. Unlike supervised learning, there are no predefined labels, and the algorithm independently uncovers meaningful insights. Common tasks include clustering, dimensionality reduction, and anomaly detection. Algorithms such as k-means and PCA are used to group similar data points and simplify complex datasets into lower-dimensional forms. In biomedical 3D printing, unsupervised learning can support design optimisation and pattern recognition without the need for labelled datasets, making it flexible and cost-effective. For example, clustering implant data can help identify common failure modes and guide improvements in durability and performance. Although the outputs of unsupervised learning are typically more exploratory and less precise than those of supervised learning, they provide valuable early insights into complex datasets, driving innovation and enhancing the quality of medical devices and implants (**Figure 4.2**).

Semi-supervised learning combines elements of supervised and unsupervised learning by using both labelled and unlabelled data during training. This approach is particularly useful in situations where acquiring labelled data is expensive or time-consuming, but large volumes of unlabelled data are available. The model uses the small and labelled dataset to guide learning from the broader unlabelled dataset and enhance its predictive accuracy. In 3D printing for medical devices and implants, semi-supervised learning is useful when only a limited number of designs are annotated by experts, while a larger pool remains unlabelled. Algorithms like

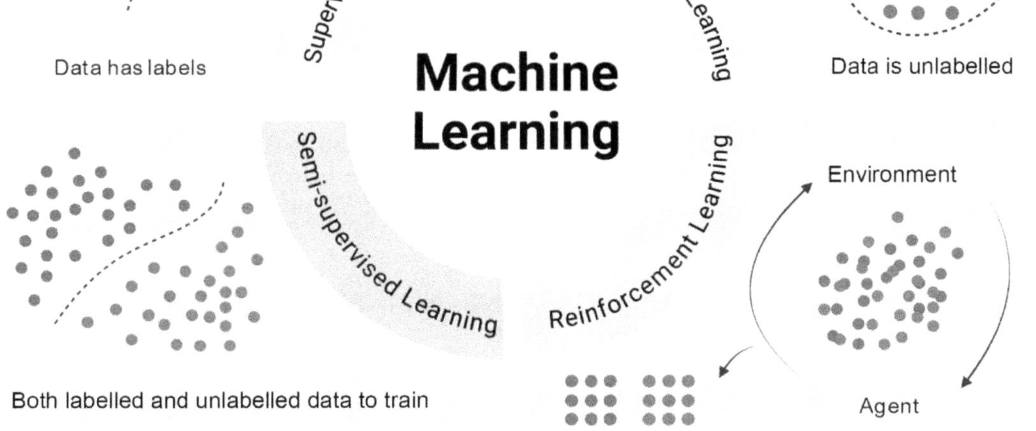

Figure 4.2 Machine learning models—supervised learning, unsupervised learning, semi-supervised learning, and reinforcement learning.

Table 4.1 Comparison of Machine Learning Methods

	Supervised Learning	*Unsupervised Learning*	*Semi-Supervised Learning*	*Reinforcement Learning*
Label Availability	Fully labelled dataset	No labels available	Partially labelled dataset	No direct labels, only rewards/punishments
Data Amount Requirement	Large	Varies	A small amount of labelled data with a large amount of unlabelled data	A large amount of interaction data
Learning Objective	Minimise error based on labelled outputs	Find hidden patterns or structures	Improve learning efficiency using limited labels	Maximise cumulative reward over time
Key Algorithms	Logistic Regression, SVM, Decision Trees, Neural Networks	K-means, PCA, autoencoders	Semi-supervised SVM, graph-based methods	Q-Learning, Deep Q-Networks (DQN), policy gradient
Challenges	High cost of data labelling	Difficult to evaluate results	Noise pseudo-labels	Difficult to train
Strengths	High accuracy	Less human effort	Can improve accuracy with less labelled data	Learns complex decision-making tasks dynamically

graph-based methods and semi-supervised Support Vector Machines (SVMs) effectively utilise this data to optimise the design and functionality of implants. For instance, these models can help identify optimal material compositions and structural configurations that improve implant durability and biocompatibility. By balancing supervision and autonomy, semi-supervised learning offers a cost-efficient approach to improving the quality and performance of 3D-printed medical devices and implants, even with limited labelled data (**Table 4.1**).

Reinforcement learning (RL) is a machine learning paradigm in which an agent learns to interact with an environment and receive feedback in the form of rewards or penalties. The goal is to develop a policy that maximises cumulative rewards over time. Unlike supervised learning, RL does not require labelled data; instead, it learns through trial and error guided by its reward system. In the field of medical devices, RL is playing an increasingly important role in optimising 3D printing processes to improve the quality of final products. Key algorithms such as Q-learning, Deep Q-Networks (DQNs), and policy gradient methods are used to fine-tune 3D printer parameters to achieve precise, reliable outcomes. While RL offers high adaptability, it often requires extensive training time and significant computational resources, which make it challenging to implement for real-world applications. Despite these limitations, RL continues to drive innovation in the customisation and performance of 3D-printed medical devices.

4.4 MACHINE LEARNING APPLICATIONS IN 3D PRINTING FOR MEDICAL DEVICES AND IMPLANTS

4.4.1 Pre-3D Printing Optimisation

Machine learning has played a crucial role in advancing the design of 3D-printed medical devices and implants, enabling the optimisation of mechanical properties, cost efficiency, and manufacturability. For instance, a convolutional neural network (CNN) has been employed to predict the stiffness and toughness of 2D composites, achieving high accuracy even with limited training data, small batch sizes, and minimal training loops (Gu et al. 2018). Generative models have further enhanced multi-material 3D printing by simplifying the exploration of complex material combinations. For example, variational autoencoders (VAEs) coupled with Bayesian optimisation have been used to design composite materials with targeted elastic properties. Xue et al. applied a similar generative strategy to automatically design multi-material composite solids with distinct elastic moduli, and successfully fabricated these optimised structures, which confirmed superior mechanical properties as predicted (Xue et al. 2020). In the domain of soft material printing, hierarchical machine learning (HML) has been effectively applied to the freeform reversible embedding (FRE) method, optimising material formulations and processing parameters. This technique not only accelerated printing speed by 2.5 times while maintaining fidelity but also identified novel silicone compositions and print settings beyond traditional trial-and-error approaches (Menon et al. 2019). Machine learning has also improved FDM by optimising layer thickness, infill percentage, and extruder temperature to enhance the toughness and cost-effectiveness of PLA components. Response surface methodology (RSM) experiments demonstrated that honeycomb fill patterns yielded lightweight yet strong specimens, while artificial neural networks (ANN) and hybrid ANN-genetic algorithm (ANN-GA) models provided accurate predictions, with the latter improving estimation accuracy by up to 11.5% (Meiabadi et al. 2021). In addition, by monitoring in-process signals such as nozzle pressure, temperature, and near-infrared (NIR) spectra during extrusion, regression models can predict the resulting polymer's mechanical strength in real time. Mulrennan et al. demonstrated that incorporating NIR spectroscopy data into a multivariate regression enabled accurate forecasting of the tensile yield stress of bioresorbable polymer filaments as they were being extruded, facilitating immediate process control to ensure each filament met quality standards (Mulrennan et al. 2022).

Beyond materials, machine learning significantly enhances part design for 3D-printed medical devices by uncovering complex relationships between design parameters and performance. Traditional design approaches often rely on empirical knowledge and iterative prototyping, which can be time-consuming and resource-intensive. In contrast, machine learning enables data-driven optimisation by uncovering complex relationships between process parameters and final part characteristics, thus facilitating more efficient and precise design modifications. One key area where machine learning has demonstrated significant impact is in predicting and optimising structural integrity based on printing parameters. ANNs have been extensively used to model the mechanical properties of 3D-printed components. For example, Bayrakter et al. developed an ANN model to predict the tensile strength of FDM-printed PLA parts based on parameters such as melt temperature, layer thickness, and raster orientation. The model achieved exceptionally high accuracy (R^2 values above 0.999), enabling precise control over mechanical performance during part fabrication (Bayraktar et al. 2017). Similarly, research on compressive strength optimisation for FDM-printed ABS components demonstrated that layer thickness, raster angle, and build orientation significantly influence mechanical behaviour. By integrating ANN predictions with quantum-behaved particle swarm optimisation (QPSO), the study identified an optimal set of parameters that maximised compressive stress, offering a systematic approach to enhance part strength while minimising anisotropic effects (Sood et al. 2012).

In addition to mechanical performance, machine learning has been applied to improve geometric accuracy and compensate for distortions caused by thermal effects during additive manufacturing. One study introduced an ANN-based compensation method that corrects geometric deformations in STL files before printing, effectively reducing dimensional inaccuracies caused by thermal expansion and contraction (Chowdhury and Anand 2016). Similarly, a multi-scale neural network approach has been developed to accelerate the evaluation of micro- and nanoscale structures in complex lattice-based biomedical implants. By combining offline training and parameter sensitivity analysis, the model reduced computational evaluation times from hours to milliseconds, making it feasible for real-time topology optimisation in large-scale medical devices (Koeppe et al. 2018).

The optimisation of printability and material efficiency is another area where machine learning has provided substantial benefits. Jiang et al. applied neural networks to optimise the printable bridge length (PBL) in FDM, considering print speed, cooling fan speed, and temperature. Their model accurately predicted the maximum PBL, enabling an increase in unsupported span length and a reduction in support material use, which is particularly beneficial for medical implants requiring intricate geometries with minimal material waste (Jiang et al. 2019). Furthermore, in SLS, machine learning has been utilised to enhance dimensional accuracy and material density. Li et al. improved dimensional prediction by incorporating fuzzy C-means clustering into a radial basis function (RBF) neural network, effectively refining process parameter selection for better geometric fidelity (Li et al. 2009). Moreover, an ANN model trained with a genetic algorithm was employed to optimise SLS process parameters, such as layer thickness, hatch spacing, laser power, and scanning speed, ultimately reducing shrinkage and improving part consistency (Rong-Ji et al. 2009).

Machine learning-driven design has also extended into bio-functional optimisation. For tissue-engineering scaffolds, aside from mechanical properties, one must consider factors such as pore size for nutrient flow and surface topology for cell attachment. In a pioneering study, Z. Jiang et al. combined a genetic algorithm with a neural network to optimise the strand layout in a bioprinted scaffold, aiming to achieve uniform pore distribution and adequate strength (Jiang et al. 2014). The ML model captured the nonlinear relationships between print parameters (such as nozzle speed and material viscosity) and the resulting filament width and pore size in the scaffold, enabling the algorithm to find settings that produced a highly uniform grid conducive to cell growth. In summary, ML inspires the design phase of medical 3D printing by providing predictive models and optimisation algorithms that handle the complexity of modern implant requirements. Designers can now rely on ANNs, SVMs, and evolutionary algorithms as collaborative tools to predict outcomes of design choices.

One of the primary challenges in AM is manufacturability analysis, especially when printing complex geometries. Machine learning is proving invaluable in this context by tackling problems such as manufacturability analysis, build time reduction, and multi-objective process tuning with greater sophistication than traditional empirical or rule-based methods. Shi et al. developed a feature-based method using the Heat Kernel Signature to analyse geometric constraints and identify potential manufacturing issues before printing. This approach aids in redesigning and process planning, reducing the likelihood of printing failures and enhancing overall efficiency (Shi et al. 2018). Additionally, deep learning models have been employed to estimate key manufacturing metrics, including build time, material usage, and structural integrity. Williams et al. applied 3D CNNs to predict these metrics from voxel-based geometries, outperforming traditional linear regression models in accuracy. Their findings indicated that orientation-independent metrics were more predictable, providing valuable insights for optimising print setup in FDM (Williams et al. 2019).

Beyond manufacturability, machine learning has been used to optimise printing parameters for different AM processes. He et al. explored various machine learning models to predict feasible printing speeds in SLA, with a Siamese network achieving the highest accuracy of 82.67%, demonstrating its effectiveness in guiding process parameter selection for improved efficiency (He et al. 2019). Similarly, Deng et al. combined Orthogonal Experimental Design (OED) with BP neural networks to optimise extrusion parameters in ceramic 3D printing. Their approach identified key influencing factors, improving extrusion efficiency by 34% and reducing errors in parameter predictions to less than 3% (Deng et al. 2018). These studies illustrate how machine learning can refine process parameters across different AM techniques, leading to higher precision and material efficiency. In summary, machine learning-driven process optimisation has markedly enhanced the efficiency, reliability, and scalability of 3D printing for medical applications. By leveraging predictive modelling, adaptive control strategies, and hybrid physics-ML methods, engineers can achieve more precise control over the printing process.

4.4.2 3D Printing Process

Real-time process monitoring during the 3D printing of medical devices is crucial for ensuring print quality, detecting defects early, and optimising process parameters on the fly. Machine learning-based monitoring approaches typically fall into two categories: acoustic and optical. Each modality provides unique insights into the printing process, and ML algorithms can analyse complex sensor data to flag anomalies or predict part quality in situ. Acoustic monitoring leverages sound emissions to detect irregularities during printing. Shevchik et al. employed acoustic emission sensors and spectral convolutional neural networks (SCNN) to classify part quality based on porosity levels in LPBF, achieving 83–89% accuracy (Shevchik et al. 2018). Another study by Ye et al. focused on using simpler acoustic data (collected with a standard microphone) to detect defects in LPBF processes using a deep belief network (DBN) model. The DBN could classify acoustic patterns without requiring extensive signal preprocessing or complex feature engineering, simplifying the deployment of the monitoring system. This approach showed that even relatively straightforward acoustic setups combined with deep learning can be effective for detecting issues such as spatter, balling, and layer delamination in metal printing (Ye et al. 2018).

Optical monitoring techniques capture visual data to assess process stability and detect defects. Gobert et al. used high-resolution imaging and supervised learning to identify defects in PBF. A linear support vector machine (SVM) trained on high-resolution CT scan data achieved over 80% accuracy in defect detection (Gobert et al. 2018). Meanwhile, infrared imaging has also been explored for process stability assessment. Grasso et al. combined infrared thermal imaging with statistical process monitoring to detect unstable melting conditions in metal AM, demonstrating effectiveness in LPBF of zinc powder (Grasso et al. 2018). Semi-supervised and unsupervised ML methods have also been applied to optical monitoring to reduce the dependency on large labelled datasets. Okaro et al. developed a semi-supervised approach for fault detection in LPBF, using photodiode sensor data and clustering techniques. They showed that even without explicit labels for every type of defect, the algorithm could effectively separate normal vs. anomalous prints by identifying patterns that did not conform to the majority of good data (Okaro et al. 2019).

Machine vision systems have also been explored for defect identification. Grasso et al. applied PCA and K-means clustering to automate defect detection in LPBF, demonstrating applicability across different geometries (Grasso et al. 2016). Lyu and Manoochehri introduced an online laser-based monitoring and control system to enhance geometric accuracy and in-plane surface quality in additive manufacturing (Lyu and Manoochehri 2021). Using a 3D laser scanner, a CNN with a pre-trained VGG16 model classifies in-plane anomalies. Two PID-based closed-loop control systems significantly reduce height deviation errors and correct surface anomalies. Machine learning has also been applied to monitor the bioprinting process. Jin et al. developed an anomaly detection system for transparent hydrogel-based bioprinted materials using layer-by-layer sensor images and machine learning algorithms (Jin et al. 2023). High detection accuracy was achieved with CNN and advanced image processing techniques. The system can categorise and localise anomalies, enabling real-time autonomous correction of process parameters for high-quality 3D-bioprinted tissue constructs.

4.4.3 Post-3D Printing Treatments

Post-printing processes play a crucial role in additive manufacturing, ensuring that printed parts meet the required quality, mechanical performance, and surface characteristics. Machine learning-based monitoring techniques facilitate defect detection, mechanical property prediction, and process optimisation, improving the efficiency and reliability of post-processing workflows (**Figure 4.3**).

Automated vision systems powered by deep learning have significantly improved post-printing quality inspection, particularly in vision-based defect detection. Lim et al. introduced a fully automated vision pipeline that integrates CAD models with deep neural networks to detect and localise powder-based 3D-printed parts, enabling automated post-processing tasks such as sorting and cleaning. The system demonstrated high precision across various powder-bed fusion technologies, including SLS, binder jetting, and HP MJF (Lim and Pham 2021).

In terms of surface optimisation, Badogu et al. applied classification and regression trees (CART) to predict the tensile properties of zirconia (ZrO_2)-reinforced ABS composite filaments used for post-processing applications, specifically in roller burnishing (Badogu et al. 2024). The study demonstrated that heat treatment played a dominant role in improving the tensile strength of these filaments, with an R^2 of 0.8503 for peak strength and 0.8558 for break strength. These optimised composite filaments were subsequently used in roller burnishing, significantly reducing the surface roughness of 3D-printed PLA parts, showcasing how ML can assist in fine-tuning post-processing conditions to enhance final part quality.

Intelligent post-processing optimisation strategies are transforming post-processing treatments such as coatings and structural modifications. Sharma et al. utilised genetic algorithms

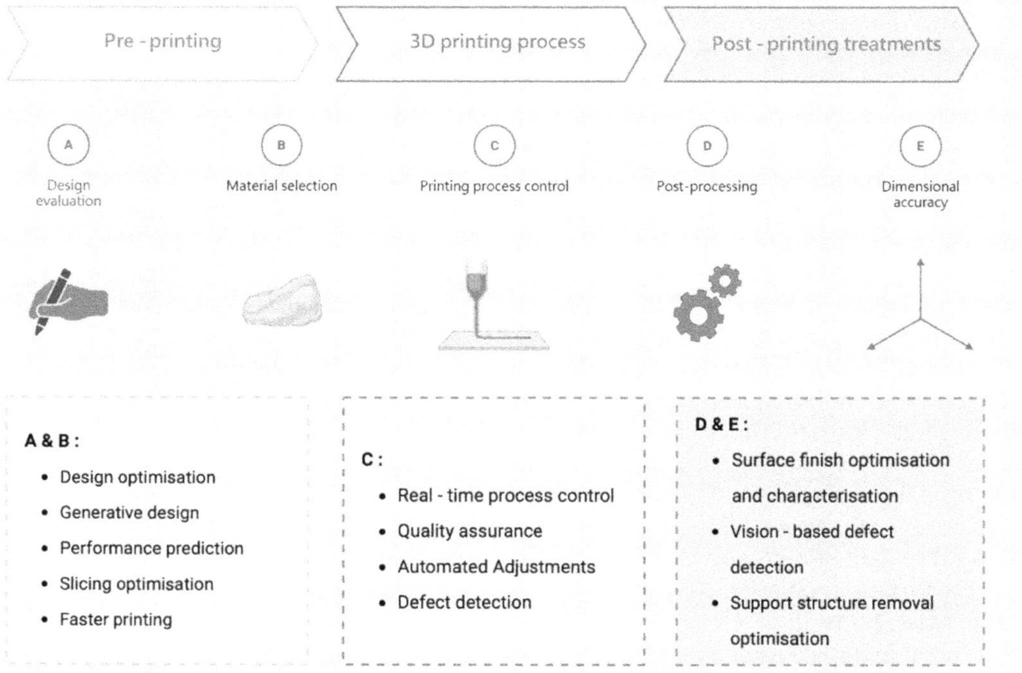

Figure 4.3 Integration of AI/ML methods across the 3D printing workflow.

(GA), particle swarm optimisation (PSO), and other ML techniques to refine polydopamine (PDM) coating parameters on 3D-printed PLA bone plates, significantly improving mechanical strength and surface characteristics (Sharma et al. 2024). Such data-driven optimisation ensures more uniform coatings and enhances the durability of biomedical implants, contributing to their long-term reliability and performance.

4.5 PROSPECT OF MACHINE LEARNING IN 3D PRINTING OF MEDICAL DEVICES AND IMPLANTS

The integration of machine learning into the 3D printing of medical devices and implants is poised to revolutionise the industry by enhancing precision, automation, and material innovation. As we move forward, ML-driven advancements will likely reshape the entire manufacturing process (from intelligent design algorithms and self-optimising printers to autonomous quality control), leading to highly customised, patient-specific implants with superior performance. This future aligns closely with the vision of Industry 4.0, where smart, connected systems collaborate to continuously improve manufacturing outcomes.

- Pre-Printing: Intelligent Material and Design Optimisation
  ML is already driving significant improvements in material selection and part design for medical applications, but its future potential lies in the development of self-optimising systems that can autonomously generate novel biomaterials. By leveraging generative adversarial networks (GANs) and physics-informed machine learning, researchers could accelerate the discovery of next-generation bioresorbable polymers, ceramic composites, and hybrid materials optimised for both mechanical strength and biological integration. Beyond material selection, ML will further refine the design process through AI-driven generative modelling, allowing for implants with complex, patient-specific geometries that maximise bone growth and vascularisation. In the near future, closed-loop AI systems could seamlessly integrate pre-printing optimisation with real-time feedback from the printing process, dynamically adjusting CAD models and process parameters to account for variations in material behaviour or anatomical constraints.
- Printing: Real-Time Process Monitoring for Precision Manufacturing
  While ML-driven acoustic and optical monitoring techniques already enhance defect detection, the next step is full autonomy in process control. Future ML models will not only identify printing anomalies but also enable self-correcting mechanisms that dynamically adjust process parameters in real time. By integrating reinforcement learning and predictive analytics, printers could autonomously modify laser power, extrusion rates, or scan paths to compensate for inconsistencies in material behaviour. This level of autonomy would significantly reduce waste, improve print consistency, and make high-precision implants more accessible, even in decentralised healthcare settings. Furthermore, the combination of AI and digital twin technology will enable virtual simulations to predict part performance before fabrication, further reducing trial-and-error and enhancing production efficiency.
- Post-Printing: Automated Quality Control and Functional Enhancement
  Post-processing will shift towards AI-driven automation, reducing reliance on manual inspection and empirical parameter tuning. In medical applications, ML-based vision systems will streamline the detection of surface irregularities, internal defects, and structural weaknesses in printed implants, allowing for automated classification and correction. Additionally, ML-guided optimisation of post-processing techniques, such as heat treatment, mechanical polishing, and bioactive coatings, will enhance implant

durability, wear resistance, and osseointegration. In the future, self-learning systems could analyse vast datasets of clinical outcomes and manufacturing conditions to dynamically adjust post-processing protocols, tailoring treatments to specific patient needs. As regulatory frameworks evolve, AI-powered predictive models will also play a critical role in accelerating certification processes by providing virtual validation of mechanical and biological performance, significantly reducing time-to-market for innovative medical implants.

By integrating machine learning across all stages of 3D printing, the medical device industry is moving towards an era of autonomous, intelligent manufacturing. The combination of AI-driven material discovery, adaptive process monitoring, and smart post-processing will lead to highly personalised, high-performance implants, ultimately transforming patient care through enhanced precision, reduced costs, and broader accessibility.

4.6 CONCLUSION

Integrating cutting-edge machine learning (ML) techniques with additive manufacturing (AM) processes such as FDM, SLA, SLS, and LPBF is fundamentally advancing the fabrication of medical devices and implants by enhancing precision, automation, and the ability to produce highly customised, patient-specific products. Across the pre-printing, printing, and post-processing stages of 3D printing, ML algorithms can be incorporated in a complementary role. In summary, ML-driven additive manufacturing will ensure more consistent production quality by reducing process variability and defects, while also enabling unprecedented customisation, resulting in devices with superior fit and functionality tailored to individual patients. The convergence of ML and 3D printing technologies is transformative. It advances an intelligent, highly automated production workflow aligned with the Industry 4.0 paradigm, paving the way for highly personalised, high-performance medical devices.

REFERENCES

Badogu, Ketan, Vishal Thakur, Raman Kumar, Ranvijay Kumar, and Sunpreet Singh. 2024. “Acrylonitrile Butadiene Styrene-ZrO2 Composites for Roller Burnishing as Post-Processing of 3D Printed Parts: Machine Learning Modeling Using Classification and Regression Trees.” *Journal of Materials Engineering and Performance* 33 (18):9522–9533. https://doi.org/10.1007/s11665-023-08620-y.

Bayraktar, Ömer, Gültekin Uzun, Ramazan Çakiroğlu, and Abdulmecit Guldas. 2017. “Experimental Study on the 3D-Printed Plastic Parts and Predicting the Mechanical Properties Using Artificial Neural Networks.” *Polymers for Advanced Technologies* 28 (8):1044–1051. https://doi.org/10.1002/pat.3960.

Chowdhury, Sushmit, and Sam Anand. 2016. “Artificial Neural Network Based Geometric Compensation for Thermal Deformation in Additive Manufacturing Processes.” In *ASME 2016 11th International Manufacturing Science and Engineering Conference*. https://doi.org/10.1115/MSEC2016-8784

Deng, Lina, Bo Feng, and Yue Zhang. 2018. “An Optimization Method for Multi-Objective and Multi-Factor Designing of a Ceramic Slurry: Combining Orthogonal Experimental Design with Artificial Neural Networks.” *Ceramics International* 44 (13):15918–15923. https://doi.org/10.1016/j.ceramint.2018.06.010.

Gobert, Christian, Edward W. Reutzel, Jan Petrich, Abdalla R. Nassar, and Shashi Phoha. 2018. “Application of Supervised Machine Learning for Defect Detection During Metallic Powder Bed Fusion

Additive Manufacturing Using High Resolution Imaging." *Additive Manufacturing* 21:517–528. https://doi.org/10.1016/j.addma.2018.04.005.

Grasso, M., A. G. Demir, B. Previtali, and B. M. Colosimo. 2018. "In Situ Monitoring of Selective Laser Melting of Zinc Powder Via Infrared Imaging of the Process Plume." *Robotics and Computer-Integrated Manufacturing* 49:229–239. https://doi.org/10.1016/j.rcim.2017.07.001.

Grasso, Marco, Vittorio Laguzza, Quirico Semeraro, and Bianca Maria Colosimo. 2016. "In-Process Monitoring of Selective Laser Melting: Spatial Detection of Defects Via Image Data Analysis." *Journal of Manufacturing Science and Engineering* 139 (5). https://doi.org/10.1115/1.4034715.

Gu, Grace X., Chun-Teh Chen, and Markus J. Buehler. 2018. "De Novo Composite Design Based on Machine Learning Algorithm." *Extreme Mechanics Letters* 18:19–28. https://doi.org/10.1016/j.eml.2017.10.001.

Han, Wei, Lingbao Kong, and Min Xu. 2022. "Advances in Selective Laser Sintering of Polymers." *International Journal of Extreme Manufacturing* 4 (4):042002. https://doi.org/10.1088/2631-7990/ac9096.

He, Haiyang, Yang Yang, and Yayue Pan. 2019. "Machine Learning for Continuous Liquid Interface Production: Printing Speed Modelling." *Journal of Manufacturing Systems* 50:236–246. https://doi.org/10.1016/j.jmsy.2019.01.004.

Inayathullah, Sirajudeen, and Raviteja Buddala. 2025. "Review of Machine Learning Applications in Additive Manufacturing." *Results in Engineering* 25:103676. https://doi.org/10.1016/j.rineng.2024.103676.

Jiang, Jingchao, Guobiao Hu, Xiao Li, Xun Xu, Pai Zheng, and Jonathan Stringer. 2019. "Analysis and Prediction of Printable Bridge Length in Fused Deposition Modelling Based on Back Propagation Neural Network." *Virtual and Physical Prototyping* 14 (3):253–266. https://doi.org/10.1080/17452759.2019.1576010.

Jiang, Zhenglong, Yuan Yuan Liu, Haiping Chen, and Qingxi Hu. 2014. "Optimization of Process Parameters for Biological 3D Printing Forming Based on BP Neural Network and Genetic Algorithm." CE. https://doi.org/10.3233/978-1-61499-440-4-351

Jin, Zeqing, Zhizhou Zhang, Xianlin Shao, and Grace X. Gu. 2023. "Monitoring Anomalies in 3D Bioprinting with Deep Neural Networks." *ACS Biomaterials Science & Engineering* 9 (7):3945–3952. https://doi.org/10.1021/acsbiomaterials.0c01761.

Koeppe, Arnd, Carlos Alberto Hernandez Padilla, Maximilian Voshage, Johannes Henrich Schleifenbaum, and Bernd Markert. 2018. "Efficient Numerical Modeling of 3D-Printed Lattice-Cell Structures Using Neural Networks." *Manufacturing Letters* 15:147–150. https://doi.org/10.1016/j.mfglet.2018.01.002.

Li, X. F., J. H. Dong, and Y. Z. Zhang. 2009. "Modeling and Applying of RBF Neural Network Based on Fuzzy Clustering and Pseudo-Inverse Method." In *2009 International Conference on Information Engineering and Computer Science*, 19–20 December. https://doi.org/10.1109/ICIECS.2009.5362683

Lim, Joyce Xin-Yan, and Quang-Cuong Pham. 2021. "Automated Post-Processing of 3D-Printed Parts: Artificial Powdering for Deep Classification and Localisation." *Virtual and Physical Prototyping* 16 (3):333–346. https://doi.org/10.1080/17452759.2021.1927762.

Lyu, Jiaqi, and Souran Manoochehri. 2021. "Online Convolutional Neural Network-based Anomaly Detection and Quality Control for Fused Filament Fabrication Process." *Virtual and Physical Prototyping* 16 (2):160–177. https://doi.org/10.1080/17452759.2021.1905858.

Mamo, Hana Beyene, Marcin Adamiak, and Anil Kunwar. 2023. "3D Printed Biomedical Devices and Their Applications: A Review on State-of-the-Art Technologies, Existing Challenges, and Future Perspectives." *Journal of the Mechanical Behavior of Biomedical Materials* 143:105930. https://doi.org/10.1016/j.jmbbm.2023.105930.

Meiabadi, Mohammad Saleh, Mahmoud Moradi, Mojtaba Karamimoghadam, Sina Ardabili, Mahdi Bodaghi, Manouchehr Shokri, and Amir H. Mosavi. 2021. "Modeling the Producibility of 3D Printing in Polylactic Acid Using Artificial Neural Networks and Fused Filament Fabrication." *Polymers* 13 (19):3219.

Menon, Aditya, Barnabás Póczos, Adam W. Feinberg, and Newell R. Washburn. 2019. "Optimization of Silicone 3D Printing with Hierarchical Machine Learning." *3D Printing and Additive Manufacturing*. https://doi.org/10.1089/3dp.2018.0088

Mulrennan, Konrad, Nimra Munir, Leo Creedon, John Donovan, John G. Lyons, and Marion McAfee. 2022. "NIR-Based Intelligent Sensing of Product Yield Stress for High-Value Bioresorbable Polymer Processing." *Sensors* 22 (8):2835.

Okaro, Ikenna A., Sarini Jayasinghe, Chris Sutcliffe, Kate Black, Paolo Paoletti, and Peter L. Green. 2019. "Automatic Fault Detection for Laser Powder-Bed Fusion Using Semi-Supervised Machine Learning." *Additive Manufacturing* 27:42–53. https://doi.org/10.1016/j.addma.2019.01.006.

Rong-Ji, Wang, Li Xin-hua, Wu Qing-ding, and Wang Lingling. 2009. "Optimizing Process Parameters for Selective Laser Sintering Based on Neural Network and Genetic Algorithm." *The International Journal of Advanced Manufacturing Technology* 42 (11):1035–1042. https://doi.org/10.1007/s00170-008-1669-0.

Savini, Antonio, and Gianluca Savini. 2015. "A Short History of 3D Printing, a Technological Revolution Just Started." In *2015 ICOHTEC/IEEE International History of High-Technologies and Their Socio-Cultural Contexts Conference (HISTELCON)*:1–8. https://api.semanticscholar.org/CorpusID:19609972

Shanthar, Rajinth, Kun Chen, and Chamil Abeykoon. 2023. "Powder-Based Additive Manufacturing: A Critical Review of Materials, Methods, Opportunities, and Challenges." *Advanced Engineering Materials* 25 (19):2300375. https://doi.org/10.1002/adem.202300375.

Sharma, Shrutika, Vishal Gupta, Deepa Mudgal, and Vishal Srivastava. 2024. "Optimization of Polydopamine Coating Process for Poly Lactic Acid-Based 3D Printed Bone Plates Using Machine Learning Approaches." *Polymer Engineering & Science* 64 (1):279–295. https://doi.org/10.1002/pen.26546.

Shevchik, S. A., C. Kenel, C. Leinenbach, and K. Wasmer. 2018. "Acoustic Emission for in Situ Quality Monitoring in Additive Manufacturing Using Spectral Convolutional Neural Networks." *Additive Manufacturing* 21:598–604. https://doi.org/10.1016/j.addma.2017.11.012.

Shi, Yang, Yicha Zhang, Steven Baek, Wout De Backer, and Ramy Harik. 2018. "Manufacturability Analysis for Additive Manufacturing Using a Novel Feature Recognition Technique." *Computer-Aided Design and Applications* 15 (6):941–952. https://doi.org/10.1080/16864360.2018.1462574.

Singh, Mahima, and Sriramakamal Jonnalagadda. 2020. "Advances in Bioprinting Using Additive Manufacturing." *European Journal of Pharmaceutical Sciences* 143:105167. https://doi.org/10.1016/j.ejps.2019.105167.

Sood, Anoop K., Raj K. Ohdar, and Siba S. Mahapatra. 2012. "Experimental Investigation and Empirical Modelling of FDM Process for Compressive Strength Improvement." *Journal of Advanced Research* 3 (1):81–90. https://doi.org/10.1016/j.jare.2011.05.001.

Williams, Glen, Nicholas A. Meisel, Timothy W. Simpson, and Christopher McComb. 2019. "Design Repository Effectiveness for 3D Convolutional Neural Networks: Application to Additive Manufacturing." *Journal of Mechanical Design* 141 (11). https://doi.org/10.1115/1.4044199.

Xue, Tianju, Thomas J. Wallin, Yigit Menguc, Sigrid Adriaenssens, and Maurizio Chiaramonte. 2020. "Machine Learning Generative Models for Automatic Design of Multi-Material 3D Printed Composite Solids." *Extreme Mechanics Letters* 41:100992. https://doi.org/10.1016/j.eml.2020.100992.

Ye, Dongsen, Geok Soon Hong, Yingjie Zhang, Kunpeng Zhu, and Jerry Ying Hsi Fuh. 2018. "Defect Detection in Selective Laser Melting Technology by Acoustic Signals with Deep Belief Networks." *The International Journal of Advanced Manufacturing Technology* 96 (5):2791–2801. https://doi.org/10.1007/s00170-018-1728-0.

Chapter 5

Recent Developments and Challenges in Bioprinting and Tissue Engineering

Kartikeya Dixit, Hitendra Kumar, Keekyoung Kim, and Niraj Sinha

5.1 INTRODUCTION TO BIOPRINTING AND TISSUE ENGINEERING

Bioprinting and tissue engineering represent transformative technologies in biomedical sciences, enabling the creation of complex tissue constructs for therapeutic, diagnostic, and research applications. These disciplines have evolved to address critical challenges in organ shortages, tissue repair, and disease modelling. By combining principles from biology, engineering, and material science, bioprinting and tissue engineering offer innovative approaches to mimic native tissue architecture and functionality. Tissue engineering emerged as a multidisciplinary field in the late 20th century, aiming to repair or replace damaged tissues and organs. Early research focused on scaffold-based approaches, where biocompatible materials were seeded with cells to promote tissue regeneration. For instance, Langer and Vacanti (1993) laid the foundation for tissue engineering by describing the potential of combining scaffolds, cells, and signalling molecules to engineer functional tissues [1]. Advances in cell biology, biomaterials, and biotechnology have since propelled the field towards more sophisticated designs and applications. Bioprinting is a subset of additive manufacturing that utilises layer-by-layer deposition of bioinks, comprising cells, biomaterials, and bioactive molecules, to fabricate three-dimensional (3D) tissue constructs. This technology enables precise spatial control over cell placement, mimicking the structural complexity of native tissues. Commonly employed bioprinting methods include inkjet, extrusion-based, and laser-assisted bioprinting, each offering distinct advantages and limitations [2].

Bioprinting and tissue engineering success rely on the interplay between cells, biomaterials, and bioactive factors. The first essential component in bioprinting is bioink, as mentioned earlier. A typical bioink composition includes cells, biomaterials, and bioactive molecules. A diverse array of cell types can be used, including stem cells (e.g., mesenchymal stem cells, induced pluripotent stem cells), primary cells, and cell lines. Stem cells are particularly advantageous due to their potential for differentiation and regenerative properties [3]. Biomaterials serve as scaffolds or matrices, providing structural support and influencing cell behaviour. Natural polymers (e.g., collagen, alginate) and synthetic polymers (e.g., polycaprolactone (PCL), polyethene glycol (PEG)) are commonly used biomaterials. These are often combined to complement each other's properties and overcome their respective limitations [4]. Bioactive molecules include growth factors, cytokines, and other signalling molecules that are crucial in guiding cell proliferation, differentiation, and tissue maturation. Controlled delivery of these molecules is essential for achieving functional outcomes [5]. Bioprinting and tissue engineering have made significant progress in regenerative medicine over the years. For example, researchers have created skin grafts for burn victims and cartilage replacements for joint repair [6, 7]. The bioprinting of vascularised tissues is a critical milestone, enabling the creation of constructs that closely mimic native organs [8]. Traditional drug discovery methods often rely on animal models, which may not accurately predict human responses. Bioprinted tissue models offer a

DOI: 10.1201/9781003590774-5

more relevant platform for testing drug efficacy and toxicity, reducing the reliance on animal studies and improving clinical outcomes [9]. Bioprinted tumour models allow researchers to study the tumour microenvironment, drug resistance, and metastatic behaviour. For instance, bioengineered 3D models of glioblastoma have been used to test novel therapeutic strategies [10]. Bioprinted organ-on-a-chip systems emerged with advances in 3D and 2D *in vitro* models. These integrate multiple cell types and microfluidic devices to simulate organ-level functions. These systems are valuable for studying disease mechanisms and evaluating therapeutic interventions with precision and high throughput [11]. **Figure 5.1** illustrates the bioprinting process for tissue models, outlining the key stages from cell sourcing to biomedical applications. The process begins with obtaining specific cells from organs, which are cultured in a laboratory setting to increase their numbers and ensure their viability. These cultured cells are then combined with natural or synthetic biomaterials to create a specialised bioink that provides the necessary structural support and biological environment for cell growth and function.

The resulting bioprinted tissue models have a wide range of applications in the biomedical field. They are commonly used for drug discovery, allowing researchers to test new medications in a controlled environment that mimics human tissues. Additionally, these models play a crucial role in drug screening, helping to identify the most effective compounds with minimal side effects. Another significant application is disease modelling, where bioprinted tissues provide insights into disease onset and progression and enable the development of personalised treatment strategies.

Bioprinting and tissue engineering are at the forefront of biomedical innovation, with the potential to revolutionise healthcare and research. Despite the challenges, advances in

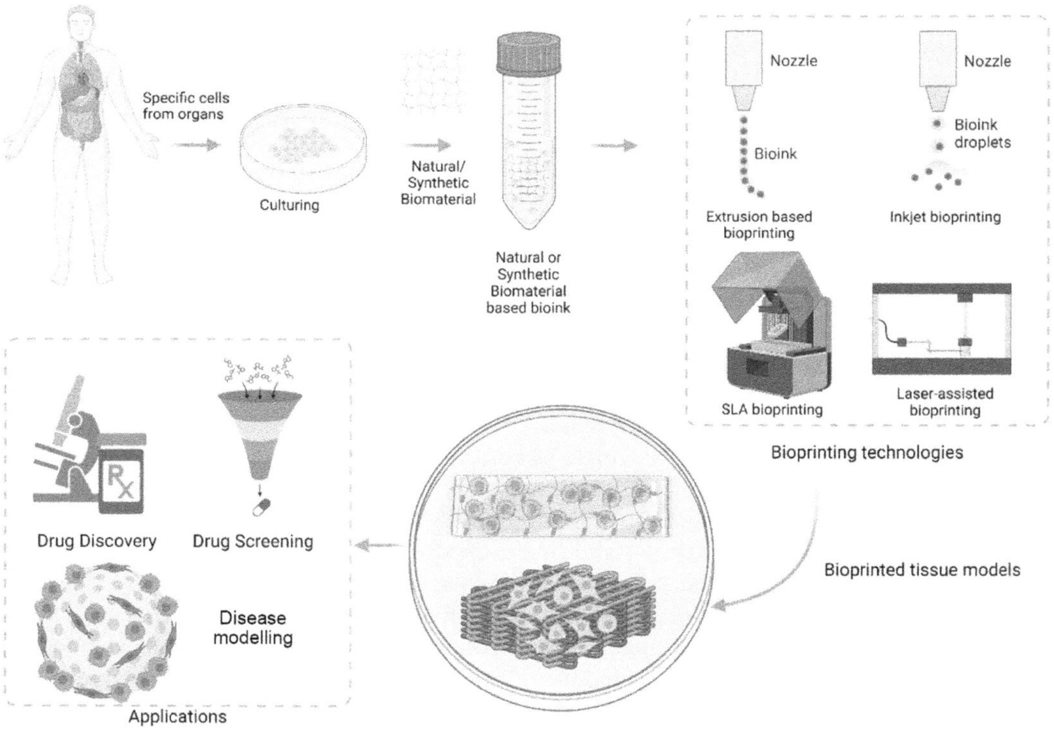

Figure 5.1 A schematic representation of the bioprinting process, illustrating the steps from cell culture and bioink preparation to various bioprinting technologies and their applications in drug discovery, screening, and disease modelling. (Created in BioRender. Kim, K. (2025) https://BioRender.com/o87b765).

biomaterials, bioprinting technologies, and interdisciplinary collaboration continue to drive progress. As these technologies mature, their applications in regenerative medicine, drug discovery, and personalised healthcare are poised to have a transformative impact on society. We will delve into these advancements and challenges in the upcoming sections.

5.2 RECENT DEVELOPMENTS IN BIOPRINTING

5.2.1 Advancements in Bioinks

Recent advances in bioinks for tissue engineering have focused on developing dynamic and tailorable materials to enhance printability and support cell function. Supramolecular and dynamic covalent chemistry are being explored to create more precise microenvironments for tissue maturation [12]. Researchers are combining various biocompatible materials, including hydrogels, cells, and biosignals, to produce complex tissue structures [13]. The incorporation of carbon-based nanomaterials, clay, ceramics, and growth factors has led to hybrid and multi-composite hydrogels with unique functionalities [14–17]. These innovations have resulted in bioinks that improve cell viability, nutrient diffusion, and mechanical properties of printed constructs (see **Figure 5.2**). *In vivo* studies have demonstrated the potential of bioprinted tissues to support cellular activity and regeneration, signifying advancements in clinical viability. However, continued research is needed to overcome current limitations and expand the range of engineered tissues for widespread clinical application.

5.2.2 Natural Polymer-Based Bioinks

Collagen, alginate, gelatin, and hyaluronic acid are examples of natural polypeptides, polysaccharides, and glycan biopolymers that have been extensively studied due to their inherent biocompatibility and bioactivity. One major component of the extracellular matrix (ECM), collagen-based bioinks, for instance, has undergone chemical modification to improve their printability and mechanical characteristics. Li et al. highlighted collagen-based bioinks as versatile materials with exceptional biocompatibility, printability, and cell-supporting capabilities for 3D bioprinting [18]. These bioinks are effectively applied in both soft tissue repair (e.g., skin, cartilage, heart, blood vessels) and hard tissue regeneration (e.g., skull, femur, teeth, spine). For soft tissues, structural fidelity and functional biomimicry are essential, while for hard tissues, mechanical strength and mineralisation support are crucial. In another study, Lee et al. introduced a freeform reversible embedding of suspended hydrogels (FRESH) 3D printing technique using collagen-based bioinks to fabricate human heart components, which bridges the multiscale gap in tissue engineering, ranging from microvasculature to whole organs [19]. They also recreated intricate trabecular structures by modelling a scanned chicken embryo heart [20]. This method demonstrated high spatial fidelity and was extended to other complex tissues, including the femur, brain, and coronary vessels.

Chawla et al. summarised silk-based bioinks, particularly from silkworm and spider silk, as promising materials for 3D bioprinting due to their tunable rheology, biocompatibility, and structural fidelity [21]. These bioinks exhibited shear-thinning behaviour, cytocompatible gelation, and the ability to form β-sheet structures, enabling stable construct formation without harsh crosslinking. Applications of silk-based biomaterials span cartilage, bone, and skin tissue engineering, with constructs showing high cellular functionality and matrix remodelling. In another study, Rodriguez et al. developed silk-gelatin-glycerol bioinks optimised for extrusion-based 3D printing to reconstruct soft tissues with patient-specific geometries [22]. These bioinks exhibited shear-thinning behaviour, rapid physical crosslinking via β-sheet formation, and tunable mechanical properties suitable for native tissue mimicry. A study by

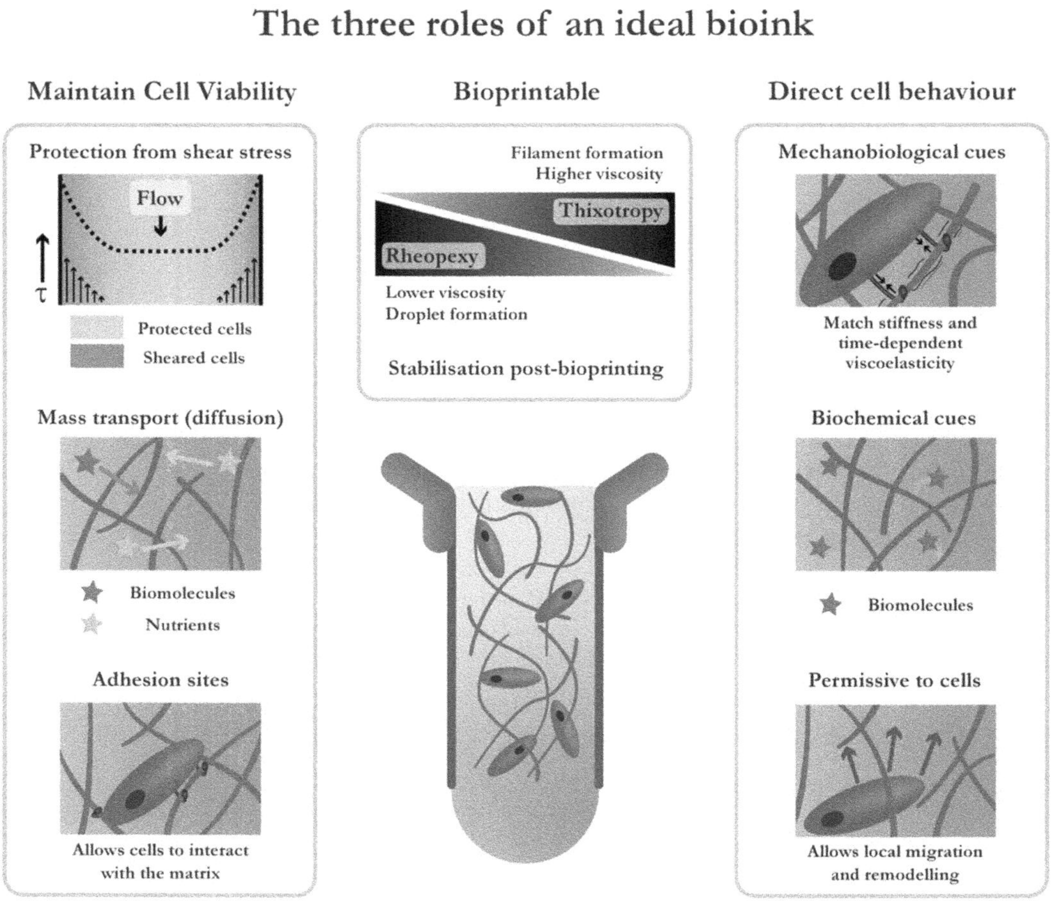

Figure 5.2 An ideal bioink fulfils three essential roles. First, it must support cell viability by protecting cells from mechanical stress, enabling nutrient diffusion, and providing adhesion sites. Second, it requires suitable mechanical properties, such as optimal viscosity and shear-thinning behaviour, to ensure printability for the selected bioprinting method. Lastly, it should actively guide cellular behaviour by releasing biomolecules in a controlled manner, offering appropriate mechanical cues for mechanotransduction, and maintaining a structure that supports cell migration, interaction, and matrix remodelling (Adapted from Morgan et al., Advanced Healthcare Materials, 9, 2020) [12].

Burdick et al. reported distinctive biological and mechanical properties of hyaluronic acid (HA) that make it particularly suitable for hydrogel formation in tissue engineering and drug delivery systems [23]. The authors discussed various chemical modification techniques used to tune HA hydrogel characteristics, such as degradability, mechanical strength, and bioactivity, thereby enabling adaptation to specific biomedical needs. Alginate, another widely used natural polymer, is popular for its ability to crosslink rapidly in the presence of calcium ions. However, its lack of cell adhesion properties has been a significant limitation. Jeon et al. addressed this by functionalising alginate with arginyl-glycyl-aspartic acid (RGD) peptides, thereby improving cell adhesion and proliferation [24]. Similarly, gelatin methacrylate (GelMA) has emerged as a versatile material due to its tunable mechanical properties and photocrosslinking capabilities [25]. The ability to tailor GelMA's stiffness and degradation rate has made it suitable for various tissue engineering applications, including cartilage and vascular constructs. Further, a study by Duong et al. reported a cytocompatible Gelatin-Norbornene (GelNB) bioink for digital light processing (DLP) 3D bioprinting, enabling the precise fabrication of perfusable, vascularised soft tissue constructs [26]. Its tunable stiffness and post-printing functionalisation make it highly suitable for regenerative applications involving complex vascular architectures.

Gomes et al. reported the versatility of gellan gum (GG) as a bioink in tissue engineering due to its biocompatibility, shear-thinning behaviour, and ionic crosslinking capability [27]. GG-based bioinks can be physically or chemically modified and combined with polymers like gelatin, silk methacrylate, or graphene oxide to enhance mechanical strength, printability, and bioactivity. Studies show that GG supports cell adhesion, proliferation, and differentiation, and even has anti-inflammatory and antimicrobial properties [28]. Furthermore, Taghizadeh et al. critically reviewed chitosan-based bioinks as versatile, biocompatible materials ideal for 3D bioprinting in tissue engineering, drug delivery, and biosensing [29]. These bioinks offer good printability, antimicrobial properties, tunable degradation, and favourable cell viability and adhesion. Chitosan's chemical versatility enables structural and rheological tailoring through functionalisation or blending with polymers such as gelatin and HA. Despite their promise, challenges include slow gelation, limited mechanical strength, and regulatory hurdles for clinical translation. In a recent work, Li et al. engineered a highly self-supporting chitosan-based hydrogel ink (CMAL) combining carboxymethyl chitosan and nanoclay, enabling in situ 3D printing of complex diabetic wound dressings [30]. The ink exhibits excellent printability, antibacterial activity (via silver nanoparticles, AgNPs), and pro-angiogenic effects (via a Von Hippel-Lindau (VHL) protein antagonist, VH298), accelerating wound healing in vivo. A study by Shpichka et al. reviewed fibrin-based bioinks as highly biocompatible, biodegradable platforms with tunable mechanical and biological properties suitable for 3D bioprinting [31]. Fibrin's printability and cell-guiding features are enhanced via blending with polymers (e.g., gelatin, collagen) or by covalently linking growth factors [32]. These bioinks facilitate wound healing, angiogenesis, and tissue-specific applications, including skin, cardiac, neural, and tumour models. Despite its rapid gelation and limited shape fidelity, structural tuning and smart modifications have broadened its applicability. For instance, Hangad et al. developed 3D-printed fibrin constructs incorporating all-trans retinoic acid (atRA)-loaded PCL microparticles produced via microfluidics for enhanced drug delivery and scaffold modulation [33]. The microparticles achieved high encapsulation efficiency (85.9%) and slowed atRA release in fibrin compared to buffer, while also accelerating scaffold degradation. Additionally, their inclusion increased the constructs' solid-like mechanical properties, offering a route to bioinks with tailored therapeutic and structural profiles.

5.2.3 Synthetic Polymer-Based Bioinks

Synthetic polymers offer unmatched tunability and reproducibility, making them ideal candidates for creating mechanically robust bioinks. Among these, PEG has garnered significant attention due to its non-immunogenic and hydrophilic properties. Sun et al. reviewed PEG-based hydrogels as multifunctional platforms for bone regeneration, highlighting their biocompatibility, tunable degradation, and drug-delivery capacity [34]. These hydrogels support sustained release of growth factors, nucleic acids, and osteogenic drugs, and can be engineered for antibacterial and osteoinductive functions. Structural variants, such as linear, four-arm, and eight-arm PEGs, enable precise control over mechanical strength, drug loading, and in vivo performance for localised bone defect repair. Polylactic-co-glycolic acid (PLGA) has also been extensively studied, especially in composite formulations with natural polymers. For instance, Choe et al. developed an osteoinductive bioink combining alginate and BMP-2-loaded PLGA nanoparticles to support mesenchymal stem cell (MSC) bioprinting for bone tissue engineering [35]. This bioink exhibited enhanced printability, sustained BMP-2 release, and promoted osteogenic differentiation, as shown by elevated ALP activity, calcium deposition, and osteogenic gene expression. Further, Muscolino et al. investigated κ-Carrageenan and polyvinyl alcohol (PVA) blends to fabricate 3D scaffolds for cartilage regeneration [36]. The authors reported that cell viability was maintained regardless of the freeze-thaw cycle, ranging from −20°C to +20°C, and that cells expressed the chondrogenic SOX9 marker, confirming their differentiation. Ullah et al. developed a highly reproducible hydrogel-based bioink composed of polyethene oxide-chitosan-polymethyl methacrylate (PEO-CS-PMMA), enhanced with nicotinamide, for skin tissue regeneration via 3D bioprinting [37]. The bioink demonstrated excellent printability, shear-thinning, and self-healing behaviour at physiological conditions and supported over 90% cell viability post-printing. In another study, Oh et al. developed a biocompatible, highly conductive poly(3,4-ethylenedioxythiophene): polystyrene sulfonate (PEDOT:PSS) ionic liquid colloidal bioink (PILC) that enables high-resolution 3D printing without post-processing [38]. The PILC ink supports exceptional cellular compatibility, with NIH/3T3 fibroblasts exhibiting high viability, and facilitates direct interfacing with biological tissues for electrophysiological applications. Its viscoplastic behaviour and structural integrity enable rapid fabrication of implantable and on-skin bioelectronics with stable signal recordings and stimulation, highlighting its promise for next-generation biointerfacing systems.

5.2.4 Composite Bioinks

Composite bioinks combine natural and synthetic polymers, enabling a balance between mechanical robustness and biological functionality. Recent research has focused on enhancing their performance by incorporating bioactive nanoparticles, growth factors, and stimuli-responsive components. In one study, Cai et al. assessed cell-laden nanocomposite bioinks as multifunctional platforms that enhance the mechanical strength, printability, and bioactivity of 3D bioprinted constructs [39]. They highlighted how nanomaterials, such as nanosilicates, bioactive ceramics, nanocellulose, and carbon-based nanostructures, reinforce hydrogels by improving viscosity, shear-thinning behaviour, and structural fidelity. In another study, Potere et al. developed a composite bioink combining alginate, gelatin, and decellularised extracellular matrix (dECM) with a novel double-crosslinking strategy to fabricate self-supporting, multi-layered aortic-like constructs [40]. Further, researchers have reported bioinks containing nanoparticles for the controlled release of therapeutic agents, offering potential in drug delivery and tissue regeneration. For example, Fu et al. developed a standardised bioprinting protocol for minocycline-loaded nanoparticles (MH-DS) incorporated GelMA bioinks, optimising flow behaviour, printability, and cell viability through precise time-temperature control [41]. The optimised bioink demonstrated high structural fidelity and biocompatibility, supporting applications in wound-healing and drug-delivery models. Other studies, such as those by Yeh et al., have developed a conductive bioink composed of Silk fibroin Methacrylate (SilMA), pectin, and MXene modified with soybean phospholipids (MXene-SP), optimised for 3D bioprinting of neural stem cell (NSC) spheroids with potential in neural tissue engineering [42]. In another study, Sakr et al. developed a novel electrically conductive bioink by combining GelMA with zwitterionic ornithine methacryloyl (OrnMA), enabling tunable stiffness, high transparency, and rapid visible-light photocrosslinking [43]. OrnMA-incorporated GelMA bioink showed significant promise for fabricating functional neural constructs via high-resolution bioprinting. Further, Meenakshy et al. reviewed polypyrrole (PPy)-based bioinks as promising materials for 3D printing due to their high electrical conductivity, biocompatibility, and rheological tunability [44]. PPy-based bioinks also support drug delivery, wearable biosensors, and cardiac patches with electroactive therapeutic properties. Despite challenges in mechanical strength and long-term stability, PPy inks offer scalable, functional platforms for next-gen biomedical devices. Further, Cheng et al. reported a 3D-printed double-network hydrogel scaffold comprising GelMA, chitosan, and PPy to promote peripheral nerve regeneration [45]. The conductive GelCS-PPy scaffold enhanced electrical signalling, supported PC-12 and HT-22 cell adhesion, differentiation, and neurite outgrowth, especially under electrical stimulation. Integrating natural and synthetic polymers in composite bioinks has allowed researchers to leverage the strengths of both material types. For instance, nanoengineered bioinks incorporating nanoparticles such as hydroxyapatite and carbon nanotubes have enhanced mechanical strength and promoted osteogenic differentiation [46, 47].

5.2.5 dECM Bioinks

dECM bioinks represent a significant advancement in tissue-specific engineering. These bioinks retain the biochemical composition of native tissues, providing essential cues for cell behaviour. In one study, Pati et al. investigated a bioink derived from the dECM of adipose, cartilage, and heart tissues to create tissue-specific, cell-laden 3D constructs [4]. The dECM bioinks were optimised for extrusion-based printing by adjusting pH and temperature, forming gels at 37 °C with rheological properties tailored to each tissue type. Another work by Jang et al. developed a 3D bioprinted pre-vascularised stem cell patch using a heart-derived dECM as a tissue-specific bioink to promote cardiac repair [48]. The bioink supported high cell viability, enhanced cardiomyogenic and vasculogenic differentiation, and showed effective sol-gel transition and crosslinking behaviour under physiological conditions. *In vivo* implantation of the dual-cell patterned patch significantly improved vascularisation, reduced fibrosis, and enhanced cardiac function post-myocardial infarction by mimicking the native cardiac microenvironment. Rezaei et al. studied a photocrosslinkable hybrid hydrogel of GelMA and liver dECM methacrylate (LdECMMA) that enabled tunable mechanics, supported hepatocellular carcinoma cell proliferation, and showed promise for drug screening and liver cancer studies [49]. Khati et al. studied a multi-material bioink based on decellularised liver matrix (dLM) reinforced with gelatin and PEG, enabling 3D bioprinting at a physiological temperature (37 °C) [50]. By enzymatic crosslinking with mushroom tyrosinase, the printed constructs demonstrated significantly enhanced mechanical strength, rheological stability, and high HepG2 cell viability (85–93%), making them promising for liver tissue engineering, drug testing, and cancer model applications.

5.2.6 Crosslinking Strategies

Crosslinking is crucial for stabilising bioink structures post-printing. Recent innovations in this area have focused on achieving a balance between mechanical stability and biocompatibility (see **Figure 5.3**). Ionic crosslinking involves rapid gelation via electrostatic interactions, typically between negatively charged polymers, such as alginate, and divalent cations, such as Ca^{2+}. It is mild and cell-friendly, making it ideal for extrusion-based and embedded hydrogel bioprinting, though it offers limited mechanical strength. Next, enzymatic crosslinking, for example, employs mild, cell-friendly conditions, as reported by Li et al., using transglutaminase to crosslink hydrogel-based bioinks [51]. In some cases, DNA hybridisation is used as a non-covalent crosslinking mechanism in 3D bioprinting to form hydrogels under mild, physiologically relevant conditions. This method involves mixing polypeptide-based bioinks with complementary double-stranded DNA sequences, resulting in rapid gelation via base-pair recognition. Further, one of the most commonly used methods is photocrosslinking, which utilises light-activated reactions and has been significantly improved through the

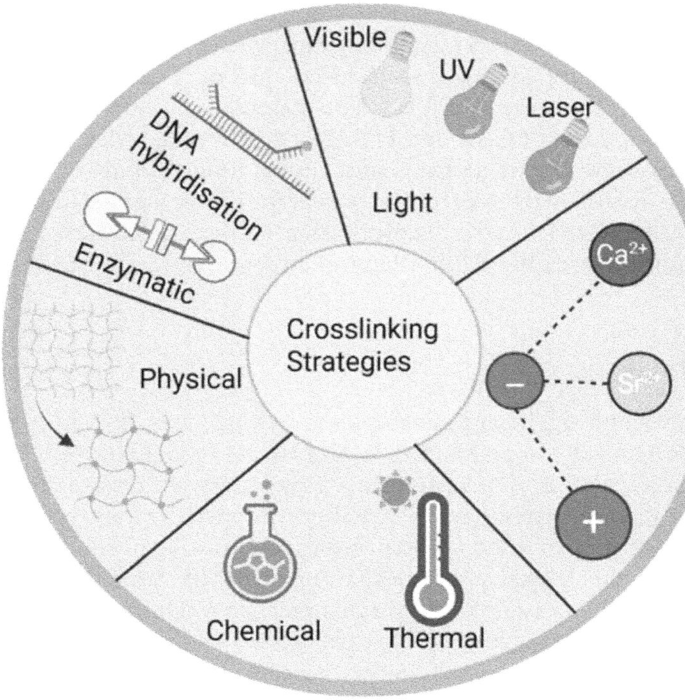

Figure 5.3 Various crosslinking strategies for bioinks (Created in BioRender. Kim, K. (2025) https://BioRender.com/xida34n).

development of biocompatible photoinitiators [52]. Dual crosslinking, as explored by Kim et al., combines calcium ions with secondary agents to enhance the structural stability of alginate-tyramine hydrochloride hydrogels [53]. Additionally, in chemical crosslinking, covalent networks are formed using reagents such as glutaraldehyde or genipin, yielding robust hydrogels. However, potential cytotoxicity from residual chemicals should be taken care of. Moreover, during thermal crosslinking, bioinks undergo a sol-gel transition with increasing temperature. However, in this case, mechanical properties are often limited, and hence these are combined with other crosslinking strategies. In a recent work, Li et al. developed a novel bioink based on PVA functionalised with styrylpyridinium groups (PVA-SbQ), thereby enabling visible-light laser direct writing of high-resolution hydrogels within 30 seconds [54]. The PVA-SbQ hydrogel's performance is attributed to photodimerisation-driven covalent crosslinking, abundant free hydroxyl groups, and cationic SbQ moieties, making it highly promising for rapid biofabrication and biomedical engineering applications. Additional examples, including diffusion-based crosslinking, are discussed in Section 5.2.4.

5.2.7 Physical Functional Additives and Enhancements

Incorporating functional additives into bioinks has opened new avenues for creating constructs with specialised properties. Conductive bioinks, for instance, have been developed by integrating conductive polymers like PPy, enabling applications in neural and cardiac tissue engineering [45]. Similarly, antimicrobial bioinks infused with silver nanoparticles or antimicrobial peptides have shown promise for wound-healing applications by preventing infections. Controlled-release systems have also gained traction, with growth factor-loaded nanoparticles embedded in bioinks to facilitate sustained delivery of bioactive molecules. Huang et al. demonstrated that such systems significantly enhance tissue regeneration by providing localised and prolonged growth factor delivery [55]. Advancements in bioink have enabled the development of high-throughput, multi-material printing platforms. Coaxial bioprinting, which enables the fabrication of hollow, vascularised structures, has benefited from bioinks tailored for dual-layer printing [56]. Additionally, multi-material bioinks have enabled the simultaneous deposition of diverse materials, creating constructs with spatially defined mechanical and biological properties [57]. The advancements in bioink technology have revolutionised the field of bioprinting, enabling the fabrication of complex, functional constructs with applications in regenerative medicine, disease modelling, and personalised healthcare. By addressing challenges related to printability, mechanical stability, and bioactivity, researchers have paved the way for bioinks to play a transformative role in modern biomedicine.

5.3 ADVANCES IN BIOPRINTING TECHNOLOGIES

Traditional bioprinting methods involve fabricating tissue constructs in controlled laboratory environments and then implanting them. While effective, this approach often faces challenges, including delayed healing, the risk of immune rejection, and poor integration with native tissues. Unlike traditional approaches, where tissues are fabricated externally and later implanted, *in situ* bioprinting enables the precise deposition of bioinks directly onto the patient's body at the targeted site [58]. This approach offers several advantages, including improved cell viability, reduced contamination risks, better tissue integration, and more efficient treatment procedures, particularly in clinical and emergency settings, such as battlefield wound healing.

Several methodologies have been explored for *in situ* bioprinting. One approach involves modifying conventional bioprinters for direct application during surgical procedures. Another involves miniaturising printheads to create handheld devices that enable surgeons to manually apply bioinks with precision [59]. Additionally, catheter-based robotic systems have been developed to deliver therapeutic cells in a structured manner [60]. Initial studies have shown promising results, particularly with biomaterial-enriched bioinks. For example, the use of microgels with high cell densities has yielded better healing outcomes than traditional cell-laden hydrogel applications. Combining in situ bioprinting with biomaterial-free cell-dense bioinks, such as spheroids and organoids, is anticipated to revolutionise regenerative medicine by providing enhanced cellular interactions, self-assembled tissue-specific extracellular matrices, and improved functional outcomes. However, challenges remain in achieving proper cell adhesion to surrounding tissues, maintaining cohesion within the bioprinted structures, and ensuring minimally invasive application methods. Ultimately, *in situ* bioprinting holds great promise for personalised and rapid tissue regeneration with applications across various medical fields. The following sections discuss key bioprinting methods that form the basis of fabricating tissue engineering and have evolved over time towards advanced bioprinting processes capable of *in situ* fabrication.

5.3.1 Extrusion-Based Bioprinting

Extrusion-based bioprinting is one of the most widely used bioprinting techniques, offering versatility and scalability for a range of tissue engineering applications. This method extrudes bioink through a nozzle to form continuous filaments, enabling the creation of complex structures with high cell densities. Recent studies have focused on improving the resolution, scalability, and cell viability of this technique [61]. Ozbolat et al. highlighted the role of low-shear-stress extrusion mechanisms in enhancing cell viability during bioprinting [62]. The study emphasised the importance of nozzle design in mitigating cell damage. Gao et al. introduced a simultaneous multi-material, embedded-extrusion-based printing technique that enables precise fabrication of heterogeneous 3D structures with improved structural integrity [63]. By employing individually controlled nozzles and rheology-tuned embedding media, the method enables high-resolution deposition and strong interfacial bonding between materials. Their approach was validated by printing multi-layered organ models, demonstrating superior performance compared to conventional sequential printing. Another significant breakthrough was the development of microfluidic-based coaxial extrusion systems by Maharjan et al., which enabled the printing of vascularised tissue constructs with perfusable channels [64]. Advanced extrusion-based bioprinting is seen as a key approach for fabricating large, cell-laden constructs directly at defect sites. In a recent study, Yang et al. developed a novel autonomous in situ bioprinting robotic framework for the treatment of volumetric muscle loss (VML) using generic bioinks [65]. The system integrates a seven-degree-of-freedom (DoF) robotic arm, structured-light 3D vision, and a quantitative evaluation module to optimise bioink deposition and scaffold uniformity. Through ninety experiments, they identified optimal printing parameters and demonstrated enhanced geometric precision and reproducibility critical for functional tissue regeneration.

5.3.2 Inkjet Bioprinting

Inkjet bioprinting, characterised by the deposition of tiny droplets of bioink, offers high resolution and cost-effectiveness. This technique has been extensively studied for applications in high-throughput drug screening, tissue modelling, and cell patterning. However, challenges such as limited bioink viscosity range and potential cell damage due to shear stress persist. Xu et al. demonstrated the potential of inkjet bioprinting in fabricating viable cell-laden constructs with precise spatial control, providing a foundation for drug discovery applications [66]. Park et al. introduced piezoelectric nozzle systems that minimised thermal damage, thereby increasing cell viability and uniform droplet formation [67]. Ng and Shkolnikov extensively analysed the interplay between bioink rheology and inkjet bioprinting performance, showing that cell concentration significantly affects viscosity, surface tension, and droplet behaviour [68]. Both thermal and piezoelectric chambers subject cells to high shear stress for short durations, making bioink tuning critical to preserving cell viability. Shear-thinning and viscoelastic bioinks reduce droplet rebound and splashing, enhancing cell placement accuracy and print resolution. Additives like polyvinylpyrrolidone (PVP) improve droplet impact dynamics and viability by cushioning cells and stabilising jet formation. Additionally, Kim et al. developed a lung-on-a-chip platform using high-resolution inkjet bioprinting to fabricate ultra-thin, three-layered alveolar barrier tissues integrated into a microfluidic device [69]. This modular system enables air–liquid interface culture and holds promise for high-throughput lung disease modelling.

5.3.3 Laser-Assisted Bioprinting (LAB)

LAB employs laser pulses to deposit bioink with exceptional precision, making it suitable for applications requiring fine resolution, such as vascular tissue engineering, neural constructs, and complex microenvironments. This process is advantageous for maintaining high cell viability and achieving submicron resolution. Guillotin et al. explored LAB's utility in patterning cells and biomolecules, emphasising its capability to achieve high-resolution deposition [70]. Dou et al. comprehensively reviewed LAB, highlighting its nozzle-free, high-resolution, and high-viability capabilities for depositing bioinks, including highly viscous and cell-dense formulations [71]. The authors also discussed the laser-induced forward transfer (LIFT) mechanism, which enables precise single-cell deposition and the successful patterning of various cell types and biological materials. Further, authors reported innovations such as biodegradable or reusable energy-absorbing layers (EALs) improved cell viability (up to ~95–100%) and minimised contamination during printing. LAB was shown to outperform inkjet and extrusion methods in cell survival and droplet precision, supporting its potential for vascular, skin, and bone tissue models. However, LAB remains limited by low material transfer volume per pulse, high equipment costs, and complex bioink layer preparation, which restricts its scalability for large-volume tissue fabrication. Furthermore, Zennifer et al. assessed the design requirements of bioinks for LAB-based tissue regeneration [72]. The authors concluded that bioinks must exhibit a viscosity range of 1–300 mPa to ensure stable droplet formation without jet breakup. Additionally, surface tension should typically lie between 35 and 65 mN/m, balancing energy transfer efficiency and print resolution. Also, the thickness of the bioink layer on the ribbon (typically 10–100 μm) is critical for maintaining consistent energy absorption and droplet size. The bioinks must be optimised for low volatility to reduce evaporation under laser irradiation and should incorporate energy-absorbing materials (e.g., gelatin, ethyl acetate or triazene polymers) in the sacrificial layer. Additionally, the inclusion of stabilisers like glycerol or methylcellulose can modulate rheological behaviour and prevent thermal denaturation of encapsulated cells.

5.3.4 Advanced Bioprinting Processes

Advanced bioprinting techniques have revolutionised the fabrication of complex, functional biological structures. Among these, electrowriting, volumetric bioprinting (VBP) (utilising light and microwave energy), ultrasound-mediated bioprinting, xolography, and multiphoton lithography stand out for their precision and versatility [73–75]. Electrowriting, particularly melt electrowriting (MEW), employs high-voltage electric fields to control the deposition of molten polymer fibres, thereby enabling the creation of micro- to nanoscale scaffolds with high precision. This technique excels in fabricating structures with fine architectural features, making it ideal for applications requiring detailed scaffolding, such as vascular grafts and nerve regeneration conduits. Recent advancements have integrated MEW with VBP to fabricate multi-material constructs that combine the mechanical strength of electrowritten fibres with the biological functionality of cell-laden hydrogels. For instance, Größbacher et al. demonstrated the convergence of VBP with MEW to create tubular hydrogel-based composites with enhanced mechanical properties, suitable for engineering complex vascular structures [76].

VBP represents a paradigm shift by enabling the rapid fabrication of three-dimensional structures in a single step, overcoming the limitations of traditional layer-by-layer approaches [73, 77]. This method utilises dynamic light patterns projected into a rotating volume of photosensitive bioresin, simultaneously initiating polymerisation throughout the entire volume. Specifically, the tomographic process, depicted in **Figure 5.4**, involves rotating a cylindrical resin container while exposing it to patterned light from the side. These light patterns are generated using a DLP modulator and are synchronised with the container's rotation. Each pattern corresponds to a projection of the target 3D object from a specific angle, calculated using a Radon transform analogous to X-ray computed tomography [78]. The result is the swift creation of complex geometries with high cell viability, as the rapid printing process minimises exposure to potentially harmful conditions. Thijssen et al. provide a comprehensive review of the mechanistic aspects of volumetric 3D printing, highlighting its potential in fabricating intricate biological constructs [74]. Mukherjee et al. investigated the use of focused microwave radiation for localised curing of polymer resins, aiming to expand volumetric additive manufacturing (VAM) beyond optically transparent materials [75, 79]. They developed an X-band microwave applicator array with a time-reversal beam-steering algorithm to focus energy into specific 3D regions of opaque resins, achieving spot sizes as small

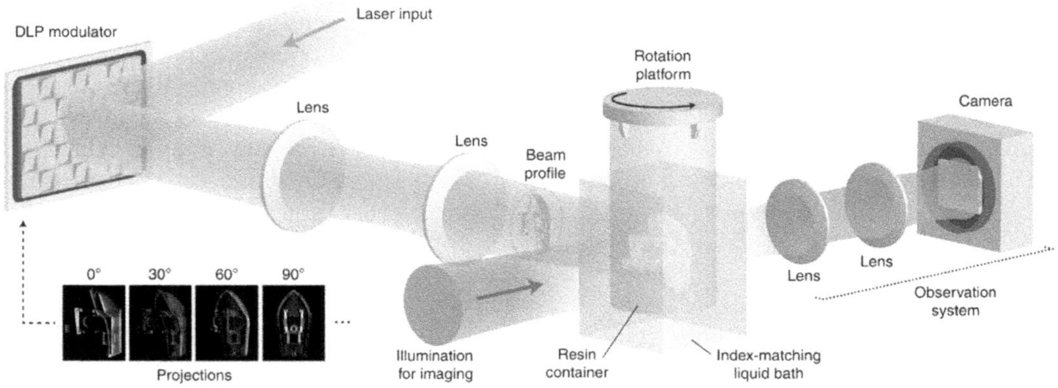

Figure 5.4 An experimental setup for volumetric printing developed by Loterie et al. (Adapted from Loterie et al., Nature Communications, 11, 2020) [77].

as 1 cm. Ultrasound-mediated bioprinting leverages focused ultrasound waves to induce localised polymerisation within a bioresin. This technique offers significant advantages in terms of penetration depth and spatial control, allowing for the fabrication of structures deep within tissue-mimicking environments. Zheng et al. have discussed the opportunities and challenges of ultrasound volumetric bioprinting, emphasising its potential for non-invasive, in situ tissue-engineering applications [80].

Further, xolography is another emerging volumetric 3D printing technique that utilises dual-colour photopolymerisation within a continuously flowing resin [81, 82]. This method enables the fabrication of complex geometries with smooth surfaces and high resolution. In its continuous version, called *xolography in flow* developed by Stuwe et al., the resin flows vertically through a static light sheet, enabling rapid, layerless fabrication of complex geometries with resolutions down to ~10 μm laterally and 25 μm axially, and production rates up to 1.75 mm^3/s [83]. Multiphoton lithography employs femtosecond laser pulses to induce localised polymerisation via nonlinear absorption processes, allowing the fabrication of structures with sub-diffraction-limited features [84, 85]. This technique is particularly suited for creating micro- and nanoscale scaffolds with high precision. Yang et al. developed a parallel two-photon lithography method capable of generating thousands of individually controlled foci, achieving uniform sub-200 nm features, thereby enhancing the throughput and scalability of this approach [86]. Additionally, Riffe et al. investigated a multi-material volumetric additive manufacturing (VAM) strategy using gelatin as both a sacrificial viscosity modifier and a shear-yielding suspension bath to enable the printing of low-viscosity hydrogel precursors [87]. By incorporating gelatin and high-molecular-weight hyaluronic acid, they successfully printed multi-material structures with distinct regions of GelMA and methacrylated hyaluronic acid (MeHA) using VAM, achieving high cell viability and spatial control over cellular responses.

Further, Cai et al. introduced diffusion-based 3D bioprinting as a powerful method to fabricate complex, functional tissue constructs by manipulating molecular diffusion during and after printing [88]. They categorised strategies into inward, outward, and intra-construct diffusion, each enabling modulation of gelation, mechanical properties, and biochemical gradients. These methods support the fabrication of multi-material constructs, perfusable networks, and spatially controlled morphogen gradients. Integration with extrusion printing, coaxial bioprinting, and photopolymerisation broadens their applicability to engineered tissues. These advanced bioprinting processes offer diverse capabilities in fabricating complex, functional biological structures. For example, 4D bioprinting introduces the dimension of time, enabling printed constructs to dynamically change their shape or function in response to environmental stimuli [89]. This innovation relies on smart materials, such as shape-memory polymers and hydrogels, that can respond to triggers such as temperature, moisture, pH, or light. However, 4D printing faces challenges, including limited availability of biocompatible smart materials and the need for complex computational models to design post-printing transformations. Further, 5D printing advances additive manufacturing by introducing rotational axes, enabling the fabrication of complex, curved geometries. Unlike linear 3D printing, 5D printing enhances mechanical integrity by eliminating weak points and is particularly suited for patient-specific, load-bearing scaffolds. In bone tissue engineering, it improves anatomical accuracy and structural strength, though it requires advanced computational systems and specialised materials. 6D printing represents a further evolution by integrating real-time adaptability and smart materials. This technology combines robotics, imaging, and stimuli-responsive materials to create scaffolds that dynamically respond to physiological conditions, such as pH or enzymatic changes. It also allows embedded sensors to monitor cell activity, degradation, or tissue growth, making scaffolds functionally active upon implantation. However, 6D printing faces challenges, including complex system integration, high costs, and technical demands. Overall, 4D, 5D, and 6D printing are transforming tissue engineering by enabling the production of precise, adaptive, and multifunctional scaffolds.

5.3.5 Design for Bioprinting

Although often underemphasised, design plays a pivotal role in tissue engineering by influencing cell viability, tissue integration, and long-term mechanical stability. As highlighted by Ozbolat et al., effective bioprinting begins with acquiring patient-specific anatomical data via imaging techniques such as magnetic resonance imaging (MRI) and computed tomography (CT) (see **Figure 5.5**) [90]. These inputs help create external geometries and internal architectures that support cellular activities, vascularisation, and tissue regeneration. Beyond structural information, medical imaging modalities such as elastography, angiography, and diffusion tensor imaging can also provide functional information about various tissues and significantly improve engineered tissue fabrication. The need for dynamic, patient-specific constructs highlights the critical importance of design in overcoming the limitations of traditional fabrication techniques and meeting clinical demands. To replicate functional tissues, bioprinted constructs must balance biological, mechanical, and transport properties. The porosity, interconnectivity, and surface chemistry of the internal architecture significantly influence cell attachment, nutrient flow, and tissue remodelling over time. Hollister et al. demonstrated that optimising porosity and stiffness for bone tissue constructs enables effective cell migration and mechanical load tolerance, thus emphasising the synergy between biological and mechanical design parameters [91]. Spatially varying porosity is especially critical in tissues such as bone and cartilage, where functional gradients play a vital role [92]. However, many constructs still use uniform porosity due to design and manufacturing limitations, highlighting the need for advanced modelling techniques. Dynamic mechanical requirements are also critical. Constructs for load-bearing applications, such as bone or cartilage, must withstand mechanical stresses during tissue formation. Techniques that incorporate gradient stiffness within a single construct are being explored to better replicate tissue interfaces, such as the cartilage-bone junction. Bioprinted tissues must also adapt to biochemical and mechanical stimuli to support long-term integration within the host, requiring innovative approaches in material design and computational modelling [90].

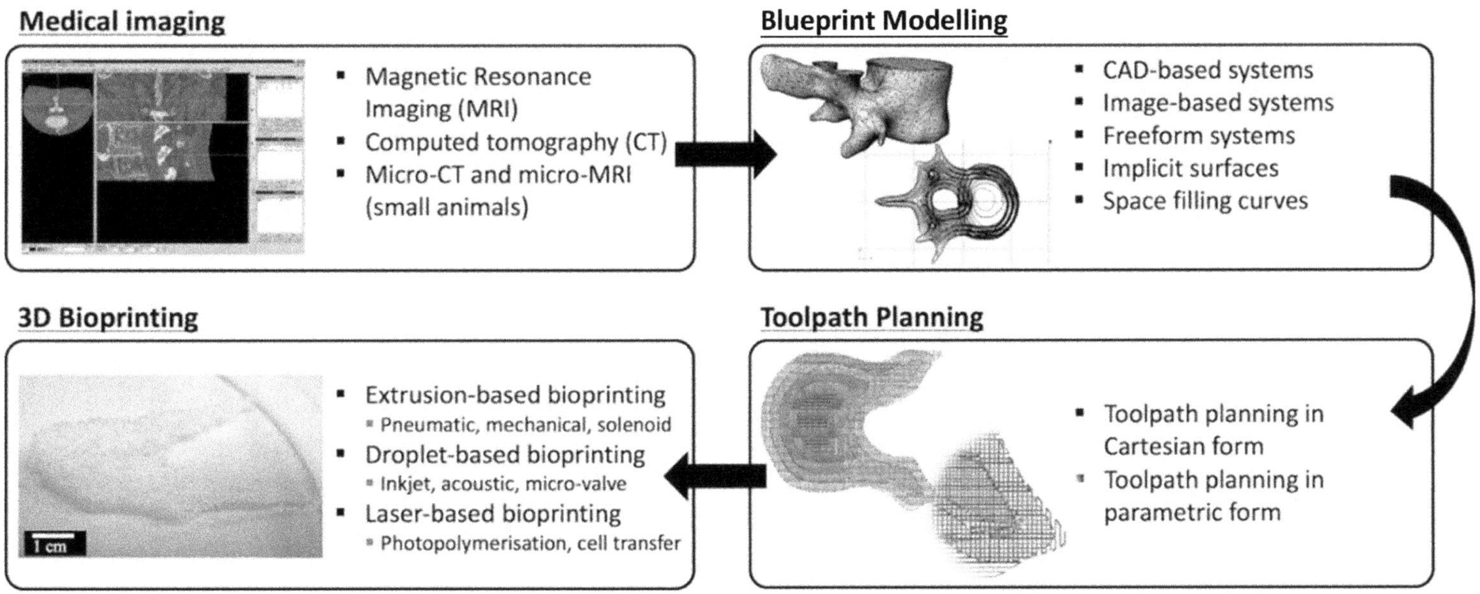

Figure 5.5 Bioprinting design's main steps comprise medical imaging, blueprint modelling, toolpath planning, and 3D bioprinting (Adapted from Ozbolat et al., Bioprinting, 3–4, 2016) [90].

5.3.6 Imaging for Bioprinting Design

MRI offers high-resolution imaging for soft tissues, making it a preferred tool for acquiring patient-specific anatomical data. With resolutions of 5–200 μm in μMRI systems, it enables detailed mapping of tissue hydration and cellular organisation [93]. Recent advancements in contrast agents, such as magnetic nanoparticles, have further improved the imaging of vascular and cellular structures, making MRI invaluable for soft-tissue applications. However, its limitations include susceptibility to motion artefacts, longer scan times, and restrictions on imaging patients with metallic implants. Despite constraints, MRI is one of the most widely used imaging modalities. Based on MRI datasets, Kumar et al. developed a closed-loop framework for designing vascular networks in bioprinted scaffolds by extracting organ geometry and tissue properties [94]. Using segmented liver MR images, they guided vasculature generation using a venation algorithm, enabling patient-specific, anatomically constrained vascular designs. This method ensures nutrient delivery throughout solid liver tissues by simulating mass transport and iteratively refining the vascular layout based on diffusion-limited regions. CT, on the other hand, excels at imaging hard tissues such as bone and provides resolution across various length scales [95]. It is more tolerant of patient movement than MRI and is compatible with metallic implants. However, the ionising radiation limits its use, particularly in paediatric and pregnant patients. Recent innovations, such as micro-CT (μCT), have enabled high-resolution imaging for small-scale constructs and animal models, providing detailed anatomical information for scaffold design. Based on μCT, researchers have developed a framework wherein structural properties (porosity, pore size distribution, pore throat size distribution, tortuosity, interconnectivity, and pore network classification) can be evaluated for any pore type [96, 97]. Such frameworks aid in characterising the anatomy of diseased bone tissue and, accordingly, guide the scaffold design process. After image acquisition, the imaging data undergoes filtration and segmentation to create 3D models. Tools such as ITK-Snap enable precise delineation of tissue structures, but segmentation accuracy significantly affects construct fidelity [98]. For example, adaptive segmentation techniques can improve the modelling of heterogeneous tissues like bones or vascular networks. The integration of AI in image segmentation has further streamlined this process, enabling the automated generation of anatomically accurate models.

5.3.7 Modelling for Bioprinting

Modelling involves designing the internal and external geometries of bioprinted constructs, which are then used to simulate various biophysical and biochemical phenomena. Several approaches are used for modelling the constructs. Computer-aided design (CAD) tools, such as SOLIDWORKS, Autodesk, and CATIA, provide flexibility in designing constructs. For instance, Cheah et al. (2004) used CAD to create porous femoral bone models with tunable porosity and geometry. However, these systems often struggle to efficiently represent complex, biomimetic architectures. Recent developments include libraries of pre-designed unit cells that can be assembled into larger constructs, improving the representation of hierarchical structures. Image-based modelling leverages medical imaging data to create anatomically accurate models. Hollister et al. developed scaffolds for craniofacial reconstruction by integrating CT scans with voxel-based designs [99]. This approach ensures anatomical precision but is computationally intensive. Advances in hybrid modelling, which combines image-based and parametric methods, have improved the scalability and efficiency of these systems. Freeform modelling partitions constructs into regions with distinct material and porosity distributions. For example, Ozbolat et al. created wound-healing devices with region-specific biomolecule delivery [100]. This method supports heterogeneous designs but demands advanced software capabilities. Recent studies have expanded freeform modelling to incorporate dynamic, time-dependent material properties, paving the way for more adaptive tissue constructs.

Implicit surface modelling, on the other hand, generates highly complex non-Euclidean geometries using mathematical equations. Triply periodic minimal surfaces (TPMS), such as gyroid structures, enhance mechanical strength and nutrient diffusion within scaffolds [101]. Space-filling curves, such as the Hilbert curve, provide continuous toolpaths compatible with extrusion bioprinting, thereby reducing fabrication errors [102]. Additionally, bioprinting modelling includes effective toolpath planning to ensure precision and efficiency. Two primary approaches are Cartesian and parametric. Cartesian toolpaths use raster patterns to fill constructs layer by layer. While simple and widely adopted, they often fail to account for geometrical complexities and material transitions. Innovations in adaptive rastering, which dynamically adjust path parameters based on construct geometry, have improved the accuracy and efficiency of Cartesian toolpaths. Parametric approaches generate toolpaths that align with the geometry of the constructs. For instance, ruling-line-based paths proposed for the fabrication of vascular structures enhanced continuity and material distribution in the fabricated constructs [103]. Recent advancements include spiral and hybrid toolpath strategies that integrate multiple printing modalities, enabling the fabrication of more intricate and heterogeneous constructs. In a work by Zhao et al., a helical tool path generation method for a rotary four-DOF 3D printer is proposed to overcome the limitations of traditional planar slicing, such as seam lines and poor surface continuity. This method improves surface finish, geometric fidelity, and mechanical properties, especially for thin-walled and curved structures. Various case studies and preliminary experiments confirmed its effectiveness using both simulated G-code and physical prints [104].

5.3.8 Challenges with Design for Bioprinting

The rheological properties of bioinks significantly affect printability and structural fidelity. Innovations in rheological tuning, such as the incorporation of shear-thinning materials, are critical for fabricating complex designs. Integrating nanoscale and microscale features into constructs remains a challenge. Hybrid approaches may offer solutions by combining extrusion bioprinting with advanced processes such as electrospinning. For instance, Fazal et al. developed a low-cost hybrid 3D bioprinting–electrospinning system by modifying a low-cost 3D printer to fabricate layered vascular grafts that mimic native blood vessels [105]. The setup includes a bioprinting head and dual electrospinning heads, enabling the deposition of hydrogel and nanofiber layers around a rotating needle for enhanced mechanical and biological functionality. This approach enables the fabrication of tubular constructs with improved structural integrity, demonstrating promise for vascular tissue-engineering applications. Computational models that simulate multi-scale interactions, including cell behaviour and material degradation, are increasingly being used to guide design decisions. Gill et al. reviewed the challenges of multiscale bioprinting in replicating complex cellular microenvironments, focusing on integrating features across nanoscale, microscale, and macroscale [106]. They highlighted that current techniques struggle to simultaneously achieve high shape fidelity, biologically relevant stiffness, and microstructural cues, due to limitations in hydrogel mechanics and printing resolution. Combining multiple fabrication strategies, such as extrusion printing, electrospinning, and two-photon polymerisation, is essential to mimic the biphasic architecture of native ECMs. Soft materials, such as hydrogels, often deform after printing due to swelling or dehydration. Developing predictive compensation strategies can improve the accuracy of printed constructs [107]. ML algorithms have been employed to predict deformation patterns and adjust toolpaths in real time, thereby enhancing construct fidelity. Designs for bioprinting must align with regulatory standards to ensure safety and efficacy in clinical applications.

5.4 INTEGRATION OF BIOREACTORS AND TISSUE MICROENVIRONMENTS

A critical challenge in the bioprinting of artificial constructs is ensuring they maintain viability, functionality, and scalability post-fabrication. In this section, the roles of bioreactors and microenvironments in tissue engineering are highlighted, with a focus on recent advancements and their implications for bioprinting. Bioreactors are engineered systems designed to provide controlled environments for cell and tissue culture [108]. They regulate critical parameters such as oxygen concentration, temperature, pH, and mechanical stimuli, effectively mimicking the natural conditions of the human body. Bioreactors can be classified into various types based on their function and design, including the spinner flask, rotating wall vessels, hollow fibre, direct perfusion, and dynamic bioreactors. However, with advancements in bioprinting technologies, 3D-printed, modular, and microfluidics-integrated bioreactors are now available.

Spinner flask bioreactors are typically cylindrical glass containers in which tissue is suspended, and a stirring system ensures mixing of the cell culture medium. These are commonly used for growing articular cartilage, while other tissues can be cultured [109]. Rotating wall vessels (RWVs)-based bioreactors create a microgravity-like environment, promoting three-dimensional tissue assembly and uniform cell growth. Studies have utilised RWVs to culture various tissue types, resulting in enhanced functionality and long-term viability compared to static cultures. Hollow fibre bioreactors (HFBs) are advanced perfusion-based systems widely used in tissue engineering, regenerative medicine, and high-density cell culture applications such as antibody production and stem cell expansion [110]. These bioreactors consist of semipermeable hollow fibres that separate the cell compartment from the medium reservoir, enabling efficient exchange of nutrients, gases, and waste products via diffusion or convection. Cells are typically seeded in the extracapillary space, while media flows through the lumens, mimicking capillary-like microenvironments for sustained viability and function. Such bioreactors support 3D cell culture for scalable production under physiological conditions. Despite their high complexity, HFBs offer precise control over oxygenation, shear stress, and metabolic gradients, making them suitable for long-term culture of various cell types. Direct perfusion-based systems ensure uniform nutrient and oxygen delivery to large tissue constructs, addressing significant challenges in tissue engineering, such as vascularisation [111]. Perfusion bioreactors typically enhance cell viability and functionality in engineered tissues. Dynamic bioreactors that apply mechanical or fluid shear stress have been shown to enhance the maturation of musculoskeletal tissues.

Recent advancements in bioreactor technology focus on improving automation, scalability, and real-time monitoring capabilities. The development of smart bioreactors equipped with integrated biosensors allows continuous measurement of key culture parameters, including pH, glucose, lactate, and dissolved oxygen. Such real-time feedback enables precise control over culture conditions, enhancing the reproducibility and quality of tissue constructs. Additive manufacturing has also enabled rapid prototyping of bioreactors with complex geometries and integrated functionalities. For instance, 3D printing has been utilised to fabricate bioreactors with embedded vascular-like channels, enhancing nutrient distribution in large tissue constructs [112]. Modular designs offer customisation and scalability, allowing researchers to tailor bioreactor systems to specific tissue types. Developments in modular bioreactor systems with interchangeable chambers have been optimised for bone, cartilage, and vascular tissue engineering [113]. Advances in microfluidic technologies have enabled the creation of miniaturised bioreactors that simulate organ-specific microenvironments. These systems, often called "organ-on-a-chip," provide high-resolution control over fluid flow and nutrient gradients, making them ideal for studying cellular interactions and drug responses [114].

5.5 MICROENVIRONMENTS IN BIOPRINTING

The microenvironment within a bioprinted construct is a crucial factor in regulating cell behaviour, influencing processes such as proliferation, differentiation, and migration [115]. These microenvironments are characterised by their biochemical, mechanical, and topographical properties, creating a niche for cellular activities. Designing and controlling these microenvironments is critical for achieving tissue-specific functionality. Recent advancements in bioactive hydrogels have transformed microenvironment engineering. These hydrogels, enriched with growth factors and signalling molecules, provide a dynamic platform for cell interaction. The use of gradient hydrogels to mimic the heterogeneity of native tissues, such as cartilage-bone interfaces, has significantly enhanced tissue-specific differentiation. Engineering microenvironments involves designing constructs with gradients of mechanical stiffness, biomolecular concentrations, and topographical features. Gradient stiffness in hydrogels has been shown to direct stem cell differentiation into specific lineages [116]. For example, stiffness gradients in bioprinted constructs have facilitated the zonation of hepatocytes, mimicking the liver's functional architecture [117]. The integration of bioreactors and microenvironments offers synergistic benefits for tissue engineering. Bioreactors provide external stimuli, such as mechanical loading and perfusion, while microenvironments deliver localised cues that regulate cellular function. Together, these systems create a dynamic and physiologically relevant culture environment that closely mimics *in vivo* conditions. For instance, using a perfusion bioreactor combined with gradient hydrogels has facilitated enhanced osteogenic differentiation and mineralisation, highlighting the importance of integrating systemic and localised cues for successful tissue engineering. Integrating bioreactors with bioprinted constructs presents challenges such as ensuring compatibility between the construct and the bioreactor system, maintaining sterility, and optimising nutrient delivery. Advances in computational modelling of bioreactors and collective biosystems with a multiphysics approach have provided valuable insights for addressing these challenges.

5.6 CHALLENGES IN BIOPRINTING AND TISSUE ENGINEERING

5.6.1 Vascularisation and Integration of Complex Tissues

One of the primary challenges in bioprinting and tissue engineering is achieving vascularisation within complex tissue constructs. Vascularisation is essential for the delivery of oxygen, nutrients, and signalling molecules, as well as the removal of metabolic waste. Without proper vascular networks, cells within larger constructs undergo hypoxia and necrosis, limiting the functionality and longevity of engineered tissues. Several strategies have been proposed to address this challenge. One approach involves incorporating prevascularised networks using sacrificial bioinks. These bioinks, such as Pluronic F127, are printed alongside the primary construct and subsequently removed, creating hollow channels that can later be seeded with endothelial cells. Kolesky et al. demonstrated the utility of sacrificial bioinks in creating perfusable vascular networks that supported long-term cell viability and tissue maturation

[118]. Another promising approach is coaxial bioprinting, which enables the simultaneous deposition of endothelial and stromal cells to form tubular structures mimicking blood vessels [119]. Beyond technological innovations, biological factors also play a critical role in vascularisation. The incorporation of angiogenic growth factors, such as VEGF and fibroblast growth factor (FGF), has been shown to stimulate the formation of capillary-like structures within bioprinted tissues. Lu et al. utilised VEGF-loaded hydrogels to promote endothelial sprouting, resulting in functional vascular networks [120]. The integration of bioprinted constructs with host tissues remains another significant hurdle. For example, matching the mechanical and biochemical properties of the engineered tissue to those of native tissue is crucial for successful integration. Advances in dynamic bioreactors have shown promise in facilitating this process by providing biomechanical stimuli that mimic *in vivo* conditions, thereby guiding vascularisation and tissue maturation [121].

5.6.2 Scale-Up and Reproducibility

Scaling up bioprinting processes to produce clinically relevant tissue constructs presents another major hurdle. Transitioning from small-scale research models to large-scale manufacturing requires robust methods that maintain cell viability, structural fidelity, and functionality. Moreover, reproducibility across multiple batches is critical for ensuring clinical applicability and regulatory approval. One key issue in scaling up is the uniform distribution of cells and bioactive molecules within large constructs (see **Figure 5.6**). In this context, adherence to current Good Manufacturing Practices (cGMP) is essential to guarantee

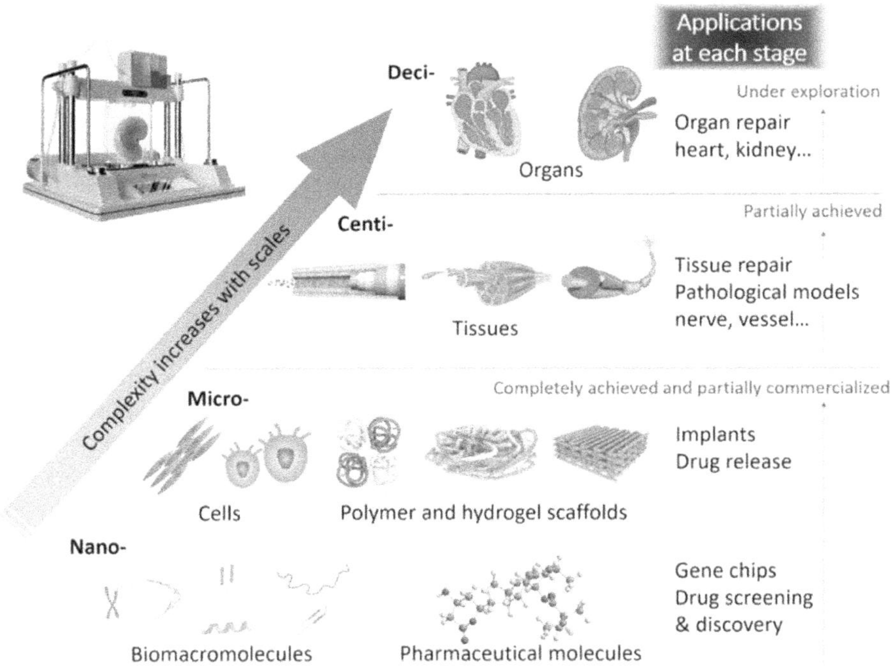

Figure 5.6 Schematic showing increasing assembly complexity level with scales via 3D bioprinting. Components of the upper-stage organs are shown in the lower stages with their usual applications (Reprinted (adapted) with permission from Li et al. [124]. Copyright {2020} American Chemical Society).

product quality, safety, and consistency, while Good Documentation Practices (GDP) enable traceability and standardised workflows. Together, cGMP and GDP frameworks form the foundation for scalable and compliant bioprinting processes. Traditional extrusion-based bioprinting techniques often lead to cell sedimentation and aggregation, thereby compromising construct quality. Advances in nozzle design and microfluidic systems have mitigated some of these issues [122]. For example, using microfluidic systems, microfibres with cells encapsulated from one to a few cells have been produced [123].

Factors such as viscosity, crosslinking efficiency, and cell encapsulation density must be precisely controlled to ensure reproducibility. Standardised protocols for bioink preparation and characterisation are urgently needed to minimise batch-to-batch variability. Kesti et al. proposed a framework for bioink standardisation, highlighting the importance of rheological profiling and pre-printing quality checks [125]. Automation and real-time monitoring systems are emerging as solutions to reproducibility challenges. Smart bioprinters equipped with integrated sensors can continuously monitor printing parameters, such as extrusion pressure and flow rate, ensuring consistency across multiple prints.

5.6.3 Mechanical Strength and Functionality

Achieving the appropriate mechanical strength and functionality in bioprinted constructs is required for their application in load-bearing tissues, such as bone and cartilage, as well as dynamic tissues like the heart and lungs. The mechanical properties of bioprinted tissues must closely match those of native tissues to ensure structural stability and proper integration with the host. A major challenge lies in balancing the mechanical properties of bioinks with their biocompatibility. Natural hydrogels, such as collagen and alginate, are highly biocompatible but often lack the mechanical strength required for load-bearing tissues. On the other hand, synthetic polymers, such as PEG and PCL, offer superior mechanical properties but may lack the bioactivity necessary for cell adhesion and proliferation. Composite bioinks that combine natural and synthetic materials have shown promise in addressing this trade-off. For instance, Yu et al. developed a hybrid bioink comprising alginate and PCL nanoparticles, achieving constructs with enhanced stiffness and bioactivity [7]. Another critical aspect is the anisotropy of mechanical properties in native tissues. For example, muscle and ligament tissues exhibit directionally dependent mechanical behaviours that are challenging to replicate in bioprinted constructs. Advances in multi-material and gradient printing techniques have enabled the fabrication of constructs with spatially varying mechanical properties, thereby mimicking the anisotropy of native tissues. Sun et al. utilised a dual-factor-releasing MSC-laden hydrogel to fabricate cartilage constructs for anisotropic cartilage regeneration [126]. Along with mechanical properties, functional integration of the fabricated constructs is equally important. Dynamic bioreactors that apply mechanical stimulation, such as compression, tension, or shear stress, have been shown to enhance the functionality of engineered tissues. For example, Song et al. reported that cyclic stretching conditions improve contractile properties and cardiomyocyte alignment [127]. Additionally, incorporating conductive materials, such as graphene and carbon nanotubes, into bioinks has shown potential to improve the electrical conductivity and functionality of cardiac and neural tissues [128].

5.6.4 Regulatory and Ethical Considerations

The clinical translation of bioprinted constructs is fraught with regulatory and ethical challenges. Regulatory frameworks for bioprinting and tissue engineering are still in their infancy, often lagging behind technological advancements [129]. This creates uncertainty for researchers and manufacturers seeking to bring bioprinted products to clinical applications. One of the primary regulatory hurdles is the classification of bioprinted constructs. These products often fall into a grey area between medical devices, biologics, and combination products, requiring input from multiple regulatory bodies. Although the U.S. Food and Drug Administration (FDA) has previously approved various additively manufactured medical devices for clinical applications such as patient-specific implants, dental bridges, surgical guides, and orthopaedic devices [130], it has yet to establish specific guidelines for bioprinting, instead evaluating constructs on a case-by-case basis [131]. Similarly, the European Medicines Agency (EMA) faces challenges in harmonising regulations across member states. The Department of Science and Technology in India has taken a significant step by including biomaterials and fabricated constructs in medical devices. However, as with other international regulatory bodies, clear distinctions and guidelines for bioprinted constructs remain under development. Ethical considerations also play a significant role in engineered tissues. The use of patient-derived cells raises concerns about donor consent, privacy, and data security. Researchers must navigate these ethical dilemmas while ensuring compliance with established ethical guidelines. Standardisation and quality control are essential for regulatory approval. Robust preclinical studies and clinical trials are needed to demonstrate the safety, efficacy, and reproducibility of bioprinted constructs. Advanced computational models and *in vitro* testing platforms can complement traditional preclinical studies, providing additional data to support regulatory submissions [132]. To address these challenges, collaborative efforts between academia, industry, and regulatory agencies are crucial. Furthermore, public engagement and education initiatives can help address ethical concerns and build societal trust in these transformative technologies.

5.7 CONCLUSION AND FUTURE DIRECTIONS

The development of effective vascularisation strategies remains a cornerstone of bioprinting, particularly for thick, metabolically active tissues. Future research will emphasise the generation of perfusable vascular networks using technologies such as co-axial bioprinting, sacrificial bioinks, and microfluidic integration. Concurrently, the integration of microstimulators and bioelectronic systems is gaining momentum, as electrical stimulation has shown promise in enhancing angiogenesis and tissue regeneration. Future systems may embed electroactive elements directly into bioprinted matrices to deliver localised stimuli in engineered tissues. Hybrid bioprinting platforms that combine extrusion, droplet-based, and laser-assisted methods are expected to transform construct complexity and functionality. These systems will support the deposition of multiple bioinks with varying viscosities, cell types, and densities, enabling hierarchical, biomimetic architectures. For example, combining extrusion for bulk material deposition with laser-assisted bioprinting for high-resolution patterning allows fabrication of functionally diverse and heterogeneous tissues. Advances in hybrid printing will also facilitate spatial control over nanoparticles, growth factors, and mechanical reinforcements, enabling gradient-based tissue properties. Bioink development is evolving towards "smart" materials responsive to stimuli such as temperature, pH, and light. Shape-memory hydrogels, capable of undergoing preprogrammed transformations, are particularly suited for 4D bioprinting. Additionally, integrating bioactive molecules, such as microRNAs and exosomes, can significantly enhance cell differentiation, angiogenesis, and immunomodulation, supporting more precise and functional tissue regeneration. AI and ML are expected to revolutionise bioprinting and tissue engineering by optimising design, manufacturing, and evaluation processes. AI models can predict cellular behaviours, optimise printing parameters, and simulate tissue maturation, thereby reducing experimental trial-and-error. Additionally, ML models can analyse data from biosensors embedded in bioreactors, providing real-time feedback for adaptive control of culture conditions. Another promising application of AI is in the design of bioinks. AI can suggest optimal combinations of polymers, crosslinkers, and bioactive agents for specific applications by analysing databases of material properties and biological outcomes. This approach will accelerate the development of tailored bioinks for complex tissues like the liver, kidney, heart, and brain. Current studies are exploring how ML algorithms, including supervised, unsupervised, and reinforcement learning, can be applied across pre-printing, printing, and post-printing stages to predict printability, detect anomalies, and simulate tissue maturation. The use of bioprinted organoids and tissue-on-chip systems for disease modelling and drug screening is also gaining traction. Future systems will incorporate immune and neural components to improve physiological relevance, enabling personalised therapies and mechanistic studies of diseases such as Alzheimer's and cancer. Scaffold-free bioprinting and in situ regeneration approaches, such as cardiac and cartilage repair, are emerging as viable strategies. Long-term success will require scalable production, automation, standardised protocols, and strong regulatory frameworks to ensure clinical translation and accessibility.

REFERENCES

[1] R. Langer and J. P. Vacanti, "Tissue engineering," *Science (80-.)*, vol. 260, no. 5110, pp. 920–926, 1993 [Online]. Available: http://www.jstor.org/stable/2885618

[2] S. V Murphy and A. Atala, "3D bioprinting of tissues and organs," *Nat. Biotechnol.*, vol. 32, no. 8, pp. 773–785, 2014, https://doi.org/10.1038/nbt.2958.

[3] S. Mao et al., "Bioprinting of in vitro tumor models for personalized cancer treatment: A review," *Biofabrication*, vol. 12, no. 4, p. 042001, 2020, https://doi.org/10.1088/1758-5090/ab97c0.

[4] F. Pati et al., "Printing three-dimensional tissue analogues with decellularized extracellular matrix bioink," *Nat. Commun.*, vol. 5, no. 1, p. 3935, 2014, https://doi.org/10.1038/ncomms4935.

[5] R. Langer and D. A. Tirrell, "Designing materials for biology and medicine," *Nature*, vol. 428, no. 6982, pp. 487–492, 2004, https://doi.org/10.1038/nature02388.

[6] A. Pappalardo et al., "Engineering edgeless human skin with enhanced biomechanical properties," *Sci. Adv.*, vol. 9, no. 4, pp. 1–16, 2023, https://doi.org/10.1126/sciadv.ade2514.

[7] J. Yu et al., "Fabrication of a polycaprolactone/alginate bipartite hybrid scaffold for osteochondral tissue using a three-dimensional bioprinting system," *Polymers (Basel)*, vol. 12, no. 10, pp. 1–16, 2020, https://doi.org/10.3390/polym12102203.

[8] B. Grigoryan et al., "Multivascular networks and functional intravascular topologies within biocompatible hydrogels," *Science (80-.)*, vol. 364, no. 6439, pp. 458–464, 2019, https://doi.org/10.1126/science.aav9750.

[9] K. Yang, L. Wang, S. Vijayavenkataraman, Y. Yuan, E. C. K. Tan, and L. Kang, "Recent applications of three-dimensional bioprinting in drug discovery and development," *Adv. Drug Deliv. Rev.*, vol. 214, no. September, p. 115456, 2024, https://doi.org/10.1016/j.addr.2024.115456.

[10] C. Wang, X. Tong, and F. Yang, "Bioengineered 3D brain tumor model to elucidate the effects of matrix stiffness on glioblastoma cell behavior using peg-based hydrogels," *Mol. Pharm.*, vol. 11, no. 7, pp. 2115–2125, 2014, https://doi.org/10.1021/mp5000828.

[11] N. Tabatabaei Rezaei, H. Kumar, H. Liu, S. S. Lee, S. S. Park, and K. Kim, "Recent advances in organ-on-chips integrated with bioprinting technologies for drug screening," *Adv. Healthc. Mater.*, vol. 12, no. 20, 2023, https://doi.org/10.1002/adhm.202203172.

[12] F. L. C. Morgan, L. Moroni, and M. B. Baker, "Dynamic bioinks to advance bioprinting," *Adv. Healthc. Mater.*, vol. 9, no. 15, 2020, https://doi.org/10.1002/adhm.201901798.

[13] J. Gopinathan and I. Noh, "Recent trends in bioinks for 3D printing," *Biomater. Res.*, vol. 22, no. 1, pp. 1–15, 2018, https://doi.org/10.1186/s40824-018-0122-1.

[14] M. Mobaraki, M. Ghaffari, A. Yazdanpanah, Y. Luo, and D. K. Mills, "Bioinks and bioprinting: A focused review," *Bioprinting*, vol. 18, no. January, p. e00080, 2020, https://doi.org/10.1016/j.bprint.2020.e00080.

[15] Q. Fu, E. Saiz, M. N. Rahaman, and A. P. Tomsia, "Bioactive glass scaffolds for bone tissue engineering: State of the art and future perspectives," *Mater. Sci. Eng. C. Mater. Biol. Appl.*, vol. 31, no. 7, pp. 1245–1256, 2011, https://doi.org/10.1016/j.msec.2011.04.022.
[16] K. Dixit, A. Vishwakarma, H. Kumar, K. Kim, and N. Sinha, "Material extrusion additive manufacturing of graphene oxide reinforced 13–93B1 bioactive glass scaffolds for bone tissue engineering applications," *Addit. Manuf.*, vol. 94, p. 104481, 2024, https://doi.org/10.1016/j.addma.2024.104481.
[17] F. Luo et al., "PCL-CNT nano-composites for bone tissue engineering application," in *Polymer Science: Research Advances, Practical Applications and Educational Aspects*, 2016, pp. 369–378 [Online]. Available: http://www.formatex.info/polymerscience1/book/369-378.pdf
[18] Z. Li, C. Ruan, and X. Niu, "Collagen-based bioinks for regenerative medicine: Fabrication, application and prospective," *Med. Nov. Technol. Devices*, vol. 17, no. 37, p. 100211, 2023, https://doi.org/10.1016/j.medntd.2023.100211.
[19] A. Lee et al., "3D bioprinting of collagen to rebuild components of the human heart," *Science (80-.)*, vol. 365, no. 6452, pp. 482–487, 2019, https://doi.org/10.1126/science.aav9051.
[20] T. J. Hinton et al., "Three-dimensional printing of complex biological structures by freeform reversible embedding of suspended hydrogels," *Sci. Adv.*, vol. 1, no. 9, 2015, https://doi.org/10.1126/sciadv.1500758.
[21] S. Chawla, S. Midha, A. Sharma, and S. Ghosh, "Silk-based bioinks for 3D bioprinting," *Adv. Healthc. Mater.*, vol. 7, no. 8, pp. 1–23, 2018, https://doi.org/10.1002/adhm.201701204.
[22] M. J. Rodriguez, J. Brown, J. Giordano, S. J. Lin, F. G. Omenetto, and D. L. Kaplan, "Silk based bioinks for soft tissue reconstruction using 3-dimensional (3D) printing with in vitro and in vivo assessments," *Biomaterials*, vol. 117, pp. 105–115, 2017, https://doi.org/10.1016/j.biomaterials.2016.11.046.
[23] J. A. Burdick and G. D. Prestwich, "Hyaluronic acid hydrogels for biomedical applications," *Adv. Mater.*, vol. 23, no. 12, pp. 41–56, 2011, https://doi.org/10.1002/adma.201003963.
[24] O. Jeon and E. Alsberg, "Photofunctionalization of alginate hydrogels to promote adhesion and proliferation of human mesenchymal stem cells," *Tissue Eng.—Part A*, vol. 19, no. 11–12, pp. 1424–1432, 2013, https://doi.org/10.1089/ten.tea.2012.0581.
[25] H. Kumar, K. Sakthivel, M. G. A. Mohamed, E. Boras, S. R. Shin, and K. Kim, "Designing gelatin methacryloyl (GelMA)-based bioinks for visible light stereolithographic 3D biofabrication," *Macromol. Biosci.*, vol. 21, no. 1, p. 2000317, 2021, https://doi.org/10.1002/mabi.202000317.
[26] V. T. Duong and C. C. Lin, "Digital light processing 3D bioprinting of gelatin-norbornene hydrogel for enhanced vascularization," *Macromol. Biosci.*, vol. 23, no. 12, pp. 1–12, 2023, https://doi.org/10.1002/mabi.202300213.
[27] D. Gomes, J. P. Batista-Silva, A. Sousa, and L. A. Passarinha, "Progress and opportunities in Gellan gum-based materials: A review of preparation, characterization and emerging applications," *Carbohydr. Polym.*, vol. 311, no. December 2022, p. 120782, 2023, https://doi.org/10.1016/j.carbpol.2023.120782.
[28] X. Peng, G. Liu, L. Zhu, K. Yu, K. Qian, and X. Zhan, "In vitro and in vivo study of novel antimicrobial gellan–polylysine polyion complex fibers as suture materials," *Carbohydr. Res.*, vol. 496, no. February, p. 108115, 2020, https://doi.org/10.1016/j.carres.2020.108115.
[29] M. Taghizadeh et al., "Chitosan-based inks for 3D printing and bioprinting," *Green Chem.*, vol. 24, no. 1, pp. 62–101, 2022, https://doi.org/10.1039/d1gc01799c.
[30] S. Li et al., "High self-supporting chitosan-based hydrogel ink for in situ 3D printed diabetic wound dressing," *Adv. Funct. Mater.*, vol. 35, no. 5, pp. 1–17, 2025, https://doi.org/10.1002/adfm.202414625.
[31] A. Shpichka et al., "Fibrin-based bioinks: New tricks from an old dog," *Int. J. Bioprinting*, vol. 6, no. 3, pp. 1–14, 2020, https://doi.org/10.18063/IJB.V6I3.269.
[32] C. Benwood et al., "Natural biomaterials and their use as bioinks for printing tissues," *Bioengineering*, vol. 8, no. 2, pp. 1–19, 2021, https://doi.org/10.3390/bioengineering8020027.
[33] M. V. Hangad, A. Forigua, K. Scheck, S. M. Willerth, and K. S. Elvira, "Investigating how all-trans retinoic acid polycaprolactone (atRA-PCL) microparticles alter the material properties of 3D printed fibrin constructs," *Macromol. Biosci.*, vol. 2400464, pp. 1–13, 2025, https://doi.org/10.1002/mabi.202400464.

[34] S. Sun et al., "Drug delivery systems based on polyethylene glycol hydrogels for enhanced bone regeneration," *Front. Bioeng. Biotechnol.*, vol. 11, no. January, pp. 1–17, 2023, https://doi.org/10.3389/fbioe.2023.1117647.
[35] G. Choe et al., "Three-dimensional bioprinting of mesenchymal stem cells using an osteoinductive bioink containing alginate and BMP-2-loaded PLGA nanoparticles for bone tissue engineering," *Biomater. Adv.*, vol. 136, no. December 2021, p. 212789, 2022, https://doi.org/10.1016/j.bioadv.2022.212789.
[36] E. Muscolino et al., "κ-Carrageenan and PVA blends as bioinks to 3D print scaffolds for cartilage reconstruction," *Int. J. Biol. Macromol.*, vol. 222, no. PB, pp. 1861–1875, 2022, https://doi.org/10.1016/j.ijbiomac.2022.09.275.
[37] F. Ullah et al., "Development of highly-reproducible hydrogel based bioink for regeneration of skin-tissues via 3-D bioprinting technology," *Int. J. Biol. Macromol.*, vol. 230, no. January, p. 123131, 2023, https://doi.org/10.1016/j.ijbiomac.2022.123131.
[38] B. Oh et al., "3D printable and biocompatible PEDOT:PSS-ionic liquid colloids with high conductivity for rapid on-demand fabrication of 3D bioelectronics," *Nat. Commun.*, vol. 15, no. 1, pp. 1–13, 2024, https://doi.org/10.1038/s41467-024-50264-6.
[39] Y. Cai, S. Y. Chang, S. W. Gan, S. Ma, W. F. Lu, and C. C. Yen, "Nanocomposite bioinks for 3D bioprinting," *Acta Biomater.*, vol. 151, pp. 45–69, 2022, https://doi.org/10.1016/j.actbio.2022.08.014.
[40] F. Potere et al., "Double-crosslinked dECM bioink to print a self-sustaining 3D multi-layered aortic-like construct," *Bioprinting*, vol. 44, no. August, p. e00368, 2024, https://doi.org/10.1016/j.bprint.2024.e00368.
[41] Z. Fu, N. Hai, Y. Zhong, and W. Sun, "Printing GelMA bioinks: A strategy for building in vitro model to study nanoparticle-based minocycline release and cellular protection under oxidative stress," *Biofabrication*, vol. 16, no. 2, 2024, https://doi.org/10.1088/1758-5090/ad30c3.
[42] Y. C. Yeh, P. Y. Chen, K. T. Chen, and I. C. Lee, "Innovative MXene/SilMA-based conductive bioink for three dimensional bioprinting of neural stem cell spheroids in neural tissue engineering," *ACS Appl. Mater. Interfaces*, 2025, https://doi.org/10.1021/acsami.4c19373.
[43] M. A. Sakr et al., "A novel GelMA-OrnMA electrically conductive bioink for developing engineered neural tissues," *Cold Spring Harbor Laboratory*, 2024. https://doi.org/10.1101/2024.10.06.616594.
[44] S. Meenakshy, J. Jesslyn, and S. Anas, "Development and applications of polypyrrole-based conductive inks: An overview," *Adv. Mater. Technol.*, p. 2401216, 2024, https://doi.org/10.1002/admt.202401216.
[45] R. Cheng et al., "Peripheral nerve regeneration with 3D printed bionic double-network conductive scaffold based on GelMA/chitosan/polypyrrole," *Int. J. Biol. Macromol.*, vol. 304, no. P1, p. 140746, 2025, https://doi.org/10.1016/j.ijbiomac.2025.140746.
[46] K. Dixit and N. Sinha, "Additive manufacturing of carbon nanotube reinforced bioactive glass scaffolds for bone tissue engineering," *ASME J. Eng. Sci. Med. Diagnostics Ther.*, vol. 4, no. 4, pp. 1–8, Nov. 2021, https://doi.org/10.1115/1.4051801.
[47] S. Zheng et al., "Surface-modified nano-hydroxyapatite uniformly dispersed on high-porous GelMA scaffold surfaces for enhanced osteochondral regeneration," *Int. J. Nanomedicine*, vol. 18, pp. 5907–5923, 2023, https://doi.org/10.2147/IJN.S428965.
[48] J. Jang et al., "3D printed complex tissue construct using stem cell-laden decellularized extracellular matrix bioinks for cardiac repair," *Biomaterials*, vol. 112, pp. 264–274, 2017, https://doi.org/10.1016/j.biomaterials.2016.10.026.
[49] N. Tabatabaei Rezaei et al., "3D bioprinting of liver microenvironment model using photocrosslinkable decellularized extracellular matrix based hydrogel," *bioRxiv*, p. 2024, 2024.
[50] V. Khati, H. Ramachandraiah, F. Pati, H. A. Svahn, G. Gaudenzi, and A. Russom, "3D bioprinting of multi-material decellularized liver matrix hydrogel at physiological temperatures," *Biosensors*, vol. 12, no. 7, 2022, https://doi.org/10.3390/bios12070521.
[51] Z. Li, F. Lu, and Y. Liu, "A review of the mechanism, properties, and applications of hydrogels prepared by enzymatic cross-linking," *J. Agric. Food Chem.*, vol. 71, no. 27, pp. 10238–10249, 2023, https://doi.org/10.1021/acs.jafc.3c01162.
[52] K. S. Lim, J. H. Galarraga, X. Cui, G. C. J. Lindberg, J. A. Burdick, and T. B. F. Woodfield, "Fundamentals and applications of photo-cross-linking in bioprinting," *Chem. Rev.*, vol. 120, no. 19, pp. 10662–10694, 2020, https://doi.org/10.1021/acs.chemrev.9b00812.

[53] E. Kim, M. H. Kim, J. H. Song, C. Kang, and W. H. Park, "Dual crosslinked alginate hydrogels by riboflavin as photoinitiator," *Int. J. Biol. Macromol.*, vol. 154, pp. 989–998, 2020, https://doi.org/10.1016/j.ijbiomac.2020.03.134.
[54] Z. Li et al., "Visible light laser direct-writing of high-resolution, biocompatible, super-multifunctional and tough hydrogels without photoinitiators in 30 s," *Biomater. Adv.*, vol. 147, no. February, p. 213318, 2023, https://doi.org/10.1016/j.bioadv.2023.213318.
[55] H. Huang et al., "Intra-articular drug delivery systems for osteoarthritis therapy: Shifting from sustained release to enhancing penetration into cartilage," *Drug Deliv.*, vol. 29, no. 1, pp. 767–791, 2022, https://doi.org/10.1080/10717544.2022.2048130.
[56] J. Chen et al., "A scalable coaxial bioprinting technology for mesenchymal stem cell microfiber fabrication and high extracellular vesicle yield," *Biofabrication*, vol. 14, no. 1, 2022, https://doi.org/10.1088/1758-5090/ac3b90.
[57] H. Zhou et al., "Simultaneous multimaterial multimethod bioprinting," *Bio-Design Manuf.*, vol. 5, no. 3, pp. 433–436, 2022, https://doi.org/10.1007/s42242-022-00203-5.
[58] Y. S. Zhang, A. Dolatshahi-Pirouz, and G. Orive, "Regenerative cell therapy with 3D bioprinting," *Science (80-.)*, vol. 385, no. 6709, pp. 604–606, 2024, https://doi.org/10.1126/science.add8593.
[59] W. Li et al., "An adhesive bioink toward biofabrication under wet conditions," *Small*, vol. 19, no. 50, pp. 1–12, 2023, https://doi.org/10.1002/smll.202205078.
[60] C. Zhou et al., "Ferromagnetic soft catheter robots for minimally invasive bioprinting," *Nat. Commun.*, vol. 12, no. 1, pp. 1–12, 2021, https://doi.org/10.1038/s41467-021-25386-w.
[61] S. Zoghi, "Advancements in tissue engineering: A review of bioprinting techniques, scaffolds, and bioinks," *Biomed. Eng. Comput. Biol.*, vol. 15, 2024, https://doi.org/10.1177/11795972241288099.
[62] I. T. Ozbolat and M. Hospodiuk, "Current advances and future perspectives in extrusion-based bioprinting," *Biomaterials*, vol. 76, pp. 321–343, 2016, https://doi.org/10.1016/j.biomaterials.2015.10.076.
[63] Z. Gao et al., "Simultaneous multi-material embedded printing for 3D heterogeneous structures," *Int. J. Extrem. Manuf.*, vol. 5, no. 3, 2023, https://doi.org/10.1088/2631-7990/acd285.
[64] S. Maharjan, J. J. He, L. Lv, D. Wang, and Y. S. Zhang, "Microfluidic coaxial bioprinting of hollow, standalone, and perfusable vascular conduits," in *Methods in Molecular Biology (Clifton, N.J.)*, vol. 2375, 2022, pp. 61–75. https://doi.org/10.1007/978-1-0716-1708-3_6.
[65] S. Yang et al., "Development and quantitative evaluation of a novel autonomous in situ bioprinting surgical robotic framework for treatment of volumetric muscle loss injuries," *IEEE Trans. Autom. Sci. Eng.*, vol. 22, pp. 2565–2576, 2025, https://doi.org/10.1109/TASE.2024.3382200.
[66] T. Xu, W. Zhao, J.-M. Zhu, M. Z. Albanna, J. J. Yoo, and A. Atala, "Complex heterogeneous tissue constructs containing multiple cell types prepared by inkjet printing technology," *Biomaterials*, vol. 34, no. 1, pp. 130–139, 2013, https://doi.org/10.1016/j.biomaterials.2012.09.035.
[67] J. A. Park, Y. Lee, and S. Jung, "Inkjet-based bioprinting for tissue engineering," *Organoid*, vol. 3, p. e12, 2023, https://doi.org/10.51335/organoid.2023.3.e12.
[68] W. L. Ng and V. Shkolnikov, "Optimizing cell deposition for inkjet-based bioprinting," *Int. J. Bioprinting*, vol. 10, no. 2, p. 2135, 2024, https://doi.org/10.36922/ijb.2135.
[69] W. Kim, Y. Lee, D. Kang, T. Kwak, H. R. Lee, and S. Jung, "3D inkjet-bioprinted lung-on-a-chip," *ACS Biomater. Sci. Eng.*, vol. 9, no. 5, pp. 2806–2815, 2023, https://doi.org/10.1021/acsbiomaterials.3c00089.
[70] B. Guillotin and F. Guillemot, "Cell patterning technologies for organotypic tissue fabrication," *Trends Biotechnol.*, vol. 29, no. 4, pp. 183–190, 2011, https://doi.org/10.1016/j.tibtech.2010.12.008.
[71] C. Dou, V. Perez, J. Qu, A. Tsin, B. Xu, and J. Li, "A state-of-the-art review of laser-assisted bioprinting and its future research trends," *ChemBioEng Rev.*, vol. 8, no. 5, pp. 517–534, 2021, https://doi.org/10.1002/cben.202000037.
[72] A. Zennifer, A. Subramanian, and S. Sethuraman, "Design considerations of bioinks for laser bioprinting technique towards tissue regenerative applications," *Bioprinting*, vol. 27, no. April, p. e00205, 2022, https://doi.org/10.1016/j.bprint.2022.e00205.
[73] P. N. Bernal et al., "Volumetric bioprinting of complex living-tissue constructs within seconds," *Adv. Mater.*, vol. 31, no. 42, 2019, https://doi.org/10.1002/adma.201904209.

[74] Q. Thijssen, J. Toombs, C. C. Li, H. Taylor, and S. Van Vlierberghe, "From pixels to voxels: A mechanistic perspective on volumetric 3D-printing," *Prog. Polym. Sci.*, vol. 147, p. 101755, 2023, https://doi.org/10.1016/j.progpolymsci.2023.101755.
[75] S. Mukherjee, J. Schwartz, T. Chang, E. Baluyot, J. Tringe, and M. Shusteff, "Focused microwave radiation for the localized curing of polymer resins," in *2022 IEEE International Symposium on Antennas and Propagation and USNC-URSI Radio Science Meeting (AP-S/URSI 2022 Proceedings)*, pp. 159–160, 2022, https://doi.org/10.1109/AP-S/USNC-URSI47032.2022.9886029.
[76] G. Größbacher et al., "Volumetric printing across melt electrowritten scaffolds fabricates multi-material living constructs with tunable architecture and mechanics," *Adv. Mater.*, vol. 35, no. 32, pp. 1–18, 2023, https://doi.org/10.1002/adma.202300756.
[77] D. Loterie, P. Delrot, and C. Moser, "High-resolution tomographic volumetric additive manufacturing," *Nat. Commun.*, vol. 11, no. 1, pp. 1–6, 2020, https://doi.org/10.1038/s41467-020-14630-4.
[78] A. C. Kak and M. Slaney, *Principles of Computerized Tomographic Imaging*, vol. 27. Society for Industrial and Applied Mathematics, 2001. https://doi.org/10.1137/1.9780898719277.
[79] S. Mukherjee et al., "Towards microwave volumetric additive manufacturing: Generation of a computational multi-physics model for localized curing," *Addit. Manuf. Lett.*, vol. 10, no. April, 2024, https://doi.org/10.1016/j.addlet.2024.100209.
[80] Z. Zheng, L. Lian, and M. Xie, "Ultrasound volumetric bioprinting: Opportunities and challenges," *Innov. Life*, vol. 2, no. 1, p. 100053, 2024, https://doi.org/10.59717/j.xinn-life.2024.100053.
[81] L. Stoecker et al., "Xolography for biomedical applications: Dual-color light-sheet printing of hydrogels with local control over shape and stiffness," *Adv. Mater.*, vol. 2410292, pp. 1–16, 2025, https://doi.org/10.1002/adma.202410292.
[82] M. Regehly et al., "Xolography for linear volumetric 3D printing," *Nature*, vol. 588, no. 7839, pp. 620–624, 2020, https://doi.org/10.1038/s41586-020-3029-7.
[83] L. Stüwe et al., "Continuous volumetric 3d printing: Xolography in flow," *Adv. Mater.*, vol. 36, no. 4, pp. 1–10, 2024, https://doi.org/10.1002/adma.202306716.
[84] P. Somers, Z. Liang, J. E. Johnson, B. W. Boudouris, L. Pan, and X. Xu, "Rapid, continuous projection multi-photon 3D printing enabled by spatiotemporal focusing of femtosecond pulses," *Light Sci. Appl.*, vol. 10, no. 1, 2021, https://doi.org/10.1038/s41377-021-00645-z.
[85] M. Bouzin et al., "Multiphoton laser fabrication of hybrid photo-activable biomaterials," *Sensors*, vol. 21, no. 17, 2021, https://doi.org/10.3390/s21175891.
[86] S. Yang et al., "Parallel two-photon lithography achieving uniform sub-200 nm features with thousands of individually controlled foci," *Opt. Express*, vol. 31, no. 9, pp. 14174–14184, 2023, https://doi.org/10.1364/OE.483524.
[87] M. B. Riffe et al., "Multi-material volumetric additive manufacturing of hydrogels using gelatin as a sacrificial network and 3D suspension bath," *Adv. Mater.*, vol. 36, no. 34, pp. 1–10, 2024, https://doi.org/10.1002/adma.202309026.
[88] B. Cai, D. Kilian, D. Ramos Mejia, R. J. Rios, A. Ali, and S. C. Heilshorn, "Diffusion-based 3D bioprinting strategies," *Adv. Sci.*, vol. 11, no. 8, 2024, https://doi.org/10.1002/advs.202306470.
[89] D. K. Gara, R. Gujjala, P. S. Prasad, N. Madaboosi, and S. Ojha, "4D bioprinting: A review on smart bio-adaptable technology to print stimuli-responsive materials," *Prog. Addit. Manuf.*, no. 0123456789, 2024, https://doi.org/10.1007/s40964-024-00876-7.
[90] I. Ozbolat and H. Gudapati, "A review on design for bioprinting," *Bioprinting*, vol. 3–4, no. November, pp. 1–14, 2016, https://doi.org/10.1016/j.bprint.2016.11.001.
[91] S. J. Hollister and C. Y. Lin, "Computational design of tissue engineering scaffolds," *Comput. Methods Appl. Mech. Eng.*, vol. 196, no. 31–32, pp. 2991–2998, 2007, https://doi.org/10.1016/j.cma.2006.09.023.
[92] V. Karageorgiou and D. Kaplan, "Porosity of 3D biomaterial scaffolds and osteogenesis," *Biomaterials*, vol. 26, no. 27, pp. 5474–5491, 2005, https://doi.org/10.1016/j.biomaterials.2005.02.002.
[93] E. Terreno, D. D. Castelli, A. Viale, and S. Aime, "Challenges for molecular magnetic resonance imaging," *Chem. Rev.*, vol. 110, no. 5, pp. 3019–3042, 2010, https://doi.org/10.1021/cr100025t.
[94] H. Kumar, K. Dixit, R. Sharma, E. MacDonald, N. Sinha, and K. Kim, "Closed-loop vasculature network design for bioprinting large, solid tissue scaffolds," *Biofabrication*, 2023, https://doi.org/10.1088/1758-5090/acb73c.

[95] M. P. Akhter and R. R. Recker, "High resolution imaging in bone tissue research-review," *Bone*, vol. 143, no. August 2020, p. 115620, 2021, https://doi.org/10.1016/j.bone.2020.115620.
[96] K. Dixit, P. Gupta, S. Kamle, and N. Sinha, "Structural analysis of porous bioactive glass scaffolds using micro-computed tomographic images," *J. Mater. Sci.*, vol. 55, no. 27, pp. 12705–12724, 2020, https://doi.org/10.1007/s10853-020-04850-w.
[97] K. Dixit, A. Raichur, and N. Sinha, "Polymer-coated and nanofiber-reinforced functionally graded bioactive glass scaffolds fabricated using additive manufacturing," *IEEE Trans. Nanobioscience*, vol. 21, no. 3, pp. 380–386, 2022, https://doi.org/10.1109/TNB.2021.3083278.
[98] P. A. Yushkevich, Y. Gao, and G. Gerig, "ITK-SNAP: An interactive tool for semi-automatic segmentation of multi-modality biomedical images," *Proc. Annu. Int. Conf. IEEE Eng. Med. Biol. Soc. EMBS*, vol. 2016, pp. 3342–3345, 2016, https://doi.org/10.1109/EMBC.2016.7591443.
[99] S. J. Hollister, R. A. Levy, T. M. Chu, J. W. Halloran, and S. E. Feinberg, "An image-based approach for designing and manufacturing craniofacial scaffolds," *Int. J. Oral Maxillofac. Surg.*, vol. 29, no. 1, pp. 67–71, 2000, https://doi.org/10.1034/j.1399-0020.2000.290115.x.
[100] I. T. Ozbolat and B. Koc, "3D hybrid wound devices for spatiotemporally controlled release kinetics," *Comput. Methods Programs Biomed.*, vol. 108, no. 3, pp. 922–931, 2012, https://doi.org/10.1016/j.cmpb.2012.05.004.
[101] S. C. Kapfer, S. T. Hyde, K. Mecke, C. H. Arns, and G. E. Schröder-Turk, "Minimal surface scaffold designs for tissue engineering," *Biomaterials*, vol. 32, no. 29, pp. 6875–6882, 2011, https://doi.org/10.1016/j.biomaterials.2011.06.012.
[102] B. Starly, L. Shor, W. Sun, and A. Darling, "Space filling curves: Its design and fabrication for extrusion based SFF systems," in *Manufacturing Engineering and Materials Handling, Parts A and B*, ASMEDC, 2005, pp. 99–104. https://doi.org/10.1115/IMECE2005-79742.
[103] I. T. Ozbolat and A. K. M. B. Khoda, "Design of a new parametric path plan for additive manufacturing of hollow porous structures with functionally graded materials," *J. Comput. Inf. Sci. Eng.*, vol. 14, no. 4, 2014, https://doi.org/10.1115/1.4028418.
[104] D. Zhao, J. He, G. Zhu, Y. Han, and W. Guo, "Helical tool path generation based on the triangle mesh model for a rotary four-DOF 3D printer," *Rapid Prototyp. J.*, vol. 29, no. 4, pp. 709–719, 2023, https://doi.org/10.1108/RPJ-03-2022-0091.
[105] F. Fazal, F. J. Diaz Sanchez, M. Waqas, V. Koutsos, A. Callanan, and N. Radacsi, "A modified 3D printer as a hybrid bioprinting-electrospinning system for use in vascular tissue engineering applications," *Med. Eng. Phys.*, vol. 94, no. October 2020, pp. 52–60, 2021, https://doi.org/10.1016/j.medengphy.2021.06.005.
[106] E. L. Gill, X. Li, M. A. Birch, and Y. Y. S. Huang, "Multi-length scale bioprinting towards simulating microenvironmental cues," *Bio-Design Manuf.*, vol. 1, no. 2, pp. 77–88, 2018, https://doi.org/10.1007/s42242-018-0014-1.
[107] H. Wang, H. Al Shraida, and Y. Jin, "Predictive modeling for online in-plane shape deviation inspection and compensation of additive manufacturing," *Rapid Prototyp. J.*, vol. 30, no. 2, pp. 350–363, 2024, https://doi.org/10.1108/RPJ-03-2023-0076.
[108] N. Ferencik, R. Hudak, V. Rajtukova, M. Kohan, T. Breskovic, and J. Zivcak, "Overview article: Bioreactors designed for 3D bioprinted tissue and process parameters," in *2021 IEEE 19th World Symposium on Applied Machine Intelligence and Informatics (SAMI)*, IEEE, Jan. 2021, pp. 000131–000134. https://doi.org/10.1109/SAMI50585.2021.9378697.
[109] P. Sucosky, D. F. Osorio, J. B. Brown, and G. P. Neitzel, "Fluid mechanics of a spinner-flask bioreactor," *Biotechnol. Bioeng.*, vol. 85, no. 1, pp. 34–46, 2004, https://doi.org/10.1002/bit.10788.
[110] M. P. Storm et al., "Hollow fiber bioreactors for In Vivo-like mammalian tissue culture," *J. Vis. Exp.*, vol. 2016, no. 111, pp. 1–12, 2016, https://doi.org/10.3791/53431.
[111] B.-N. B. Nguyen, H. Ko, R. A. Moriarty, J. M. Etheridge, and J. P. Fisher, "Dynamic bioreactor culture of high volume engineered bone tissue," *Tissue Eng. Part A*, vol. 22, no. 3–4, pp. 263–271, 2016, https://doi.org/10.1089/ten.tea.2015.0395.
[112] J. Nie et al., "Vessel-on-a-chip with hydrogel-based microfluidics," *Small*, vol. 14, no. 45, pp. 1–14, 2018, https://doi.org/10.1002/smll.201802368.
[113] S. Zakhireh, A. Mesgari-Shadi, J. Barar, Y. Omidi, Y. Beygi-Khosrowshahi, and K. Adibkia, "Current researches in modular biofabrication: Tissue building blocks and bioreactors," *Korean J. Chem. Eng.*, vol. 41, no. 6, pp. 1567–1585, 2024, https://doi.org/10.1007/s11814-024-00059-8.

[114] L. A. Low, C. Mummery, B. R. Berridge, C. P. Austin, and D. A. Tagle, "Organs-on-chips: Into the next decade," *Nat. Rev. Drug Discov.*, vol. 20, no. 5, pp. 345–361, 2021, https://doi.org/10.1038/s41573-020-0079-3.

[115] R. Martin and D. Joung, "The promise and challenges of bioprinting in tissue engineering," *Micromachines*, vol. 15, no. 12, p. 1529, 2024, https://doi.org/10.3390/mi15121529.

[116] M. Kuzucu et al., "Extrusion-based 3D bioprinting of gradients of stiffness, cell density, and immobilized peptide using thermogelling hydrogels," *ACS Biomater. Sci. Eng.*, vol. 7, no. 6, pp. 2192–2197, 2021, https://doi.org/10.1021/acsbiomaterials.1c00183.

[117] T. Xia, W. Liu, and L. Yang, "A review of gradient stiffness hydrogels used in tissue engineering and regenerative medicine," *J. Biomed. Mater. Res.—Part A*, vol. 105, no. 6, pp. 1799–1812, 2017, https://doi.org/10.1002/jbm.a.36034.

[118] D. B. Kolesky, R. L. Truby, A. S. Gladman, T. A. Busbee, K. A. Homan, and J. A. Lewis, "3D bioprinting of vascularized, heterogeneous cell-laden tissue constructs," *Adv. Mater.*, vol. 26, no. 19, pp. 3124–3130, 2014, https://doi.org/10.1002/adma.201305506.

[119] Y. C. Hou et al., "Three-dimensional bioprinting of artificial blood vessel: Process, bioinks, and challenges," *Int. J. Bioprinting*, vol. 9, no. 4, 2023, https://doi.org/10.18063/ijb.740.

[120] D. Lu et al., "VEGF loading heparinized hyaluronic acid macroporous hydrogels for enhanced 3D endothelial cell migration and vascularization," *Biomater. Adv.*, vol. 167, no. June 2024, p. 214094, 2025, https://doi.org/10.1016/j.bioadv.2024.214094.

[121] N. Sarkar, S. Bhumiratana, L. Geris, I. Papantoniou, and W. L. Grayson, "Bioreactors for engineering patient-specific tissue grafts," *Nat. Rev. Bioeng.*, vol. 1, no. 5, pp. 361–377, 2023, https://doi.org/10.1038/s44222-023-00036-6.

[122] E. Kang, S. J. Shin, K. H. Lee, and S. H. Lee, "Novel PDMS cylindrical channels that generate coaxial flow, and application to fabrication of microfibers and particles," *Lab Chip*, vol. 10, no. 14, pp. 1856–1861, 2010, https://doi.org/10.1039/c002695f.

[123] E. Davoodi et al., "Extrusion and microfluidic-based bioprinting to fabricate biomimetic tissues and organs," *Adv. Mater. Technol.*, vol. 5, no. 8, 2020, https://doi.org/10.1002/admt.201901044.

[124] X. Li et al., "Inkjet bioprinting of biomaterials," *Chem. Rev.*, vol. 120, no. 19, pp. 10793–10833, 2020, https://doi.org/10.1021/acs.chemrev.0c00008.

[125] M. Kesti, P. Fisch, M. Pensalfini, E. Mazza, and M. Zenobi-Wong, "Guidelines for standardization of bioprinting: A systematic study of process parameters and their effect on bioprinted structures," *BioNanoMaterials*, vol. 17, no. 3–4, pp. 193–204, 2016, https://doi.org/10.1515/bnm-2016-0004.

[126] Y. Sun, Y. You, W. Jiang, B. Wang, Q. Wu, and K. Dai, "3D bioprinting dual-factor releasing and gradient-structured constructs ready to implant for anisotropic cartilage regeneration," *Sci. Adv.*, vol. 6, no. 37, 2020, https://doi.org/10.1126/sciadv.aay1422.

[127] M. Song, Y. Jang, S. J. Kim, and Y. Park, "Cyclic stretching induces maturation of human-induced pluripotent stem cell-derived cardiomyocytes through nuclear-mechanotransduction," *Tissue Eng. Regen. Med.*, vol. 19, no. 4, pp. 781–792, 2022, https://doi.org/10.1007/s13770-021-00427-z.

[128] D. Moura, R. F. Pereira, and I. C. Gonçalves, "Recent advances on bioprinting of hydrogels containing carbon materials," *Mater. Today Chem.*, vol. 23, 2022, https://doi.org/10.1016/j.mtchem.2021.100617.

[129] T. Mladenovska, P. F. Choong, G. G. Wallace, and C. D. O'connell, "The regulatory challenge of 3D bioprinting," *Regen. Med.*, vol. 18, no. 8, pp. 659–674, 2023, https://doi.org/10.2217/rme-2022-0194.

[130] S. V. Murphy, P. De Coppi, and A. Atala, "Opportunities and challenges of translational 3D bioprinting," *Nat. Biomed. Eng.*, vol. 4, no. 4, pp. 370–380, 2020, https://doi.org/10.1038/s41551-019-0471-7.

[131] L. M. Ricles, J. C. Coburn, M. Di Prima, and S. S. Oh, "Regulating 3D-printed medical products.," *Sci. Transl. Med.*, vol. 10, no. 461, 2018, https://doi.org/10.1126/scitranslmed.aan6521.

[132] X. Wang et al., "Progress in organ bioprinting for regenerative medicine—article," *Engineering*, vol. 42, pp. 121–142, 2024, https://doi.org/10.1016/j.eng.2024.04.023.

Chapter 6

3D Printing of MEMS Devices

Brahmansh Kaushik, Akshay Pathania, Prathamesh Rajan Korgaonkar, Kunal Singh, and S Anand Kumar

6.1 INTRODUCTION

Microelectromechanical systems (MEMS) have revolutionised the landscape of miniaturised devices, enabling seamless integration of sensing, actuation, and signal processing at the microscale. Over the past few decades, MEMS devices have become indispensable components in a diverse range of applications, including automotive safety systems, biomedical diagnostics, aerospace instrumentation, consumer electronics, and environmental monitoring systems [1, 2]. By enabling the manipulation of micro-scale features with remarkable precision, MEMS technology has facilitated the development of high-performance components, including accelerometers, gyroscopes, pressure sensors, and microfluidic chips [3, 4].

Traditionally, the fabrication of MEMS has relied heavily on well-established semiconductor manufacturing techniques, including photolithography, chemical vapour deposition, dry and wet etching, and thin-film deposition. While these methods have proven highly effective for producing planar and batch-fabricated devices, they impose inherent limitations on the geometric complexity, material diversity, and integration flexibility required for next-generation microsystems. The dependence on cleanroom environments, the significant costs associated with photomask-based workflows, and the restricted scope for rapid design modifications further constrain innovation, particularly in rapidly evolving application areas. In response to the growing demand for personalised, miniaturised, and multifunctional devices, especially in fields such as healthcare, environmental monitoring, and wearable technology, researchers have increasingly turned to alternative fabrication strategies. Among these, 3D printing, or additive manufacturing, has emerged as a transformative platform, offering enhanced design freedom, faster prototyping cycles, and greater cost-effectiveness, thereby addressing several key limitations of conventional MEMS fabrication [5, 6].

Additive manufacturing (AM), or 3D printing, is a layer-by-layer fabrication process where materials are precisely deposited or solidified to create complex three-dimensional structures. This method offers significant advantages over traditional MEMS fabrication by enabling intricate geometries, freeform designs, and multi-material integration without relying on photomasks or elaborate etching processes [7]. Various 3D printing techniques, such as inkjet printing, stereolithography, digital light processing, fused deposition modelling, selective laser sintering, and two-photon polymerisation, are increasingly being adopted for microscale applications. Each method offers distinct advantages in terms of resolution, material compatibility, and scalability. Among these, high-resolution techniques like two-photon polymerisation, inkjet printing, and aerosol jet printing have demonstrated particular success in fabricating MEMS devices with precise microscale features. Alongside these process innovations, the introduction of functional materials, including nanoparticle-laden resins, piezoelectric polymers, and conductive inks, has further expanded MEMS capabilities, supporting

DOI: 10.1201/9781003590774-6

their integration into energy-harvesting systems, microfluidic devices, implantable biomedical sensors, and flexible electronic platforms [8].

Given the rapidly expanding body of literature and experimental work in this domain, it is imperative to systematically explore and critically evaluate the progress of 3D printing technologies in MEMS device fabrication. This chapter aims to provide a comprehensive overview of the evolution of MEMS devices, outline the fundamentals of MEMS operation and fabrication, and delve into recent technological advancements enabled by 3D printing. Special emphasis is placed on emerging applications, material innovations, and the practical challenges of integrating additive manufacturing into established MEMS fabrication workflows. By synthesising insights from recent research, this chapter seeks to elucidate the transformative role of 3D printing in shaping the next generation of MEMS devices.

6.2 FUNDAMENTALS OF MEMS DEVICES

Microelectromechanical systems (MEMS) are highly integrated microscale devices that combine mechanical structures with electrical functionalities to perform sensing, actuation, and signal processing at small dimensions. They are widely employed across industries, including biomedical diagnostics, industrial automation, aerospace systems, and portable electronics, due to their compact size and multifunctionality. The primary strength of MEMS lies in their ability to convert various physical stimuli, such as mechanical displacement, acceleration, pressure, or chemical interactions, into precise electrical signals. This integration facilitates the development of complex devices capable of real-time interaction with their environment within a minimal footprint [7].

MEMS devices operate using well-defined transduction mechanisms. Capacitive transduction detects changes in capacitance due to the movement of microstructures, while piezoresistive systems exploit changes in resistance under mechanical strain. Piezoelectric mechanisms generate electrical outputs through the deformation of specific materials, making them effective in energy harvesting and dynamic sensing. Thermal MEMS use temperature changes to create actuation or sensing effects. These mechanisms are implemented using mechanical microcomponents, such as diaphragms, cantilevers, and springs, which are often integrated with signal-processing electronics to form efficient microsystems [7].

The fabrication of MEMS has traditionally relied on established semiconductor manufacturing processes such as photolithography, chemical vapour deposition (CVD), and etching. While these methods enable high-precision microsystems, recent advancements in additive manufacturing have expanded design capabilities by allowing non-planar geometries and multi-material integration. Hybrid processes that merge 3D printing with conventional micromachining have been particularly effective in realising complex MEMS devices for emerging applications such as flexible biosensing, autonomous micro-robotics, and environmental monitoring, where performance metrics such as sensitivity, resolution, and reliability are crucial [9].

6.3 OVERVIEW OF 3D PRINTING TECHNOLOGY

3D printing, or additive manufacturing (AM), has emerged as a transformative approach in MEMS fabrication, enabling rapid prototyping, design flexibility, and the integration of complex geometries. Unlike traditional planar microfabrication, AM facilitates out-of-plane structures and non-silicon materials, broadening the functional scope of MEMS devices. The schematic diagrams of laser-based additive manufacturing processes are shown in **Figure 6.1**.

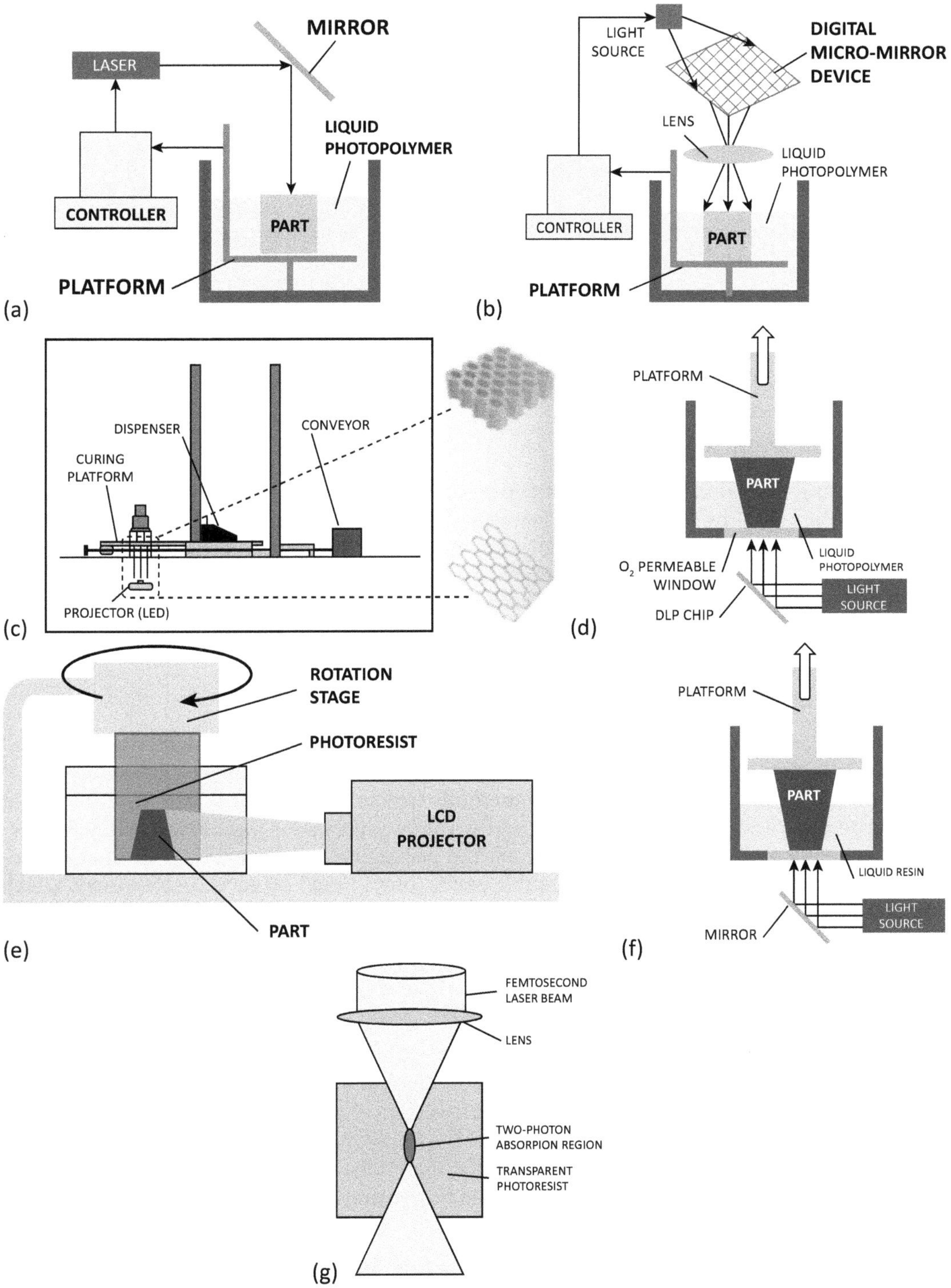

Figure 6.1 Other laser-based processes for AM of microstructures: (a) Scanning micro-stereolithography (μ-SLA), (b) Projection micro-stereolithography (μ-SLA), (c) Mask-image-projection stereolithography with 3D mask (MIPSLA) [3], (d) Continuous liquid interphase printing (CLIP), (e) Computed axial lithography (CAL), (f) Digital light processing (DLP), (g) Two-photon polymerisation (TPP) or direct laser writing (DLW) [7].

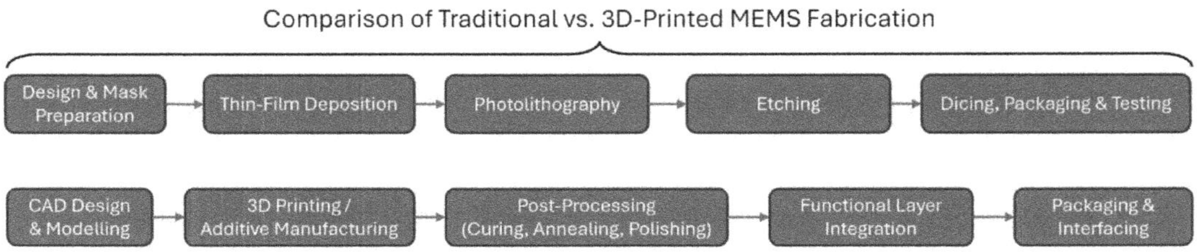

Figure 6.2 Comparison of MEMS fabrication workflows: top shows the traditional process, while bottom illustrates the 3D printing approach.

Further, a Comparison of MEMS fabrication workflows: The top shows the traditional process, while the bottom illustrates the 3D printing approach, as shown in **Figure 6.2.**

6.3.1 Inkjet Printing

Inkjet printing has gained significant attention due to its compatibility with a broad range of functional inks and low-temperature processing. This method is frequently employed for fabricating microfluidic devices, microelectrodes, and energy harvesters, achieving feature sizes of 16 to 50 micrometres. Inkjet printing primarily utilises conductive nanoparticle inks, such as silver nanoparticle suspensions, polymer-based dielectric inks, and piezoelectric functional inks to produce microscale functional components. [10, 11].

6.3.2 Stereolithography (SLA) and DLP

Stereolithography (SLA) and digital light processing (DLP) are photopolymerisation-based techniques widely used in MEMS fabrication for their ability to achieve high-dimensional accuracy and complex geometries. These methods typically offer lateral resolutions of 10 to 50 micrometres and are particularly suited for the production of mechanical MEMS structures, including accelerometers, microvalves, and packaging components. The primary materials used in SLA and DLP are photocurable resins, often based on acrylate chemistry, which can be further functionalised with nanoparticles to introduce specific mechanical or electrical properties [12].

6.3.3 Two-Photon Polymerisation (TPP)

Two-photon polymerisation (TPP) provides the highest resolution among 3D printing techniques, achieving lateral feature sizes below 200 nanometres. TPP is especially valuable for applications requiring ultra-fine structures, such as optical MEMS, micromechanical springs, and microfluidic mixers. The process utilises photopolymerisable organic resins and can incorporate semiconducting or biocompatible materials through resin functionalisation, enabling applications in flexible electronics and biosensing platforms.

6.3.4 Aerosol Jet and Electrohydrodynamic Jet Printing

Aerosol jet and electrohydrodynamic jet printing have emerged as flexible techniques for printing high-resolution features on complex, non-planar MEMS substrates. These methods enable the conformal deposition of nanoparticle-loaded inks with feature sizes approaching 10 micrometres. Common materials include metal nanoparticles such as silver and copper, dielectric materials, and polymeric inks, enabling these techniques to be applied to the fabrication of energy-harvesting devices, flexible interconnects, and printed electronic circuits [13].

6.3.5 Fused Deposition Modelling (FDM)

Fused deposition modelling (FDM), though limited by a relatively coarse resolution of approximately 100 micrometres, remains useful for structural MEMS components and rapid prototyping applications. FDM is valued for its accessibility and cost-effectiveness, utilising thermoplastic materials such as PLA and ABS, as well as advanced composite filaments loaded with conductive fillers, piezoelectric particles, or metal oxides like copper oxide (CuO). This versatility allows FDM to contribute to both structural and semi-functional MEMS prototyping, particularly in educational and exploratory research contexts [14].

6.4 RECENT TECHNOLOGICAL ADVANCEMENTS IN 3D PRINTING FOR MEMS

Recent advancements in 3D printing have significantly expanded the possibilities for fabricating complex, high-resolution MEMS devices, pushing beyond conventional material and resolution limitations. These developments span novel materials, hybrid fabrication strategies, and process innovations.

6.4.1 Sub-Micron Resolution Techniques

Two-photon polymerisation (TPP) and multiphoton lithography are among the most advanced high-resolution 3D printing techniques used in MEMS fabrication, capable of producing intricate three-dimensional microstructures with lateral feature sizes below 200 nanometres. These techniques enable the direct fabrication of mechanically and optically functional components such as springs, cantilevers, microfluidic mixers, and optical waveguides with exceptional precision. Their key advantage lies in the ability to create freeform geometries without the limitations of planar processing, making them ideal for applications in flexible electronics, bioMEMS, and high-frequency microsystems. By incorporating doped photopolymers, these methods also enable the integration of functional properties, such as electrical conductivity or optical activity, into sub-micron-scale devices, positioning them as powerful tools for next-generation microscale systems [15, 16].

6.4.2 Advanced Material Systems

Advances in functional material formulations have significantly expanded the capabilities of 3D-printed MEMS devices. Nanoparticle-infused inks containing materials such as graphene, Fe_3O_4, and CuO have demonstrated improved performance in microelectrodes, energy harvesters, and microsensors. These materials contribute enhanced electrical conductivity, magnetic response, and thermal stability, making them highly suitable for multifunctional MEMS applications. For instance, rGO-Fe_3O_4 composite inks have shown promising electrochemical properties for organic semiconductor applications, while CuO-loaded thermoplastic composites have been effectively utilised in printing semiconducting MEMS components with stable functional properties [17, 18].

6.4.3 Integration of Structural and Functional Elements

To overcome the inherent limitations of resolution and material functionality in standalone AM processes, researchers have turned to hybrid manufacturing. This approach combines 3D printing with traditional MEMS techniques such as photolithography, etching, and mould replication. For instance, in the work by Camposeo et al., CAD/CAM-enabled systems were employed to create scalable hybrid architectures using multiple toolpaths for high-resolution MEMS fabrication. Another study describes the fabrication of 3D carbon MEMS (C-MEMS) using patterned photoresist followed by pyrolysis to create conductive carbon microstructures with complex geometries [9, 19].

6.4.4 Hybrid Manufacturing Strategies

Hybrid fabrication approaches that combine additive manufacturing with subtractive techniques or soft lithography are gaining increasing attention. Notable examples include the use of 3D-printed sacrificial templates for moulding microfluidic channels in polydimethylsiloxane (PDMS) and the integration of printed optical elements with silicon-based substrates. These strategies offer a practical pathway to bridge the design flexibility of additive manufacturing with the high precision and material compatibility of conventional MEMS fabrication processes.

6.4.5 Towards Flexible and Autonomous Microsystems

A growing frontier in MEMS research is the creation of autonomous, flexible, and wearable microsystems using 3D printing. Researchers have reported the development of a 3D-printable self-powered piezoelectric smart stent designed for wireless endoleak sensing, enabling real-time health monitoring without batteries. Similarly, direct laser 3D printing has been used to fabricate organic semiconductor microdevices for flexible biosensing applications, combining structural adaptability with electronic functionality [20, 21].

6.5 APPLICATIONS OF 3D PRINTING IN MEMS

The integration of 3D printing with MEMS technology has opened a vast array of applications across biomedical systems, energy harvesting, environmental monitoring, microfluidics, and wearable devices. **Figure 6.3** The additive nature of 3D printing allows complex geometries, rapid iteration, and material heterogeneity, making it ideal for fabricating multifunctional, application-specific MEMS structures.

6.5.1 MEMS Devices for Biomedical Applications

3D printing has enabled innovative advances in the fabrication of MEMS devices for biomedical applications, particularly by enabling the production of highly customised geometries and the integration of functional materials suitable for biological environments. A prominent example involves the development of self-powered MEMS devices, such as piezoelectric-based smart stents, capable of wireless physiological monitoring without reliance on external power sources. These devices leverage 3D-printed biocompatible materials to create compact, minimally invasive platforms for patient care.

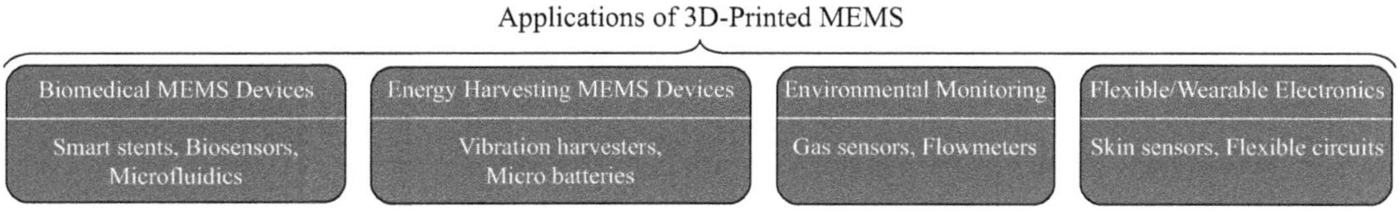

Figure 6.3 Application domains of 3D-printed MEMS devices, including biomedical systems, energy harvesting, environmental monitoring, and flexible/wearable electronics.

Further, additive manufacturing has enabled the development of miniaturised biosensors for real-time diagnostics. Electrochemical sensing systems fabricated via precise 3D printing techniques offer high sensitivity and portability, making them promising candidates for point-of-care testing in clinical settings.

In addition to healthcare applications, 3D-printed MEMS devices have been utilised in biological research, where the integration of microfluidic architectures enables detailed observation and manipulation of biological processes. Such systems enable direct studies of cellular interactions and tissue mechanics, broadening the scope of MEMS technology in life sciences.

6.5.2 Energy Harvesting and Integrated Power Supplies

Energy harvesting is a strategic application area for MEMS, especially for autonomous and wireless sensor systems. Recent advances in 3D printing have enabled the development of vibrational electromagnetic energy harvesters, in which both the electromagnetic coil and the mechanical suspension are fabricated via inkjet-based AM. Such devices are compact and capable of generating energy from ambient mechanical motion, making them suited for powering low-power MEMS sensors in IoT environments [10].

Additionally, microbattery systems fabricated directly on flexible or polymer substrates provide localised energy storage within MEMS packages. These systems use 3D-printed interdigitated electrode architectures and novel hybrid chemistries (e.g., Zn-based and Fe-based formulations) to provide miniaturised power sources for standalone operation [12, 18].

6.5.3 Microfluidics and Lab-on-Chip Platforms

3D printing has significantly advanced the development of microfluidic MEMS devices by enabling the fabrication of intricate fluidic pathways, integrated mixers, valves, and thermal control elements within compact microsystems. The ability to directly print complex fluidic architectures reduces the need for multi-step assembly processes and facilitates rapid prototyping of application-specific designs. Recent developments have demonstrated the fabrication of self-regulating microfluidic platforms incorporating embedded heating elements, allowing precise temperature control and faster thermal responses, making them suitable for biochemical assays and on-chip diagnostic applications [22].

Additionally, additive manufacturing has expanded the functional scope of microfluidic MEMS by integrating sensing capabilities. Venturi-based microflowmetres fabricated by inkjet printing have shown high sensitivity and low pressure drop, making them suitable for biomedical and industrial flow monitoring [23]. Furthermore, layer-by-layer printing of silicon and polymer microstructures has facilitated the integration of fluidic channels with electrical pathways, supporting compact analytical platforms for chemical detection, health diagnostics, and environmental sensing [6].

6.5.4 Flexible, Wearable, and Conformal MEMS

Flexible and wearable MEMS devices are increasingly benefiting from 3D printing technologies, particularly through the use of materials with inherent mechanical compliance and the ability to conform to non-planar surfaces. These advancements have enabled the fabrication of skin-compatible biosensors and flexible electronic circuits that integrate seamlessly with human tissue or dynamic mechanical structures. High-resolution additive manufacturing methods, such as direct laser writing and multiphoton lithography, have been successfully applied to produce organic semiconductor devices suitable for real-time physiological monitoring and adaptable wearable electronics [8].

Beyond sensing, additive manufacturing has also advanced the production of flexible MEMS actuators, including microgrippers and droplet dispensers. Devices fabricated from photocurable polymer resins have demonstrated controlled mechanical actuation at the microscale, offering practical applications in soft robotics and microfluidic manipulation [24]. The convergence of mechanical flexibility, miniaturisation, and material functionality underscores the expanding role of 3D printing in the development of next-generation wearable MEMS technologies.

6.5.5 Structural Health and Environmental Monitoring

Additive manufacturing has expanded the applicability of MEMS devices for structural health monitoring and environmental sensing by enabling the integration of compact sensing architectures with customised geometries. Inkjet-printed vibrational energy-harvesting devices incorporating embedded microsensors have been utilised for real-time monitoring of mechanical vibrations in infrastructure systems such as bridges, industrial machinery, and building frameworks. These devices leverage the design flexibility of inkjet printing to produce sensitive, low-profile energy-harvesting modules suitable for distributed monitoring applications [10].

In parallel, gas-sensing MEMS devices fabricated via 3D printing have shown promise in environmental monitoring and industrial safety diagnostics. In particular, sensors constructed from copper oxide (CuO)-infused thermoplastic composites via fused deposition modelling (FDM) exhibit improved gas adsorption characteristics, benefiting from the increased surface area and porosity inherent to additive manufacturing methods. These developments highlight the growing role of 3D-printed MEMS in enabling low-cost, scalable, and adaptable sensor systems for critical environmental and structural monitoring tasks [14].

6.6 CHALLENGES AND LIMITATIONS

Additive manufacturing (AM) of microelectromechanical systems (MEMS) offers several advantages, including greater design flexibility and rapid prototyping. However, despite its transformative potential, integrating 3D printing into MEMS fabrication workflows is constrained by several technical and practical limitations. These challenges merit rigorous academic attention, particularly as the field seeks to transition from proof-of-concept prototypes to fully functional, reliable, and scalable MEMS devices.

6.6.1 Resolution Limitations and Feature Definition

A central limitation in the additive fabrication of MEMS is the resolution of prevalent 3D printing techniques. Conventional AM processes such as fused deposition modelling (FDM), inkjet printing, and stereolithography (SLA) often fail to achieve the sub-micron resolution demanded by MEMS components. While multiphoton lithography and two-photon polymerisation offer higher precision, their throughput is limited, and they remain cost-prohibitive for broad applications. This trade-off between resolution and scalability continues to constrain the feasibility of 3D printing for fabricating mechanically and electrically functional microdevices.

6.6.2 Material Incompatibilities

The functional heterogeneity of MEMS devices, which must incorporate conductors, insulators, semiconductors, and piezoelectric materials, poses a significant challenge for selecting suitable materials for 3D printing. Although research has advanced the development of polymer nanocomposite inks and hybrid resins, persistent issues such as incomplete curing, uneven filler distribution, and limited electrical conductivity continue to constrain device performance. In addition, the scarcity of printable materials that offer biocompatibility, high-temperature resistance, or chemical inertness further limits the applicability of 3D-printed MEMS in biomedical devices and under harsh environmental conditions.

6.6.3 Surface Quality and Dimensional Fidelity

MEMS devices are highly sensitive to surface topography and geometric precision. Additively manufactured surfaces frequently exhibit roughness due to the discrete layering process, which can adversely affect fluidic resistance, optical alignment, or electrical interfacing. Although post-processing techniques such as polishing or coating can mitigate these issues, they introduce additional complexity and potential dimensional deviation. This limitation can be solved with novel process parameters, in situ monitoring and smoothing techniques.

6.6.4 Integration with Traditional Microsystems Technologies

The hybrid integration of 3D-printed structures with traditional MEMS platforms, which are commonly fabricated on substrates such as silicon, glass, or quartz, poses a range of complex challenges. Differences in thermal expansion coefficients, bonding compatibility, and lithographic alignment can lead to interface failures and degrade overall device performance. Additionally, the integration of printed materials with CMOS-compatible processes remains relatively underexplored, contributing to a disconnect between additive microfabrication approaches and established microelectronics manufacturing infrastructure.

6.6.5 Reliability, Repeatability, and Metrology

Ensuring device reliability and manufacturing repeatability is fundamental for MEMS applications in safety-critical domains. However, AM techniques are often sensitive to environmental variables such as temperature, humidity, and resin ageing. Additionally, layer-wise deposition may introduce anisotropic mechanical properties, porosity, or internal stresses. The lack of standardised metrology tools to assess microscale defects and performance further complicates the validation and certification of 3D-printed MEMS components.

6.6.6 Design Standardisation and CAD Toolchains

The absence of standardised design guidelines, material libraries, and process models for MEMS-specific 3D printing inhibits streamlined development. Unlike the well-established MEMS foundry model, AM lacks universally accepted design rules for feature size, interconnect tolerances, and multi-material printing. Moreover, current CAD tools are not inherently optimised for the anisotropic and multi-physics behaviour of printed microstructures, necessitating further development of physics-based simulation platforms.

6.6.7 Economic and Scalability Constraints

While additive manufacturing is often lauded for its low-cost prototyping capabilities, its application to MEMS remains economically constrained. High-resolution systems such as TPP or electrohydrodynamic jet printing entail significant capital and operational costs. Moreover, batch processing of microscale components via AM has yet to achieve the throughput and cost-efficiency of photolithographic or wafer-level manufacturing processes.

6.7 FUTURE TRENDS AND RESEARCH DIRECTIONS

Emerging research on 3D-printed MEMS is advancing to enhance both functional capabilities and manufacturing efficiency. Technological advances in high-resolution printing, advanced materials development, and hybrid fabrication methods are unlocking new opportunities for creating complex, integrated microsystems. Alongside these advancements, there is growing recognition of the need to address scalability, reliability, and process standardisation challenges. The following points highlight the most prominent trends and research directions likely to influence the next generation of 3D-printed MEMS devices.

- **Advancement of High-Resolution Additive Manufacturing Techniques:** Future research will focus on enhancing the resolution and scalability of additive manufacturing processes such as two-photon polymerisation (TPP), micro-stereolithography, and direct ink writing (DIW). This includes improving printing speeds, optimising optical systems, and enabling batch production of complex three-dimensional MEMS components with sub-micron precision for practical industrial applications.
- **Development of Advanced Multi-Functional Materials:** Expanding the library of printable materials remains a key direction, with an emphasis on incorporating advanced functional properties. Graphene-based inks, iron oxide (Fe_3O_4) nanocomposites, copper oxide (CuO) reinforced thermoplastics, and piezoelectric polymer blends will enable MEMS devices with enhanced conductivity, magnetic responsiveness, biocompatibility, and mechanical flexibility suitable for diverse microsystem applications.
- **Hybrid Manufacturing Strategies Combining 3D Printing with Traditional Processes:** A promising trend involves the integration of additive manufacturing with established MEMS fabrication techniques like photolithography and etching. This hybrid approach aims to combine the geometric freedom of 3D printing with the precision of silicon-based micromachining, enabling highly complex, multifunctional MEMS devices, such as accelerometers, microfluidic chips, and flexible sensors [25].
- **Expansion into Flexible and Wearable MEMS Applications:** The proliferation of soft, flexible materials in additive manufacturing is expected to accelerate the development of MEMS devices tailored for wearable technology, soft robotics, and implantable

biomedical systems. Future efforts will address the challenges of robust encapsulation, high-density interconnects, and stretchable sensor integration for real-time physiological monitoring and therapeutic applications.

- **Emergence of Autonomous, Self-Powered MEMS Systems:** Integrating energy harvesting capabilities directly into MEMS devices through 3D printing will be a significant research focus. Self-powered systems incorporating vibrational, piezoelectric, and thermoelectric mechanisms are expected to enable wireless, maintenance-free MEMS suitable for Internet of Things (IoT) applications and distributed sensing networks.
- **Incorporation of Artificial Intelligence for Design and Process Optimisation:** Machine learning and AI-driven design tools are anticipated to revolutionise MEMS fabrication by optimising geometry, minimising material waste, predicting failure points, and accelerating design iterations. AI will also contribute to real-time process monitoring and adaptive control during 3D printing, improving device consistency and performance (**Figure 6.4**).
- **Focus on Standardisation, Reliability Testing, and Industrial Scalability:** Establishing standardised protocols for performance validation, mechanical reliability, and long-term stability is essential for transitioning 3D-printed MEMS from research prototypes to commercial products. Future work will address durability testing, especially for biomedical and aerospace environments, alongside efforts to develop scalable, cost-effective production workflows.

6.8 CONCLUSION

The integration of 3D printing into MEMS fabrication represents a pivotal shift in microsystem design and manufacturing. Moving beyond the constraints of conventional two-dimensional microfabrication, 3D printing offers the ability to realise intricate geometries, complex three-dimensional architectures, and customised device configurations. This capability addresses longstanding limitations in the field, particularly those related to rigid design constraints, limited material versatility, and the high costs associated with traditional cleanroom-dependent fabrication processes.

The emergence of high-resolution additive manufacturing techniques, such as two-photon polymerisation, micro-stereolithography, and inkjet printing, has enabled the direct fabrication of MEMS devices with enhanced functional integration and design freedom. These methods facilitate the incorporation of diverse material systems, including conductive, piezoelectric, and biocompatible composites, supporting the development of advanced MEMS devices for applications ranging from biomedical sensors and flexible electronics

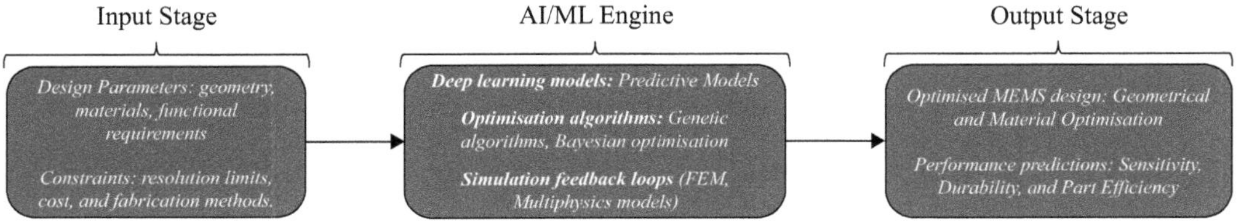

Figure 6.4 Workflow of AI-driven MEMS design optimisation, integrating design parameters and constraints with predictive models, optimisation algorithms, and simulations to generate optimised device designs and performance predictions.

to microfluidics, energy harvesters, and autonomous microsystems. Such advancements are particularly impactful in emerging sectors like personalised medicine, soft robotics, and the Internet of Things, where device adaptability and rapid customisation are critical.

Despite these technological strides, the field of 3D-printed MEMS remains in a dynamic phase of development, with ongoing research required to overcome challenges related to resolution scalability, mechanical reliability, multi-material printing fidelity, and process standardisation. Addressing these gaps through continued interdisciplinary collaboration and materials innovation will be essential to fully realise the potential of 3D-printed MEMS devices. Looking forward, these advancements promise to expand the scope of microsystem applications and redefine the accessibility of complex, multifunctional microdevices in both research and industry.

REFERENCES

[1] Blachowicz, Tomasz, and Andrea Ehrmann. 2020. "3D Printed MEMS Technology—Recent Developments and Applications." *Micromachines* 11 (4): 434. https://doi.org/10.3390/mi11040434.

[2] French, Paddy J., Gijs J. M. Krijnen, Sten Vollebregt, and Massimo Mastrangeli. 2022. "Technology Development for MEMS: A Tutorial." *IEEE Sensors Journal* 22 (11): 10106–10125. https://doi.org/10.1109/JSEN.2021.3104715.

[3] Pagliano, Simone, David E. Marschner, Damien Maillard, et al. 2022. "Micro 3D Printing of a Functional MEMS Accelerometer." *Microsystems & Nanoengineering* 8 (1): 105. https://doi.org/10.1038/s41378-022-00440-9.

[4] Khorsandi, Danial, Mehrab Nodehi, Tayyab Waqar, et al. 2021. "Manufacturing of Microfluidic Sensors Utilizing 3D Printing Technologies: A Production System." *Journal of Nanomaterials* 2021 (August): 1–16. https://doi.org/10.1155/2021/5537074.

[5] Gendreau, D., A. Mohand-Ousaid, P. Rougeot and M. Rakotondrabe. "3D-Printing: A Promising Technology to Design Three-Dimensional Microsystems." In *2016 International Conference on Manipulation, Automation and Robotics at Small Scales (MARSS)*, Paris, France, 2016, pp. 1–5, https://doi.org/10.1109/MARSS.2016.7561724.

[6] Fischer, Andreas C., Kristinn B. Gylfason, Lyubov M. Belova, et al. 2012. "Layer-by-Layer 3D Printing of Si Micro- and Nanostructures by Si Deposition, Ion Implantation and Selective Si Etching." In *2012 12th IEEE International Conference on Nanotechnology (IEEE-NANO)*, August, 1–4. https://doi.org/10.1109/nano.2012.6322048.

[7] De Pasquale, Giorgio. 2021. "Additive Manufacturing of Micro-Electro-Mechanical Systems (MEMS)." *Micromachines* 12 (11): 1374. https://doi.org/10.3390/mi12111374.

[8] Dadras-Toussi, Omid, Milad Khorrami, Anto Sam Crosslee Louis Sam Titus, Sheereen Majd, Chandra Mohan, and Mohammad Reza Abidian. 2022. "Multiphoton Lithography of Organic Semiconductor Devices for 3D Printing of Flexible Electronic Circuits, Biosensors, and Bioelectronics." *Advanced Materials* 34 (30). https://doi.org/10.1002/adma.202200512.

[9] Kawa, Bartosz, Krzysztof Śliwa, Vincent Lee, Qiongfeng Shi, and Rafał Walczak. 2020. "Inkjet 3D Printed MEMS Vibrational Electromagnetic Energy Harvester." *Energies* 13 (11): 2800. https://doi.org/10.3390/en13112800.

[10] Walczak, Rafał, Bartosz Kawa, and Krzysztof Adamski. 2019. "Inkjet 3D Printed Microfluidic Device for Growing Seed Root and Stalk Mechanical Characterization." *Sensors and Actuators A: Physical* 297 (October): 111557. https://doi.org/10.1016/j.sna.2019.111557.

[11] Cohen, E., S. Menkin, M. Lifshits, et al. 2018. "Novel Rechargeable 3D-Microbatteries on 3D-Printed-Polymer Substrates: Feasibility Study." *Electrochimica Acta* 265 (March): 690–701. https://doi.org/10.1016/j.electacta.2018.01.197.

[12] Seifert, Tobias, Mario Baum, Frank Roscher, Maik Wiemer, and Thomas Gessner. 2015. "Aerosol Jet Printing of Nano Particle Based Electrical Chip Interconnects." *Materials Today: Proceedings* 2 (8): 4262–4271. https://doi.org/10.1016/j.matpr.2015.09.012.

[13] Salea, Ahamad, Rat Prathumwan, Jedsada Junpha, and Kittitat Subannajui. 2017. "Metal Oxide Semiconductor 3D Printing: Preparation of Copper(ii) Oxide by Fused Deposition Modelling for Multi-Functional Semiconducting Applications." *Journal of Materials Chemistry C* 5 (19): 4614–4620. https://doi.org/10.1039/c7tc00990a.
[14] Pawar, Amol Ashok, Shira Halivni, Nir Waiskopf, et al. 2017. "Rapid Three-Dimensional Printing in Water Using Semiconductor–Metal Hybrid Nanoparticles as Photoinitiators." *Nano Letters* 17 (7): 4497–4501. https://doi.org/10.1021/acs.nanolett.7b01870.
[15] Pawar, Amol Ashok, Shira Halivni, Nir Waiskopf, et al. 2017. "Rapid Three-Dimensional Printing in Water Using Semiconductor–Metal Hybrid Nanoparticles as Photoinitiators." *Nano Letters* 17 (7): 4497–4501. https://doi.org/10.1021/acs.nanolett.7b01870.
[16] Fares, Mohammad M., and Samah K. Radaydeh. 2023. "3D Printing Inks of rGO/Fe3O4@Polyacrylonitrile as Organic Semiconductors." *Synthetic Metals* 297 (August): 117418. https://doi.org/10.1016/j.synthmet.2023.117418.
[17] Ferrari, Stefania, Melanie Loveridge, Shane D. Beattie, Marcus Jahn, Richard J. Dashwood, and Rohit Bhagat. 2015. "Latest Advances in the Manufacturing of 3D Rechargeable Lithium Microbatteries." *Journal of Power Sources* 286 (July): 25–46. https://doi.org/10.1016/j.jpowsour.2015.03.133.
[18] Yoon, Hae-Sung, Hyun-Taek Lee, Ki-Hwan Jang, et al. 2017. "CAD/CAM for Scalable Nanomanufacturing: A Network-Based System for Hybrid 3D Printing." *Microsystems & Nanoengineering* 3 (1). https://doi.org/10.1038/micronano.2017.72.
[19] Mineta, Takashi, Yusuke Masuda, and Jinseo Hong. 2018. "Fabrication of a 3D Nano-Printing Device." In *Nano-, Bio-, Info-Tech Sensors, and 3D Systems II*, edited by Vijay K. Varadan. SPIE. https://doi.org/10.1117/12.2317917.
[20] Islam, Sayemul, Jun Ying Tan, Trisha Das Mou, et al. 2024. "3D-Printable Self-Powered Piezoelectric Smart Stent for Wireless Endoleaks Sensing." In *2024 IEEE 37th International Conference on Micro Electro Mechanical Systems (MEMS)*, January 21, 789–792. https://doi.org/10.1109/mems58180.2024.10439508.
[21] Islam, Sayemul, Jun Ying Tan, Trisha Das Mou, et al. 2024. "3D-Printable Self-Powered Piezoelectric Smart Stent for Wireless Endoleaks Sensing." In *2024 IEEE 37th International Conference on Micro Electro Mechanical Systems (MEMS)*, January 21, 789–792. https://doi.org/10.1109/mems58180.2024.10439508.
[22] Cañada, Jorge, and Luis F. Velásquez-García. 2023. "Monolithically 3D-Printed, Self-Heating Microfluidics." In *2023 IEEE 22nd International Conference on Micro and Nanotechnology for Power Generation and Energy Conversion Applications (PowerMEMS)*, December 11, 210–13. https://doi.org/10.1109/powermems59329.2023.10417177.
[23] Courson, Rémi, Oleksii Bratash, Ali Maziz, et al. 2023. "Rapid Prototyping of a Polymer MEMS Droplet Dispenser by Laser-Assisted 3D Printing." *Microsystems & Nanoengineering* 9 (1). https://doi.org/10.1038/s41378-023-00559-3.
[24] Adamski, Krzysztof, Bartosz Kawa, and Rafal Walczak. 2018. "Inkjet 3D Printed Venturi Microflowmeter." In *2018 XV International Scientific Conference on Optoelectronic and Electronic Sensors (COE)*, June, 1–3. https://doi.org/10.1109/coe.2018.8435163.
[25] Liu, Guandong, Changhai Wang, Kexin Wang, Zhili Jia, Ruiqi Luo, and Wei Ma. 2023. "A Full 3D Printing Method for Monolithic Integration of an Accelerometer and a Force Sensor." In *2023 IEEE 36th International Conference on Micro Electro Mechanical Systems (MEMS)*, January 15, 598–601. https://doi.org/10.1109/mems49605.2023.10052173.

Chapter 7

Additive Manufacturing in the Energy Sector

Advancing Efficiency and Sustainability

Shajahan Shaik, Bhargavi Rani Anne, and VVSS Srikanth

7.1 INTRODUCTION

The rising global energy demand presents significant challenges for sustainability, efficiency, and cost-effectiveness in energy production. Between 2018 and 2050, global energy consumption is projected to increase by nearly 50%, with industries such as manufacturing, mining, and refining experiencing over a 30% rise in energy usage. This increasing demand conflicts with the urgent need to reduce carbon dioxide emissions by half within the next few decades, as emphasised in the Intergovernmental Panel on Climate Change (IPCC) report on global warming. Addressing these opposing demands requires innovative advanced manufacturing solutions, particularly those that improve efficiency, optimise material utilisation, and enable cleaner energy technologies.

Additive Manufacturing (AM), commonly known as 3D printing, has emerged as a transformative manufacturing approach, enabling the layer-by-layer fabrication of components directly from digital models. Unlike traditional methods such as casting, forging, and machining, which often result in significant material waste and high production costs, AM eliminates many of these constraints by providing greater design freedom, customisation, and material efficiency. Studies suggest that widespread adoption of AM could reduce the global energy demand by up to 27%, making it a crucial tool for achieving sustainability goals.

In the energy sector, AM technologies are being rapidly integrated into power generation, oil and gas, hydrogen production, energy storage, and nuclear energy applications (**Figure 7.1**). These technologies enable the production of high-performance turbine blades, heat exchangers, fuel cells, and advanced energy storage devices, thereby enhancing operational efficiency and durability. For instance, in nuclear power plants, AM enables on-site fabrication of reactor components, reducing lead times and enhancing performance. The oil and gas industry benefits from AM's ability to create corrosion-resistant, high-strength parts for drilling and extraction. Additionally, utilisation of AM in hydrogen production and energy storage is driving advancements in fuel cells, batteries, and electrochemical devices, paving the way for more efficient and cost-effective renewable energy solutions.

This chapter explores key AM technologies, including Powder Bed Fusion (PBF), Directed Energy Deposition (DED), and Material Extrusion, and their relevance across various energy domains. It also highlights case studies that demonstrate the transformative impact of AM in reducing costs, improving performance, and advancing sustainability across the energy sector.

7.1.1 Importance of AM in Energy

As global energy demands rise, the shift towards sustainable solutions becomes increasingly critical. AM addresses this need by promoting efficiency and sustainability across energy systems. The ability to minimise material wastage, reduce production costs, and fabricate

DOI: 10.1201/9781003590774-7

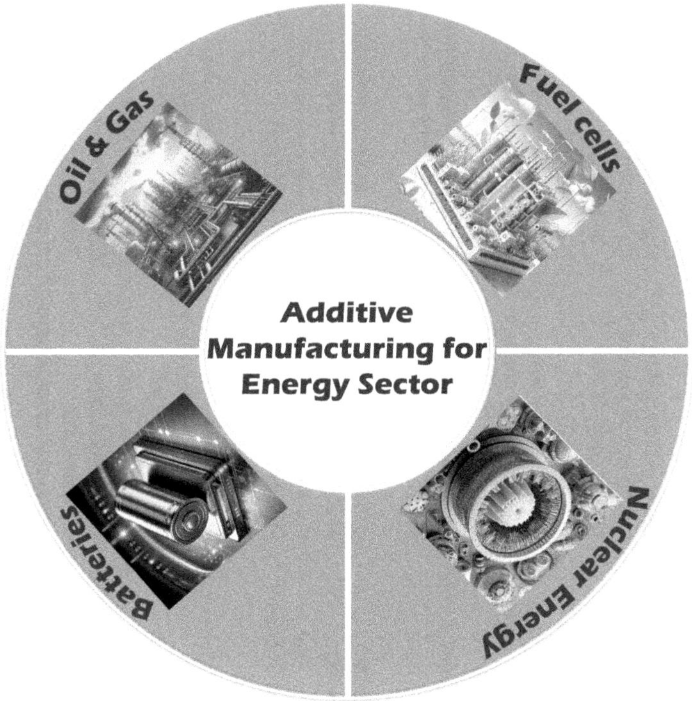

Figure 7.1 Overview of AM technologies used for various energy sectors.

complex components aligns with the industrial goals of optimising resource utilisation and minimising environmental impact.

In renewable energy systems, AM enables the production of lightweight and durable components for wind turbines, solar panels, and energy storage devices. For instance, AM technologies have been instrumental in improving the energy density and lifecycle of batteries (1, 2), making them more viable for large-scale applications. In the oil and gas sector, AM has facilitated the production of customised downhole tools and valves, enhanced operational efficiency and reduced downtime (3).

The role of AM in nuclear energy is equally significant, as it enables the fabrication of intricate components for advanced reactors and accelerates the deployment of critical systems (4). By leveraging AM capabilities, the energy sector can transition to cleaner, more efficient technologies while addressing the challenges posed by climate change.

7.2 OVERVIEW OF AM TECHNOLOGIES FOR THE ENERGY SECTOR

Table 7.1 summarises the additive manufacturing technologies and their applications in various energy sectors. Among the AM methods, the most impactful are Powder Bed Fusion (PBF), Directed Energy Deposition (DED), and Material Extrusion, each tailored to address specific challenges and opportunities within the energy sector. These technologies enable the creation of high-performance components, driving advancements in renewable and traditional energy systems.

Table 7.1 Summary of Additive Manufacturing Technologies and Their Applications in Energy Sectors

Additive Manufacturing Technology	*Key Materials*	*Applications (Examples)*	*Advantages*	*Limitations*
Powder Bed Fusion	Metal powders (Titanium, Inconel, Stainless Steel, Aluminium), Polymers (Nylon, PEEK), Ceramics (Zirconia, Alumina)	Turbine blades, Heat exchangers, Nuclear reactor components, Battery electrodes	High precision, strong mechanical properties, complex geometries	High cost, requires post-processing (heat treatment, machining)
Directed Energy Deposition	Metal powders and wires (Titanium, Nickel-based superalloys, Steel, Cobalt-chromium)	Repairing turbine blades, Pipeline maintenance, and nuclear reactor repair	High deposition rates, ability to repair and add material to existing parts	Surface roughness requires post-processing, limited resolution
Vat Photopolymerisation	Photopolymer Resins (Epoxy, Vinyl Ether, Acrylate)	Fuel cells, Microfluidic devices, Battery components	High accuracy, smooth surface quality	Limited to photopolymer materials, requires post-curing
Digital Light Processing	Epoxy, Vinyl Ether, Acrylate	Solid oxide fuel cells, Battery components	Improved speed of production	Material constraints and post-curing issues
Material Jetting	Photoreactive Materials (UV-curable resins), Metal and Ceramic Inks	Battery fabrication, Fuel cell components, Complex geometries	Multi-material printing, high resolution	Slow print speeds, high material costs
Fused Deposition Modelling	Thermoplastics (PLA, ABS, PETG, PEI, PEEK)	Biodegradable energy systems, Prototypes, Insulation components	Cost-effective, easy material handling, widely available materials	Lower resolution and weaker mechanical properties compared to other AM techniques
Binder Jetting	Metal powders (Stainless steel, Titanium, Nickel alloys), Sand, Ceramics	Lightweight energy components, Porous structures for fuel cells, Hydrogen storage systems	No need for support structures, fast production of large parts	Requires post-processing (sintering, infiltration) to improve strength
Selective Laser Sintering	Plastic powders (Nylon, Polyamide), Metal powders	Structural components for turbines and reactors, Fuel cell components	Strong and durable end-products, no need for support structures	High cost, longer processing times, rough surface finish

7.2.1 Powder Bed Fusion

PBF is one of the most widely adopted AM technologies in the energy sector. The process involves selectively melting or sintering powder materials using a high-energy laser or electron beam. The primary PBF methods include the following:

Selective Laser Melting (SLM): Uses a high-powered laser to fully melt metallic powders layer by layer, enabling the fabrication of dense, high-strength components such as turbine blades and heat exchanger parts.
Electron Beam Melting (EBM): Uses an electron beam in a vacuum environment, reducing oxidation and making it suitable for high-temperature applications such as nuclear reactor components.
Selective Laser Sintering (SLS): Utilises a laser source to sinter polymer or ceramic powders, commonly used in insulation materials and lightweight structural supports.

Material versatility of PBF is crucial for energy applications, as it supports metals (Inconel, stainless steel, titanium alloys), ceramics (alumina, zirconia), and polymers (polyamide, thermoplastic polyurethane). Its ability to produce complex geometries and high-performance materials makes it ideal for gas turbines, fuel cells, and heat exchangers in power generation and aerospace applications.

7.2.1.1 Directed Energy Deposition

DED utilises focused thermal energy (laser, electron beam, or plasma arc) to melt and deposit feedstock material (powder or wire) onto a substrate. This technology is particularly beneficial for: Producing functionally graded materials with layered compositions, such as nuclear reactor internals and turbine coatings; repairing worn components in gas turbines and oil pipelines; reducing downtime and extending service life; and fabricating large-scale structures that are difficult to manufacture using traditional methods. The ability of DED to integrate multiple materials into a single build enhances mechanical, thermal, and corrosion resistance, making it indispensable in harsh energy environments.

7.2.1.2 Material Extrusion

Material Extrusion, especially Fused Deposition Modelling (FDM), involves extruding thermoplastic or composite materials through a heated nozzle to build components layer by layer. This process is particularly relevant for battery and fuel cell components, where lightweight, customisable electrode structures are required; hydrogen storage systems, which use polymer-based pressure vessels; and solar panel structures with intricate, weight-optimised frameworks.

Low cost and design flexibility of material extrusion make it suitable for rapid prototyping and functional testing in energy applications. However, limited material strength and lower resolution restrict its use in high-performance components.

7.3 OTHER AM TECHNIQUES

Vat photopolymerisation, including stereolithography (SLA) and digital light processing (DLP), has shown significant potential in energy-related applications. These methods use UV light sources to cure liquid photopolymer resins layer by layer, forming complex geometries with high surface precision. SLA employs a laser beam, while DLP utilises a projector system, enabling faster layer exposure. Common materials include epoxy, acrylate, and vinyl ether resins, with epoxies continuing to cure even after exposure to UV light. The smooth surface finish and high precision of vat photopolymerisation make it suitable for fabricating microfluidic fuel cell devices, such as poly(dimethylsiloxane) (PDMS) components used in energy storage systems (5). Additionally, solid oxide fuel cells (SOFCs), which operate at high temperatures with a ceramic electrolyte, have been successfully manufactured using DLP-based stereolithography (6). Despite its advantages in precision and smooth surface quality, vat photopolymerisation is limited by the availability of photocurable materials, restricting its widespread adoption in certain energy applications.

Other emerging AM technologies are also being integrated into energy applications, including material jetting and binder jetting. Material jetting operates similarly to inkjet printing, where droplets of a photoreactive material are selectively deposited and solidified under UV light. This technique has been effectively utilised in lithium-ion battery fabrication (7). Another promising AM method, binder jetting, involves dispensing liquid binding agents onto powdered materials to construct 3D components with high production efficiency. This method has been explored for SOFC fabrication (8). However, binder jetting faces challenges such as significant part shrinkage, low density, and limited geometric accuracy, necessitating post-processing treatments to enhance mechanical integrity. Despite these challenges, material jetting and binder jetting hold substantial promise for scalable, high-precision AM applications in next-generation energy systems.

7.4 APPLICATION OF AM IN ENERGY SECTORS

7.4.1 Power Generation Sector

AM enhances power generation by enabling internal cooling channels and lattice structures in turbine blades and heat exchangers, improving thermal efficiency and reducing fuel consumption. Conventional manufacturing techniques, such as casting and forging, often impose design constraints and result in material waste. AM minimises material use while maintaining strength, resulting in lighter, more efficient turbine components. Processes like Laser PBF (LPBF) and DED significantly reduce material waste, which is crucial for high-cost alloys such as titanium and Inconel. By optimising design, efficiency, and sustainability, AM lowers production costs and supports resource-efficient energy solutions. The power generation sector relies heavily on high-performance components, such as turbine blades, heat exchangers, and fuel nozzles, which must withstand high temperatures, mechanical stress, and extreme operational conditions.

7.4.2 Turbine Blades

Turbine blades require intricate geometries and exceptional thermal resistance, making them prime candidates for additive manufacturing (AM). Technologies such as LPBF and DED enable the production of turbine blades with internal cooling channels, complex lattice structures, and multi-material compositions, thereby enhancing performance and longevity. These blades operate under extreme mechanical loads and temperatures exceeding 1,000°C, necessitating designs that maximise thermal efficiency, structural integrity, and durability. AM facilitates the integration of advanced cooling features and lightweight lattice structures, significantly improving operational efficiency while reducing material waste. Siemens Energy has pioneered the use of LPBF to fabricate high-temperature-resistant gas turbine blades with internal cooling channels, which enhance thermal management and extend service life.

Additionally, Siemens developed an AM-based repair strategy for turbine blade tips, enabling cost-effective restoration and functional enhancement of worn components (9). GE Aviation has leveraged DED to repair nickel-based superalloy turbine blades, extending their lifespan by restoring worn surfaces with high-performance alloys. This approach minimises downtime and optimises material properties for critical blade sections, ensuring sustained performance (10). NASA has further advanced turbine blade technology by employing PBF and Material Extrusion to optimise aerodynamic efficiency and vibrational performance, achieving a 60% reduction in vibrational stresses and significantly enhancing reliability, energy efficiency, and operational stability (11, 12). A recent study by Keshavarz et al. (13) demonstrated that laser DED, using a Rene 142–MERL 72 powder blend, can effectively restore turbine blade tips with superior oxidation resistance and mechanical properties. **Figure 7.2** (**a**) illustrates the AM repair of turbine blade tips, where a squeezer rim is selectively rebuilt using a deposited material layer, extending the component's operational life. This repair method ensures structural integrity while maintaining aerodynamic performance. As depicted in **Figure 7.2** (**b**), a bidirectional deposition approach is utilised to rebuild worn turbine blade sections, demonstrating AM's capability to restore critical components. The microhardness distribution

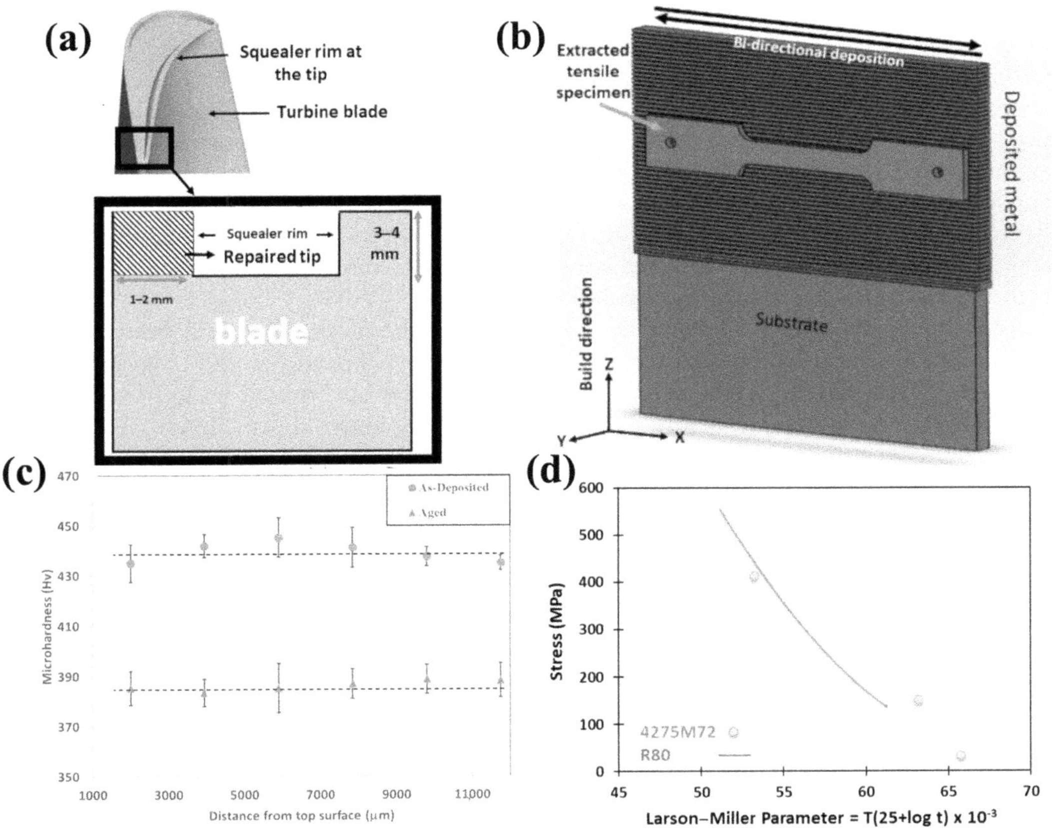

Figure 7.2 (a) Schematic of turbine blade repair using AM, showing selective material deposition at the tip. (b) Bi-directional deposition approach for AM-based repair. (c) Microhardness variation of as-deposited and aged material in repaired turbine blades. (d) Larson-Miller parameter plot illustrating high-temperature durability of AM-repaired blades (Adapted from (13), Copyright: MDPI-Switzerland. Open access under Creative Commons Attribution-CC BY-NC-ND license).

of the as-deposited and aged material in repaired turbine blades (**Figure 7.3 (c)**) highlights the mechanical integrity of the deposited regions, ensuring long-term performance stability. The deposited alloy exhibited high creep resistance (**Figure 7.3 (d)**) and oxidation behaviour comparable to that of Rene 142, making it an ideal material for high-temperature repair applications. These advancements highlight the transformative role of AM in turbine blade manufacturing, repair, and performance optimisation, driving improvements in power generation and aerospace propulsion systems.

7.4.3 Heat Exchangers

AM has revolutionised the design and production of heat exchangers, enabling intricate internal geometries that maximise heat transfer efficiency while minimising material usage. These components are essential for optimal thermal management in power plants, steam turbines, and cooling circuits, ensuring efficient energy conversion and operational stability. Traditional manufacturing methods impose constraints on heat exchanger designs, limiting their ability to incorporate complex internal pathways and high-surface-area structures. AM overcomes these limitations by enabling the fabrication of compact, lightweight, and thermally efficient heat exchangers that improve performance, reduce operational costs, and enhance durability in high-temperature environments. Linear AMS developed an air/oil heat exchanger via LPBF, incorporating intricate internal features previously unattainable with conventional techniques. This innovation resulted in a highly efficient, cost-effective heat exchanger that improved thermal performance and reduced manufacturing costs (14). Similarly, GE Additive leveraged AM to fabricate titanium heat exchangers with enhanced thermal conductivity and mechanical strength, making them particularly valuable for energy storage and industrial applications (15, 16). Research demonstrated the advantages of using AM-produced lattice structures in heat exchangers, optimising thermal efficiency while reducing component weight and size. These findings align with the needs of renewable energy systems, where lightweight, high-efficiency thermal management solutions are crucial (17–19).

Additionally, AM plays a crucial role in enhancing efficiency and sustainability in power generation. A project by the National Renewable Energy Laboratory (NREL) leveraged AM to develop lightweight turbine components, reducing the operational weight of wind turbine systems and increasing energy output. By optimising material distribution, AM enables strong yet lightweight structures, improving overall performance and reducing mechanical strain on turbine systems (20). Moreover, research conducted by ETH Zurich highlighted the environmental advantages of AM in turbine blade production, demonstrating a significant reduction in material waste and energy consumption compared to traditional manufacturing methods (21). These advancements underscore the potential of AM to drive sustainable energy solutions by enhancing efficiency, reducing resource consumption, and minimising environmental impact in power generation. These advancements highlight the critical role of AM in enhancing power generation efficiency, reducing energy consumption, and supporting the transition to more sustainable energy solutions.

7.4.4 Oil and Gas Sector

AM is gradually transforming the oil and gas industry by enhancing efficiency, reducing costs, and enabling on-demand production of complex components. Despite a slower adoption rate than in the aerospace and medical fields, AM offers significant advantages, such as design flexibility, part consolidation, and material efficiency, making it highly beneficial for fabricating downhole tools, valves, pumps, and pipeline components. One of its most impactful contributions is reducing multi-component assemblies into single, high-performance parts, cutting lead times and manufacturing costs by up to 70%. While AM is primarily used for rapid prototyping and tooling, it is increasingly being adopted for end-use parts in turbine components, centrifugal pumps, and sensor housings. The projected growth of AM in this sector is substantial, with an estimated market value of $2 billion by 2027 (22). Furthermore, AM has the potential to decentralise supply chains by enabling on-site manufacturing, reducing reliance on large spare parts inventories, and minimising downtime at remote oil and gas facilities.

In drilling and extraction operations, AM has led to the production of advanced downhole tools, lightweight valves, and optimised pumps with enhanced wear resistance and corrosion protection. Baker Hughes and Protolabs have employed PBF and SLM, respectively, to fabricate drill bits and valve components with complex internal cooling channels, improving durability and operational lifespan (23). The ability of AM to produce tailored, high-strength alloys has also improved pipeline maintenance and repair processes. DED technology enables in-situ repairs of pipeline components, reducing downtime by 30% while extending service life. Companies like GE Oil & Gas and Schlumberger have successfully integrated SLM and LPBF for gas turbines and downhole tools, enhancing fuel efficiency, operational reliability, and component performance. Research at ETH Zurich further demonstrated the benefits of AM-fabricated impellers, increasing pump efficiency by 20% while ensuring superior durability in high-pressure, high-temperature environments (24). With its ability to manufacture corrosion-resistant components, optimise material use, and support energy-efficient designs, AM is poised to revolutionise oil and gas manufacturing, enabling more cost-effective, sustainable, and reliable energy production.

7.4.5 Hydrogen Production and Fuel Cells

The global transition towards sustainable energy has made hydrogen production and fuel cells essential for clean energy generation and storage. AM is transforming this sector by enabling the fabrication of complex, high-efficiency components for fuel cells, electrolysers, and hydrogen storage systems while reducing costs, material waste, and production time. AM technologies such as PBF, DED, SLS, and material extrusion have been extensively used to manufacture bipolar plates, flow fields, membrane electrodes, and porous transport layers, all of which play a crucial role in optimising gas distribution, electrochemical performance, and heat dissipation in fuel cells. Microbial fuel cells (MFCs) and enzymatic fuel cells have benefited from AM's ability to fabricate chambers, membranes, and electrodes with high surface area, conductivity, and biocompatibility. A study by Calignano et al. (25) demonstrated a fully AM-fabricated MFC with optimised open pores, achieving an energy output of 3 kWh/m^3 per day, while Bian et al. (26) developed 3D-printed porous copper electrodes for MFC anodes, leading to a 10-fold efficiency increase over conventional copper-mesh anodes. Similarly, Zawadzki et al. (27) designed an AM-printed MFC with an exchangeable membrane, demonstrating the potential of AM in creating efficient and modular biofuel cells.

In PEM fuel cells, AM has been used to fabricate bipolar plates (BPs) and balance-of-plant components with highly efficient gas flow channels and improved water management. Bourell et al. (28) examined laser-sintered graphite BPs and highlighted how carbon fibre content affects mechanical strength and conductivity. Additionally, the Toyota Mirai fuel cell vehicle (FCV) incorporates porous media flow fields, which have been successfully produced using AM techniques, demonstrating superior permeability and gas diffusion. AM enables the fabrication of metal/carbon foams with over 90% porosity, thereby enhancing thermal/electric conductivity and corrosion resistance, which are critical for next-generation fuel cell architectures (29). Advanced AM processes such as EBM have also been utilised to fabricate Ti–6Al–4V open-cellular foams and copper lattice structures, further optimising heat transfer and durability (30). Microfuel cells with active areas below 1 cm^2 have also been successfully manufactured using AM-based rapid prototyping techniques. Scotti et al. (31) produced miniature stainless-steel fuel cells with different flow fields via laser AM, while Gould et al. (32) fabricated titanium-alloy BPs with flow channels via Direct Metal Laser Sintering, demonstrating the potential to scale up AM-produced fuel cell stacks. AM can also be used to create polymer-based microfluidic fuel cells, in which printed microchannels undergo pyrolysis carbonisation to form high-performance carbon plates (33, 34).

Beyond fuel cells, electrolysers for hydrogen production have benefited from AM innovations. Mo et al. (35) successfully fabricated perforated titanium plates as porous transport layers using EBM, optimising liquid supply and gas removal in electrolysis systems. Additionally, Ambrosi et al. (36) used SLM and FDM to produce metallic electrodes and liquid/gas handling components, improving overall hydrogen production efficiency. A case study from the Korea Advanced Institute of Science and Technology and the Korea Institute of Science and Technology demonstrated the fabrication of a 3D-printed hydrogen production catalyst, which was 20 times more efficient than conventional catalysts, providing a breakthrough for cost-effective green hydrogen production (37). Hydrogen storage has also been enhanced by AM, with PBF and SLS enabling the creation of lightweight, high-strength pressure vessels. A collaboration between industry and academia resulted in 3D-printed aluminium alloy hydrogen storage tanks, which were 20% lighter than conventional designs while maintaining high-pressure containment requirements (38). The ability of AM to integrate complex internal lattice structures into hydrogen storage systems enhances gas containment, thermal management, and mechanical stability, making hydrogen transportation safer and more efficient.

The benefits of AM in hydrogen energy systems are vast, including customised electrode designs, reduced material waste, and improved system performance. AM-driven advancements have led to a 15% increase in fuel cell stack efficiency, higher hydrogen generation rates, and cost reductions through material optimisation. However, challenges such as hydrogen embrittlement, material degradation, and high initial costs still need to be addressed. Ongoing research aims to develop AM-compatible, hydrogen-resistant materials, while international organisations like ISO and ASTM work towards AM-specific regulatory frameworks for hydrogen storage components. As AM technologies continue to evolve, their integration into fuel cell manufacturing, electrolyser design, and hydrogen storage is set to play a pivotal role in the global clean energy transition.

7.4.6 Energy Storage and Batteries

AM is transforming energy storage systems, particularly lithium-ion batteries (LIBs), by enabling the fabrication of complex architectures, customised electrode structures, and solid-state electrolytes that improve energy density, performance, and longevity. Traditional LIBs consist of five fundamental layers: anode current collector (copper), anode (active material, binder, and conductive agent), electrolyte/separator, cathode (active material, binder, and conductive agent), and cathode current collector (aluminium). These layers are typically produced through roll-to-roll printing, followed by stacking or winding to create cylindrical or orthorhombic shapes. AM overcomes these design limitations by directly printing batteries into arbitrary 3D geometries, enabling seamless integration with irregularly shaped electronics, wearable devices, and vehicle chassis. This customisation extends to printed electronics, where co-designing batteries alongside electronic components eliminates unnecessary packaging and enhances overall system efficiency. Unlike traditional LIBs, which rely on metallic foils, AM enables the fabrication of freestanding electrodes, replacing copper and aluminium layers with 3D percolating networks of carbon nanofibers or carbon nanotubes (CNT). This eliminates metal foils, reducing weight and enabling fully 3D-printed battery architectures where even the electrolyte layer can be deposited via AM techniques. However, eliminating metallic current collectors introduces challenges to electrical conductivity, as printed free-standing electrodes achieve only 10 S/cm, significantly lower than copper's 600,000 S/cm. Studies indicate that for long-range lateral electron transport, metallic support grids must be embedded at 1 cm intervals to mitigate internal resistance issues (39).

Pinilla et al. (40) investigated the influence of various AM techniques, including extrusion printing, aerosol jet printing (AJP), spray coating, and vacuum filtration, on LIB performance compared to conventional slurry casting. The study demonstrates that deposition methods significantly affect electrode microstructure (**Figure 7.3 (a–e)**), with wet techniques (slurry, filtration, extrusion) forming more continuous and segregated CNT networks, while dry methods (spray and AJP) create more dispersed structures. Electrochemical analysis confirms that wet techniques yield superior rate performance, characterised by lower τ/L_E^2 (rate performance metric) values (**Figure 7.3 (f)**), due to enhanced ionic and electronic transport, while dry techniques exhibit higher resistance and slower ion transport (**Figure 7.3 (g)**). Wet techniques enhanced out-of-plane (OOP) conductivity and charge transport, whereas dry techniques led to lower OOP conductivity and increased internal resistance (**Figure 7.3 (h)**). However, AM techniques enable precise structural control, which can be leveraged to optimise energy storage through strategic patterning and post-processing treatments. This study highlights AM's potential to tailor electrode architectures for specific applications, balancing electrochemical performance, scalability, and design flexibility.

AM has demonstrated breakthroughs in porous electrode fabrication, solid-state batteries (SSBs), and lithium-sulphur battery architectures. Research using Direct Ink Writing (DIW) to print interdigitated LIB electrodes resulted in a 25% increase in energy density, while SLS and SLA techniques successfully produced solid-state electrolytes with enhanced ionic conductivity and mechanical stability (41). PBF has been used to fabricate sulphur-carbon composite electrodes, which enhance charge retention, improve cycle stability, and enable higher sulphur loading in lithium-sulphur batteries, a promising alternative to conventional lithium-ion systems (42). Mechanical flexibility is another critical area of AM-driven battery innovation. Studies by Kohlmeyer et al. (43) revealed that increasing binder content (40–50%) in printed electrodes improved mechanical strength, preventing catastrophic failure under bending and creasing conditions. When tested under 288 creasing cycles, AM-fabricated multi-walled CNT substrates retained electrochemical function, whereas conventional metal foil electrodes suffered adhesion failures and fatigue fractures (44). Research has also demonstrated that spray printing and co-extrusion methods, rather than traditional inkjet printing, are more suitable for battery production due to their ability to handle high-viscosity inks with controlled deposition of multi-material layers (45).

Industry leaders and research institutions are leveraging AM for advanced battery applications. Tesla has integrated AM techniques in the development of its 4680 battery cells, using rapid prototyping to optimise energy density, manufacturing efficiency, and charge-discharge performance (46). Fraunhofer IPA successfully demonstrated fully 3D-printed LIBs, enabling customised shapes and direct integration into structural components, thereby reducing weight while improving energy efficiency (47). NASA has employed AM to fabricate high-capacity batteries optimised for space applications, utilising PBF-produced electrode structures that withstand extreme temperatures and mechanical stress. At Oak Ridge National Laboratory, researchers developed DIW-fabricated microbattery electrodes, which exhibited high power density, making them ideal for portable electronics and medical devices (48). In the development of SSBs, AM techniques such as SLA and SLS have been used to produce lithium-stuffed garnet electrolytes, which offer improved ionic conductivity, mechanical stability, and safety by eliminating liquid electrolytes (49).

Despite these advancements, challenges remain. The limited availability of AM-compatible battery materials, high production costs, and slow scalability hinder widespread adoption. Additionally, ensuring uniformity and consistency in 3D-printed battery components remains

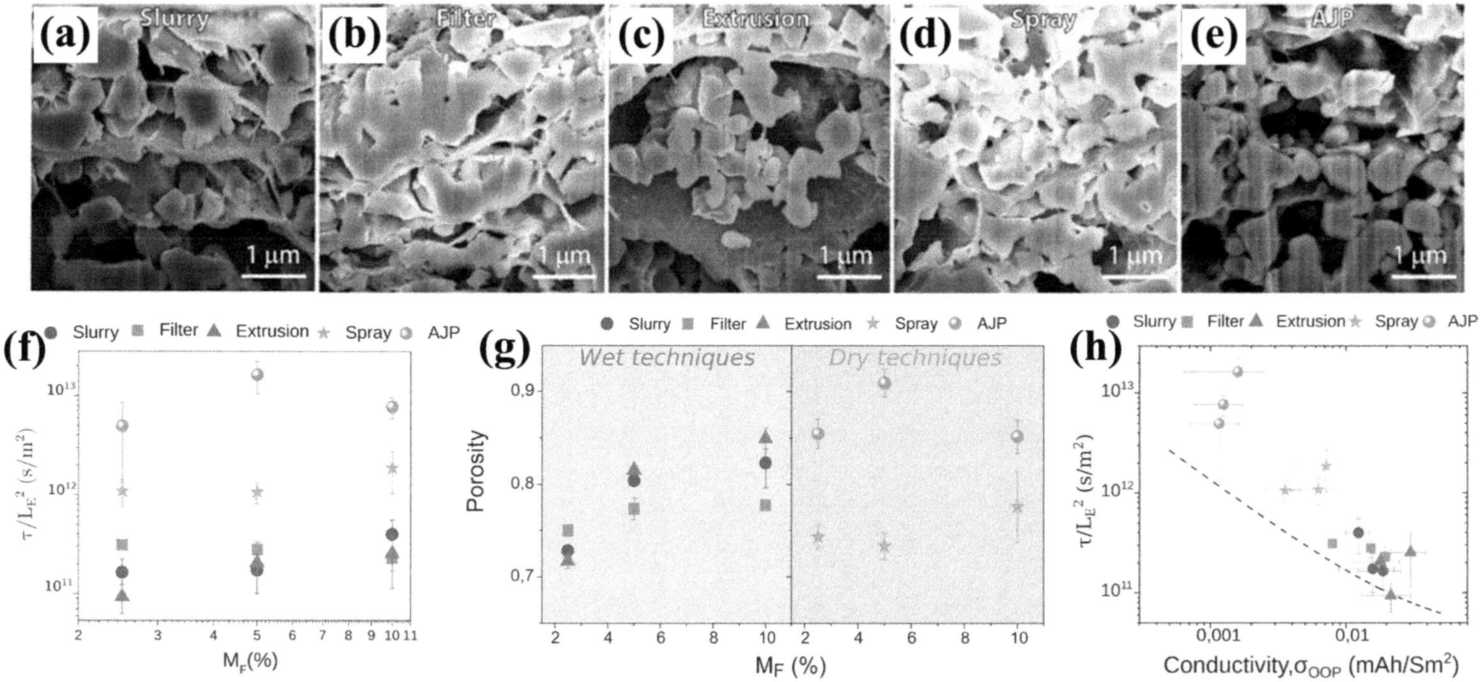

Figure 7.3 Influence of deposition techniques on LIB electrode microstructure and electrochemical performance. (a–e) SEM images show microstructural differences among electrodes fabricated using slurry casting (a), filtration (b), extrusion printing (c), spray coating (d), and aerosol jet printing (AJP) (e). Wet techniques (a–c) yield more interconnected structures, while dry techniques (d–e) result in more dispersed microstructures. (f) Rate performance metric as a function of filler mass fraction (M_F), showing significantly higher values for dry techniques (spray, AJP) due to greater internal resistance. (g) Porosity differences between wet and dry techniques, with wet techniques exhibiting more compact structures that enhance conductivity. (h) Correlation between out-of-plane (OOP) conductivity and rate performance, demonstrating that higher conductivity in wet-deposited electrodes leads to better charge transport and rate capability (Adapted from (40), Copyright: Wiley-VCH GmbH. Open access under Creative Commons Attribution-CC BY-NC-ND license).

a technical hurdle. Future research is focused on developing novel printable electrode materials, optimising hybrid AM techniques, and improving AI-driven process controls to make AM a viable large-scale solution for energy storage. With continued advancements, AM is set to redefine battery manufacturing, enhance renewable energy storage solutions, and drive innovation in electric vehicles, aerospace, and grid-scale energy storage systems.

7.4.7 Nuclear Energy

AM is emerging as a transformative technology in the nuclear energy sector, offering significant advantages in design flexibility, cost reduction, and accelerated production timelines for reactor components, radiation-resistant materials, and waste management solutions. While powder metallurgy techniques such as hot isostatic pressing have traditionally been used for nuclear components, AM offers new opportunities for enhanced process monitoring, improved microstructure control, and functionally graded material transitions. One of the primary benefits of AM is its ability to enable layer-by-layer process monitoring, ensuring high precision and collecting detailed data on the material properties of the final component. Additionally, AM enables the seamless integration of dissimilar materials through gradual compositional transitions, reducing residual stresses and minimising heat-affected zones compared to conventional joining techniques such as welding, brazing, or soldering. Given the harsh operating conditions in nuclear power plants, including high radiation levels, extreme temperatures, and corrosive environments, materials must be carefully selected to withstand these conditions. AM has demonstrated success in fabricating functionally graded materials such as Inconel-steel, copper-steel, titanium-steel, and titanium-carbide composites, which improve thermal resistance, mechanical strength, and radiation tolerance. Advances in modelling and simulation have further optimised AM processes for nuclear applications, enhancing the feasibility of 3D-printed micro-composites and complex reactor components.

Kang et al. (50) manufactured a 3-inch nuclear safety class 1 valve, a vital safety-grade component in pressurised light water reactors using L-DED, demonstrating the technology's capability to produce high-integrity parts with complex geometries. The L-DED process enhances mechanical properties, including improved high-temperature strength, corrosion resistance, and fatigue performance, as evident from **Figure 7.4**. The yield strength and ultimate tensile strength of L-DED-fabricated materials, presented in **Figures 7.4 (a)** and **(b)**, show a stable response across a wide range of test temperatures, with post-processing treatments like hot isostatic pressing further refining these properties. Additionally, the fatigue performance of L-DED components, as analysed in **Figure 7.4 (c)**, highlights their long-term durability under cyclic loading, ensuring reliability in demanding nuclear reactor applications.

Additionally, the United States Department of Energy's Office of Nuclear Energy (US DOE-NE) is actively funding AM research through the Advanced Methods for Manufacturing (AMM) initiative under the Nuclear Energy Enabling Technologies (NEET) programme, which aims to reduce costs, shorten construction timelines, and improve component reliability. Several industry leaders, including Westinghouse, GE, Rolls-Royce, Novatech, BWXT, and NuScale, are integrating AM into nuclear reactor designs, spanning from light-water reactors to small modular reactors. The Nuclear Regulatory Commission has also begun addressing standardisation challenges, convening a public meeting in 2017 that brought together over 20 companies, research labs, and government agencies to discuss the certification of AM-produced reactor materials. The adoption of AM in structural nuclear components is still in its early stages due to the stringent requirements of the American Society of Mechanical Engineers (ASME) Boiler and Pressure Vessel Code, which governs pressure-retaining structures. However, committees such as ASTM F42 and the ASME special committee on AM for pressure-retaining equipment are working to integrate AM into regulatory frameworks. AM has seen greater progress

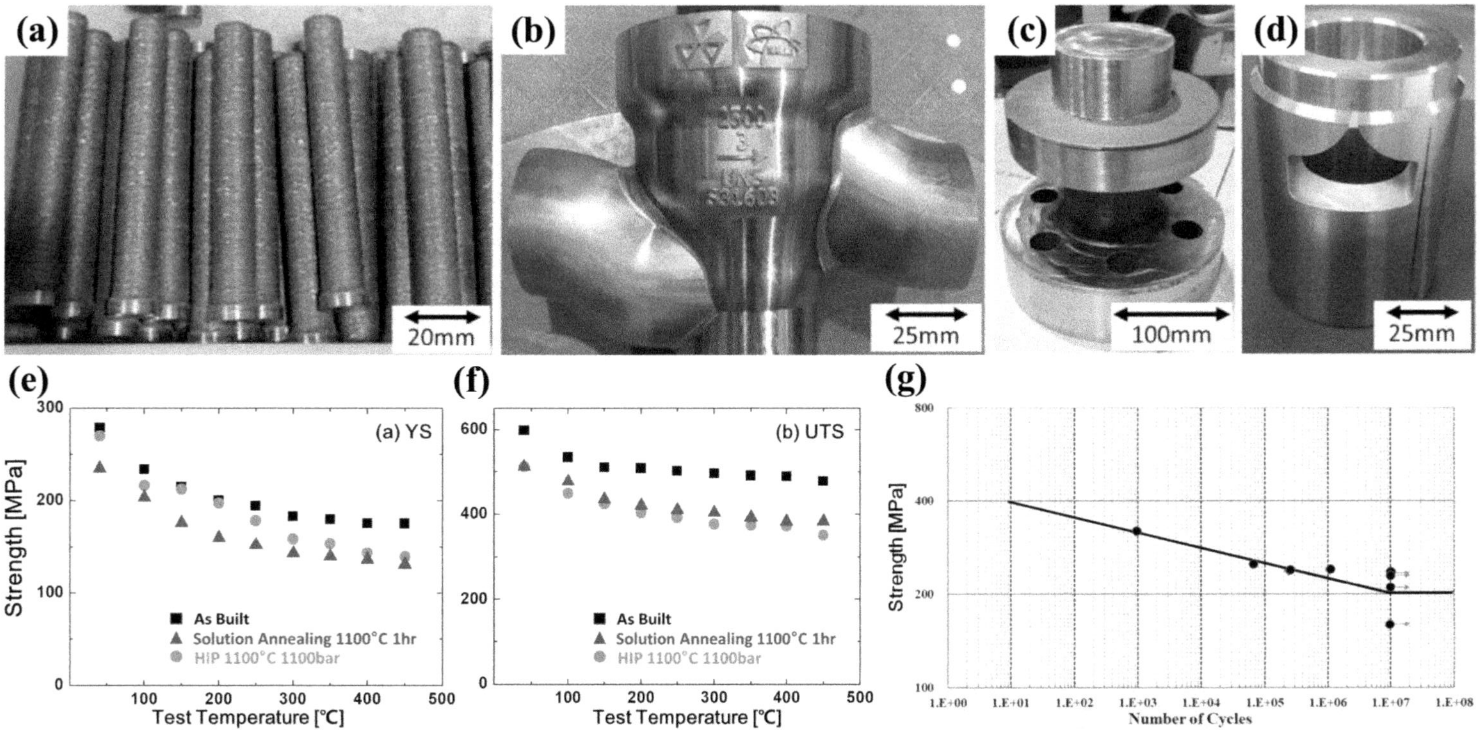

Figure 7.4 L-DED-Fabricated Nuclear Components and Mechanical Properties: (a) cylindrical specimens for mechanical testing, (b) 3-inch nuclear safety class I valve, (c) Nuclear-grade component for fatigue testing, (d) Machined and sectioned L-DED component showing internal features, (e) Yield Strength and (f) Ultimate Tensile Strength vs. temperature, with the as-built condition outperforming heat-treated samples, (g) Fatigue strength vs. cycles, demonstrating durability under cyclic loading (Adapted from (50), Copyright: Elsevier B.V., Open access under Creative Commons Attribution-CC BY-NC-ND license).

in reactor internals, where pressure constraints are less stringent, leading to advancements in reactor core internals, control elements, and fuel cladding. Westinghouse pioneered the use of AM-produced reactor components, installing an AM-fabricated thimble plugging device made from 316L stainless steel in an operational reactor (51). Additionally, Oak Ridge National Laboratory has explored the potential of AM in manufacturing control elements for the High-Flux Isotope Reactor (52), while Idaho National Laboratory has successfully utilised AM to fabricate test capsules for the Advanced Test Reactor (53).

AM is also playing a crucial role in next-generation nuclear reactor designs, including small modular reactors (SMRs) and molten salt reactors (MSRs). SMRs, which emphasise compact reactor designs, benefit from AM's ability to manufacture highly intricate, space-efficient components, such as compact heat exchangers and reactor core structures. Siemens Energy has leveraged AM to produce turbine blades with advanced cooling channels, significantly enhancing thermal efficiency in SMRs. Meanwhile, MSRs, which operate at elevated temperatures with highly corrosive molten salts, require specialised materials such as nickel-based superalloys (e.g., Inconel 625). Researchers have successfully applied PBF techniques to fabricate MSR-compatible components, demonstrating improved thermal resistance, oxidation resistance, and structural integrity under extreme conditions. Beyond power generation, AM is making significant contributions to nuclear waste management by improving the design and integrity of storage containers and encapsulation systems. Similarly, Idaho National Laboratory has applied AM to fabricate geometrically optimised capsules for intermediate-level waste storage, minimising storage footprints while improving containment efficiency (54).

One of the most promising areas of AM in nuclear energy is its potential in fabricating radiation-resistant materials. Nuclear reactors demand materials that can withstand prolonged exposure to neutron irradiation, extreme heat, and corrosive environments. AM enables the production of oxide dispersion-strengthened steels, which exhibit superior radiation tolerance and mechanical strength (55). MIT researchers have successfully fabricated titanium-zirconium-molybdenum alloys using AM, demonstrating excellent high-temperature performance and resistance to neutron-induced swelling. Additionally, AM-fabricated high-temperature alloys such as Hastelloy and Inconel are being integrated into nuclear heat exchangers and core components (56).

Despite its transformative potential, AM in nuclear energy faces several challenges, including rigorous testing and certification requirements, the limited availability of AM-compatible radiation-resistant materials, and scalability concerns for large-scale reactor components. Standardising AM processes for nuclear applications is a key focus for ASTM and ASME, as certification and safety compliance remain major hurdles. Additionally, research efforts are dedicated to expanding the material selection for AM to ensure that printed components maintain structural integrity under extreme nuclear environments. While AM has proven effective for prototyping and manufacturing small-to-medium-sized components, scaling up for full-scale reactor structures presents both technical and economic challenges. However, as AM technology advances, regulatory frameworks evolve, and hybrid AM techniques are refined, its role in nuclear energy is expected to expand significantly. With continued investment in next-generation reactor designs, radiation-resistant materials, and advanced waste containment solutions, AM is set to play a critical role in the future of nuclear energy, improving efficiency, safety, and sustainability across the industry.

7.5 CHALLENGES AND FUTURE PROSPECTS

AM has demonstrated its transformative potential across multiple energy sectors, from power generation to hydrogen storage and nuclear applications. However, several technical, economic, and regulatory challenges must be addressed to fully harness its capabilities. Simultaneously, rapid advancements in materials science, process optimisation, and hybrid manufacturing techniques are paving the way for a future where AM becomes an integral part of energy-efficient, sustainable, and cost-effective industrial practices.

7.5.1 Challenges

One of the primary challenges in AM for energy applications is the limited availability of materials compatible with AM. While AM supports a variety of metals, polymers, and ceramics, many advanced energy-sector materials, such as high-temperature superalloys and functionally graded composites, pose processing challenges due to thermal distortion, residual stress formation, cracking, and inconsistent powder properties. For instance, in nuclear and gas turbine applications, where materials must withstand extreme environments, oxide dispersion-strengthened steels and nickel-based superalloys require further development to achieve printability, mechanical integrity, and long-term stability.

Another significant limitation is the high cost of AM systems and materials, which remains a barrier to widespread adoption, particularly for small- and medium-sized enterprises SMEs. The cost of metal powders, advanced polymer resins, and ceramic feedstocks is significantly higher than that of their conventionally manufactured counterparts. Additionally, post-processing steps, such as heat treatments, machining, and surface finishing, contribute to the overall expense. This is particularly relevant in sectors such as nuclear energy, where safety-critical components must meet stringent quality and regulatory requirements, thereby adding complexity to the certification process.

Post-processing remains a major bottleneck in AM for energy applications. Components intended for turbines, reactors, and heat exchangers demand high precision, defect-free surfaces, and enhanced mechanical properties, necessitating additional treatments such as hot isostatic pressing, chemical etching, and multi-stage machining. PBF parts, for instance, require extensive surface smoothing and internal defect mitigation before they can be deployed in high-performance applications. These additional steps increase production time and costs, limiting the full potential of AM as a rapid manufacturing technology.

Another critical limitation is scalability. While AM excels at producing small, intricate components, its ability to fabricate large-scale infrastructure elements remains constrained by build-volume limitations, deposition rates, and challenges with layer adhesion. This limits its applicability to large wind turbine blades, high-capacity heat exchangers, and structural reactor components, where traditional manufacturing techniques such as casting, forging, and extrusion still dominate due to their efficiency in handling large parts.

Qualification and certification remain key hurdles, particularly in nuclear and aerospace applications, where safety and reliability are paramount. Currently, AM-fabricated components lack well-established, standardised qualification protocols, making regulatory approval a lengthy and complex process. For instance, DED-produced nuclear reactor components must undergo extensive microstructural analysis, radiation exposure tests, and mechanical performance assessments before gaining regulatory acceptance. This lack of standardisation delays the adoption of AM in mission-critical energy applications, requiring collaborative efforts from industry, academia, and regulatory bodies to develop clear certification pathways.

Despite AM's reputation as a sustainable technology, its energy consumption and environmental footprint remain areas of concern. High-powered lasers, electron beams, and sintering processes used in PBF, DED, and EBM are highly energy-intensive, potentially offsetting the sustainability benefits of material efficiency and waste reduction. Moreover, the handling and disposal of unused metal powders, polymer residues, and process by-products necessitate careful management to ensure compliance with environmental regulations.

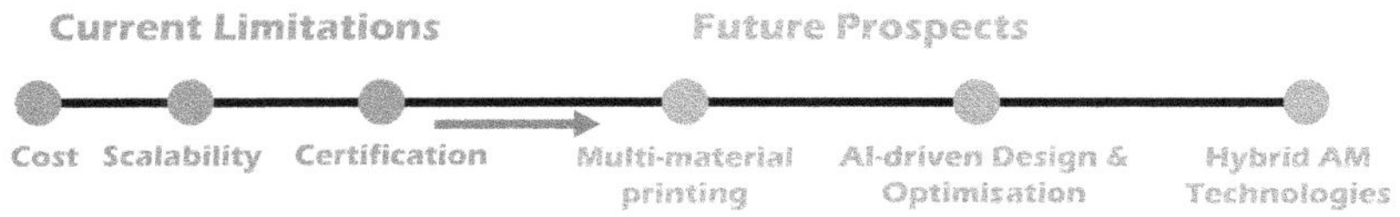

Figure 7.5 Schematic of current challenges and future prospects of AM in energy sector.

7.5.2 Future Prospects

Despite these challenges, AM is rapidly evolving, with new materials, hybrid manufacturing approaches, and process optimisations shaping its future in energy applications. Key challenges and future prospects have been presented in **Figure 7.5**. Material innovations are a key area of progress, with researchers developing oxide dispersion-strengthened steels, advanced nickel-based superalloys, and functionally graded materials tailored for AM. These next-generation materials offer superior high-temperature performance, corrosion resistance, and radiation tolerance, making them ideal for turbines, nuclear reactors, and hydrogen storage systems.

Hybrid manufacturing systems are emerging as a viable solution to overcome limitations of AM in surface finish, scalability, and cost efficiency. By integrating AM with traditional manufacturing methods, hybrid systems can leverage AM's design flexibility while maintaining the precision, scalability, and material properties of conventional processes. For example, hybrid AM-CNC machining systems allow for intricate near-net-shape part fabrication followed by precision finishing, reducing post-processing complexity and enhancing part quality.

The adoption of automation and digital twin technologies is also expected to revolutionise AM in energy applications. Automated powder handling, real-time build monitoring, and AI-driven defect detection are minimising process variability and enhancing reproducibility. Digital twins which creates a virtual simulations of AM- components, enable predictive modelling, real-time performance assessment, and optimised maintenance schedules, ensuring that AM-fabricated parts meet long-term operational requirements.

A major focus for the future of AM is sustainability. Researchers are actively exploring low-energy AM processes, recycled metal powders, and bio-based polymer resins to improve the environmental credentials of AM. Additionally, binder jetting and material extrusion methods are being adapted for lightweight, high-strength energy components, such as photovoltaic mounts and fuel cell architectures, further driving AM's role in renewable energy sectors.

Advances in multi-material printing are unlocking new design possibilities for energy storage and thermal management systems. The ability to print functionally graded materials with tailored thermal, mechanical, and electrical properties is particularly promising for heat exchangers, battery electrodes, and solid-state hydrogen storage solutions. These developments are expected to significantly enhance energy efficiency, durability, and overall system performance.

Standardisation and certification frameworks are also evolving to facilitate the widespread adoption of AM in energy applications. Organisations such as ASTM International, ISO, and ASME are actively developing guidelines for AM materials, processes, and part qualification. These efforts aim to streamline certification protocols, reduce regulatory barriers, and accelerate the integration of AM into critical energy infrastructure.

Collaboration between industry, research institutions, and government bodies is playing a crucial role in advancing AM technologies. Initiatives like the U.S. Department of Energy's AM Energy programme are funding research into AM applications in turbines, nuclear reactors, and energy storage. Public-private partnerships are driving breakthroughs in high-performance AM materials, scalable production methods, and real-world validation of AM-fabricated energy components.

7.6 CONCLUSIONS

This chapter has provided a comprehensive exploration of the transformative role of Additive Manufacturing (AM) in advancing the energy sector. Starting with the fundamental principles of AM and its growing significance in the energy sector, the discussion covered key AM technologies, including Powder Bed Fusion (PBF), Directed Energy Deposition (DED), and Material Extrusion, emphasising their capabilities in fabricating intricate, high-performance components. In power generation, AM has enabled the development of turbine blades, heat exchangers, and fuel nozzles with optimised geometries, reducing fuel consumption and improving thermal efficiency, as demonstrated by Siemens Energy. The oil and gas sector has benefited from AM's ability to produce corrosion-resistant tools and customised components, enhancing durability and reducing downtime, with industry leaders like BP and Chevron integrating AM for pipeline maintenance and subsea applications. Hydrogen production and storage have seen advancements through AM-enabled fabrication of high-efficiency fuel cell components and lightweight, high-pressure storage systems, accelerating the transition to a hydrogen economy. In energy storage, AM has revolutionised battery architecture by enabling intricate anode and cathode structures, improving energy density, lifecycle, and thermal management, which are crucial for electric vehicles and grid storage. Despite challenges such as high costs, material limitations, and certification hurdles, ongoing advancements in hybrid manufacturing, automation, and material science are addressing these barriers. Looking ahead, AM is poised to play a pivotal role in advancing sustainability, improving efficiency, and driving innovation in energy applications. By integrating AM into mainstream energy production and storage systems, the sector can achieve greater resilience, reduced environmental impact, and long-term cost-effectiveness, positioning AM as a cornerstone of future energy solutions.

This chapter underscores indispensable role of AM in shaping the future of energy, combining technological advancements with the imperative of sustainability to address the challenges of modern energy systems.

List of Abbreviations

3D	Three-Dimensional
ABS	Acrylonitrile Butadiene Styrene
AJP	Aerosol Jet Printing
AM	Additive Manufacturing
ASME	American Society of Mechanical Engineers
ASTM	American Society for Testing and Materials
BP	Bipolar Plate
CNC	Computer Numerical Control
CNT	Carbon NanoTube
DED	Directed Energy Deposition
DIW	Direct Ink Writing
DLP	Digital Light Processing
DOE	Department of Energy
EBM	Electron Beam Melting
FCV	Fuel Cell Vehicle
FDM	Fused Deposition Modelling
IPCC	Intergovernmental Panel on Climate Change
L-DED	Laser-Directed Energy Deposition

LIB	Lithium-Ion Battery
LPBF	Laser Powder Bed Fusion
MFC	Microbial Fuel Cell
MSR	Molten Salt Reactor
NEET	Nuclear Energy Enabling Technologies
NREL	National Renewable Energy Laboratory
OOP	Out-Of-Plane
PBF	Powder Bed Fusion
PEEK	Polyether Ether Ketone
PEM	Proton Exchange Membrane
PLA	Polylactic Acid
PVA	PolyVinyl Alcohol
SLA	Stereolithography
SLM	Selective Laser Melting
SLS	Selective Laser Sintering
SMR	Small Modular Reactor
SOFC	Solid Oxide Fuel Cell
SSB	Solid-State Battery
UV	Ultraviolet

REFERENCES

1. Cao D, Xing Y, Tantratian K, Wang X, Ma Y, Mukhopadhyay A, et al. 3D printed high-performance lithium metal microbatteries enabled by nanocellulose. *Advanced Materials*. 2019;31(14).
2. Kohlmeyer RR, Blake AJ, Hardin JO, Carmona EA, Carpena-Núñez J, Maruyama B, et al. Composite batteries: A simple yet universal approach to 3D printable lithium-ion battery electrodes. *Journal of Materials Chemistry A*. 2016;4(43):16856–16864.
3. Sireesha M, Lee J, Kranthi Kiran AS, Babu VJ, Kee BBT, Ramakrishna S. A review on additive manufacturing and its way into the oil and gas industry. *RSC Advances. Royal Society of Chemistry*. 2018;8:22460–22468.
4. Balbaud F, Cabet C, Cornet S, Dai Y, Gan J, Hernández Mayoral M, et al. A NEA review on innovative structural materials solutions, including advanced manufacturing processes for nuclear applications based on technology readiness assessment. *Nuclear Materials and Energy*. 2021;27.
5. Au AK, Lee W, Folch A. Mail-order microfluidics: Evaluation of stereolithography for the production of microfluidic devices. Lab on a Chip. 2014;14(7):1294–1301.
6. Wei L, Zhang J, Yu F, Zhang W, Meng X, Yang N, et al. A novel fabrication of yttria-stabilized-zirconia dense electrolyte for solid oxide fuel cells by 3D printing technique. *International Journal of Hydrogen Energy*. 2019;44(12):6182–6191.
7. Singh R. Process capability study of polyjet printing for plastic components. *Ecology and Evolution*. 2011;25(4):1011–1015.
8. Manogharan G, Kioko M, Linkous C. Binder jetting: A novel solid oxide fuel-cell fabrication process and evaluation. *JOM*. 2015;67(3):660–667.
9. Material solutions & siemens—innovative project outstanding results [Internet]. https://www.eos.info/industries/manufacturing/turbomachinery-turbines
10. Angel NM, Basak A. On the fabrication of metallic single crystal turbine blades with a commentary on repair via additive manufacturing. *Journal of Manufacturing and Materials Processing*. 2020;4(4):101.
11. Shyy W. *Optimization of Turbine Blade Design for Reusable Launch Vehicles* (Technical Report No. NAG8-1251). University of Florida, Washington; 1998.
12. Integral Tuned Mass Absorber for Turbine Blades (MFS-TOPS-111). USA; U.S. Patent No. 11,808,166, 2021.

13. Keshavarz MK, Gontcharov A, Lowden P, Chan A, Kulkarni D, Brochu M. Turbine blade tip repair by laser directed energy deposition additive manufacturing using a rene 142–merl 72 powder blend. *Journal of Manufacturing and Materials Processing*. 2021;5(1).
14. Case study: Additive manufacturing creates next-generation heat exchangers. 2017 [Internet]. https://www.engineering.com/resources/case-study-additive-manufacturing-creates-next-generation-heat-exchangers/
15. Careri F, Khan RHU, Todd C, Attallah MM. Additive manufacturing of heat exchangers in aerospace applications: A review. *Applied Thermal Engineering*. 2023;235.
16. Galati M, Iuliano L. A literature review of powder-based electron beam melting focusing on numerical simulations. *Additive Manufacturing*. 2018;19:1–20.
17. Maloney KJ, Fink KD, Schaedler TA, Kolodziejska JA, Jacobsen AJ, Roper CS. Multifunctional heat exchangers derived from three-dimensional micro-lattice structures. *International Journal of Heat and Mass Transfer*. 2012;55(9–10):2486–2493.
18. Lebaal N, SettaR A, Roth S, Gomes S. Conjugate heat transfer analysis within in lattice-filled heat exchanger for additive manufacturing. *Mechanics of Advanced Materials and Structures*. 2022; 29(10):1361–1369.
19. Kaur I, Singh P. Critical evaluation of additively manufactured metal lattices for viability in advanced heat exchangers. *International Journal of Heat and Mass Transfer*. 2021;168.
20. DOE advanced manufacturing office funds next-gen turbine blades and more. 2021 [Internet]. https://www.nrel.gov/news/program/2021/doe-advanced-manufacturing-office-funds-next-gen-turbine-blades-and-more.html
21. Eisenbarth D. *Buildup Strategies for Additive Manufacturing by Direct Metal Deposition*. Doctoral Thesis. DISS. ETH NO. 26907. Technische Universit¨at M¨unchen, Germany; 2020.
22. Research and Markets. Additive manufacturing opportunities in oil and gas markets—an opportunity analysis and ten-year forecast. PR Newswire, 17 November 2017. https://www.prnewswire.com/news-releases/additive-manufacturing-opportunities-in-oil--gas-markets-report-2017-an-opportunity-analysis-and-ten-year-forecast-300557696.html
23. Stephanie Hendrixson. 2024. Additive Manufacturing Versus Cavitation. https://www.valvemagazine.com/articles/additive-manufacturing-versus-cavitation-.
24. Huber M, Hartmann M, Ess J, Loeffel PD, Kränzler T, Rettberg R. Process method for manufacturing impellers by Selective Laser Melting (SLM). In: *Proceedings of the International Conference on Additive Manufacturing in Products and Applications (AMPA)*. Zürich, Switzerland: ETH; 2017, pp. 13–15.
25. Calignano F, Tommasi T, Manfredi D, Chiolerio A. Additive manufacturing of a microbial fuel cell—a detailed study. *Scientific Reports*. 2015;5.
26. Bian B, Wang C, Hu M, Yang Z, Cai X, Shi D, et al. Application of 3D printed porous copper anode in microbial fuel cells. *Frontiers in Energy Research*. 2018;6(June).
27. Zawadzki D, Pędziwiatr P, Michalska K. A novel microbial fuel cell with exchangeable membrane—application of additive manufacturing technology for device fabrication. *Acta Innovations*. 2018; 28:20–31.
28. Alayavalli K, Bourell DL. Fabrication of modified graphite bipolar plates by indirect selective laser sintering (SLS) for direct methanol fuel cells. *Rapid Prototyping Journal*. 2010;16(4):268–274.
29. Migita, S., Watanabe, Y., Ota, H., Nabatame, T., Toriumi, A. *Structural metastability and size scalability of phase-controlled HfO2 formed through cap-PDA. Electrochemical Society Transactions*. 2009;19(2):563–575.
30. Murr LE, Gaytan SM, Medina F, Martinez E, Martinez JL, Hernandez DH, et al. Characterization of Ti-6Al-4V open cellular foams fabricated by additive manufacturing using electron beam melting. *Materials Science and Engineering: A*. 2010;527(7–8):1861–1868.
31. Scotti G, Matilainen V, Kanninen P, Piili H, Salminen A, Kallio T, et al. Laser additive manufacturing of stainless steel micro fuel cells. *Journal of Power Sources*. 2014;272:356–361.
32. Gould BD, Rodgers JA, Schuette M, Bethune K, Louis S, Rocheleau R, et al. Performance and limitations of 3D-printed bipolar plates in fuel cells. *ECS Journal of Solid State Science and Technology*. 2015;4(4):P3063–P3068.

33. Wang Y, Pham L, Vasconcellos GPS de, Madou M. Fabrication and characterization of micro PEM fuel cells using pyrolyzed carbon current collector plates. *Journal of Power Sources*. 2010; 195(15):4796–4803.
34. Jiang S, Shi T, Zhan X, Xi S, Long H, Gong B, et al. Scalable fabrication of carbon-based MEMS/NEMS and their applications: A review. *Journal of Micromechanics and Microengineering*. 2015; 25(11):113001.
35. Mo J, Dehoff RR, Peter WH, Toops TJ, Green JB, Zhang FY. Additive manufacturing of liquid/gas diffusion layers for low-cost and high-efficiency hydrogen production. *International Journal of Hydrogen Energy*. 2016;41(4):3128–3135.
36. Ambrosi A, Pumera M. Multimaterial 3D-printed water electrolyzer with earth-abundant electrodeposited catalysts. *ACS Sustainable Chemistry & Engineering*. 2018;6(12):16968–16975.
37. Kim YJ, Lim A, Kim JM, Lim D, Chae KH, Cho EN, et al. Highly efficient oxygen evolution reaction via facile bubble transport realized by three-dimensionally stack-printed catalysts. *Nature Communications*. 2020;11(1).
38. Dong L, Zhu Q, Liu X, Xing Y, Zhang S, Jia Q, et al. 3D printing Al porous metamaterials with triply periodic minimal surfaces (TPMS) for hydrogen generation from Al-water reaction. *International Journal of Hydrogen Energy*. 2024;49:1426–1435.
39. Ho CC, Evans JW, Wright PK. Direct write dispenser printing of a zinc microbattery with an ionic liquid gel electrolyte. *Journal of Micromechanics and Microengineering*. 2010;20(10):104009.
40. Pinilla S, Ryan S, McKeon L, Lian M, Vaesen S, Roy A, et al. Additive manufacturing of li-ion batteries: A comparative study between electrode fabrication processes. *Advanced Energy Materials*. 2023;13(15).
41. Pang Y, Cao Y, Chu Y, Liu M, Snyder K, MacKenzie D, et al. Additive manufacturing of batteries. *Advanced Functional Materials*. 30;2020.
42. Shen K, Mei H, Li B, Ding J, Yang S. 3D printing sulfur copolymer-graphene architectures for Li-S batteries. *Advanced Energy Materials*. 2018;8(4).
43. Kohlmeyer RR, Blake AJ, Hardin JO, Carmona EA, Carpena-Núñez J, Maruyama B, et al. Composite batteries: A simple yet universal approach to 3D printable lithium-ion battery electrodes. *Journal of Materials Chemistry A*. 2016;4(43):16856–16864.
44. Blake AJ, Kohlmeyer RR, Drummy LF, Gutiérrez-Kolar JS, Carpena-Núñez J, Maruyama B, et al. Creasable batteries: Understanding failure modes through dynamic electrochemical mechanical testing. *ACS Applied Materials & Interfaces*. 2016;8(8):5196–5204.
45. Cobb CL, Ho CC. Additive manufacturing: Rethinking battery design. *Interface Magazine*. 2016; 25(1):75–78.
46. Tesla gives update on its game-changing 4680 battery cell. Fred Lambert; 2023. https://electrek.co/2023/04/21/tesla-update-4680-battery-cell/.
47. Printsolid—printable solid-state lithium-ion batteries. Andreas Georg. 2021. https://www.ise.fraunhofer.de/en/research-projects/printsolid.html.
48. Lai W, Erdonmez CK, Marinis TF, Bjune CK, Dudney NJ, Xu F, et al. Ultrahigh-energy-density microbatteries enabled by new electrode architecture and micropackaging design. *Advanced Materials*. 2010;22(20).
49. McOwen DW, Xu S, Gong Y, Wen Y, Godbey GL, Gritton JE, et al. 3D-printing electrolytes for solid-state batteries. *Advanced Materials*. 2018;30(18).
50. Kang SH, Suh J, Lim SY, Jung S, Jang YW, Jun IS. Additive manufacture of 3 inch nuclear safety class 1 valve by laser directed energy deposition. *Journal of Nuclear Materials*. 2021;547.
51. Carlota V. First installation of a 3D printed thimble plugging device in nuclear plant. 2020. https://www.3dnatives.com/en/3d-printed-thimble-plugging-device-110520205/#!
52. Burns JR, Petrovic B, Chandler D, Terrani KA. Reactor physics phenomena in additively manufactured control elements for the High Flux Isotope Reactor. *Annals of Nuclear Energy*. 2018;115: 403–414.
53. Wagner AR, Rufner JF, Jokisaari AM, McMurtrey MD. *Advanced Manufacturing for Harsh Environments at Idaho National Laboratory* [Internet]; 2024. http://www.inl.gov
54. 3D printing could transform nuclear industry, INL results indicate. https://inl.gov/nuclear-energy/3d-printing-could-transform-nuclear-industry-inl-results-indicate/?__cf_chl_f_tk=4i1cncQmyi6r

HGbEjWpI4n08W_uON3.qhxvIV.UIJcM-1740828694-1.0.1.1-nhVYoOvrDzOHAhBunLi.YSwXr6csF7l3fW9pJiJVgMA
55. Arkhurst BM, Park J-J, Lee C-H, Kim JH. Direct laser deposition of 14Cr oxide dispersion strengthened steel powders using Y2O3 and HfO2 dispersoids. *Korean Journal of Metals and Materials*. 2017;55(8).
56. Brinkman CR, Rittenhouse PL, Corwin WR, Strizak JP, Lystrup A, DiStefano JR. *Application of Hastelloy X in Gas-Cooled Reactor Systems*. Oak Ridge, TN (United States); 1976.

Chapter 8

From Cradle to Grave

A Holistic View of Environmental and Economic Sustainability of Sophisticated Additively Manufactured Metal Products

Shadi Samemaleki, Mikael Collan, and Jyrki Savolainen

8.1 INTRODUCTION

This chapter presents and examines the lifecycle of sophisticated additively manufactured metal products, from the initial stages of metal mining through abandonment, disassembly, and recycling. The point of view taken is that of economic and environmental sustainability. Economic sustainability is measured by a business's ability to maintain financially stable metal additive manufacturing (AM) operations with long-term profitability, and environmental sustainability is evaluated holistically by considering materials, energy consumption, material waste, and emissions.

Additive manufacturing, also known as 3D printing, is a technology in which a product is constructed from a 3D digital model via a layer-by-layer deposition process [1]. Conventional manufacturing (CM) includes traditional production methods such as machining, which typically involve removing material from a solid to achieve the final shape and dimensions, along with other processes like casting, forging, and stamping, which are widely utilised in industrial production due to their scalability and precision. Economically, CM benefits from well-established infrastructure and economies of scale, which reduce per-unit costs in large production runs [2, 3].

Compared to conventional manufacturing, AM typically benefits less from economies of scale, primarily due to slower production speeds and a limited production capacity that can be expanded only by producing multiple similar products in a single run or by acquiring additional additive manufacturing equipment. As AM equipment is relatively expensive, there is typically no economic sustainability in such endeavours. This means that AM is seldom the most suitable method for the mass production of simple components [4, 5]. For the mass manufacturing of products such as simple standard metal components, the traditional technologies are likely to remain the go-to solution [5, 6].

AM offers substantial benefits over traditional manufacturing, such as flexibility and customisation, particularly in the ability to create advanced designs that reduce material use and offer cost-of-use benefits, for example, due to the comparatively lightweight nature of the produced geometries, during the product lifetime [7, 8]. On the other hand, excessive design complexity can lead to inefficiencies in production, necessitating careful optimisation of the design-production cycle for practical applications [9, 10]. This observation indicates a trade-off between complexity and cost. Notably, additive manufacturing (AM) enables the production of intricate geometric components with functionalities that are challenging or impossible to achieve using conventional manufacturing (CM) [11]. In such cases, AM can offer entirely new solutions, redefining cost-benefit relationships. Metal additive manufacturing has already partially replaced traditional manufacturing processes in the production of complex, high-value parts for industries, where customisation and flexibility are crucial [12]. Thus, it is the sophisticated, not simple, components that are relevant from the point of

DOI: 10.1201/9781003590774-8

view of AM in terms of economic viability. By sophistication, we refer to products that offer functionality or usability beyond what can be easily achieved through conventional manufacturing, that incorporate significant design investments, and that typically provide valuable benefits to users. It is through these benefits, for example, complex geometries, weight advantages, and reduced use of expensive raw materials, that AM production can achieve economic sustainability. Beyond economic considerations, the environmental implications of CM and AM also warrant attention. The bottom line in analysing environmental sustainability is to consider material use and waste, in addition to energy consumption, in the production process. Also, recycling materials and the refurbishment of worn components [13] are of the essence from an environmental sustainability point of view.

The subtractive nature of conventional manufacturing does not easily allow the construction of cavities within solid parts, and the subtractive nature of material removal typically results in significant metal waste, for example, in CNC machining. Layer-by-layer techniques used in AM allow for considerable savings in material use without compromising structural strength. These advantages are particularly substantial in aerospace applications, where weight is crucial for carrying more payload. Moreover, when working with costly materials such as titanium, material conservation directly reduces production costs [14, 15].

Carbon emissions in manufacturing arise from multiple sources, including electricity consumption at the machine and system levels, the production and disposal of materials, tools, and liquids, and general material transportation [16]. From an environmental perspective, CM processes such as machining and casting are energy-intensive, consuming significant amounts of electricity and fuel. This trend is driven by the need to produce precise, complex, and high-quality components, particularly when working with high-strength alloys, which increase energy consumption per unit volume processed [17, 18].

For example, modern machining processes often allocate less than 15% of their total energy demand to the actual cutting, with the majority consumed in auxiliary operations [19]. The use of cutting fluids in machining presents major environmental challenges due to the management and disposal of waste products. Duflou et al. [20] and Oliveira et al. [21] analysed the environmental impact of CO_2 laser cutting and highlighted the substantial contributions of gases such as nitrogen and oxygen, as well as waste materials, to the overall environmental footprint during the use phase of laser cutting machines.

Selective laser melting (SLM), an additive manufacturing (AM) process, has demonstrated a 10% to 23% reduction in environmental impacts compared to CNC machining [14]. SLM and similar AM processes achieve these environmental benefits primarily through reduced material usage, improved energy efficiency in localised heating, and the elimination of extensive tooling requirements. Sand casting is a metal-shaping method characterised by high consumption of virgin materials and energy, along with significant hazardous gas emissions and the generation of solid waste [22]. Sand moulds also generate solid waste unless recycling measures can be effectively implemented.

Hybrid approaches that integrate CM with AM are being explored to reduce waste and enhance process flexibility while retaining the precision advantages of CM [23, 24]. For instance, hybrid CNC-AM machines can combine precise material removal with AM's ability to build complex geometries layer by layer, reducing waste and enabling components that are difficult or impossible to manufacture with traditional methods alone [24, 25].

Building on the economic and environmental sustainability of conventional and additive manufacturing, we analyse both aspects across the lifecycle of additively manufactured metal products, highlighting key differences. We examine the business models underlying these manufacturing techniques, noting that they must be "win-win" for both producers and buyers/users to be viable. Additionally, we explore the potential of organising additive manufacturing as a network-based operation, with components produced locally to order, considering both economic and environmental perspectives.

In the following section, we present a Life Cycle Assessment (LCA) of the environmental and economic sustainability of additive manufacturing and compare it with conventional manufacturing at relevant life-cycle stages. In such cases, AM can provide completely new solutions with new cost-benefit relationships.

8.2 THE ENVIRONMENTAL AND ECONOMIC SUSTAINABILITY OF METAL AM

We analyse the environmental and economic sustainability of metal additive manufacturing (AM) by examining the entire lifecycle of an additively manufactured metal product—from raw material extraction to end-of-life disposal or, as discussed here, material recycling. Our approach incorporates Life Cycle Assessment (LCA), a method focused on evaluating the environmental impact of industrial processes. Additionally, we integrate economic analysis to assess the costs and benefits of AM and explore suitable business models. Our analysis partially resembles life-cycle costing (LCC), a method that typically focuses only on the cost side of product life-cycle economics. We do not include an analysis of social impacts, so we are not conducting a life-cycle sustainability assessment (LCSA), which is a triple-bottom-line analysis of the impacts of metal AM.

Typically, LCA approaches are classified as cradle-to-grave or cradle-to-gate. A cradle-to-grave analysis examines the entire product supply chain, from raw material production (cradle) to end-of-life (grave). In contrast, a cradle-to-gate approach focuses on the portion from the raw material to the end of product production (gate), as shown in **Figure 8.1.** Our approach, as discussed, is cradle-to-grave (and beyond).

A literature review by Kokare et al. [26] analysed 77 articles on life cycle assessment (LCA) in additive manufacturing published between 2011 and 2022. Their findings revealed that only one study incorporated the triple bottom line, and very few focused on life cycle costing (LCC). Regarding system boundaries, the authors identified 42 cradle-to-gate studies, 16 cradle-to-grave studies, and 14 studies that evaluated a cradle-to-grave scope plus the end-of-life phase while omitting the use phase. Furthermore, only a single LCC study was found to utilise a cradle-to-grave approach.

Our view (**Figure 8.1**) of the production process, cradle-to-grave, is quite similar to what is presented by Pusateri and others [27]; however, we have added a feedback loop from the end-of-life waste disposal to “metal scrap” and further to the production of metal powder/wire that is the direct input into the additive manufacturing process. Additionally, we have included a feedback loop from the project disassembly to a new process part titled “parts refurbishing.” These feedback loops indicate the potential for reusing and recycling materials

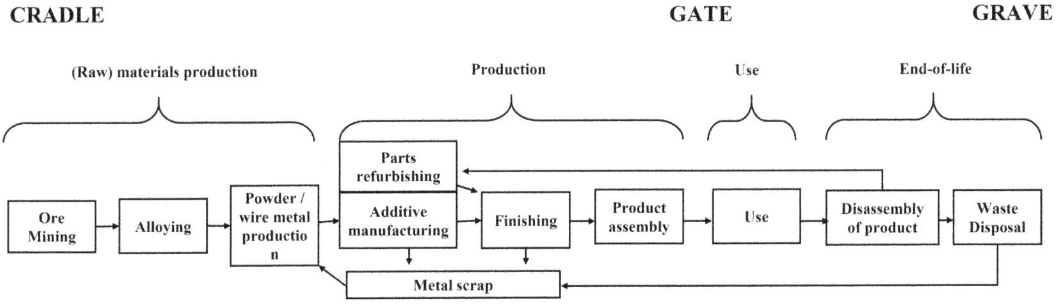

Figure 8.1 The lifecycle of an additively manufactured metal component, from cradle to grave, including recycling of scrap and end-of-life waste.

not present in the process described by Pusateri and others [27]. In the materials production part of the life cycle, we group, in addition to the mining of ore, the alloying and the powder/wire metal production as these three steps are needed to produce the input material for the production process.

Our changes and additions to the lifecycle of metal AM-produced products take into account the important benefits in LCA that metal AM offers, including the recyclability of the raw materials used and the components of the products that can be refurbished. The integration of recycling and refurbishment feedback mechanisms into the cradle-to-grave life cycle model moves us closer to a circular economy, where material efficiency is increased and environmental impact is lower. These feedback loops create a partially closed system that reduces landfill waste and may positively affect energy consumption. A recycling-based material economy with a sufficient quantity of raw material in the loop may also be very resilient from a supply chain perspective, because the system is not dependent on access to raw material in the same way as a linear production system.

Next, we go through the process, phase by phase and step by step, discussing both the environmental and economic issues involved in metal additive manufacturing. Importantly, we also discuss the business models involved and specific to metal additive manufacturing.

8.2.1 (Raw) Material Production Phase

The phase includes mining (virgin) ore, separating and enriching the metal, producing usable metal alloys for metal additive manufacturing, and preparing suitable powders and wire metal for the machinery used. There is a feedback loop from the end-of-life phase via the recycling of metal scrap. See **Figure 8.2.**

The broader outlook for metallic raw materials indicates that demand for critical and scarce metals used in AM is rising significantly rather than diminishing. For example, the growth rate of demand for metals is rather high, which means there is a push for more extraction. If demand continues while high-metal-content deposits are depleted, the costs of extraction and the prices of raw materials will rise. The lower the metal content in a ton of ore, the larger the environmental impact of extraction is likely to be [28].

Thus, from an environmental perspective, there are challenges associated with the extraction of some metals used as input materials for additive manufacturing. Metals like titanium and molybdenum are geologically scarce, and other metals like antimony and gold are predicted to face the risk of depletion [29]. Mining and sourcing of materials, such as titanium, contribute significantly to the overall CO_2 emissions of the metal industry, accounting for

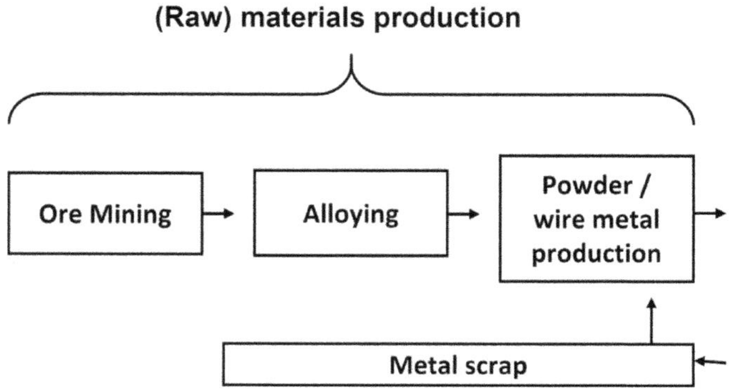

Figure 8.2 (Raw) materials production phase.

approximately 12% of global CO_2 emissions [11]. It is, however, clear that there is no significant difference in the environmental issues associated with metal AM and conventional metal manufacturing, as both technologies require the use of extracted metals.

When the environmental assessment of metal extraction is made, it must be observed that metal ore may not be a material containing only one metal or relevant substance but a mixture of many substances of importance. Many ores contain so-called companion metals, such as arsenic and cadmium, by-products of carrier metals like copper and zinc, which can add considerable complexity to environmental assessments [30]. It is common for the side flows of metal mines that concentrate on one main metal to constitute both an important source of revenue and environmental challenges for the extractors. The scarcer certain metals are, the deeper one may have to go to mine them, which also influences the sustainability of metal mining operations.

In evaluating the profitability of mining operations (also of metals relevant for metal AM), one must note that metal mining is a cyclical business and that the value of a deposit is determined mainly by the size of the deposit, the richness of the ore, access to the ore, the present market price of the mined metals, and the technical setup used in the mining operations. The world market price is the most volatile of these main determinants of the value of a metal mining operation and thus drives the value of most deposits. We know from previous literature that the financial choices made when operations are designed may determine the survivability of a mining operation and will influence the flexibility that the owners must maximise their profits by way of utilising, for example, temporary mine closures [31, 32].

Technical choices made in terms of, for example, separation of metal from the ore, may also affect the environmental and economic sustainability of mining operations. The quality of the ore affects the energy consumption, water use, and waste, adding to the overall environmental footprint of mining activities [29]. Choosing hydrometallurgical enrichment of metals entails foregoing the option to execute temporary shutdowns of operations, which means a constant, continuous profile of cash outflows, while more conventional methods of metal separation (crushing and sifting) allow for flexibility. On the other hand, the environmental profiles of hydrometallurgical and conventional ore extraction techniques may vary greatly, as conventional extraction typically uses a lot of electrical energy, while hydrometallurgical techniques consume water. Planning plays a key role in achieving a good balance between the long-term economic and environmental sustainability of metal mining operations. An example of this is the planning of mine-water-related investments that affect both types of mining sustainability [33].

After the extraction and enrichment steps, the metal ore is processed to create alloys usable in production. It is well known that a great selection of alloys is used in metal AM. This is a testament to the fact that there are no de facto standards in place for metal AM at this time, which is not necessarily beneficial to industry and may also put pressure on the prices of speciality alloys. From an economic sustainability perspective, using very special alloys may pose risks, as supply-related risks increase with the speciality of the raw materials.

Here, we focus on the production of metal powders and wire metal as raw materials for metal AM. Metal powders are typically produced by atomisation methods that break down input materials into fine particles. The process is energy-intensive and requires electricity and gas, both of which contribute to the environmental impacts of feedstock material preparation [11, 18]. There are also more energy-efficient methods for producing metal powders, for example, the metalysis process, which enables less energy-intensive processing of titanium powder [34, 35].

AM powder production is less well documented in lifecycle inventory (LCI) terms than conventional raw materials (such as sheets or billets), partly due to the still-proprietary (and non-standardised) practices in this industry [18]. According to the literature, producing

Ti-6Al-4V powder by atomisation requires approximately 560.6 MJ of energy per kilogram [36]. According to Serres et al. [37], the atomisation process is conducted under argon at 25 bar, with gas flow rates ranging from 0.5 to 6 m^3/h. Steel powder production via direct or indirect atomisation requires 15–26 MJ/kg, depending on the method [38]. AM powder-based production processes often involve nickel-based superalloy powders.

Wire-based deposition systems employ metal wire as the feedstock to construct components. Wires of various alloys and diameters are produced through wire drawing and are generally more cost-effective than their powdered counterparts. Primarily manufactured for the welding industry, these wires are available in spooled form and come in a wide range of alloys. Spooled weld wire offers advantages, including a well-established history of alloy development, well-studied weld metal deposit properties, and standardised, certified production processes [39, 40].

Wire-based techniques are particularly suitable for manufacturing larger parts with lower geometric complexity and accuracy than can be achieved with powder-based metal AM. Wire-based metal AM leaves a cleaner working environment, offers higher deposition rates, and provides even more efficient material utilisation than powder-based techniques. Nearly all the feed wire is incorporated into the build, making the technology material-efficient [41, 42]. Metal wire is cheaper, more cost-effective, and more readily available than powder [43]. Wire-feed AM is considered a cost-efficient option for producing medium to large metal components, offering advantages in productivity, cost-effectiveness, and energy efficiency [44].

Material cost is an important component of overall metal AM expenses and varies depending on the type of metal used, its properties, and availability. Thus, material selection is critical to optimising both costs and performance [2]. One point of discussion is the utilisation of replacement metals, that is, using good enough, cheaper, or from the material property point of view inferior, metals to replace (very) expensive, but "optimal" alloys [29]. As discussed earlier, the prices of scarce metals (alloys) are typically high and may be volatile, posing an availability risk. At the same time, a less scarce, cheaper material may also allow price competition, as price is typically an important issue for customers. Only when physical properties must adhere to strict specifications that can be achieved with scarce metal alloys does it make sense to use them. The aforementioned issues make the use of scarce alloys less desirable from an economic sustainability perspective.

Other important approaches to reducing the use of scarce virgin raw materials include increasing product lifetimes and improving the recycling processes of once-used raw materials [29]. Increasing product lifetimes means designing more durable products and adopting a modular approach that allows parts to be used for multiple purposes. These design issues are not relevant only to metal AM but also to conventionally manufactured metal products. Enhancing recycling processes for metal AM and conventionally manufactured metal products is a way to save economically and reduce the use of virgin metal raw materials. Recycling related to metal AM has not attracted much attention, as evidenced by previous articles describing the lifespan of metal AM products that do not include recycling as part of the process charting [27]. In **Figure 8.1**, we present feedback loops from product end-of-life to the production of metal AM raw materials from recycled materials that describe the potential of recycling (scarce) raw materials for metal AM. As metal AM production is a well-documented process in terms of what one knows (exactly) about the product in terms of the metals and the specific alloys that have been used in production, it is possible to create systems that work towards the high level of recycling of scarce and valuable metals relevant to metal AM. It is plausible to imagine that a business case and a model could be constructed around the idea of recycling scarce alloys used in products manufactured by metal AM technologies.

8.2.2 The Production Phase

The production phase includes the actual manufacturing of the metal parts, their finishing, and the assembly of the final product. In addition to previous representations of the production phase, we include the refurbishing step, which is connected to the feedback loop from the product's end-of-life phase. Each one of these steps carries environmental and economic considerations for metal AM. See **Figure 8.3.**

Material efficiency achieved with metal AM compared to conventional manufacturing is one of the main environmental advantages of metal AM. Unlike traditional manufacturing, which typically uses billets and sheets, AM relies on powdered and wire feedstocks, offering greater material efficiency but requiring substantial energy inputs [18]. The material efficiency advantage of AM may come from two distinct sources:

I) AM allows the creation of sophisticated geometries that include, for example, closed cavities, which means that when designs are made in a way that minimises raw material usage, while keeping the structural integrity of the produced components at the required level, there is an environmental effect. Interestingly and importantly, if the alloy used in production is made from scarce metals and an expensive raw material, there is a win-win: it is both environmentally friendly and more cost-effective. This is especially true for metal AM jewellery, where the product's weight may be an additional "negative."

II) Material efficiency advantage is created by the fact that the AM process inherently creates less metal waste. It has been claimed that with laser beam melting (LBM), over 99% of nickel powder can be used (>1% waste), meaning the production is a near "closed loop" in terms of material waste [45]. There are also opposite experiences with powder bed processing, particularly when using unsintered powders, which have caused recycling challenges and inefficiencies in AM processes [18].

Recycling of raw material waste generated during the manufacturing process typically involves reprocessing the waste. In conventional manufacturing, this means melting and remelting suitable solids, and in metal AM, using the waste as raw material for powder and wire production. In the recycling of waste raw materials from production, the benefit is that the waste is typically "clean," the alloy used is well known, and it can be reused directly without the need for identification and/or separation of raw materials.

Energy consumption can be considered the environmental Achilles' heel of metal AM and a challenge to the production phase [2]. From an energy consumption perspective, metal AM

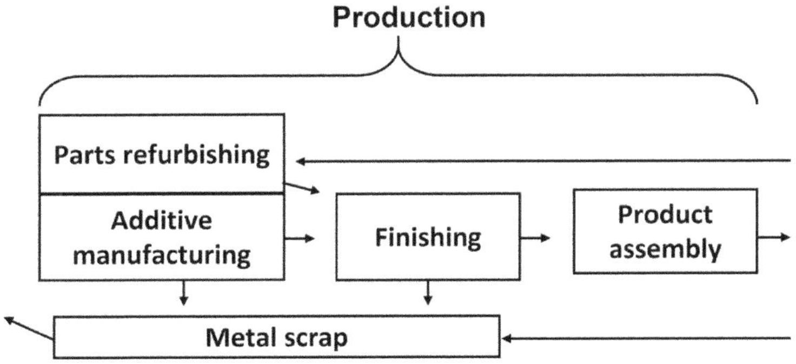

Figure 8.3 Production phase.

is not competitive with conventional manufacturing methods. Depending on energy costs, there is also an economic aspect to energy use.

Where continuous laser operation is used with chilling units and inert gas supplies, Specific Energy Consumption (SEC), the amount of energy required to produce a unit mass or volume of a product, can be very high for metal AM [18]. The shielding and cooling gases, like argon, necessary for maintaining inert environments in AM processes that use Powder Bed Fusion (PBF), and Laser Metal Deposition (LMD), contribute to their overall environmental impact [11, 37, 46, 47]. All in all, it should be noted that some AM methods exhibit significantly higher SEC than conventional manufacturing [17, 48, 49]. The energy consumption of AM processes can be influenced by factors such as building height and the mass of support structures, both of which contribute to the overall environmental footprint [11]. There are differences between metal AM technologies in terms of energy use; for example, electron beam melting (EBM) systems use less energy because of their shorter build times and higher deposition rates compared to direct metal laser sintering (DMLS) systems [2]. All in all, technological development will reduce energy consumption in metal AM and enhance sustainability [50].

A little-discussed issue in the literature on metal AM is the potential to transform the logistics of the manufacturing industry. One aspect of this concerns the possibility of setting up manufacturing as a network, rather than what we typically call a factory [51]. The idea, shortly, is that if a network of local manufacturing "cells" were used instead of concentrated manufacturing locations and there was a digital marketplace for matching the demand and supply of AM, then the manufacturing could potentially take place as close as possible to the demand and thus even considerably cut the distance and time the physical components produced would have to travel [52]. Such logistical benefits would also be environmentally beneficial and could yield economic benefits. It can be imagined that such a system would work in the context of metal AM for the supply of spare parts for maintenance and, perhaps, for the refurbishment of worn parts, in which case the logistics benefits would also include shipping the refurbishable part. Refurbishing itself reduces the need to extract raw materials.

Metal AM often requires post-processing to improve surface quality and remove support structures used during manufacturing. Support structures are an additional form of waste specific to AM, and their use entails energy consumption in addition to waste—these issues are a focus for further development to enhance sustainability [53]. Surface quality improvements can be achieved through thermal energy treatments, which increase the energy demand of AM systems [11].

Finishing processes, such as wire erosion and ultrasonic cleaning, can increase energy consumption and thus should also be considered when evaluating sustainability [54]. Accordingly, the post-processing step is a central criterion when selecting a suitable AM technique. For instance, post-processing requirements in Direct Metal Laser Sintering (DMLS) systems add both time and cost, while Electron Beam Melting (EBM) systems typically require less post-processing, making them more efficient overall [2]. Combining AM with subtractive techniques can enhance surface finishing and reduce the need for extensive post-processing [53, 55, 56].

In the assembly of metal AM-produced versus conventionally produced components, the differences may not be very large. We note that through design, one can make assembly easier, and if products are specifically designed for fast, easy assembly, there are sustainability advantages to be had from increased assembly speed and reduced assembly errors.

8.2.3 Use Phase

The use phase encompasses the product's economic life in use and can, in many cases, be long or very long, even tens of years. As we are here discussing what we call sophisticated metal AM-produced products, the use-phase benefits that accrue to the user originate from this sophistication and constitute a difference between AM-produced and conventionally produced products. Typically, the benefits include the ability to produce lighter products, which leads to lower fuel consumption and, as a result, higher payload capacity, in cases where transport is considered in the aerospace and automotive industries. The benefits have a long payback time, as the savings of some thousands of kilos per aircraft translate into more payload per flight over the economic life of the aircraft—such long-term benefits are not necessarily the easiest sell in terms of being able to add to the cost of the product. It is, however, clear that as real-world experience with beneficial AM-produced parts accumulates, there will be greater acceptance of the higher costs involved. For example, AM-designed brackets, seats, and other aerospace components have demonstrated weight reductions that have demonstrably resulted in lower fuel consumption and lifecycle emissions [57, 58]. Additionally, better product design can enhance product performance, making them more competitive for longer and thereby extending their economic life [53]. Flow-through designs can include geometries that accelerate fluid flow and thus enable higher throughputs, and similarly, nozzles for burning fuels can be designed to optimise combustion while minimising, for example, smoke.

The usability of the above-listed advantages depends on overcoming challenges related to the durability and stress resistance of metal AM parts in high-performance applications. There is a place for further research into the validation and standardisation of material properties to ensure consistency in AM applications [53]. There is a trade-off between the high energy consumption during the production phase and the energy savings achieved during the use phase [45].

Metal AM can be used for the repair and refurbishment of metal components and to extend product life. The use of hybrid machines enables servicing, for example, metal dies and other components with wearable parts [13]. Refurbishing product components typically allows the use of partially new parts or products and entails savings in raw materials and energy consumption [59].

8.2.4 End-of-Life Phase

The phase includes the final disassembly of the product and waste disposal steps. Both have linkages via feedback loops to component refurbishing and to metal recycling. The end-of-life phase of metal AM is most likely the least-discussed phase in terms of environmental and economic sustainability (see **Figure 8.4**). This is somewhat interesting because there are quite easily identifiable opportunities for firms in the metal AM ecosystem regarding both aspects.

The end-of-life stage can reveal both opportunities and challenges for AM in achieving circularity and waste reduction. Opportunities may stem from the fact that metal AM is

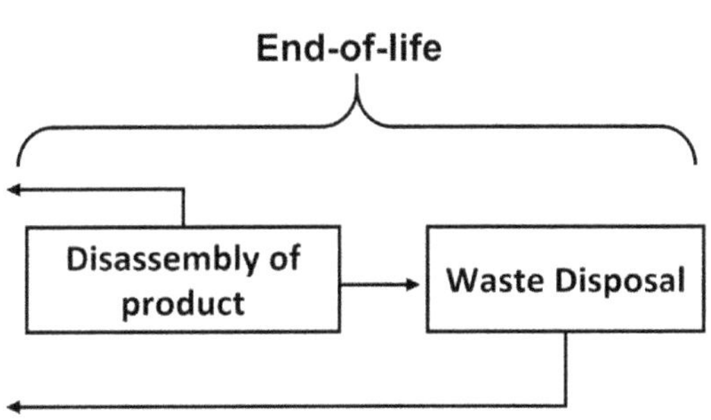

Figure 8.4 The end-of-life phase.

highly "documented," in that producers know very precisely which alloys are used and the contents of various metals in products. If this information is used in the end-of-life phase to capture the metals known to reside in the product, the phase of identifying the alloy/metal and sorting can be omitted. This would, in theory, make recycling easier and more economically viable. For expensive raw materials, such as precious metals, the case is clear: metals like gold and zinc are easier to recycle, reducing the need for primary extraction [29].

The issue of documentation has also been identified in previous literature, and a call has been made for more standardised specifications and control methods for powder quality to enhance the recyclability of products [18, 60]. Two recycling methods for assessing material recovery efficiency are Equal Quality Recycling, which maintains material quality by avoiding cross-contamination, and Down-Cycling, which can result in lower-quality alloys and reduced recycling benefits [45]. Therefore, standardised quality control measures and advancements in recycling methods, such as equal-quality recycling to avoid cross-contamination, can be critical to improve the sustainability potential of AM [45].

The volume of recyclable products per known alloy is important in determining the possible profitability of recycling as a business. If a critical mass/volume of an alloy can be salvaged with a low cost of identification, a business case is easier to form around it. There is clearly a place for multi-producer information on the alloys used in products to make recycling a more attractive business opportunity.

Refurbishing worn components by metal AM is a viable option that makes business sense for expensive-to-manufacture components that can be refurbished to original geometry at a reasonable cost. For a maintenance business servicing a fleet of similar machines that use the same components that wear and can be refurbished, there is an opportunity to save costs by refurbishing them. Again, the critical volume issue comes into play: because the equipment is relatively expensive, there needs to be a large enough set of refurbishable parts and a steady volume to keep an operation running profitably. There is potential for growth in the refurbishing business through metal AM.

8.3 DISCUSSION AND CONCLUSIONS

This paper discussed the environmental and economic aspects of metal additive manufacturing and framed these issues by examining the full economic life of metal AM-produced products. A new cradle-to-grave framework for the product's lifetime was presented that includes feedback loops from the product's end-of-life phase back to raw material production and other production phases. These feedback loops present the recycling options available for metal AM-manufactured products, as well as the possibility to refurbish products by adding new material to worn, but otherwise intact, products via hybrid metal AM technologies. Such feedback loops have not been discussed actively in the literature previously. As metal AM is digitally controlled and the materials used in the process are exactly known, it may be possible to construct a business case for the recycling of products produced from expensive speciality alloys. Such a process would include the "repowering" or "rewiring" of the alloy and would be information-intensive, as the number of products using the same speciality alloy must be large enough to sustain the business. At this time, such a business case is potential, not a tangible opportunity. On the other hand, refurbishing expensive additively manufactured special equipment used in, for example, production facilities may be a suitable add-on for maintenance service providers responsible for keeping machinery operational. This kind of "production as a service" provider could build a business case around reusing worn and refurbished products of a fleet of serviced machines. If the economic benefit of using refurbished parts were significant, this could create a competitive advantage for the service provider.

For each phase of the product lifecycle, environmental and economic sustainability issues can be identified, and the differences between these issues in metal AM and conventional metal manufacturing can be observed. In each phase, benefits can be achieved through additive manufacturing. It is, however, important to recognise that economic benefits can typically justify additive manufacturing only for products that are not very simple in geometry, manufactured with scarce and expensive raw materials (alloys), and cannot be mass-produced very cheaply by conventional manufacturing.

The gap between the costs of metal AM and conventional manufacturing for simple geometries can be made smaller by standardisation, by using cheaper good-enough metals instead of maximally optimal, but expensive alloys, by putting in place recycling of metal waste from production and end-of-life phases, and by creating a system of refurbishing for components that can be made close-enough to "as new" as is required for complete functionality and reuse.

Metal AM uses more energy and is more expensive per component manufactured than conventional manufacturing, but for sophisticated components with, for example, weight advantages, the energy disadvantage may be offset by an energy and cost advantage over the economic life of the component. There is a selling challenge associated with more expensive components, but as the life-of-product environmental and economic sustainability factors weigh in, metal AM is making headway and replacing conventional manufacturing in sophisticated component manufacturing.

Financing may play a role in making metal AM win ground if leasing and other options are made available to the owner of assets for which more expensive metal AM-produced components are a possibility. Paying as one goes rather than making a large one-time investment may be a suitable model for metal AM when the lifetime benefits considerably outweigh the higher lifetime costs.

There are many avenues for further research on the lifetime analysis of metal AM products. The design issues have not been exhaustively studied, nor have the business model opportunities been fully explored. The financial aspects and opportunities in financial engineering for the business of metal AM manufacturing remain almost uncharted.

REFERENCES

[1] Výtisk, Jan, Vladimír Kočí, Stanislav Honus, and Mojmír Vrtek. 2019. "Current Options in the Life Cycle Assessment of Additive Manufacturing Products." *Open Engineering* 9 (1): 674–682. https://doi.org/10.1515/eng-2019-0073

[2] Baumers, Martin, Phill Dickens, Chris Tuck, and Richard Hague. 2016. "The Cost of Additive Manufacturing: Machine Productivity, Economies of Scale and Technology-Push." *Technological Forecasting and Social Change* 102: 193–201. https://doi.org/10.1016/j.techfore.2015.02.015.

[3] Hegab, Hussien, Navneet Khanna, Neamat Monib, and Amr Salem. 2023. "Design for Sustainable Additive Manufacturing: A Review." *Sustainable Materials and Technologies* 35: e00576. https://doi.org/10.1016/j.susmat.2023.e00576.

[4] Berman, Barry. 2012. "3-D Printing: The New Industrial Revolution." *Business Horizons* 55 (2): 155–162. https://doi.org/10.1016/j.bushor.2011.11.003.

[5] Weller, Christian, Robin Kleer, and Frank T. Piller. 2015. "Economic Implications of 3D Printing: Market Structure Models in Light of Additive Manufacturing Revisited." *International Journal of Production Economics* 164: 43–56. https://doi.org/10.1016/j.ijpe.2015.02.020.

[6] Busa chi, Alessandro, John Erkoyuncu, Paul Colegrove, Filomeno Martina, Chris Watts, and Richard Drake. 2017. "A Review of Additive Manufacturing Technology and Cost Estimation

Techniques for the Defence Sector." *CIRP Journal of Manufacturing Science and Technology* 19: 117–128. https://doi.org/10.1016/j.cirpj.2017.07.001.

[7] Mami, Fares, Jean-Pierre Revéret, Sophie Fallaha, and Manuele Margni. 2017. "Evaluating Eco-Efficiency of 3D Printing in the Aeronautic Industry." *Journal of Industrial Ecology* 21 (S1): S37–S48. https://doi.org/10.1111/jiec.12693.

[8] Huang, Samuel H., Peng Liu, Abhiram Mokasdar, and Liang Hou. 2013. "Additive Manufacturing and Its Societal Impact: A Literature Review." *The International Journal of Advanced Manufacturing Technology* 67 (5–8): 1191–1203. https://doi.org/10.1007/s00170-012-4558-5.

[9] Baumers, Martin, Chris Tuck, and Richard Hague. 2011. "Realised Levels of Geometric Complexity in Additive Manufacturing." *International Journal of Product Development* 13 (3): 222–244. https://doi.org/10.1504/IJPD.2011.040268.

[10] Hopkinson, Neil, and Philip Dicknes. 2003. "Analysis of Rapid Manufacturing—Using Layer Manufacturing Processes for Production." *Proceedings of the Institution of Mechanical Engineers, Part C: Journal of Mechanical Engineering Science* 217 (1): 31–39. https://doi.org/10.1243/095440603762554596.

[11] Monteiro, Helena, Gabriel Carmona-Aparicio, Inês Lei, and Mélanie Despeisse. 2022. "Energy and Material Efficiency Strategies Enabled by Metal Additive Manufacturing—A Review for the Aeronautic and Aerospace Sectors." *Energy Reports* 8: 298–305. https://doi.org/10.1016/j.egyr.2022.01.035.

[12] Savolainen, Jyrki, and Mikael Collan. 2020. "How Additive Manufacturing Technology Changes Business Models?—Review of Literature." *Additive Manufacturing* 32: 101070. https://doi.org/10.1016/j.addma.2020.101070.

[13] Urbani, Michele, and Mikael Collan. 2020. "Additive Manufacturing Cases and a Vision for a Predictive Analytics and Additive Manufacturing Based Maintenance Business Model." In *Technical, Economic and Societal Effects of Manufacturing 4.0*, 131–148. Springer International Publishing. https://doi.org/10.1007/978-3-030-46103-4_7.

[14] Peng, Tao, Yanan Wang, Yi Zhu, Yang Yang, Yiran Yang, and Renzhong Tang. 2020. "Life Cycle Assessment of Selective-Laser-Melting-Produced Hydraulic Valve Body with Integrated Design and Manufacturing Optimization: A Cradle-to-Gate Study." *Additive Manufacturing* 36: 101530. https://doi.org/10.1016/j.addma.2020.101530.

[15] Campbell, Thomas. 2011. "Could 3D Printing Change the World?" In *Technologies, Potential, and Implications of Additive Manufacturing*. Strategic Foresight Report of the Atlantic Council of the United States. Washington, DC: Atlantic Council. https://www.atlanticcouncil.org/in-depth-research-reports/issue-brief/could-3d-printing-change-the-world/.

[16] Panagiotopoulou, Vasiliki Christina, Panagiotis Stavropoulos, and George Chryssolouris. 2022. "A Critical Review on the Environmental Impact of Manufacturing: A Holistic Perspective." *The International Journal of Advanced Manufacturing Technology* 118 (1–2): 603–625. https://doi.org/10.1007/s00170-021-07980-w.

[17] Gutowski, Timothy G., Matthew S. Branham, Jeffrey B. Dahmus, Alissa J. Jones, Alexandre Thiriez, and Dusan P. Sekulic. 2009. "Thermodynamic Analysis of Resources Used in Manufacturing Processes." *Environmental Science & Technology* 43 (5): 1584–1590. https://doi.org/10.1021/es8016655.

[18] Kellens, Karel, Martin Baumers, Timothy G. Gutowski, William Flanagan, Reid Lifset, and Joost R. Duflou. 2017. "Environmental Dimensions of Additive Manufacturing: Mapping Application Domains and Their Environmental Implications." *Journal of Industrial Ecology* 21 (S1): S49–S68. https://doi.org/10.1111/jiec.12629.

[19] Dahmus, Jeffrey B., and Timothy G. Gutowski. 2004. "An Environmental Analysis of Machining." *ASME International Mechanical Engineering Congress and Exposition* 47136: 643–652. https://doi.org/10.1115/IMECE2004-62600.

[20] Duflou, Joost R., Karel Kellens, Tom Devoldere, Wim Deprez, and Wim Dewulf. 2010. "Energy Related Environmental Impact Reduction Opportunities in Machine Design: Case Study of a Laser Cutting Machine." *International Journal of Sustainable Manufacturing* 2 (1): 80–98. https://doi.org/10.1504/IJSM.2010.031621.

[21] Oliveira, Marta, João P. Santos, Fernando G. Almeida, Ana Reis, João P.G.T. Pereira, and Augusto Barata Da Rocha. 2011. "Impact of Laser-Based Technologies in the Energy-Consumption of Metal Cutters: Comparison between Commercially Available Systems." *Key Engineering Materials* 473: 809–815. https://doi.org/10.4028/www.scientific.net/KEM.473.809.

[22] Dalquist, Stephanie, and Timothy Gutowski. 2004. "Life Cycle Analysis of Conventional Manufacturing Techniques: Sand Casting." *ASME International Mechanical Engineering Congress and Exposition* 47136: 631–641. https://doi.org/10.1115/IMECE2004-62599.

[23] Pragana, Joao, Rui F. Sampaio, I. Bragança, Carlos A. Silva, and P. Martins. 2021. "Hybrid Metal Additive Manufacturing: A State–of–the-Art Review." *Advances in Industrial and Manufacturing Engineering* 2: 100032. https://doi.org/10.1016/j.aime.2021.100032.

[24] Lorenz, K. A., J. B. Jones, D. I. Wimpenny, and M. R. Jackson. 2015. "A Review of Hybrid Manufacturing." *Proceedings of the 26th Annual International Solid Freeform Fabrication Symposium*, 96–108. University of Texas at Austin. https://www.semanticscholar.org/paper/A-REVIEW-OF-HYBRID-MANUFACTURING-Lorenz-Jones/dd7aedd46b8b9339bc46b0065e1e1f63f525701e

[25] Merklein, Marion, Daniel Junker, Adam Schaub, and Franziska Neubauer. 2016. "Hybrid Additive Manufacturing Technologies—An Analysis Regarding Potentials and Applications." *Physics Procedia* 83: 549–559. https://doi.org/10.1016/j.phpro.2016.08.057.

[26] Kokare, Samruddha, J. P. Oliveira, and Radu Godina. 2023. "Life Cycle Assessment of Additive Manufacturing Processes: A Review." *Journal of Manufacturing Systems* 68: 536–559. https://doi.org/10.1016/j.jmsy.2023.05.007.

[27] Pusateri, Valentina, Michael Zwicky Hauschild, Sami Kara, Constantinos Goulas, and Stig Irving Olsen. 2024. "Quantitative Sustainability Assessment of Metal Additive Manufacturing: A Systematic Review." *CIRP Journal of Manufacturing Science and Technology* 49: 95–110. https://doi.org/10.1016/j.cirpj.2023.12.005.

[28] Gordon, R. Bertram, Marlen Bertram, and Thomas E. Graedel. 2006. "Metal Stocks and Sustainability." *Proceedings of the National Academy of Sciences* 103 (5): 1209–1214. https://doi.org/10.1073/pnas.0509498103.

[29] Henckens, M. L. C. M., P. P. J. Driessen, and E. Worrell. 2014. "Metal Scarcity and Sustainability, Analyzing the Necessity to Reduce the Extraction of Scarce Metals." *Resources, Conservation and Recycling* 93: 1–8. https://doi.org/10.1016/j.resconrec.2014.09.012.

[30] Gunn, Gus. 2011. *Mineral Scarcity—a Non-Issue*. British Geological Survey.

[31] Savolainen, Jyrki, Mikael Collan, Kalevi Kyläheiko, and Pasi Luukka. 2017. "On the Trade-off between the Leverage Effect and Real Options Thinking: A Simulation-Based Model on Metal Mining Investment." *International Journal of Production Economics* 194: 43–51. https://doi.org/10.1016/j.ijpe.2017.06.002.

[32] Savolainen, Jyrki, Mikael Collan, and Pasi Luukka. 2017. "Analyzing Operational Real Options in Metal Mining Investments with a System Dynamic Model." *The Engineering Economist* 62 (1): 54–72. https://doi.org/10.1080/0013791X.2016.1167988.

[33] Savolainen, Jyrki, Daniele Pedretti, and Mikael Collan. 2019. "Incorporating Hydrologic Uncertainty in Industrial Economic Models: Implications of Extreme Rainfall Variability on Metal Mining Investments." *Mine Water and the Environment* 38: 447–462. https://doi.org/10.1007/s10230-019-00600-w.

[34] Lubik, Sarah, and Elizabeth Garnsey. 2016. "Early Business Model Evolution in Science-Based Ventures: The Case of Advanced Materials." *Long Range Planning* 49 (3): 393–408. https://doi.org/10.1016/j.lrp.2015.03.001.

[35] Mellor, Ian, Lucy Grainger, Kartik Rao, James Deane, Melchiorre Conti, Greg Doughty, and Dion Vaughan. 2015. "Titanium Powder Production via the Metalysis Process." *Titanium Powder Metallurgy*: 51–67. https://doi.org/10.1016/B978-0-12-800054-0.00004-6.

[36] Baumers, Martin, Chris Tuck, Ricky Wildman, Ian Ashcroft, and Richard Hague. 2017. "Shape Complexity and Process Energy Consumption in Electron Beam Melting: A Case of Something for Nothing in Additive Manufacturing?" *Journal of Industrial Ecology* 21 (S1): S157–S167. https://doi.org/10.1111/jiec.12397.

[37] Serres, Nicolas, Dorian Tidu, Simon Sankare, and Françoise Hlawka. 2011. "Environmental Comparison of MESO-CLAD® Process and Conventional Machining Implementing Life Cycle Assessment." *Journal of Cleaner Production* 19 (9–10): 1117–1124. https://doi.org/10.1016/j.jclepro.2010.12.010.

[38] Morrow, W. R., H. Qi, I. Kim, J. Mazumder, and S. J. Skerlos. 2007. "Environmental Aspects of Laser-Based and Conventional Tool and Die Manufacturing." *Journal of Cleaner Production* 15 (10): 932–943. https://doi.org/10.1016/j.jclepro.2005.11.030.

[39] Stecker, S., K. W. Lachenberg, H. Wang, and R. C. Salo. 2006. "Advanced Electron Beam Free Form Fabrication Methods & Technology." *Proceedings of the AWS Welding Show/Annual Meeting (Session 2, Presentation 12)*. Miami, FL: American Welding Society.

[40] Syed, Waheed Ul Haq, Andrew J. Pinkerton, and Lin Li. 2006. "Combining Wire and Coaxial Powder Feeding in Laser Direct Metal Deposition for Rapid Prototyping." *Applied Surface Science* 252 (13): 4803–4808. https://doi.org/10.1016/j.apsusc.2005.08.118.

[41] Baufeld, Bernd, Erhard Brandl, and Omer Van Der Biest. 2011. "Wire Based Additive Layer Manufacturing: Comparison of Microstructure and Mechanical Properties of Ti–6Al–4V Components Fabricated by Laser-Beam Deposition and Shaped Metal Deposition." *Journal of Materials Processing Technology* 211 (6): 1146–1158. https://doi.org/10.1016/j.jmatprotec.2011.01.018.

[42] Heralić, Almir, Anna-Karin Christiansson, and Bengt Lennartson. 2012. "Height Control of Laser Metal-Wire Deposition Based on Iterative Learning Control and 3D Scanning." *Optics and Lasers in Engineering* 50 (9): 1230–1241. https://doi.org/10.1016/j.optlaseng.2012.03.016.

[43] Schmidt, Michael, Marion Merklein, David Bourell, Dimitri Dimitrov, Tino Hausotte, Konrad Wegener, Ludger Overmeyer, Frank Vollertsen, and Gideon N. Levy. 2017. "Laser Based Additive Manufacturing in Industry and Academia." *CIRP Annals* 66 (2): 561–583. https://doi.org/10.1016/j.cirp.2017.05.011.

[44] Ding, Donghong, Zengxi Pan, Dominic Cuiuri, and Huijun Li. 2015. "Wire-Feed Additive Manufacturing of Metal Components: Technologies, Developments and Future Interests." *The International Journal of Advanced Manufacturing Technology* 81 (1): 465–481. https://doi.org/10.1007/s00170-015-7077-3.

[45] Walachowicz, Frank, Ingo Bernsdorf, Ulrike Papenfuss, Christine Zeller, Andreas Graichen, Vladimir Navrotsky, Noorie Rajvanshi, and Christoph Kiener. 2017. "Comparative Energy, Resource and Recycling Lifecycle Analysis of the Industrial Repair Process of Gas Turbine Burners Using Conventional Machining and Additive Manufacturing." *Journal of Industrial Ecology* 21 (S1): S203–S215. https://doi.org/10.1111/jiec.12637.

[46] Le Bourhis, Florent, Olivier Kerbrat, Lucas Dembinski, Jean-Yves Hascoet, and Pascal Mognol. 2014. "Predictive Model for Environmental Assessment in Additive Manufacturing Process." *Procedia CIRP* 15: 26–31. https://doi.org/10.1016/j.procir.2014.06.031.

[47] Colomo, Alberto Garcia, Dudley Wood, Filomeno Martina, and Stewart W. Williams. 2020. "A Comparison Framework to Support the Selection of the Best Additive Manufacturing Process for Specific Aerospace Applications." *International Journal of Rapid Manufacturing* 9 (2–3): 194–211. https://doi.org/10.1504/IJRAPIDM.2020.107736.

[48] Duflou, Joost R., John W. Sutherland, David Dornfeld, Christoph Herrmann, Jack Jeswiet, Sami Kara, Michael Hauschild, and Karel Kellens. 2012. "Towards Energy and Resource Efficient Manufacturing: A Processes and Systems Approach." *CIRP Annals* 61 (2): 587–609. https://doi.org/10.1016/j.cirp.2012.05.002.

[49] Yoon, Hae-Sung, Jang-Yeob Lee, Hyung-Soo Kim, Min-Soo Kim, Eun-Seob Kim, Yong-Jun Shin, Won-Shik Chu, and Sung-Hoon Ahn. 2014. "A Comparison of Energy Consumption in Bulk Forming, Subtractive, and Additive Processes: Review and Case Study." *International Journal of Precision Engineering and Manufacturing-Green Technology* 1 (3): 261–279. https://doi.org/10.1007/s40684-014-0033-0.

[50] Peng, Tao, Karel Kellens, Renzhong Tang, Chao Chen, and Gang Chen. 2018. "Sustainability of Additive Manufacturing: An Overview on Its Energy Demand and Environmental Impact." *Additive Manufacturing* 21: 694–704. https://doi.org/10.1016/j.addma.2018.04.022.

[51] Michelsen, Karl-Erik, Mikael Collan, Jyrki Savolainen, and Paavo Ritala. 2021. "Changing Manufacturing Landscape: From a Factory to a Network." In *Handbook of Smart Materials,*

Technologies, and Devices: Applications of Industry 4.0, 1–21. Springer International Publishing. https://doi.org/10.1007/978-3-030-58675-1_2-1.

[52] Collan, Mikael, Jyrki Savolainen, Veli-Matti Virolainen, and Pasi Luukka. 2023. "A Vision for a Highly Automated Digital Local Manufacturing Network—Solutions and Challenges." In *Intelligent and Transformative Production in Pandemic Times: Proceedings of the 26th International Conference on Production Research*, 377–384. Springer International Publishing. https://doi.org/10.1007/978-3-031-18641-7_36.

[53] Ford, Simon, and Mélanie Despeisse. 2016. "Additive Manufacturing and Sustainability: An Exploratory Study of the Advantages and Challenges." *Journal of Cleaner Production* 137: 1573–1587. https://doi.org/10.1016/j.jclepro.2016.04.150.

[54] Baumers, Martin, Tuck Christopher, Ricky Wildman, Ian Ashcroft, Eric Rosamond, and Richard Hague. 2012. "Combined Build-Time, Energy Consumption and Cost Estimation for Direct Metal Laser Sintering." *International Solid Freeform Fabrication Symposium*. University of Texas at Austin, 1–14.

[55] Colwell, L. V. 1956. "The Effects of High-Frequency Vibrations in Grinding." *Journal of Fluids Engineering* 78 (4): 837–844. https://doi.org/10.1115/1.4013840.

[56] Markov, Alekseĭ Ivanovich, and E. A. Neppiras. 1966. "Ultrasonic Machining of Intractable Materials." *Proceedings of the Institution of Mechanical Engineers* 181 (1): 135–148.

[57] Munsch, Maximilian, Eric Wycisk, Jannis Kranz, Vanessa Seyda, and E. Claus. 2012. "Functional Products through Laser Additive Manufacturing of TiAl6V4." *Proceedings of the 4th Laser Additive Manufacturing Workshop (LAM 2012)*. Orlando, FL: Laser Institute of America.

[58] Krailling, G., and M. Novi. 2014. "EOS and Airbus Group Innovations Team on Aerospace Sustainability Study for Industrial 3D Printing." http://www.eos.info/eos_airbusgroupinnovationteam_aerospace_sustainability_study.

[59] Gisario, Annamaria, Michele Kazarian, Filomeno Martina, and Mehrshad Mehrpouya. 2019. "Metal Additive Manufacturing in the Commercial Aviation Industry: A Review." *Journal of Manufacturing Systems* 53: 124–149. https://doi.org/10.1016/j.jmsy.2019.08.005.

[60] Dawes, Jason, Robert Bowerman, and Ross Trepleton. 2015. "Introduction to the Additive Manufacturing Powder Metallurgy Supply Chain." *Johnson Matthey Technology Review* 59 (3): 243–256. https://doi.org/10.1595/205651315X688686.

Chapter 9

Recent Developments in Modelling Strategies of Laser Powder Bed Fusion

Rakesh Kumar, Shubhra Kamal Nandi, and Anupam Agrawal

9.1 INTRODUCTION

In Laser Powder Bed Fusion (LPBF), the melt pool is the transient molten region formed when a laser strikes a metal powder bed composed of particles ranging in size from 20 to 50 μm. The metal powder melts as the laser moves along a predefined path, forming a liquid pool that rapidly cools and solidifies. This entire process occurs within an extremely short timeframe, ranging from microseconds to milliseconds, and within a confined spatial domain of a few hundred microns in volume. Melt pool behaviour is governed by several complex and interrelated physical phenomena that ultimately influence the final properties of the fabricated part. The interaction between the laser and the material determines how energy is absorbed and distributed, leading to localised melting and the coalescence of powder particles into a continuous liquid pool. Surface tension and wetting effects are crucial in shaping the molten region, influencing its stability and interaction with the partially melted powder bed and the underlying substrate. Within the melt pool, fluid flow is driven by thermal gradients, surface tension variations (Marangoni effect), and buoyancy forces, all of which affect heat and mass transfer. Additionally, the evaporation of the molten metal generates recoil pressure at the liquid surface, contributing to additional heat loss and influencing melt pool dynamics. As solidification occurs, the release or absorption of latent heat affects microstructural evolution, residual stress formation, and overall mechanical properties. Understanding these interdependent phenomena is essential for optimising process parameters and achieving the desired material characteristics in LPBF-manufactured components.

Advanced experimental techniques, such as operando X-ray tomography, have revolutionised understanding of dynamic events in the LPBF process by capturing high-resolution, real-time data. With temporal resolutions on the order of tenths of a microsecond and spatial resolutions reaching just a few micrometres, these methods enable precise observation of melt pool dynamics, powder behaviour, and defect formation. Despite their potential, widespread application remains challenging due to the complexities associated with integrating them across diverse process parameters and material systems. To overcome these limitations, computational simulations have become indispensable for studying and optimising LPBF. Simulations play a crucial role in enhancing process stability and improving part quality by providing a deeper understanding of the fundamental physics governing melt pool behaviour (Cook and Murphy 2020).

Built on mathematical formulations derived from fluid dynamics and structural mechanics, LPBF simulations have evolved significantly, incorporating advanced computational approaches to model laser-material interactions, multi-material printing, and complex multiphysics phenomena. In recent years, these developments have expanded the predictive capabilities of simulations, making them a powerful tool for guiding experimental efforts and refining process parameters to achieve superior manufacturing outcomes. The increasing

 DOI: 10.1201/9781003590774-9

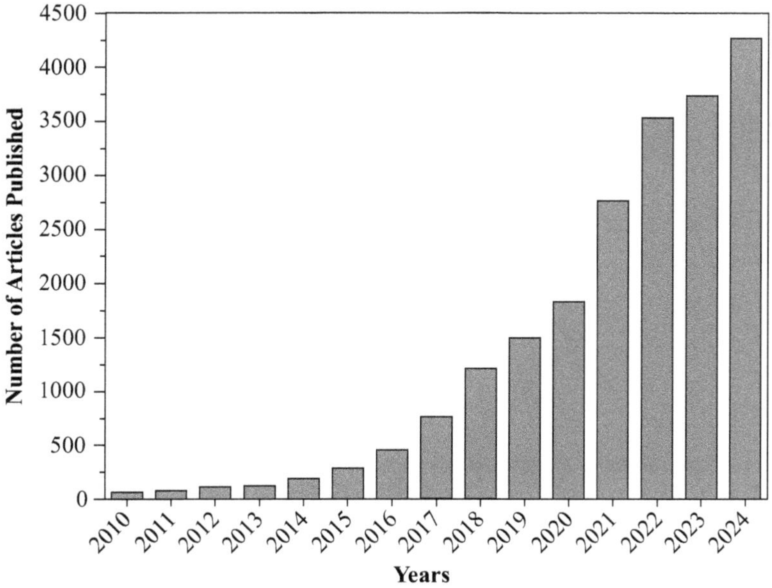

Figure 9.1 Growing volumes of publications on LPBF simulations to predict melt pool characteristics (Source: Google Scholar search, with keywords of Melt Pool, Simulation, LPBF, etc., data collected on February 2025).

volume of publications (**Figure 9.1**) on LPBF simulations over the last 15 years is a testament to the research community's growing interest in exploring and refining melt pool simulations.

A key objective of melt pool modelling is to predict temperature variations with high temporal and spatial resolution, as these govern several secondary phenomena, including geometrical characteristics, remelting, fluid velocity, porosity formation, and spattering. Furthermore, temperature distribution can be integrated with structural modelling to predict residual stress and coupled with phase-field physics to simulate microstructural evolution. This chapter presents a detailed discussion of the methodological aspects of melt-pool simulation in LPBF, focusing on their advantages and challenges.

9.2 COMPUTATIONAL FRAMEWORKS

Modelling approaches can be broadly classified into physics-based and data-driven methods. The physics-based approach relies on solving the governing equations of energy, mass, and momentum conservation. Depending on the computational domain, the governing partial differential equations can be solved via mesh-based methods such as the Finite Element Method (FEM) and Finite Volume Method (FVM) or mesh-free techniques like the Discrete Element Method (DEM), Lattice Boltzmann Method (LBM), and Smoothed Particle Hydrodynamics (SPH). Data-driven methods, such as machine learning (ML) algorithms, can handle complex data and a broader range of process parameters than traditional models. They can be trained using real-time experimental data or physics-based simulations. Once trained, ML models are computationally efficient, adaptable, and can be updated with new data, making them efficient for manufacturing. Although each strategy addresses the transient melt pool phenomenon, their differences in temporal and spatial resolution, material management, and

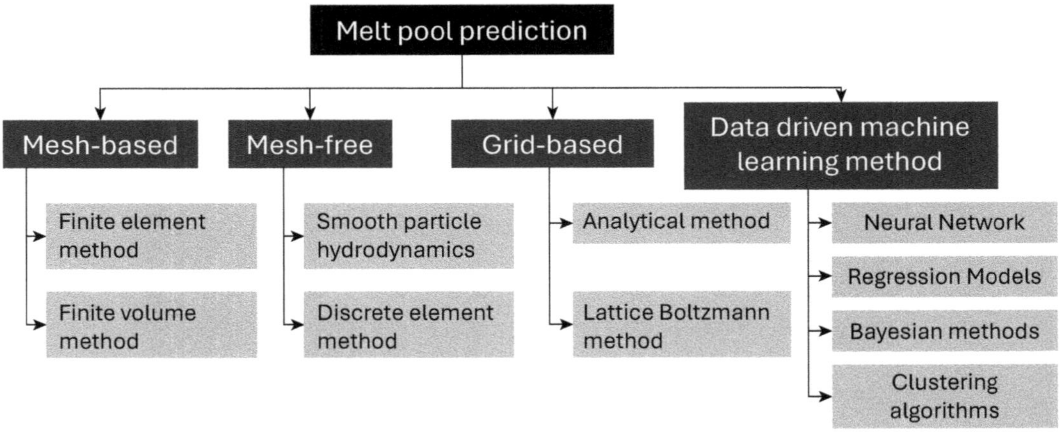

Figure 9.2 Different approaches to predict the melt-pool characteristics in laser powder bed fusion.

computational complexity influence user preferences. Recently, researchers have extensively used data- and image-based neural networks to predict various LPBF process characteristics, which require substantial physics-based and experimental process information. Each computational framework has unique features and innovations in incorporating complex physics and solving intricate phenomena, and researchers choose them based on convenience and problem-specific requirements. The subsequent sections detail the computational frameworks as classified in **Figure 9.2**.

9.2.1 Analytical Methods

The analytical route for simulations is perhaps the most computationally primitive but mathematically intensive method, operating on the classical solution steps of the partial differential equations, independent of the domain discretisation. It can provide exact, closed-form solutions for complex systems, with detailed insights into the underlying mechanisms, and often requires less computational cost, thus motivating researchers to adopt analytical methods to understand laser-material interaction, solidification behaviour, mechanical stresses, and deformations. A crucial step in the analytical framework is to formulate heat source models that effectively mimic laser irradiation in practice (Goldak et al. 1984; Van et al. 2007). Heat source models have evolved significantly over time, from line, surface, and volume models (Ebrahimi et al. 2023; Galbusera et al. 2023), as depicted in **Figure 9.3**.

Steven Y. Liang's Group at Georgia Institute of Technology has extensively contributed to this methodology by developing 3D thermal models of LPBF and studying effects of build layers, latent heat, and temperature-sensitivity of material properties, along with varying the process parameters (laser parameter, scanning parameters, layer thickness, etc.) (Mirkoohi et al. 2018; 2019). Nandi et al. (2021; 2022) proposed a thermal model based on heat conduction and Green's function by incorporating a Gaussian semi-ellipsoidal heat source and predicted melt pool dimensions for Ti6Al4V, stainless steel 316 and Inconel 718 alloy. The model's peculiarity was its lower computational cost. Building upon this study, Bombe et al. (2024) incorporated fluid flow behaviour and Marangoni effect in the solution domain, which was solved using energy and mass balance equations to estimate fluid velocity within the melt pool, improving the accuracy in melt pool dimension estimation (**Figure 9.4**). Further, a lot of solver-dependent studies have been carried out by multiple researchers to expedite the

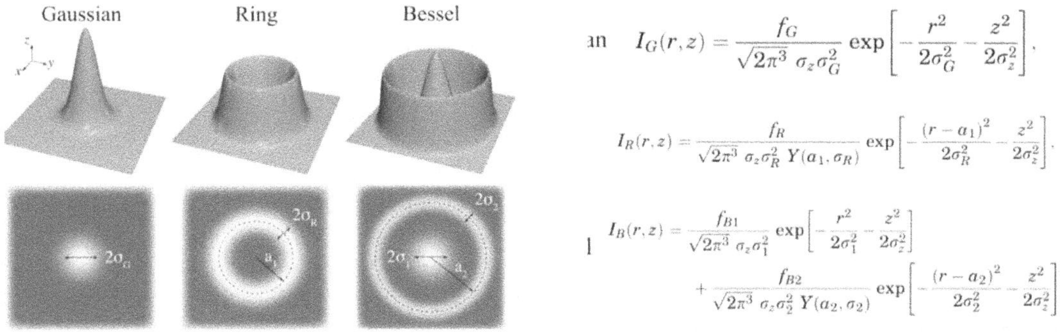

Figure 9.3 A schematic representation of laser beam shaping through different intensity profiles along with mathematical equations (Adapted from (Moore et al. 2024)).

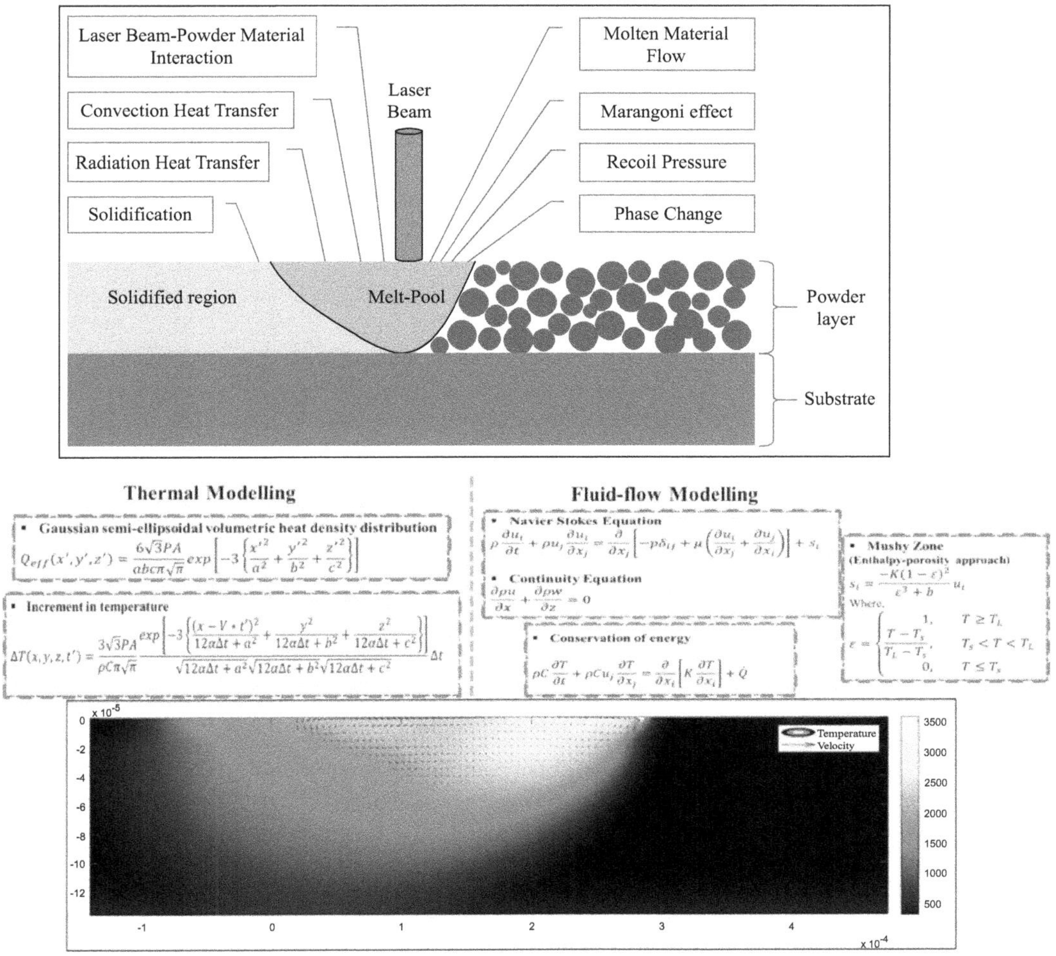

Figure 9.4 Schematic representing involved multiphysical phenomenon during laser-material interaction in LPBF process, with thermo-fluidic formulation and obtained melt pool.

simulation and achieve results at less computational cost (Kizhakkinan et al. 2023; Much et al. 2024; Zakirov et al. 2024), which increases the flexibility for the users to conduct parametric studies and even study laser additive for multi-material systems.

9.2.2 Finite Element Method

The finite element method (FEM) has been one of the most sought-after simulation techniques. FEM involves the discretisation of simulation domains into elements, and the concerned mathematical functions are reduced to their weak form and computed at every element and node. FEM of melt pool development in LPBF is a crucial area of research for understanding the thermal and mechanical behaviour of the melt pool during the additive manufacturing process. LPBF uses a laser to selectively fuse powdered metal, and the melt pool's size, shape, and dynamics play a significant role in the quality of the final part. The simulation of melt pool behaviour during this process is critical for understanding the material's thermal and mechanical characteristics, optimising manufacturing parameters, and predicting long-term material performance. FEM serves as a key tool in this effort, helping researchers model both the immediate dynamics and the long-term evolution of materials fabricated through LPBF. The application of FEM to LPBF emerged as the need for more accurate simulations in AM processes became evident. Early research into FEM for LPBF focused on understanding how laser energy interacts with powder material to form a melt pool and how different laser parameters affect its dimensions (Dong et al. 2009). Over time, this approach was refined to simulate more complex phenomena, such as thermal gradients, phase transitions, and the effects of material properties on the melt pool's size and shape (Bayat et al. 2019).

The laser is typically modelled as a moving heat source with a Gaussian distribution to approximate the laser beam's intensity (DebRoy et al. 2018). DebRoy and his research group have extensively explored heat transfer-based 3D FE simulation for metallic printing using LPBF and estimated detailed melt pool information, thermal gradients, residual stresses and solidification behaviour in the molten zone (Mukherjee et al. 2018a, 2018b). The thermal history generated by the laser profoundly affects the melt pool, influencing its size, shape, and cooling rate. The phase changes (from solid to liquid) during melting and solidification need to be carefully captured, along with temperature-dependent material properties such as thermal conductivity and the latent heat of fusion, enabling effective prediction of residual stresses (Y. Chen et al. 2022). FEM simulations can track the evolution of the melt pool's depth and width in real time, providing valuable insights into the potential for defects, such as porosity or cracks. Thermal stresses that develop as the molten material cools and solidifies can cause residual deformations or distortions, making it essential to incorporate thermal-mechanical coupling into the simulation (Li et al. 2018). By integrating time-dependent material models and validating them against experimental data, FEM simulations offer powerful tools for optimising LPBF processes and ensuring the durability of parts over a decade or more (Abhishek et al. 2024). These models consider the part dimensions and the accumulation of stresses and strains over the material's service life, which is crucial for ensuring the reliability of LPBF-manufactured components in demanding environments.

To perform FEM simulations of LPBF, commercial software such as ANSYS, ABAQUS, and COMSOL is commonly used. These software packages include advanced solvers for coupled thermal-mechanical simulations, which are essential for modelling the complex interactions during the LPBF process. The growth of FE formulations for LPBF is poised to continue accelerating as demand for high-performance parts and efficient additive manufacturing processes increases. Integrating advanced simulation tools, material modelling, and real-time adaptive control enables the next generation of LPBF technologies, making them more reliable, efficient, and applicable across a wide range of industries.

9.2.3 Finite Volume Method

The Finite Volume Method (FVM) is a numerical technique for solving partial differential equations, particularly in fluid dynamics and heat transfer. It involves discretising the governing equations (such as energy, momentum, or mass conservation) over small control volumes. For LPBF, FVM is applied to heat transfer and fluid dynamics within the melt pool, providing insight into temperature gradients, thermal cycles, and material flow. To model the heat transfer in the solid and liquid phases, laser energy absorption, thermal conduction, convection, and radiation are considered. Further, the Navier-Stokes Equation is used to simulate fluid flow in the melt pool, including convection driven by the movement of the molten material. The liquid phase in the melt pool behaves as a fluid under the influence of forces such as surface tension, the Marangoni effect, and gravity. Modelling the fluid flow inside the melt pool requires solving the Navier-Stokes equations in the liquid region. The motion of the melt pool interface due to this flow affects the final geometry of the printed layer. The transition from liquid to solid as the melt pool cools down and solidifies involves latent heat and moving interfaces. As the melt pool cools, the solidification process is influenced by the material's thermal history, which can affect the microstructure and mechanical properties of the printed part. FVM can simulate the solidification front and the cooling rates. Using proper boundary conditions, all the governing equations are solved to get the thermal history, fluid flow behaviour, and melt pool dimensions.

Researchers have performed extensive work on simulating the LPBF process using FVM techniques. Gürtler et al. (2013) have implemented the volume of fluid (VoF) approach in OpenFOAM and modified solvers to simulate the dynamics of laser beam melting for printing steel powders in the powder bed. The VoF method has been effective in reproducing the free surfaces of the multiphase system while describing melting, wetting, and solidification phenomena. Panwisawas et al. (2017) conducted thermal-fluid-dynamics and microstructural-evolution analyses for LPBF using mesoscale modelling. Through the model, thermal fluid flow and microstructure evolution have been established, which gives rise to good prediction in grain structure development during processing, considering the Marangoni effect and recoil pressure. Multi-track deposition simulations were developed using multiphysics modelling using the FVM approach to predict the thermal history and possible defect (balling effect, single-track nonuniformity and inter-track voids) formation (Yan et al. 2017; Kumar et al. 2024). With further advancements, it will be possible to improve the predictability of additive manufacturing processes, optimise production, and enhance the quality and longevity of printed components.

9.2.4 Discrete Element Method

Nonetheless, these continuum-based approaches assume homogeneity of the powder bed, which is a certain limitation in terms of spatial resolution. The discrete element method (DEM) is an extended finite-difference numerical technique grounded in Newtonian interactions among a system of particles, capable of tracking each powder particle's motion and temperature individually. To replicate the behaviour of individual particles in a granular flow, Cundall and Strack (1979) developed the DEM, which has since become one of the most effective simulation models for accurately reproducing granular flow. It is a Lagrangian method that simulates the behaviour of solid particles using Newton's laws of motion. From simulating only the behaviour of solid particles, DEM has advanced to include simulations of solid-solid contact, solid-liquid mixing, and powder flow. To execute solid-fluid coupling in the domain, DEM is currently widely utilised in LPBF to model small powder particles in powder beds and their inter-particle interactions (Chen et al. 2022). These DEM-based

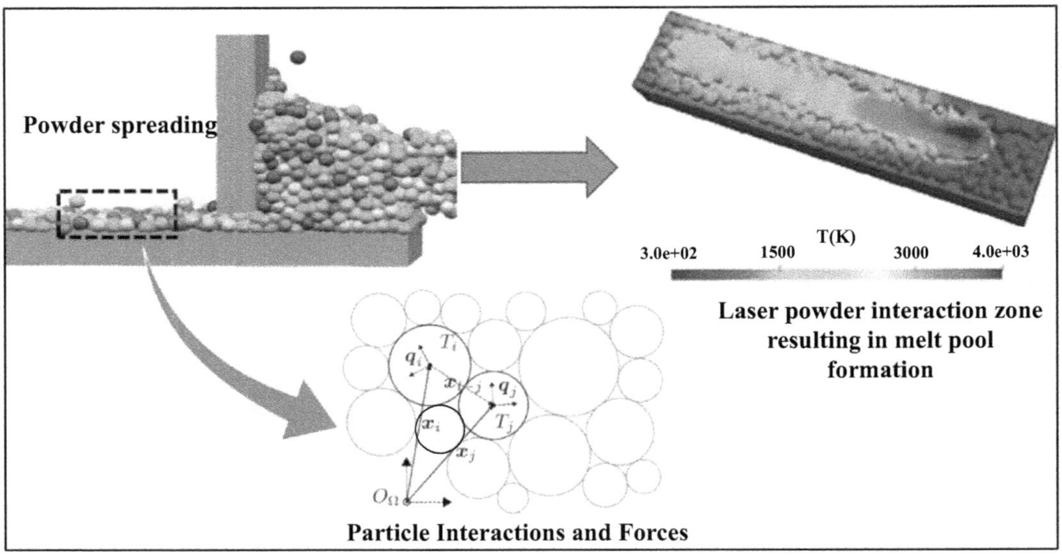

Figure 9.5 Schematic representation of DEM approach in LPBF.

powder beds are then coupled with a CFD formulation to model the actual melting and solidification phenomena using a phase-change method (**Figure 9.5**). DEM-CFD approaches are relatively computationally intensive compared to analytical and FE numerical simulations.

9.2.5 Lattice Boltzmann Method

Over the past few decades, the Lattice Boltzmann Method (LBM) has emerged as a powerful tool for fluid simulation, particularly in solving melting and solidification problems. Unlike continuum-based approaches such as the FEM and FVM, LBM is a mesoscale numerical approach that solves the incompressible Navier-Stokes (NS) equations using discrete particle distributions. The method tracks the density and velocity of these particles through a distribution function, $f(x,t)$, enabling efficient modelling of complex fluid dynamics (Zheng et al. 2019). In additive manufacturing processes such as Electron Beam Melting (EBM) and Selective Laser Melting (SLM), metal powders undergo rapid heating and cooling cycles, leading to significant thermal gradients and fluid motion. Accurately capturing these thermal effects requires coupling temperature fields with the velocity field. To achieve this, a second distribution function, $h(x,t)$, is introduced to model transient heat transfer. This advancement has led to the development of the Thermal Lattice Boltzmann Method (TLBM) (Liu and Wang 2019), which extends LBM by incorporating two distinct particle distribution functions—one for fluid dynamics and another for heat transfer. TLBM has proven highly effective in simulating single-phase and multiphase flows with complex boundary conditions and phase interfaces, making it particularly suitable for modelling powder-based metal additive manufacturing. Recent studies have successfully employed TLBM to simulate the evolution of temperature and velocity fields in EBM, providing critical insights into melt pool behaviour, thermal gradients, and defect formation. The ability of LBM and TLBM to capture microstructural evolution in melting and solidification processes makes them valuable tools for optimising process parameters and improving the quality of additively manufactured components.

The Körner group pioneered the first application of the LBM in selective beam melting (laser or electron), investigating the stochastic effects of randomly packed powder layers with identical packing density (Körner et al. 2011). Their study revealed that melt pool geometry exhibits highly chaotic behaviour, primarily influenced by local variations in bed density. They significantly advanced LBM implementation by modelling key hydrodynamic phenomena, including wetting effects, capillary forces, evaporation-induced material loss, and recoil pressure on the melt pool surface. A particular focus was placed on the localised melting of individual powder particles, providing insights into the sequential events of melting, coalescence, and melt pool growth, and demonstrating how local powder distribution irregularities affect the solidified melt pool. Building on these findings, they developed a 2D numerical model for multilayer additive manufacturing of a multi-component Ti-48Al-2Cr-2Nb alloy system (Klassen et al. 2017), emphasising melt-pool advection and selective alloy-element loss. Across their studies, they highlighted the dominant role of line energy in defect formation, demonstrating its greater influence compared to porosity formation and evaporation-induced material loss.

Dmytro Svyetlichnyy, along with his collaborators, has recently made groundbreaking progress in simulating a complete LPBF cycle, from powder bed generation to grain growth modelling, all within the scope of LBM (Krzyzanowski and Svyetlichnyy 2022). They first developed a comprehensive heating module based on a ray-tracing algorithm that appropriately accounted for refraction, scattering, and volume absorption. A holistic model was subsequently developed, comprising a powder bed generation module, followed by laser-material interaction leading to melting and solidification, and finally a powder removal step for the unmelted powders (Svyetlichnyy 2023) (**Figure 9.6**). The authors demonstrated the generation of non-uniform particle shapes in a powder bed for realistic visualisation. Multi-stage and multi-material aspects were further incorporated into their developing model, which demonstrated a diffusion model allowing a material mixture of AISI 316L austenitic stainless steel and 18Ni-300 martensitic steel to simulate the LPBF process of functionally graded materials that exhibit continuous variations in properties, offering unique multifunctional capabilities (Svyetlichnyy 2025).

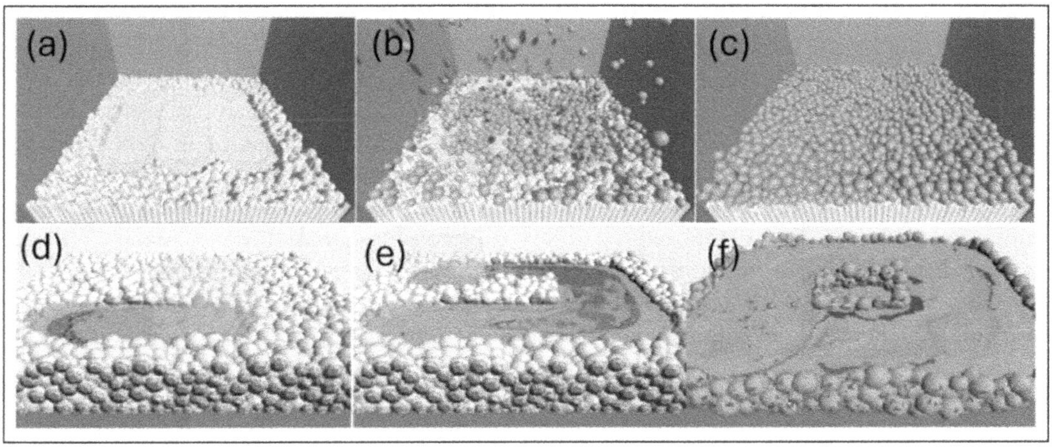

Figure 9.6 A complete cycle demonstration by LBM method, demonstrating (a) Modelling space before powder deposition, (b) Process of powder filling, and (c) Final state of deposition before the start of laser treatment. This is followed by the (d) beginning, (e) end of laser scanning, and (f) finally the solidified layer (Svyetlichnyy 2023).

Rai et al. (2016) developed an LBM model coupled with a CA-based grain growth model to predict the morphology and texture of the Inconel 718 alloy. In their 2D CA-lattice Boltzmann framework, the growth kinetics of dendritic grains were derived from the phase-field method. The grain size distribution and texture of the preheated base plate were used as inputs to determine the number of nuclei distributed in the base plate and the range of their orientations. The CA model was then employed to simulate grain-structure solidification, with phase-change information transferred back to the LB model to account for latent heat release during the solidification of liquid cells in the melt pool. A similar grain growth formulation within the TLBM framework was proposed by Liu et al. (Liu and Wang 2020), using a phase-field modelling approach for the AlSi10Mg alloy. Their model predicted the temperature field and captured compositional redistribution, which is crucial for understanding secondary arm formation. This approach significantly reduced the overestimation of microsegregation and provided a more accurate representation of dendritic growth kinetics.

9.3 SMOOTHED PARTICLE HYDRODYNAMICS

One of the most advanced methodological approaches recently adopted for melt pool simulation is the Smooth Particle Hydrodynamics (SPH) method. This inherently Lagrangian, mesh-free, particle-based numerical technique represents the simulation domain as a collection of discrete SPH particles. These particles interact through a force field derived directly from the governing equations of continuum mechanics, facilitating the exchange of mass, momentum, and energy between them (Shah and Volkov 2019). The SPH method describes the entire simulation area (or domain) using individual particles rather than a fixed grid. Each of these particles has a certain influence area, called a "smoothing length," within which its physical properties (such as temperature, velocity, or density) are spread out or averaged. This spreading is done using a mathematical function called a *kernel function*, which helps estimate the properties at any given location by considering contributions from nearby particles (Qiu et al. 2021). This approach allows SPH to smoothly approximate physical behaviours without requiring a structured grid. The primary advantages of using a Lagrangian-based numerical approach are the straightforward tracking of material points, multi-material simulation, simulation of multi-material fluids, and improved visualisation of the behaviour of the free surface.

The methodology was independently developed in the 1970s by Lucy, Gingold, and Monaghan for astrophysical problems. It was not until 2018 that Russell et al. (2018) established the foundation in laser material processing. Since then, the approach has been systematically refined to capture the complexities of the phenomenon, augment domain dimensionality, and manage computational costs. Researchers have incorporated key physical effects, including viscous forces, surface tension, and temperature-dependent material properties, to improve simulation accuracy. The methodology's ability to accommodate multiphase material models has been instrumental in simulating thermo-capillary phase change phenomena involving solid, liquid, and gaseous phases. These simulations also account for evaporation-induced recoil pressure, temperature-dependent surface tension, and wetting forces between the liquid melt and the underlying solid (Meier et al. 2021).

Building on these advancements, Meng et al. employed the SPH method to simulate spattering, in which numerous powders or partially melted particles are ejected (Meng et al. 2021). These spatters can form inclusions or remain un-melted regions within the printed part. This phenomenon, driven by recoil gas pressure and the ejection of solid-liquid material from the melt pool surface, has been experimentally observed using high-speed X-ray imaging. However, SPH remains the most suitable approach for its simulation due to its ability to model multiphase flow dynamics effectively.

The SPH-CA coupled model for the SLM process was demonstrated by Ma et al. (2024), who predicted the evolution of melt pool morphology by modelling thermal fluid flow and temperature distribution using SPH theory. The transient temperature field obtained from the SPH simulation serves as the thermal input for the CA method, which then models the solidification process to predict grain structure evolution, including grain size, shape, and spatial distribution. Compared to the FVM-CA coupled model, the SPH-CA approach offers superior accuracy in computing flow fields, temperature distributions, and data transfer. In the FVM-CA model, temperature values are interpolated from a coarse FVM grid to a finer CA grid, potentially introducing errors due to nonlinear temperature gradients in the melt pool. In contrast, using a kernel function, the SPH-CA model maintains a consistent discretisation scale and computes CA temperatures via a weighted summation of SPH particles. This approach effectively mitigates accuracy degradation and enhances numerical precision. A fully spatially adaptive 3D SPH framework for mesoscale AM simulations (**Figure 9.7**), featuring multi-level adaptivity, has been developed, with the addition of an optimised neighbour search algorithm to improve computational efficiency (Lüthi et al. 2023). By dynamically refining and coarsening particle spacing, our model reduces computational cost, requiring 80% fewer particles than a single-resolution LPBF simulation while achieving a fivefold speedup. This efficiency is crucial for simulating melt pool behaviour and material deposition, aligning with efforts to optimise mechanical properties, enhance design freedom, and mitigate fabrication defects in laser-deposited materials.

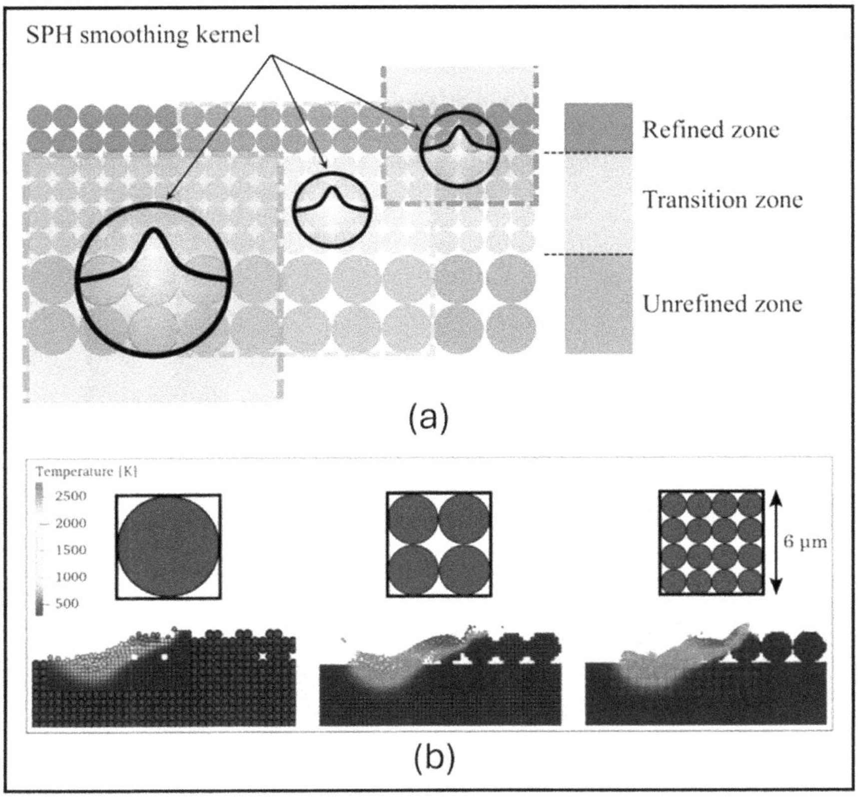

Figure 9.7 (a) Graphical illustration of SPH smoothening kernel and cell optimisation for multi-resolution SPH domains. The effect of multiple resolution and particle refinement is illustrated in (b), showing improvement in accurate melt surface in LPBF simulation (Afrasiabi et al. 2021).

9.4 MACHINE LEARNING-BASED MODELS

AI-ML has revolutionised engineering technology, and its use in manufacturing processes has accelerated the development of innovative, efficient, and successful methods for process monitoring, prediction, and control, as well as for assisting in the design of experiments (DoE) and optimising parameter selection for testing (Qi et al. 2019). The only limitation with this approach is that it requires large, high-quality datasets to learn from. There are different ML-based models, including Neural Network (NN), Regression-based Algorithms, Bayesian Algorithms, and Clustering Algorithms, that can effectively conduct the prediction of melt pool dimensions using statistical data, images, and multiphysics-based formulations. LPBF systems usually generate a vast amount of data; if that is noisy or insufficient, the performance of the models may degrade. Issues with inconsistent fabricated part quality and low process reproducibility have been an impending challenge in metal AM. Several quality improvement techniques have been put forth to overcome these obstacles and produce dependable, high-quality parts. Machine learning (ML) techniques are essential for improving LPBF quality, offering the best way to enhance the quality of manufactured parts due to the large number of parameters and complex phenomena that arise throughout the process (Zhang et al. 2024; Moradi et al. 2025). A lot of research is being done in this area to make AM even better. Because there is insufficient expertise and knowledge to integrate the complex AM process, process simulation, and closed-loop monitoring, the AM process is lagging in effectiveness and efficiency. The LPBF process has been mapped using various machine-learning-based predictive models to forecast different process attributes, and the impact of material properties has been examined (Agrawal et al. 2022). Sondagar et al. (2021) and Kumar et al. (2023) have adopted a MATLAB®-based neural network toolbox to develop data-driven models to forecast the influence of process factors on component strength and melt pool dimensions, respectively.

A convolutional neural network (CNN) is another tool that uses images/photographs as input data (**Figure 9.8**). This feature enables the use of in-situ melt-pool images, process-monitoring

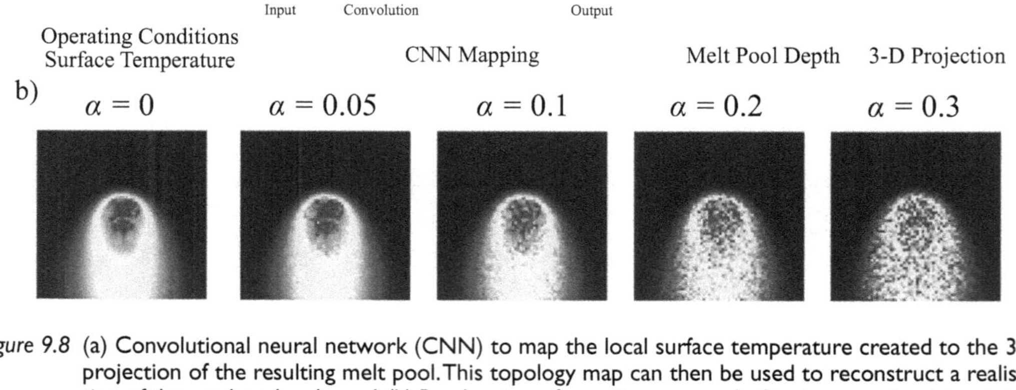

Figure 9.8 (a) Convolutional neural network (CNN) to map the local surface temperature created to the 3D projection of the resulting melt pool. This topology map can then be used to reconstruct a realisation of the produced melt pool. (b) Random transformations are applied to increase the dataset size, including rotation and noise addition (Ogoke et al. 2023).

images, part-defect features, etc., and trains an NN to detect in-process abnormalities (Yang et al. 2019). Researchers are using a physics-based deep generative model, a Convolutional Neural Network Transformer Generative Adversarial Network (CNN-TGAN), with computer vision to predict future melt pool monitoring images (Kim et al. 2024). The model's accuracy was tested by conducting experiments, comparative analyses, and quantitative and qualitative evaluations. Similarly, Cao et al. (2025) have used a DeepSort-based method to extract seven-dimensional spatter-feature images as input for the CNN model and trained the model with large data by varying the involved process parameters. The study laid the foundation for developing a relationship between melt-pool states and spatter characteristics. AI and ML have tremendous potential to enhance the efficiency, quality, and reliability of LPBF processes. By leveraging AI for process optimisation, defect/anomaly detection, material design, and real-time monitoring, manufacturers can reduce costs, improve quality, and enable the production of more complex and high-performance parts. However, integrating these technologies effectively requires careful consideration of data quality, model interpretability, and system integration.

9.5 DISCUSSION/IMPLICATION OF DIFFERENT FRAMEWORKS

Over the past decade, the LPBF process has made significant strides in the manufacturing industry and research for producing near-net-shaped metallic components with high precision. What was once primarily used for prototyping has now evolved into a method for creating complex functional components for sensitive applications. Despite its advantage of directly fabricating components from raw metallic powder, the process is complex and involves numerous thermophysical and metallurgical phenomena, making it challenging to explore novel materials. Currently, metal AM is limited to a range of materials (Fe-based, Ti-based, Ni-based, CoCr, and a few others), and researchers are striving to extend its capabilities by addressing the defects arising from the process (Chowdhury et al. 2022). This is where computational models turn out to be effective, by replicating the involved physical phenomenon in the LPBF process (Stittgen and Schleifenbaum 2021). These simulations typically involve analytical formulations, followed by FEM and FVM methods, to determine optimal process parameters for single- and multi-track processes, as well as multi-layer deposition. FE tools like COMSOL and OpenFOAM are better suited for incorporating multiple physics into the CFD simulation framework. While FE approaches have shown great potential for incorporating physical phenomena, they have become costly due to their high computational time. On the other hand, analytical methods can efficiently solve problems with lower computational costs. Additionally, implementing DEM approaches within the FEM model can improve simulation efficiency by accurately representing the powder bed rather than using a continuous bed domain, while accounting for powder spreading and particle interactions. This allows the visualisation and prediction of actual deformations and abnormalities that occur during the process by incorporating solid-mechanics formulations as subroutines within the simulation framework.

Advanced approaches, such as LBM and SPH-based techniques in conjunction with CA, are effective for simulating solidification and grain growth during the LPBF process (Lee et al. 2020). The advantage of the LBM approach is its low computational time and greater numerical stability, even with multiphase flow adaptation in the simulation (Nomoto et al. 2025). Data-driven AI-ML trends have also been among the recent trends in predicting process characteristics. The use of ML models has mechanised the process through closed-loop information systems, improving productivity by eliminating in-process defects and abnormalities. In this context, the digital twin will become a game changer by enabling minimal labour with automatic process parameter corrections and in-process improvement of part quality (Y.-P. Chen et al. 2025).

9.6 CONCLUSION

This chapter overviews the simulation techniques currently used to predict melt pool characteristics in the LPBF process. The computational approaches discussed have been instrumental for researchers to explore the process at different length scales (micro, macro, and multiscale), providing valuable insights. Numerous computational models have been developed in the past decade, with ongoing advancements that can accurately predict process and part characteristics. This enables the determination of optimal process parameters and the prediction of metallurgical properties, such as microstructure and grain growth, thereby allowing the tailoring of mechanical properties in LPBF components. The continuous development of simulation frameworks has also enabled visualisation of the entire process in real time on a computer display, facilitating process monitoring and immediate adjustments to process parameters to eliminate defects and improve part characteristics. In the near future, integrating robust, high-fidelity simulation models, online monitoring systems, and AI-ML-enabled digital twin frameworks will enable closed-loop system development, resulting in high-quality, defect-free additively manufactured metallic parts. The field of LPBF simulation is constantly evolving, with the introduction of innovative and efficient tools that provide rapid prediction of process characteristics and demonstrate high accuracy.

REFERENCES

Abhishek, Shrivastava, S. Anand Kumar, and Samrat Rao. 2024. "Multi-Scale Modelling for Optimization of Process Parameters of Laser Powder Bed Fusion Processed Inconel 718 Surrogate Part." *Engineering Failure Analysis* 155 (January): 107713. https://doi.org/10.1016/j.engfailanal.2023.107713.

Afrasiabi, Mohamadreza, Christof Lüthi, Markus Bambach, and Konrad Wegener. 2021. "Multi-Resolution SPH Simulation of a Laser Powder Bed Fusion Additive Manufacturing Process." *Applied Sciences (Switzerland)* 11 (7). https://doi.org/10.3390/app11072962.

Agrawal, Ankur K., Behzad Rankouhi, and Dan J. Thoma. 2022. "Predictive Process Mapping for Laser Powder Bed Fusion: A Review of Existing Analytical Solutions." In *Current Opinion in Solid State and Materials Science*. Elsevier. https://doi.org/10.1016/j.cossms.2022.101024.

Bayat, Mohamad, Sankhya Mohanty, and Jesper Henri Hattel. 2019. "A Systematic Investigation of the Effects of Process Parameters on Heat and Fluid Flow and Metallurgical Conditions during Laser-Based Powder Bed Fusion of Ti6Al4V Alloy." *International Journal of Heat and Mass Transfer* 139 (August): 213–230. https://doi.org/10.1016/j.ijheatmasstransfer.2019.05.017.

Bombe, Dattatraya, Rakesh Kumar, Shubhra Kamal Nandi, and Anupam Agrawal. 2024. "Semi-Analytical Formulation for Single-Track Laser Powder-Bed Fusion Process to Estimate Melt Pool Characteristics Considering Fluid-Flow and Marangoni Effect." *International Journal on Interactive Design and Manufacturing (IJIDeM)* 18 (7): 5121–5137. https://doi.org/10.1007/s12008-023-01593-1.

Cao, Longchao, Weihong Li, Yahui Zhang, Wang Cai, Zhen Wang, Wentao Guo, and Qi Zhou. 2025. "Spatter Feature-Based in-Situ Identification of Disturbances in Process Parameters during Laser Powder Bed Fusion." *Measurement Science and Technology* 36 (1): 015601. https://doi.org/10.1088/1361-6501/ad8022.

Chen, Hui, Yajing Sun, Weihao Yuan, Shengyong Pang, Wentao Yan, and Yusheng Shi. 2022. "A Review on Discrete Element Method Simulation in Laser Powder Bed Fusion Additive Manufacturing."

Chinese Journal of Mechanical Engineering: Additive Manufacturing Frontiers 1 (1): 100017. https://doi.org/10.1016/j.cjmeam.2022.100017.

Chen, Yi-Ping, Vispi Karkaria, Ying-Kuan Tsai, Faith Rolark, Daniel Quispe, Robert X. Gao, Jian Cao, and Wei Chen. 2025. "Real-Time Decision-Making for Digital Twin in Additive Manufacturing with Model Predictive Control Using Time-Series Deep Neural Networks," January, 1–19. http://arxiv.org/abs/2501.07601.

Chen, Yong, Yan Liu, Hui Chen, Ying Wu, JingQing Chen, Jun Xiong, LiSha Ren, and Jun Qian. 2022. "Multi-Scale Residual Stress Prediction for Selective Laser Melting of High Strength Steel Considering Solid-State Phase Transformation." *Optics & Laser Technology* 146 (September 2021): 107578. https://doi.org/10.1016/j.optlastec.2021.107578.

Chowdhury, Sohini, N. Yadaiah, Chander Prakash, Seeram Ramakrishna, Saurav Dixit, Lovi Raj Gupta, and Dharam Buddhi. 2022. "Laser Powder Bed Fusion: A State-of-the-Art Review of the Technology, Materials, Properties & Defects, and Numerical Modelling." *Journal of Materials Research and Technology* 20 (September): 2109–2172. https://doi.org/10.1016/J.JMRT.2022.07.121.

Cook, Peter S., and Anthony B. Murphy. 2020. "Simulation of Melt Pool Behaviour during Additive Manufacturing: Underlying Physics and Progress." *Additive Manufacturing* 31 (January): 100909. https://doi.org/10.1016/j.addma.2019.100909.

Cundall, P. A., and O. D. L. Strack. 1979. "A Discrete Numerical Model for Granular Assemblies." *Géotechnique* 29 (1): 47–65. https://doi.org/10.1680/geot.1979.29.1.47.

DebRoy, T., H. L. Wei, J. S. Zuback, T. Mukherjee, J. W. Elmer, J. O. Milewski, A. M. Beese, A. Wilson-Heid, A. De, and W. Zhang. 2018. "Additive Manufacturing of Metallic Components—Process, Structure and Properties." *Progress in Materials Science* 92: 112–224. https://doi.org/10.1016/j.pmatsci.2017.10.001.

Dong, L., A. Makradi, S. Ahzi, and Y. Remond. 2009. "Three-Dimensional Transient Finite Element Analysis of the Selective Laser Sintering Process." *Journal of Materials Processing Technology* 209 (2): 700–706. https://doi.org/10.1016/j.jmatprotec.2008.02.040.

Ebrahimi, Amin, Mohammad Sattari, Aravind Babu, Arjun Sood, Gert-Willem R. B. E. Römer, and Marcel J. M. Hermans. 2023. "Revealing the Effects of Laser Beam Shaping on Melt Pool Behaviour in Conduction-Mode Laser Melting." *Journal of Materials Research and Technology* 27 (November): 3955–3967. https://doi.org/10.1016/J.JMRT.2023.11.046.

Galbusera, Francesco, Leonardo Caprio, Barbara Previtali, and Ali Gökhan Demir. 2023. "The Influence of Novel Beam Shapes on Melt Pool Shape and Mechanical Properties of LPBF Produced Al-Alloy." *Journal of Manufacturing Processes* 85 (January): 1024–1036. https://doi.org/10.1016/J.JMAPRO.2022.12.007.

Goldak, John, Aditya Chakravarti, and Malcolm Bibby. 1984. "A New Finite Element Model for Welding Heat Sources." *Metallurgical Transactions B* 15 (2): 299–305. https://doi.org/10.1007/BF02667333.

Gürtler, F. J., M. Karg, K. H. Leitz, and M. Schmidt. 2013. "Simulation of Laser Beam Melting of Steel Powders Using the Three-Dimensional Volume of Fluid Method." *Physics Procedia* 41 (January): 881–886. https://doi.org/10.1016/J.PHPRO.2013.03.162.

Kim, Jaehyuk, Zhuo Yang, and Yan Lu. 2024. "A Physics-Guided Deep Generative Model for Predicting Melt Pool Behavior in Laser Powder Bed Fusion Additive Manufacturing." *Journal of Intelligent Manufacturing* (November): 1–21. https://doi.org/10.1007/s10845-024-02504-1.

Kizhakkinan, Umesh, Pham Luu Trung Duong, Robert Laskowski, Guglielmo Vastola, David W. Rosen, and Nagarajan Raghavan. 2023. "Development of a Surrogate Model for High-Fidelity Laser Powder-Bed Fusion Using Tensor Train and Gaussian Process Regression." *Journal of Intelligent Manufacturing* 34 (1): 369–385. https://doi.org/10.1007/s10845-022-02038-4.

Klassen, Alexander, Vera E. Forster, Vera Juechter, and Carolin Körner. 2017. "Numerical Simulation of Multi-Component Evaporation during Selective Electron Beam Melting of TiAl." *Journal of Materials Processing Technology* 247 (April): 280–288. https://doi.org/10.1016/j.jmatprotec.2017.04.016.

Körner, Carolin, Elham Attar, and Peter Heinl. 2011. "Mesoscopic Simulation of Selective Beam Melting Processes." *Journal of Materials Processing Technology* 211 (6): 978–987. https://doi.org/10.1016/j.jmatprotec.2010.12.016.

Krzyzanowski, Michal, and Dmytro Svyetlichnyy. 2022. "A Multiphysics Simulation Approach to Selective Laser Melting Modelling Based on Cellular Automata and Lattice Boltzmann Methods." *Computational Particle Mechanics*. https://doi.org/10.1007/s40571-021-00397-y.

Kumar, Rakesh, Dattatraya Bombe, and Anupam Agrawal. 2023. "A Data-Driven ANN Model for Estimation of Melt pool Characteristics in SLM Process." *Manufacturing Letters* 35 (August): 620–625. https://doi.org/10.1016/j.mfglet.2023.08.041.

Kumar, Rakesh, Abhishek Rajput, and Anupam Agrawal. 2024. "Temperature and Fluid Velocity Field Estimation in Melt pool Using Multiphysics-Based Numerical Simulation for Multi-Track Selective Laser Melting Process." In *29th International Conference on Processing and Fabrication of Advanced Materials (PFAM)—2023*, 397–407. Springer. https://doi.org/10.1007/978-981-97-5963-7_27.

Lee, Wonjoo, Yuhyeong Jeong, Jae Wook Lee, Howon Lee, Seong hoon Kang, Young Min Kim, and Jonghun Yoon. 2020. "Numerical Simulation for Dendrite Growth in Directional Solidification Using LBM-CA (Cellular Automata) Coupled Method." *Journal of Materials Science and Technology* 49 (July): 15–24. https://doi.org/10.1016/j.jmst.2020.01.047.

Li, Yingli, Kun Zhou, Pengfei Tan, Shu Beng Tor, Chee Kai Chua, and Kah Fai Leong. 2018. "Modeling Temperature and Residual Stress Fields in Selective Laser Melting." *International Journal of Mechanical Sciences* 136 (February): 24–35. https://doi.org/10.1016/j.ijmecsci.2017.12.001.

Liu, Dehao, and Yan Wang. 2019. "Mesoscale Multi-Physics Simulation of Rapid Solidification of Ti-6Al-4V Alloy." *Additive Manufacturing* 25 (October 2018): 551–562. https://doi.org/10.1016/j.addma.2018.12.005.

Liu, Dehao, and Yan Wang. 2020. "Multiphysics Simulation of Nucleation and Grain Growth in Selective Laser Melting of Alloys." *Journal of Computing and Information Science in Engineering* 20 (5): 1–11. https://doi.org/10.1115/1.4046543.

Lüthi, C., M. Afrasiabi, and M. Bambach. 2023. "An Adaptive Smoothed Particle Hydrodynamics (SPH) Scheme for Efficient Melt Pool Simulations in Additive Manufacturing." *Computers and Mathematics with Applications* 139 (October 2022): 7–27. https://doi.org/10.1016/j.camwa.2023.03.003.

Ma, Yibo, Xu Zhou, Fan Zhang, Christian Weißenfels, and Moubin Liu. 2024. "A Novel Smoothed Particle Hydrodynamics Method for Multi-Physics Simulation of Laser Powder Bed Fusion." *Computational Mechanics* 74 (5): 1009–1036. https://doi.org/10.1007/s00466-024-02465-5.

Meier, Christoph, Sebastian L. Fuchs, A. John Hart, and Wolfgang A. Wall. 2021. "A Novel Smoothed Particle Hydrodynamics Formulation for Thermo-Capillary Phase Change Problems with Focus on Metal Additive Manufacturing Melt Pool Modeling." *Computer Methods in Applied Mechanics and Engineering* 381 (August): 113812. https://doi.org/10.1016/J.CMA.2021.113812.

Meng, Lingbin, Tao Sun, Tejesh Dube, Sugrim Sagar, Xuehui Yang, Jian Zhang, and Jing Zhang. 2021. "Simulation of Spatters Sticking Phenomenon in Laser Powder Bed Fusion Process Using the Smoothed Particle Hydrodynamics Method." *ASME International Mechanical Engineering Congress and Exposition, Proceedings (IMECE)* 2A-2021: 1–7. https://doi.org/10.1115/IMECE2021-66761.

Mirkoohi, Elham, Jinqiang Ning, Peter Bocchini, Omar Fergani, Kuo Ning Chiang, and Steven Y. Liang. 2018. "Thermal Modeling of Temperature Distribution in Metal Additive Manufacturing Considering Effects of Build Layers, Latent Heat, and Temperature-Sensitivity of Material Properties." *Journal of Manufacturing and Materials Processing* 2 (3). https://doi.org/10.3390/jmmp2030063.

Mirkoohi, Elham, Daniel E. Sievers, Hamid Garmestani, Kuoning Chiang, and Steven Y. Liang. 2019. "Three-Dimensional Semi-Elliptical Modeling of Melt Pool Geometry Considering Hatch Spacing and Time Spacing in Metal Additive Manufacturing." *Journal of Manufacturing Processes* 45 (September): 532–543. https://doi.org/10.1016/j.jmapro.2019.07.028.

Moore, Robert, Giovanni Orlandi, Theron Rodgers, Daniel Moser, Heather Murdoch, and Fadi Abdeljawad. 2024. "Microstructure-Based Modeling of Laser Beam Shaping During Additive Manufacturing." *JOM* 76 (3): 1726–1736. https://doi.org/10.1007/s11837-023-06363-8.

Moradi, Alireza, Sanae Tajalli, Mohammad Hossein Mosallanejad, and Abdollah Saboori. 2025. "Intelligent Laser-Based Metal Additive Manufacturing: A Review on Machine Learning for Process

Optimization and Property Prediction." *The International Journal of Advanced Manufacturing Technology* 136 (2): 527–560. https://doi.org/10.1007/s00170-024-14858-0.

Much, Nils, Magdalena Schreter-Fleischhacker, Peter Munch, Martin Kronbichler, Wolfgang A. Wall, and Christoph Meier. 2024. "Improved Accuracy of Continuum Surface Flux Models for Metal Additive Manufacturing Melt Pool Simulations." *Advanced Modeling and Simulation in Engineering Sciences* 11 (1): 16. https://doi.org/10.1186/s40323-024-00270-6.

Mukherjee, T., H. L. Wei, A. De, and T. DebRoy. 2018a. "Heat and Fluid Flow in Additive Manufacturing—Part II: Powder Bed Fusion of Stainless Steel, and Titanium, Nickel and Aluminum Base Alloys." *Computational Materials Science* 150 (April): 369–380. https://doi.org/10.1016/j.commatsci.2018.04.027.

Mukherjee, T., H. L. L. Wei, A. De, and T. DebRoy. 2018b. "Heat and Fluid Flow in Additive Manufacturing—Part I: Modeling of Powder Bed Fusion." *Computational Materials Science* 150 (April): 304–313. https://doi.org/10.1016/j.commatsci.2018.04.022.

Nandi, Shubhra Kamal, Rakesh Kumar, and Anupam Agrawal. 2022. "Computationally Inexpensive Semi-Analytical Thermal Model to Predict Melt pool Dimensions for a Single-Track in Selective Laser Melting." *Journal of Manufacturing Processes* 80 (August): 469–479. https://doi.org/10.1016/j.jmapro.2022.06.025.

Nandi, Shubhra Kamal, Rakesh Kumar, Anubhav, and Anupam Agrawal. 2021. "Prediction of Melt pool Characteristics in SLM Process for Ti6Al4V Using a Semi-Analytical Model." In *Volume 1: Additive Manufacturing; Advanced Materials Manufacturing; Biomanufacturing; Life Cycle Engineering; Manufacturing Equipment and Automation*. American Society of Mechanical Engineers. https://doi.org/10.1115/MSEC2021-63751.

Nomoto, Sukeharu, Masahiro Kusano, Tomonori Kitashima, and Makoto Watanabe. 2025. "Numerical Simulation Method for the Laser Powder Bed Fusion Process by Lattice Boltzmann and Multi-Phase Field Methods." *Computational Materials Science* 250 (February): 113688. https://doi.org/10.1016/j.commatsci.2025.113688.

Ogoke, Francis, William Lee, Ning-Yu Kao, Alexander Myers, Jack Beuth, Jonathan Malen, and Amir Barati Farimani. 2023. "Convolutional Neural Networks for Melt Depth Prediction and Visualization in Laser Powder Bed Fusion." *The International Journal of Advanced Manufacturing Technology* 129 (7–8): 3047–3062. https://doi.org/10.1007/s00170-023-12384-z.

Panwisawas, Chinnapat, Chunlei Qiu, Magnus J. Anderson, Yogesh Sovani, Richard P. Turner, Moataz M. Attallah, Jeffery W. Brooks, and Hector C. Basoalto. 2017. "Mesoscale Modelling of Selective Laser Melting: Thermal Fluid Dynamics and Microstructural Evolution." *Computational Materials Science*. https://doi.org/10.1016/j.commatsci.2016.10.011.

Qi, Xinbo, Guofeng Chen, Yong Li, Xuan Cheng, and Changpeng Li. 2019. "Applying Neural-Network-Based Machine Learning to Additive Manufacturing: Current Applications, Challenges, and Future Perspectives." *Engineering* 5 (4): 721–729. https://doi.org/10.1016/J.ENG.2019.04.012.

Qiu, Yunji, Xiaofeng Niu, Tingting Song, Mengqing Shen, Wenqi Li, and Wenliang Xu. 2021. "Three-Dimensional Numerical Simulation of Selective Laser Melting Process Based on SPH Method." *Journal of Manufacturing Processes* 71 (September): 224–236. https://doi.org/10.1016/j.jmapro.2021.09.018.

Rai, Abha, Matthias Markl, and Carolin Körner. 2016. "A Coupled Cellular Automaton–Lattice Boltzmann Model for Grain Structure Simulation during Additive Manufacturing." *Computational Materials Science* 124 (November): 37–48. https://doi.org/10.1016/j.commatsci.2016.07.005.

Russell, M. A., A. Souto-Iglesias, and T. I. Zohdi. 2018. "Numerical Simulation of Laser Fusion Additive Manufacturing Processes Using the SPH Method." *Computer Methods in Applied Mechanics and Engineering* 341 (November): 163–187. https://doi.org/10.1016/j.cma.2018.06.033.

Shah, Deepak, and Alexey N. Volkov. 2019. "Combined Smoothed Particle Hydrodynamics—Ray Tracing Method for Simulations of Keyhole Formation in Laser Melting of Bulk and Powder Metal Targets." In *Volume 2A: Advanced Manufacturing*, 2A-2019:1–8. American Society of Mechanical Engineers. https://doi.org/10.1115/IMECE2019-11596.

Sondagar, Hardik, S. S. Bhadauria, and V. S. Sharma. 2021. "Artificial Neural Network (ANN) Based Prediction of Process Parameters in Additive Manufacturing." *IOP Conference Series: Materials Science and Engineering* 1136 (1): 012026. https://doi.org/10.1088/1757-899X/1136/1/012026.

Stittgen, Tobias, and Johannes Henrich Schleifenbaum. 2021. "Simulation of Utilization for LPBF Manufacturing Systems." *Production Engineering* 15 (1): 45–56. https://doi.org/10.1007/S11740-020-00998-1/FIGURES/5.

Svyetlichnyy, Dmytro. 2023. "Model of the Selective Laser Melting Process-Powder Deposition Models in Multistage Multi-Material Simulations." *Applied Sciences (Switzerland)*. https://doi.org/10.3390/app13106196.

Svyetlichnyy, Dmytro. 2025. "Lattice Boltzmann Modeling of Additive Manufacturing of Functionally Graded Materials." *Entropy* 27 (1). https://doi.org/10.3390/e27010020.

Van, Elsen M., M. Baelmans, P. Mercelis, and J. P. Kruth. 2007. "Solutions for Modelling Moving Heat Sources in a Semi-Infinite Medium and Applications to Laser Material Processing." *International Journal of Heat and Mass Transfer* 50 (23–24): 4872–4882. https://doi.org/10.1016/j.ijheatmasstransfer.2007.02.044.

Yan, Wentao, Wenjun Ge, Ya Qian, Stephen Lin, Bin Zhou, Wing Kam Liu, Feng Lin, and Gregory J. Wagner. 2017. "Multi-Physics Modeling of Single/Multiple-Track Defect Mechanisms in Electron Beam Selective Melting." *Acta Materialia* 134 (August): 324–333. https://doi.org/10.1016/j.actamat.2017.05.061.

Yang, Zhuo, Yan Lu, Ho Yeung, and Sundar Krishnamurty. 2019. "Investigation of Deep Learning for Real-Time Melt Pool Classification in Additive Manufacturing." *IEEE International Conference on Automation Science and Engineering* 2019-Augus (August): 640–647. https://doi.org/10.1109/COASE.2019.8843291.

Zakirov, Andrey, Sergei Belousov, Maria Bogdanova, Boris Korneev, Inna Iskandarova, Anastasia Perepelkina, and Boris Potapkin. 2024. "KiSSAM: Efficient Simulation of Melt Pool Dynamics during PBF Using GPUs." *Progress in Additive Manufacturing* m (0123456789). https://doi.org/10.1007/s40964-023-00561-1.

Zhang, Jiayi, Ce Yin, Yiyang Xu, and Swee Leong Sing. 2024. "Machine Learning Applications for Quality Improvement in Laser Powder Bed Fusion: A State-of-the-Art Review." *International Journal of AI for Materials and Design* 1 (1): 26. https://doi.org/10.36922/ijamd.2301.

Zheng, Min, Lei Wei, Jing Chen, Qiang Zhang, Chongliang Zhong, Xin Lin, and Weidong Huang. 2019. "A Novel Method for the Molten Pool and Porosity Formation Modelling in Selective Laser Melting." *International Journal of Heat and Mass Transfer* 140: 1091–1105. https://doi.org/10.1016/j.ijheatmasstransfer.2019.06.038.

Chapter 10

Recent Developments in Numerical Modelling for AM Processes

Peijie Shangguan and Swee Leong Sing

10.1 BACKGROUND

With the ongoing advancement of Industry 4.0 and the rapid development of high-performance computing, high-fidelity numerical modelling has become an essential tool for understanding and optimising advanced manufacturing processes. Compared with purely experimental approaches, numerical simulations can reduce material and equipment costs, provide access to transient physical fields, and reveal multi-scale mechanisms that are difficult to capture directly. By solving governing equations that describe heat transfer, fluid flow, phase change, and mechanical response, numerical modelling can generate quantitative insights that complement experimental observations and support process optimisation.

Powder bed fusion (PBF) is one of the most widely used metal additive manufacturing processes. This process offers significant advantages in producing geometrically complex, lightweight, and high-performance metallic parts without the need for conventional tooling. Owing to these advantages, PBF has been widely applied in the aerospace, biomedical, automotive, and energy sectors [1].

Despite its technological importance, PBF involves highly complex and strongly coupled physical phenomena. These phenomena occur over a wide range of time and length scales, making direct experimental characterisation challenging. Numerical modelling, therefore, provides a powerful framework for resolving the underlying physics of the process and predicting process-induced defects.

This chapter primarily concentrates on the numerical modelling of PBF. The content is organised as follows: Section 10.2 addresses the PBF process, from powder spreading to melt-pool dynamics, including the fundamental governing equations, key physical models, and heat-source and evaporation-induced phenomena. Section 10.3 provides an overview of the numerical techniques and software tools commonly used in AM modelling. Finally, Section 10.4 consolidates the modelling perspectives developed throughout the chapter and discusses current challenges and prospective directions.

10.2 POWDER BED FUSION SIMULATION

PBF is one of the earliest and most widely adopted AM technologies [2]. With its exceptional precision, superior material utilisation, and unparalleled design flexibility, PBF has become a critical process for producing complex geometries and high-performance components. The technology has been extensively applied to the fabrication of metallic and polymer parts and has also demonstrated considerable potential for the manufacture of ceramics and composite materials. As shown in **Figure 10.1**, PBF technologies typically involve a series of steps,

DOI: 10.1201/9781003590774-10

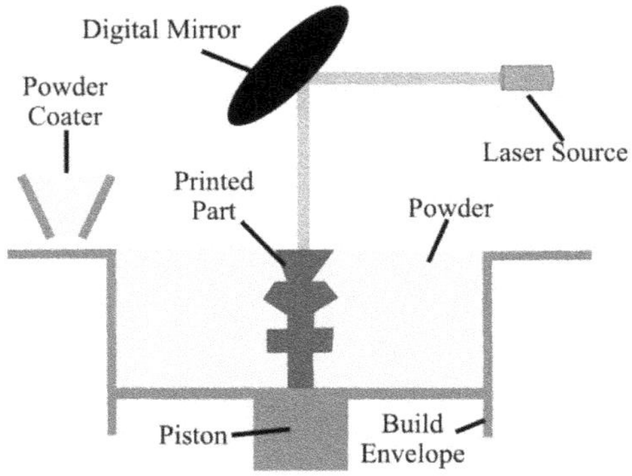

Figure 10.1 Illustration of PBF process (Adapted from [3]).

including powder spreading, energy beam scanning, and layer-by-layer deposition [3]. The energy input, delivered by either a laser or an electron beam, selectively heats the powder bed to achieve either sintering or melting. These operations are typically conducted in a vacuum or under an inert gas atmosphere to prevent oxidation of the material at elevated temperatures.

PBF processes can be further categorised into four primary technologies: Electron Beam Melting (EBM), Selective Laser Sintering (SLS), Direct Metal Laser Sintering (DMLS), and Selective Laser Melting (SLM) [4]. The SLS process is primarily utilised to produce plastic parts and is generally performed under atmospheric or inert gas conditions. EBM necessitates a vacuum environment, with an electron beam emitted from a coil-controlled tungsten filament serving as the energy source. SLM is well-suited for the rapid manufacturing of metallic parts, offering higher speeds than SLS. DMLS, primarily designed for metallic powders, is a layer-by-layer process employing computer-controlled laser technology. Similar to SLM, DMLS relies on inert gas protection and is distinguished by its ability to control the porosity of each layer, though achieving complete elimination of porosity remains challenging. Despite differences in applicable materials, energy sources, and operating environments, these technologies share the fundamental principles of PBF and provide diverse solutions for the high-precision manufacturing needs of various industries.

The PBF process involves complex multi-physics and multi-scale phenomena, such as the spreading of powder particles, the dynamic behaviour of the melt pool, heat transfer and material solidification, and the generation of residual stresses. These factors critically affect the quality and performance of the final part. Numerical simulation provides a powerful means to investigate these mechanisms, many of which are difficult to capture experimentally. Techniques such as the Discrete Element Method (DEM) for powder dynamics, Computational Fluid Dynamics (CFD) and the Lattice Boltzmann Method (LBM) for melt pool behaviour, and the Finite Element Method (FEM) for thermal stress prediction offer valuable insights. Their combined application forms a solid foundation for refining process parameters and improving part quality.

This section focuses on the numerical simulation framework of the PBF process, systematically analysing its fundamental solutions by combining foundational models with recent developments. The subsequent discussion will delve into specific processes, such as powder spreading and melt pool dynamics, as well as an in-depth exploration of state-of-the-art modelling methods and their applications.

10.2.1 Powder Spreading

Powder spreading is a critical initial step in AM, distributing powder uniformly across the build platform to form a powder bed with consistent thickness and density. This process provides a stable initial condition for subsequent laser or electron beam scanning and directly influences the microstructure and performance of the final part. In the PBF process, the packing density of the powder layer is a key factor affecting the quality of the fabricated part.

The packing density of the powder layer directly determines the interaction between the powder particles and the heat source, significantly impacting the stability of the melt pool and overall build quality. For instance, a denser powder layer can reduce defects such as balling and porosity, enhance the continuity of melting, and ultimately improve the performance of the manufactured part [5]. Therefore, a thorough understanding and careful optimisation of powder spreading directly influence the stability and quality of the PBF process.

Key factors influencing the quality of powder spreading include particle size distribution (PSD), particle morphology, flowability, and the design and operational parameters of the spreading device (e.g., rake or roller; see **Figure 10.2**) [6]. Additionally, the powder layer thickness in PBF processes typically ranges from 50 to 100 μm [7], which is comparable to the average particle size. This unique thin-layer characteristic results in more complex dynamic behaviour of the powder during spreading, making it challenging for traditional experimental methods (e.g., powder-heap experiments in containers) to accurately capture the packing characteristics under these conditions.

The study of powder spreading has increasingly relied on the synergistic use of experiments and numerical simulations to explore particle behaviour and packing mechanisms. While experiments provide valuable insights, they often face challenges in achieving a comprehensive characterisation of powder properties and ensuring reproducibility due to variability between batches in feedstock or the imperfect control of process parameters. Numerical simulations, on the other hand, offer a controlled environment for analysing powder dynamics. However, these methods are constrained by factors such as reduced domain sizes, limited simulation durations, and potential discrepancies with real-world conditions. Despite these challenges, numerical modelling remains an indispensable tool for understanding the intricate interactions within thin powder layers in PBF processes. Powder spreading phenomena in numerical simulations are predominantly captured through two modelling strategies: the continuum-based approach and particle-resolved discrete techniques [8].

The continuous approach treats the powder bed as a homogeneous medium, treating the powder particles collectively as a continuous material. This approach is typically employed to study macroscopic and part-scale phenomena, such as thermal conduction and fluid dynamics

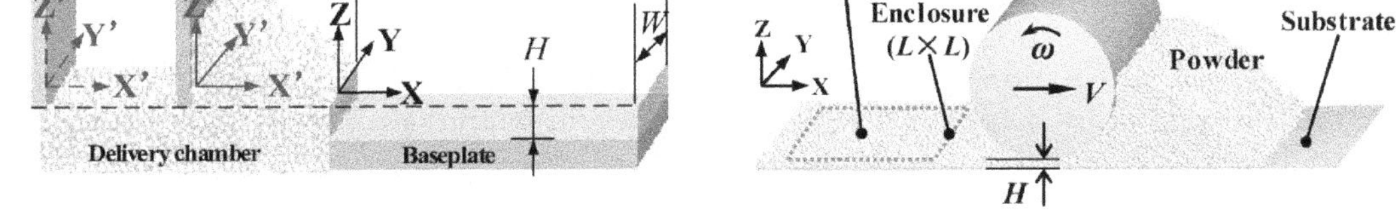

Figure 10.2 Schematic of two main powder spreading mechanisms: (a) Rake-type spreading, illustrating the delivery chamber and spreading tool's linear motion across the build platform, adapted from [8]; (b) Roller-type spreading, depicting the rotary motion of the roller and its interaction with the powder layer, adapted from [6].

within the powder bed during the printing process. Owing to its low computational cost, the continuous approach is advantageous for simulating overall powder-bed responses; however, it lacks the resolution to model particle-scale interactions and kinematics [8]. Apart from the study by Shanjani and Toyserkani (2008) [9], there have been relatively few attempts to use the continuous method for simulating powder spreading [10].

In the model proposed by Shanjani and Toyserkani, powder spreading and compaction under a counter-rotating roller are described using a continuous method. The model uses a one-dimensional slab approach under the assumption of plane strain, where stress and strain vary primarily in the rolling direction. Key governing equations include force equilibrium, mass conservation, and stress-strain relationships. The force equilibrium in the rolling direction is expressed as:

$$\frac{d\sigma_x}{d\alpha} = \frac{R}{h} \cdot \left(P_r \cdot \sin\alpha + t_r \cdot \cos\alpha - t_q \right) \tag{10.1},$$

where represents the stress in the rolling direction, α is the rolling angle, P_r is the roller contact pressure, t_r is the interfacial shear stress between loosely packed and compacted powder, t_q is the shear stress acting between the roller and loose powder, and h is the slab height.

To capture mass conservation in compressible material, the non-constant density leads to the continuity equation being formulated as [11]:

$$m = h \cdot \rho \cdot \delta \cdot b \Rightarrow \frac{\dot{\delta}}{\delta} + \frac{\dot{h}}{h} + \frac{\dot{\rho}}{\rho} = 0 \tag{10.2}$$

Here, m represents slab mass, δ represents slab width, ρ represents density.

This method provides a one-dimensional mathematical framework for analysing the powder spreading and compaction process. While it omits many finer details and operates at a macroscopic level, it offers a computationally efficient approach to understanding the powder bed's general behaviour.

The discrete approach analyses the behaviour of individual powder particles, making it particularly well-suited for studying micro- and mesoscale phenomena such as powder packing density, particle interactions, melting dynamics, and keyhole formation. Unlike the continuous approach, this method requires detailed information on the geometry and physical properties of each particle, thereby significantly increasing its computational cost. To better understand the mechanisms of laser-material interactions, recent research has increasingly adopted discrete modelling techniques, treating the powder as a system of individual particles.

The discrete element method has become a widely adopted tool for analysing the dynamic behaviour of powder spreading. Originally introduced by Cundall and Strack [12], DEM offers a particle-scale numerical framework particularly suited for systems where traditional continuum-based models fall short in capturing discrete interactions.

In the DEM model, powder particles are generated according to a specified size distribution and randomly distributed to represent the initial configuration. As the spreading tool advances and interacts with the particle bed, the trajectories of the particles are computed by solving Newtonian motion equations. At such high resolution, cohesive interactions become prominent, acting alongside tribological effects governed by contact mechanics.

For simulation purposes, powder particles are typically modelled as spheres and are characterised by the following variables: their mass m_i, vector position $\boldsymbol{r}_i$, moment of inertia I_i, and angular velocity ω_i for each particle. The translational and rotational motions of the particles are governed by the classical equations of motion:

$$m_i \frac{d^2 \boldsymbol{r}_i}{dt^2} = m_i \boldsymbol{g} + \sum_j (\boldsymbol{F}_{n,ij} + \boldsymbol{F}_{t,ij} + \boldsymbol{F}_{c,ij}) \tag{10.3}$$

$$I_i \frac{d^2 \omega_i}{dt^2} = \sum_j \boldsymbol{T}_{ij} \tag{10.4}$$

where $\boldsymbol{F}_{n,ij}$ and $\boldsymbol{F}_{t,ij}$ represent the contact forces acting normally and tangentially, $\boldsymbol{F}_{c,ij}$ corresponds to the cohesive force and $\boldsymbol{T}_{ij}$ to the torque arising from particle interactions.

The core of DEM lies in simulating interactions between particles and between particles and boundaries using contact models. These models calculate contact forces and torques based on the physical properties of the particles, including Young's modulus, Poisson's ratio, particle size, and friction behaviour. Contact forces are typically separated into normal and tangential components, as shown in **Figure 10.3** [13], which are often modelled using linear or nonlinear viscoelastic models or hysteretic models [14].

10.2.2 Melt Pool Behaviours

The melt pool plays a pivotal role in the PBF additive manufacturing process, as its behaviour governs the resulting microstructure, mechanical properties, and overall part quality. Under the influence of a laser or electron beam, powder particles melt to form a liquid metal pool, which then cools and solidifies. As illustrated in **Figure 10.4**, this process involves a combination of complex physical phenomena, including Marangoni convection within the melt pool, evaporation at the liquid surface that generates recoil pressure and heat loss, and the interplay of heat conduction, radiation, and convection [15, 16]. Furthermore, the dynamic evolution of the liquid surface profile and the release or absorption of latent heat during phase transitions contribute significantly to the complexity. These mechanisms collectively define

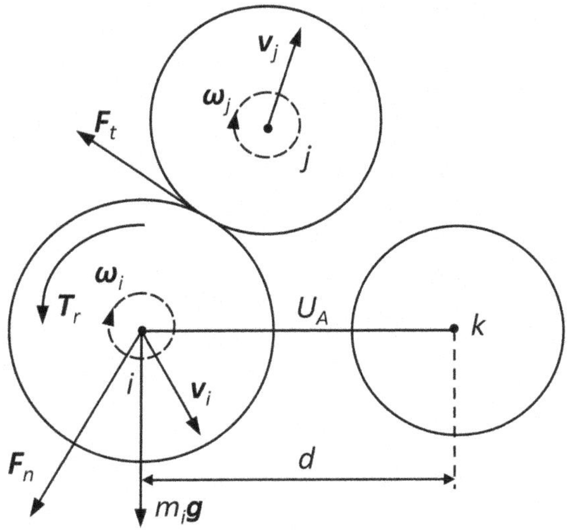

Figure 10.3 Schematic of contact forces acting between particles in the DEM model [13].

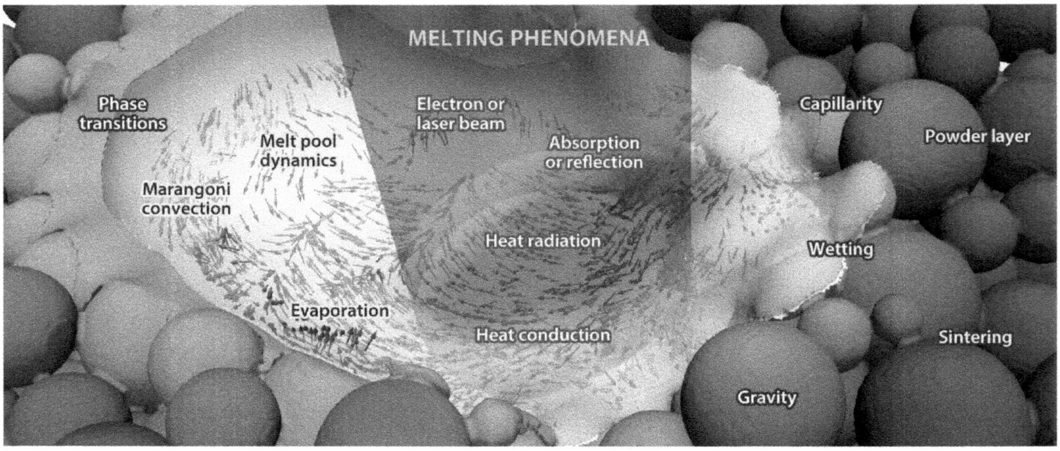

Figure 10.4 Key melt pool behaviours and associated physical processes during powder bed fusion [15] (Adapted from [16]).

the melt pool's geometry and stability, and their role in shaping the quality and performance of the final part.

Understanding the behaviour of the melt pool is critical because it directly impacts defect formation and microstructural evolution. Variations in heat source input, material properties, and boundary conditions further influence its geometry, flow patterns, and thermal characteristics. Numerical simulations have emerged as an indispensable tool for investigating melt pool dynamics, providing a means to solve governing equations describing heat and fluid flow, material evaporation, and free-surface evolution. Coupled with experimental validation, these models provide valuable insights into the mechanisms underlying melt pool behaviour and enhance predictive capability for optimising the PBF process.

To facilitate a systematic understanding of these complex phenomena, this section is organised into two parts: the first introduces the fundamental governing equations underpinning melt pool modelling, and the second discusses the key physical models that capture energy input, fluid flow, and interfacial effects.

10.2.2.1 Fundamental Governing Equations

To accurately model melt pool behaviour, a set of fundamental governing equations is formulated based on the principles of mass, momentum, and energy conservation. These equations are further supplemented by phase-change models and free-surface tracking techniques to capture the dynamic nature of the melt pool.

In melt pool modelling, the Navier-Stokes equations describe the fluid motion within the molten region, while the energy conservation equation governs heat transport and thermal diffusion, and phase-change models account for melting and solidification. Moreover, as the melt pool exhibits a distinct free surface, its evolution is typically tracked using the Volume of Fluid (VOF) method, the phase-field approach, or the level-set method to accurately represent the liquid-solid boundary dynamics:

- Continuity Equation:
  The continuity equation ensures mass conservation in the melt pool and is given by:

$$\frac{\partial \rho}{\partial t} + \nabla \cdot (\rho \boldsymbol{u}) = 0 \qquad (10.5),$$

where ρ denotes fluid density, t represents time, and $\boldsymbol{u}$ indicates the velocity vector. This equation ensures that the local mass flux remains balanced over time.

- Momentum Conservation (Navier-Stokes Equations):
  The momentum conservation equation describes the fluid motion within the melt pool, accounting for viscosity, pressure, and external forces. It is given by:

$$\rho\left(\frac{\partial \boldsymbol{u}}{\partial t}+\boldsymbol{u}\cdot\nabla\boldsymbol{u}\right)=-\nabla p+\mu\nabla^2\boldsymbol{u}+\boldsymbol{f} \quad (10.6),$$

  where p is pressure; μ is the dynamic viscosity; $\boldsymbol{f}$ represents external forces, including buoyancy force due to density variations, surface tension force affecting the free surface, Marangoni force caused by surface tension gradients, recoil pressure force due to evaporation.
  And for incompressible fluids, the divergence-free condition is often applied:

$$\nabla\cdot\boldsymbol{u}=0 \quad (10.7)$$

- Energy Conservation Equation:
  The energy conservation equation governs heat transfer within the melt pool and accounts for conduction, convection, and external heat sources:

$$\rho c_p\frac{\partial T}{\partial t}+\rho c_p\boldsymbol{u}\cdot\nabla T=\nabla\cdot(k\nabla T)+Q \quad (10.8)$$

  where T is temperature, k is the thermal conductivity, Q is the external heat source (e.g., laser or electron beam energy input), c_p is the specific heat capacity.
- Phase Change Model:
  To account for melting and solidification, the enthalpy method is often employed. The governing equation is:

$$\frac{\partial(\rho h)}{\partial t}+\nabla\cdot(\rho\boldsymbol{u}h)=\nabla\cdot(k\nabla T)+Q \quad (10.9),$$

  where the enthalpy h is defined as:

$$h=h_{ref}+\int_{T_{ref}}^{T}c_p dT+Lf_L \quad (10.10)$$

  Here h_{ref} is the reference enthalpy at temperature T_{ref}; L is the latent heat of fusion; f_L is the liquid fraction, which is a function of temperature and determines the phase transition.
  The liquid fraction is often computed using:

$$f_L=\begin{cases}0, T<T_{solidus}\\ \dfrac{T-T_{solidus}}{T_{liquidus}-T_{solidus}}, T_{solidus}\le T\le T_{liquidus}\\ 1, T>T_{liquidus}\end{cases} \quad (10.11)$$

- Free Surface Tracking:
  Since the melt pool involves a moving free surface, interface-tracking methods are essential for accurately modelling its shape and evolution. The most commonly used techniques include:

VOF Method:

$$\frac{\partial F}{\partial t} + \boldsymbol{u} \cdot \nabla F = 0 \qquad (10.12),$$

where $F = 1$ for liquid, $F = 0$ for gas, $0 < F < 1$ for interface cells.

The level-set method represents the free surface as the zero level-set of a signed distance function ϕ:

$$\frac{\partial \phi}{\partial t} + \boldsymbol{u} \cdot \nabla \phi = 0 \qquad (10.13)$$

Phase-Field Method:

The phase-field method (PFM) describes phase evolution using an order parameter ξ, which represents the phase distribution:

$$\frac{\partial \xi}{\partial t} = M \nabla^2 \xi - \frac{\partial f}{\partial \xi} \qquad (10.14),$$

where $\xi = 1$ for the solid phase, $\xi = 0$ for the liquid phase; M is the mobility coefficient, which governs the phase transition kinetics.

10.2.2.2 Physical Models

In the numerical simulation of PBF, the evolution of the melt pool relies on multiple physical models to accurately characterise energy input, molten metal flow, and interfacial effects. These models dictate how the heat source interacts with the material, direct the internal flow dynamics of the melt pool, and influence its stability through surface tension and evaporation.

To improve simulation accuracy, it is vital to establish a precise heat-source model that defines the energy-deposition profile of laser or electron beams, while integrating evaporation and recoil-pressure models to account for dynamic effects at the melt pool surface. Moreover, Marangoni convection, induced by temperature gradients, significantly influences the melt pool morphology and solidification process, making its mathematical modelling essential for accurate prediction of melt pool behaviour.

Understanding these interconnected effects is key to developing robust numerical models that represent the intricate multi-physics phenomena inherent to the PBF process, ultimately improving simulation fidelity and the predictive accuracy of material properties.

10.2.2.2.1 Heat Source Models

In thermal modelling of additive manufacturing, heat source models are essential for defining how energy is introduced and interacts with the processed material. PBF primarily utilises either a laser beam or an electron beam as the energy input mechanism. Since Rosenthal's early theoretical work on moving heat sources [17], various heat source models have been developed, ranging from simplified surface approaches to volumetric and optical-physics-based models. The following sections provide a comprehensive classification of heat source models, integrating recent developments.

10.2.2.2.1.1 ENERGY INPUT MODELS

Energy input models define the initial heat deposition from the laser or electron beam and serve as the foundation for PBF heat-source simulations. Different energy-input models can be selected based on specific process characteristics.

- Gaussian Surface Heat Source Model:
  The Gaussian surface heat source model is commonly employed due to its simplicity of implementation and low computational cost [18, 19]. It assumes that the energy distribution follows a Gaussian profile in the x-y plane, with the laser intensity peaking at the beam centre and decreasing radially. The heat flux distribution is given by:

$$q_L(x,y) = \frac{Q}{2\pi R_b^2} \exp\left(-\frac{(x-x_0)^2 + (y-y_0)^2}{2R_b^2}\right) \quad (10.15),$$

  where $q_L(x,y)$ represents the heat flux, Q is the total energy power, R_b denotes the beam radius, (x_0, y_0) corresponds to the beam centre position.
  Despite its simplicity, this model does not account for laser penetration depth, resulting in an underestimation of the melt pool depth in many scenarios. Consequently, while it provides a reasonable approximation of surface heat flux, it remains inadequate for modelling subsurface energy absorption.
- Goldak Double Ellipsoid Heat Source Model:
  The Goldak model was developed to simulate asymmetrical weld pools, providing a more flexible heat source representation than purely Gaussian-based models. The double-ellipsoid structure consists of a front ellipsoid and a rear ellipsoid, distributing heat asymmetrically between the leading and trailing regions of the melt pool [20]:

$$q_f(x,y,z) = \frac{6\sqrt{3} f_f Q}{\pi a b c_f} \exp\left(-3\frac{x^2}{a^2} - 3\frac{y^2}{b^2} - 3\frac{z^2}{c_f^2}\right) \quad (10.16)$$

$$q_r(x,y,z) = \frac{6\sqrt{3} f_r Q}{\pi a b c_r} \exp\left(-3\frac{x^2}{a^2} - 3\frac{y^2}{b^2} - 3\frac{z^2}{c_r^2}\right) \quad (10.17),$$

  where $q_f(x,y,z)$ and $q_r(x,y,z)$ represent the heat source distributions in the front and rear regions, respectively; f_f and f_r are the heat source fractions in the front and rear regions, ensuring total energy conservation with $f_f + f_r = 2$; a, b define the transverse dimensions of the heat source; c_f and c_r determine the longitudinal (z-direction) spread of the heat source in the front and rear regions.

This model is particularly useful for modelling non-axisymmetric energy deposition, making it more applicable to real-world laser- and electron-beam-based processes than traditional surface heat sources. While the Goldak model was initially developed for arc welding, it has been extended to laser processing, including PBF simulations. In fact, the standard Gaussian heat source model is a special case of the Goldak model. Cheng et al. (2024) further refined the Goldak model for keyhole-mode melting in PBF, incorporating modifications to better account for laser energy absorption dynamics and vaporisation effects, which are critical in deep penetration processes [21]. Asghari Ilani and Banad (2024) refined the Goldak

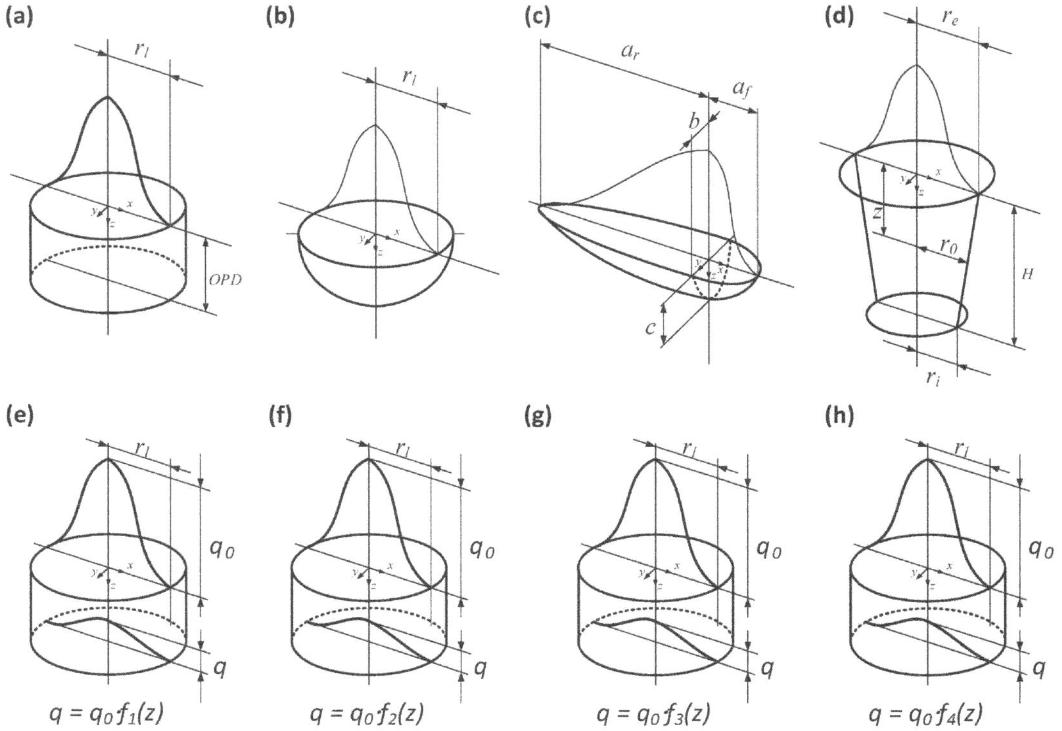

Figure 10.5 Schematic illustrations of various heat source models: (a) Cylindrical, (b) Semi-spherical, (c) Semi-ellipsoidal, (d) Conical, (e) Radiation transfer model, (f) Ray-tracing model, (g) Linearly decaying model, (h) Exponentially decaying model [22].

Semi-Ellipsoidal Heat Source model (see **Figure 10.5** (c)) to improve the prediction of melt pool geometry in metal additive manufacturing [22]. Their model integrates an analytical solution to the convection-diffusion heat transfer equation, accounting for temperature-dependent material properties and phase change effects. The validation process demonstrated high accuracy in predicting transient temperature fields, offering insights into the thermal dynamics of selective laser melting (SLM) processes [23].

Beyond the standard Gaussian and Goldak models, heat source models can also be classified based on their geometrical shapes, including cylindrical (**Figure 10.5** (**a**)), semi-spherical (**Figure 10.5** (**b**)), semi-ellipsoidal (**Figure 10.5** (**c**)), and conical (**Figure 10.5** (**d**)) models, each representing different energy distribution characteristics [22]. A comprehensive discussion of these geometrical heat source models and their applications in PBF simulations is provided in Zhang et al. (2019) [22], where different volumetric heat source models are systematically compared to improve energy deposition accuracy in additive manufacturing.

10.2.2.2.1.2 ENERGY ATTENUATION MODELS

During PBF processes, the energy source penetrates the powder bed and undergoes progressive absorption along the depth direction, rather than being solely deposited at the surface. Due to the porous nature of the powder layer and its material-dependent energy transport properties, energy transmission is influenced not only by material absorptivity but also by scattering, multiple reflections (in LPBF), or electron backscattering (in EB-PBF). To accurately characterise these effects, energy attenuation models employ different mathematical formulations to describe energy distribution within the powder bed, thereby improving predictions of melt pool morphology and the heat-affected zone (HAZ).

These models typically assume a two-dimensional Gaussian distribution at the powder bed surface, combined with a depth-dependent attenuation function. The general mathematical expression can be formulated as:

$$I(x,y,z) = \frac{P}{2\pi r_b^2} \exp\left(-\frac{x^2+y^2}{r_b^2}\right) f(z) \quad (10.18),$$

where $I(x,y,z)$ represents the energy intensity distribution, P is the power of heat source (laser or electron beam), r_b is the beam radius, $f(z)$ is a function describing energy attenuation as a function of depth z within the powder bed.

Depending on the dominant energy dissipation mechanism, different formulations of $f(z)$ are used, leading to the following classifications:

- Exponential Attenuation Model:
  A widely used approach is to model absorptivity as exponentially decreasing with depth, following the Beer–Lambert law:

  $$f(z) = e^{-\beta z} \quad (10.19),$$

  where β is the extinction coefficient, defining how strongly the material attenuates the energy.

  For instance, Mishra and Kumar (2019) applied a volumetric heat-source model based on the Beer-Lambert law to describe the energy-absorption characteristics in SLM of Ti6Al4V [24]. The model was used to simulate single- and multi-track SLM processes, and the results demonstrated good agreement with experimental melt-pool dimensions and porosity predictions. Similarly, Yin et al. (2016) utilised an exponentially decaying heat-source model, following the same depth-dependent attenuation principle, to characterise the volumetric energy distribution during Laser Micro Sintering (LMS) of metallic powders [25].

  Extended Beer-Lambert Model (Finite Powder Layer Correction):
  To account for the finite thickness L of the powder bed, a more refined expression is used [8]:

  $$f(z) = \frac{\beta \exp(-\beta z)}{1-\exp(-\beta L)} \quad (10.20),$$

  where L is the powder layer thickness.
- Linear Attenuation Model:
  The linear attenuation model assumes that energy absorption decreases linearly with depth, making it particularly suitable for scenarios where the laser penetration depth is comparable to the powder layer thickness. Unlike exponential decay models, which assume continuous energy dissipation, this model better represents finite-thickness powder layers, where absorption diminishes at a uniform rate.
  Ladani et al. [26] applied a normalised linear attenuation model, formulated as:

  $$f(z) = \frac{2\beta}{\delta}\left(1-\frac{z}{\delta}\right) \quad (10.21),$$

  where δ is the beam penetration depth.

Bayat et al. (2019) also employed a linear attenuation model, but instead of using an absorptivity coefficient, they used the optical penetration depth (OPD) to define the maximum depth of laser energy absorption in the powder bed [27]. Their Conico-Gaussian volumetric heat source model incorporates a Gaussian lateral distribution (x-y plane) while introducing a linear attenuation function along the depth (z-direction), formulated as [28]:

$$q(x,y,z) = \frac{2\eta P}{OPD \cdot \pi r_b^2} e^{-\frac{2\left(x^2+y^2\right)}{r_b^2}} \cdot \left(1 - \frac{z}{H}\right) \tag{10.22}$$

where H is the maximum penetration depth.

- Radiation Transfer Model:
  The radiation transfer equation (RTE) governs energy transport within the powder bed, treating it as an absorbing and scattering medium. The general form of the equation is:

$$f(z) = -\eta \cdot \frac{dq}{d\xi} \tag{10.23}$$

where $\xi = \eta \cdot z$ refers to the dimensionless local depth coordinate, and q represents the dimensionless net radiative energy flux density.

Gusarov et al. (2009) applied this model to simulate the interaction of laser radiation with metal powders in SLM, integrating it with thermal diffusion calculations to predict local temperature fields and melt pool behaviour. The results demonstrated that radiation scattering significantly influences energy absorption, affecting melt pool stability and morphology [29].

10.2.2.2.1.3 RAY-TRACING MODELS

Ray-tracing models provide a physically accurate method to simulate laser energy absorption, reflection, and scattering in PBF processes. Unlike purely analytical heat source models, ray-tracing methods explicitly track laser ray propagation, enabling more detailed calculations of energy deposition in powder beds. While many ray-tracing models still incorporate Gaussian-based energy distributions, they improve accuracy by modelling how laser rays interact with individual powder particles.

Based on their underlying principles, ray-tracing heat source models can be categorised into two main groups:

- Monte Carlo-Based Ray-Tracing:
  Monte Carlo ray-tracing simulates laser-powder interactions statistically, accounting for random powder packing, multiple reflections, and scattering effects. The laser beam is decomposed into discrete rays, which are traced through the powder bed while undergoing probabilistic absorption, reflection, and refraction. In such simulations, it is essential to account for the fundamental assumptions of geometrical optics [30], ensuring that laser propagation follows the principles of optics when interacting with powder particles.

  The energy absorption function in a Monte Carlo-based ray-tracing model can be expressed as [31]:

$$I(x,y,z) = \frac{P}{\pi r_b^2} \exp\left(-2\frac{x^2+y^2}{r_b^2}\right) \cdot f(z) \qquad (10.24),$$

where $f(z) = \frac{d\beta}{dz}$ denotes the absorptivity function obtained through Monte Carlo ray-tracing method.

- **Geometrical Optics-Based Ray-Tracing:**
  Geometrical-optics-based ray tracing treats the laser beam as a set of geometric rays and explicitly tracks their deterministic propagation through the powder bed. It incorporates Fresnel's equations to compute reflection and transmission coefficients when rays encounter powder particles and Snell's law to determine ray refraction at material boundaries [32].

Snell's Law (Refraction at Interfaces):

$$n_1 \sin\theta_1 = n_2 \sin\theta_2 \qquad (10.25),$$

where: n_1, n_2 are the refractive indices of two materials, θ_1, θ_2 are the angles of incidence and refraction.

Fresnel's Equations (Reflection and Transmission at Interfaces):

$$R_s = \left(\frac{n_1 \cos\theta_1 - n_2 \cos\theta_2}{n_1 \cos\theta_1 + n_2 \cos\theta_2}\right)^2 \qquad (10.26)$$

$$T_s = 1 - R_s \qquad (10.27),$$

where R_s is the reflectance of the light, which represents the fraction of incident energy that is reflected at the interface, T_s is the transmittance, which represents the fraction of energy transmitted through the interface.

Ray-tracing models have been applied in both LPBF and EB-PBF to account for energy redistribution effects that analytical heat-source models cannot fully capture. In LPBF, they have been used to study multi-material interactions and keyhole formation, enabling a more accurate representation of laser absorptivity variations and melt pool dynamics. Tang et al. (2022) developed a geometrical optics-based ray-tracing model to track laser absorption in multi-material LPBF [33], while Wang et al. (2022) integrated ray-tracing into a multi-physics model to analyse keyhole pore formation and molten metal flow behaviour [34]. Additionally, Yang et al. (2024) extended the application of ray-tracing models to polymer powder bed fusion (PBF-LB/P), explicitly capturing laser energy absorption and multiple reflections within the powder bed, as shown in **Figure 10.6** [35].

For EB-PBF, where electron scattering and backscattering significantly influence energy deposition, ray-tracing models can simulate electron penetration depth and interactions with powder particles. Yan et al. (2015) applied a Monte Carlo-based ray-tracing model to investigate electron-material interactions, improving predictions of thermal distribution and melt pool evolution [36]. Zakirov et al. (2020) further extended this approach using a detailed Monte Carlo ray-tracing model that captures multiple reflections and scattering effects to enhance understanding of energy absorption and defect formation in EB-PBF [37].

These applications highlight the role of ray-tracing models in modelling complex beam-powder interactions, improving process predictability, and supporting the optimisation of PBF techniques across both laser and electron beam-based systems.

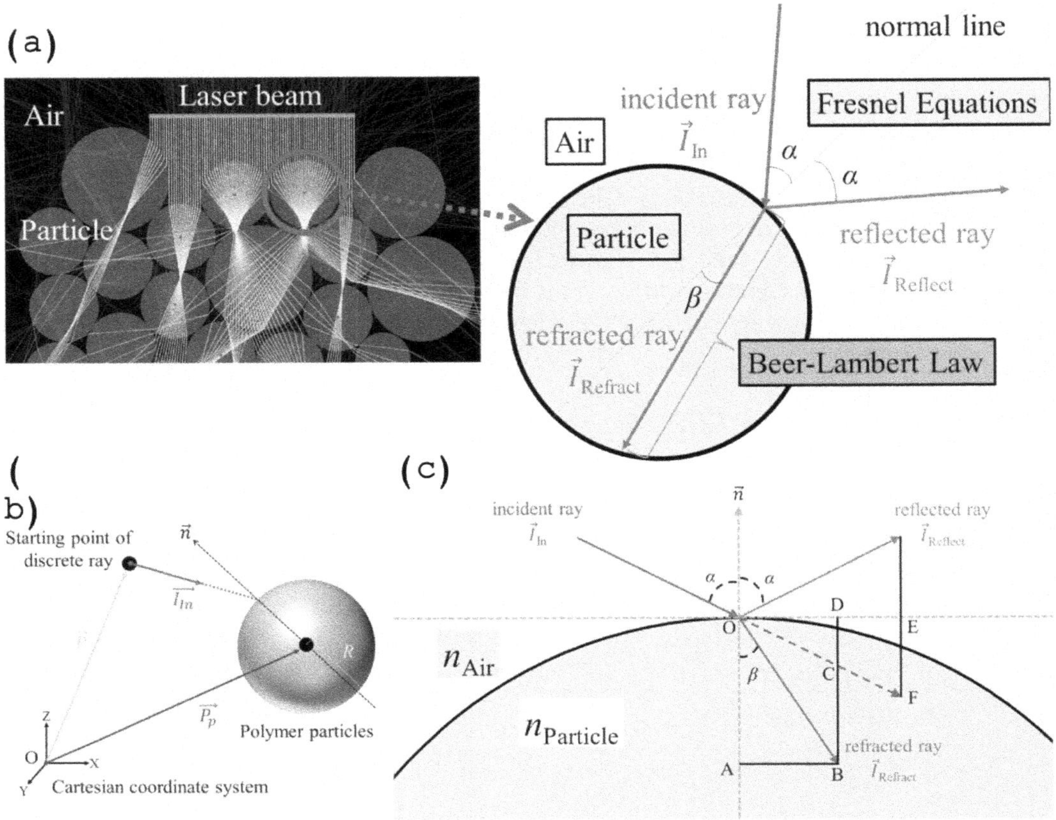

Figure 10.6 Modelling laser-powder interaction via ray-tracing method in PBF-LB/P [35]. (a) Ray propagation considering Beer-Lambert law and Fresnel equations, (b) Geometric representation of ray interaction with a single particle, (c) Reflection and refraction of a laser ray at a particle interface.

10.2.2.2.2 Evaporation and Recoil Pressure Models

In the PBF process, high-energy laser exposure rapidly heats the metal powder, leading to its evaporation. As the temperature rises, metal atoms transition from the liquid phase to the vapour phase, driven by local thermodynamic conditions, as seen in **Figure 10.7** [38]. The enthalpy required for evaporation is significantly greater than that for fusion. For example, in the case of Ti-6Al-4V, the energy required for evaporation is approximately four times the enthalpy required to heat the material from room temperature to its fully molten state [39]. This substantial energy requirement means that evaporation not only results in material loss but also strongly influences the thermal balance of the melt pool. Additionally, the escape of metal vapour can alter laser energy absorption and propagation, potentially leading to process instabilities and defect formation.

The effects of evaporation are multifaceted and can be summarised as follows:

1. Elemental Loss: Low-boiling-point elements, such as Zn, Mg, and Al, evaporate more readily, leading to compositional segregation and localised chemical variations, which in turn affect the microstructure and final material properties [40–43]. Due to the differing evaporation tendencies of individual elements, alloy compositions in PBF-produced

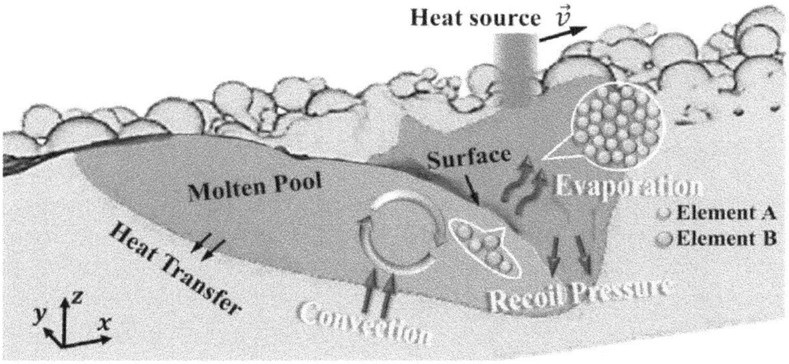

Figure 10.7 Schematic illustration of evaporation and recoil pressure in PBF (Adapted from [38]).

components may significantly deviate from the initial powder composition, particularly in alloys containing highly volatile elements.

2. Recoil Pressure: The high-velocity vapour jet generated during evaporation exerts recoil pressure on the melt pool surface. When the recoil pressure exceeds the surface tension of the liquid metal, a vapour cavity (keyhole) forms, enhancing laser absorption. Keyhole instability can lead to its collapse, entrapping gas and causing porosity in the final part [44].
3. Laser Attenuation: Metal evaporation generates a vapour plume that interacts with the laser beam, leading to attenuation due to absorption and scattering by process by-products. Insufficient shielding gas flow can cause fume accumulation, leading to condensate redeposition and spatter on surfaces exposed to the laser beam, which can degrade beam quality and contribute to defects such as porosity and lack of fusion [45].
4. Powder Denudation: The metal vapour jet expands radially and induces a vortex flow in the surrounding gas, entraining nearby powder particles and redistributing them within the powder bed. This process leads to the formation of denudation zones, where powder is cleared from the surrounding area, disrupting powder bed uniformity and contributing to defect formation [46].

To quantitatively describe these evaporation-induced effects in the PBF process, mathematical models are employed to characterise vapour pressure, evaporation flux, and recoil pressure. The following equations provide a theoretical framework for modelling these processes:

10.2.2.2.2.1 SATURATED VAPOUR PRESSURE

The driving force for metal evaporation is the saturated vapour pressure at the melt pool surface, which determines the evaporation rate over time. According to the Clapeyron-Clausius equation, the dependence of saturated vapour pressure on temperature is given by:

$$\frac{dP_s}{dT} = \frac{\Delta H_V}{T(V_G - V_L)} \tag{10.28}$$

$$P_s = P_0 \exp\left(\Delta H_V \frac{T - T_b}{RTT_b}\right) \tag{10.29},$$

where P_s is the saturated vapour pressure; ΔH_V is the latent heat of vaporisation; V_G and V_L represent the molar volumes of the gaseous and liquid phase, respectively; T_b is the boiling point of the metal, and P_0 is the reference vapour pressure.

10.2.2.2.2.2 EVAPORATION FLUX

When the vapour pressure at the melt pool surface reaches the saturation vapour pressure, metal atoms transition from the liquid phase to the vapour phase, forming an evaporation flux. The theoretical evaporation flux can be determined using the Langmuir evaporation model:

$$J_m = P_s\sqrt{\frac{M}{2\pi RT}} \tag{10.30}$$,

where J_m is the vaporisation flux in vacuum, M is molecular mass, and R is the gas constant. However, within the Knudsen layer (Region IV in **Figure 10.8**) [47], molecular collisions and vapour recondensation reduce the actual net evaporation flux. The expression for practical vapourisation flux is presented in Eq. (31), with J_{con} denoting the condensation flux and Φ representing the vaporisation ratio.

$$J_{net} = J_m - J_{con} = \Phi J_m \tag{10.31}$$

10.2.2.2.2.3 RECOIL PRESSURE

The evaporated metal atoms generate recoil pressure P_r on the melt pool surface, which expels the vapour from the molten region and concurrently distorts the interface separating the vapour and liquid phases. The magnitude of P_r is directly proportional to the saturated vapour pressure P_s and the empirical recoil coefficient α:

$$P_r = \alpha P_s \tag{10.32}$$

10.2.2.2.2.4 MULTI-COMPONENT EVAPORATION MODEL

In metallic alloys, evaporation rates vary among different elements, necessitating a model that considers the contributions of individual species. Based on Raoult's Law, the total evaporation flux for a multi-component system is expressed as:

$$J_{net}^{\Sigma} = \Sigma \gamma_i X_i J_{net}^i = \Sigma \gamma_i X_i \phi_i J_m^i \tag{10.33}$$

$$P_r^{\Sigma} = \Sigma \gamma_i X_i P_r^i = \Sigma \gamma_i X_i \alpha_i P_s^i \tag{10.34}$$,

where γ_i is the activity coefficient of element i, X_i is its molar fraction in the alloy, J_{net}^i represents its individual evaporation flux, and P_r^i is its recoil pressure contribution.

$$\phi = \sqrt{2\pi\gamma_v} \cdot Ma_{Kn}(T_S) \cdot \frac{\rho_{Kn}}{\rho_S}\sqrt{\frac{T_{Kn}}{T_S}} \tag{10.35}$$

$$\alpha = \frac{1}{2}\left[1 + \frac{1}{2}(1-\phi)\cdot\left(1 + \frac{\rho_{Kn}}{\rho_S}\frac{T_{Kn}}{T_S}\right)\right] \tag{10.36}$$,

where T_{Kn} is the vapour temperature on the outer side of the Knudsen layer; γ_v is the ratio of specific heats; Ma represents the flow Mach number at the Knudsen layer's outer boundary, expressed as a function of the surface temperature T_S.

The aforementioned models are formulated based on saturated vapour pressure, evaporation flux, and recoil pressure, providing a theoretical framework for understanding evaporation behaviour. However, these models exhibit certain limitations, particularly in their treatment of mass conservation, evaporation kinetics, and vapour-environment interactions, making it difficult to fully capture the complexity of the PBF process.

Existing numerical models provide reasonable predictions of single-component evaporation behaviour, yet metal evaporation in practical PBF applications often involves multi-component effects, leading to dynamic changes in material composition and melt pool conditions. However, many models fail to adequately account for evaporation-induced mass loss and its impact on melt pool composition, thereby reducing simulation accuracy. For instance, the model proposed by Khairallah et al. incorporates evaporation-induced cooling effects and recoil pressure, yet it does not explicitly account for mass loss due to evaporation in the mass conservation equation, leading to inconsistencies in the physical modelling [48]. Additionally, the Langmuir evaporation equation remains widely used to estimate evaporation flux, assuming that evaporation is solely governed by saturated vapour pressure. However, this approach neglects the Knudsen layer effect, vapour recondensation, and environmental influences, potentially leading to overestimated evaporation rates under realistic PBF conditions.

Another major challenge lies in the transport of vapour and its interaction with the processing atmosphere. Once evaporated from the melt pool surface, metal vapour is influenced by Knudsen layer formation, gas-phase collisions, and vapour-induced turbulence, all of which affect melt pool dynamics and vapour evolution. Nevertheless, many existing models adopt a simplified representation of these processes. For example, recoil pressure is often modelled as a function of saturated vapour pressure, without accounting for vapour flow dynamics, local temperature gradients, or external gas environments. Such simplifications can lead to inaccuracies in predicting melt pool morphology, keyhole dynamics, and porosity formation.

Figure 10.8 illustrates the distribution of different vapour-phase regions during metal evaporation, including the Knudsen layer, vapour region, expanded vapour zone, and ambient gas, along with the presence of shock waves and rarefaction waves [47]. Knight extended the evaporation model from a vacuum to an environment with back pressure, making it more applicable to laser processing conditions [49]. This refinement also incorporated the effect of condensation on evaporation flux, as depicted in **Figure 10.8**. These structures indicate that evaporation is not merely a localised surface phenomenon but rather a multiscale gas-dynamic process. He et al. developed jump conditions for the Knudsen layer, establishing state equations for calculating vapour temperature, density, velocity, pressure, and the extent of condensation [50]. These conditions provide a theoretical foundation for accurately describing vapour transport and have been widely integrated into numerical simulations. Additionally, Anisimov [51] and Ytrehus [52] further extended Knight's model by introducing dynamic relations on both sides of the Knudsen layer, enhancing the precision of evaporation flux and recoil pressure calculations.

A key aspect of evaporation modelling is its interaction with keyhole dynamics and the generation of recoil pressure. Keyhole dynamics have been widely studied in laser welding [53], with studies integrating keyhole evolution and laser energy absorption into computational frameworks to better capture the complex interplay between evaporation flux, recoil pressure, and keyhole stability. Given the similarities between keyhole dynamics in laser welding and evaporation-driven processes in LPBF, the methodologies developed for welding simulations offer valuable insights for advancing LPBF modelling. However, evaporation

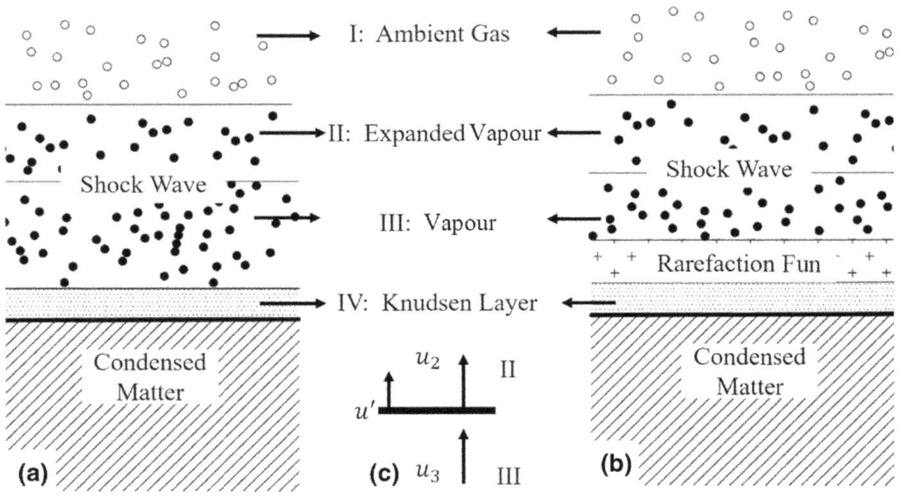

Figure 10.8 Representative flow structures of the evaporation model under different ambient conditions: (a) Flow behaviour in a standard atmospheric environment; (b) Flow patterns in a near-vacuum setting; (c) Schematic of the shock wave interface between Region II and Region III. Black solid circles denote vapour molecules, while blank circles indicate ambient gas molecules [47].

modelling in LPBF remains relatively underexplored. Its significance has often been underestimated, leading to omission or oversimplification in many numerical studies [54, 55]. As a result, only a limited number of reports have specifically addressed vapour transport, Knudsen layer effects, and multi-component evaporation dynamics in LPBF. Nevertheless, growing evidence suggests that evaporation plays a crucial role in powder spattering, compositional changes, melt pool dynamics, and defect formation [56–59]. Zhang et al. employed high-fidelity simulations to examine powder spattering and entrainment phenomena, revealing that evaporation-induced high-speed gas jets can significantly alter powder particle trajectories, contributing to melt pool instability and porosity defects [59].

Studies on recoil pressure modelling indicate that condensation reduces the net evaporation flux, thereby affecting recoil pressure [60]. Traditional models often assume that the interface reaches thermodynamic equilibrium and employ Mach number-based calculations to estimate recoil pressure (Eq. 10.35). However, in actual LPBF conditions, the melt pool surface can become superheated, invalidating this equilibrium assumption. Anisimov suggested that in vacuum conditions, the recoil pressure ratio α (Eq. 10.36) should be set at 0.54 [51], whereas in non-vacuum environments, this value may be significantly higher due to vapour flow dynamics and external gas interactions. Nevertheless, for simplicity, most LPBF studies continue to adopt an empirical value of 0.54 [15, 61].

In recent years, researchers have increasingly incorporated high-fidelity multi-physics simulations to better capture evaporation effects. For instance, Wang et al. conducted a comprehensive study integrating high-resolution X-ray imaging with multi-physics numerical simulations, providing valuable insights into the relationships among keyhole morphology, melt pool temperature, and recoil pressure. Their findings demonstrated simulation results fit well with the experimental observations, reinforcing the reliability of the improved evaporation model, as seen in **Figure 10.9** [47]. Furthermore, their study revealed that neglecting evaporation effects results in melt pool temperatures that are significantly overestimated, with traditional models predicting values far exceeding those observed experimentally. By

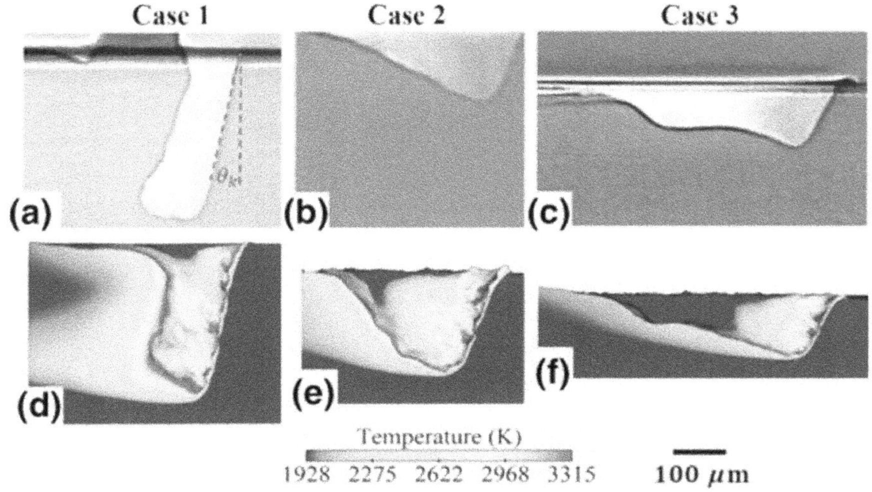

Figure 10.9 Comparison between Wang's model and experimental observations regarding keyhole geometry: (a)–(c) X-ray visualisations of the keyhole profile; (d)–(f) corresponding simulated keyhole morphologies and associated temperature fields [47].

incorporating a more refined evaporation mechanism, Wang et al. successfully reduced this discrepancy, bringing simulated melt pool temperatures closer to the measured 3500 K, thereby confirming the need to account for evaporation-induced cooling effects (see **Figure 10.10**) [47]. These results reinforce the importance of Knudsen-layer dynamics, recondensation effects, and multi-component evaporation modelling, highlighting the need for a more sophisticated approach to recoil-pressure calculations under realistic LPBF conditions.

Building on previous work, Wang et al. further proposed an evaporation-induced composition-evolution model to simulate the loss of various alloying elements during laser melting [38]. Their findings indicated that evaporation-driven compositional gradients influence solidification behaviour and may contribute to elemental segregation and cracking. Furthermore, their study highlighted that evaporation in alloy systems is not solely dictated by individual element vapour pressures but is also significantly influenced by mole fractions and activity coefficients, factors often overlooked in earlier models. As shown in **Figure 10.11**, the study provided a detailed numerical analysis of evaporation rates, velocity fields, and elemental distribution in the melt pool, illustrating the impact of evaporation-induced flow on composition evolution [38]. The velocity field visualisation highlights the complex convective patterns driven by evaporation-induced pressure gradients, while the evolution of the Zn distribution further confirms that preferential evaporation alters the local alloy composition over time.

10.3 NUMERICAL METHODS AND SOFTWARE

10.3.1 Overview of Numerical Methods

Numerical modelling plays a fundamental role in understanding and optimising powder bed fusion (PBF), providing critical insights into the complex multi-physics interactions governing powder spreading, melt pool dynamics, heat transfer, fluid flow, phase transitions, and

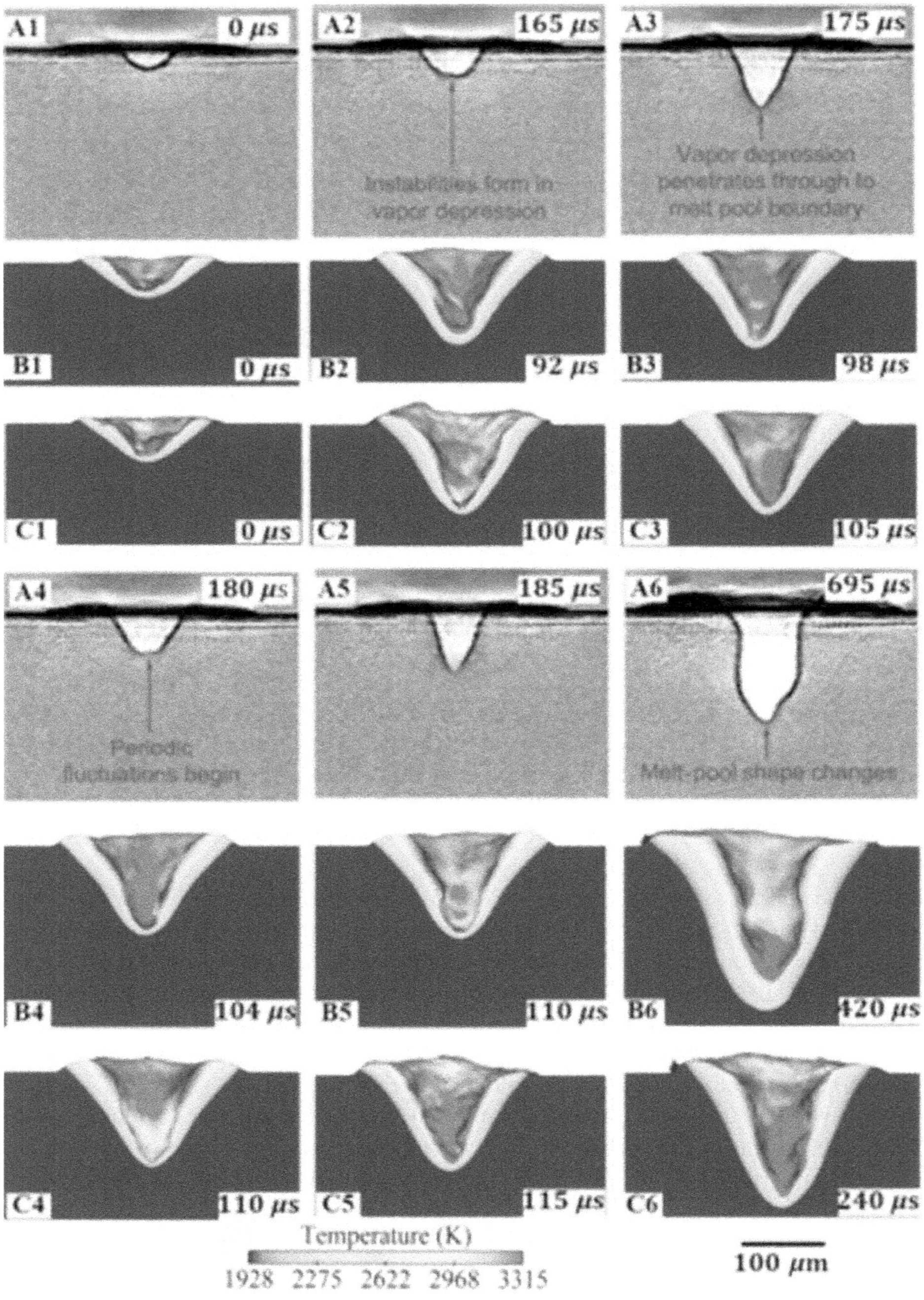

Figure 10.10 Keyhole and melt pool evolution during stationary laser melting: (A1)–(A6) X-ray images; (B1)–(B6) simulations using Wang's evaporation model; (C1)–(C6) simulations based on Anisimov's model [47].

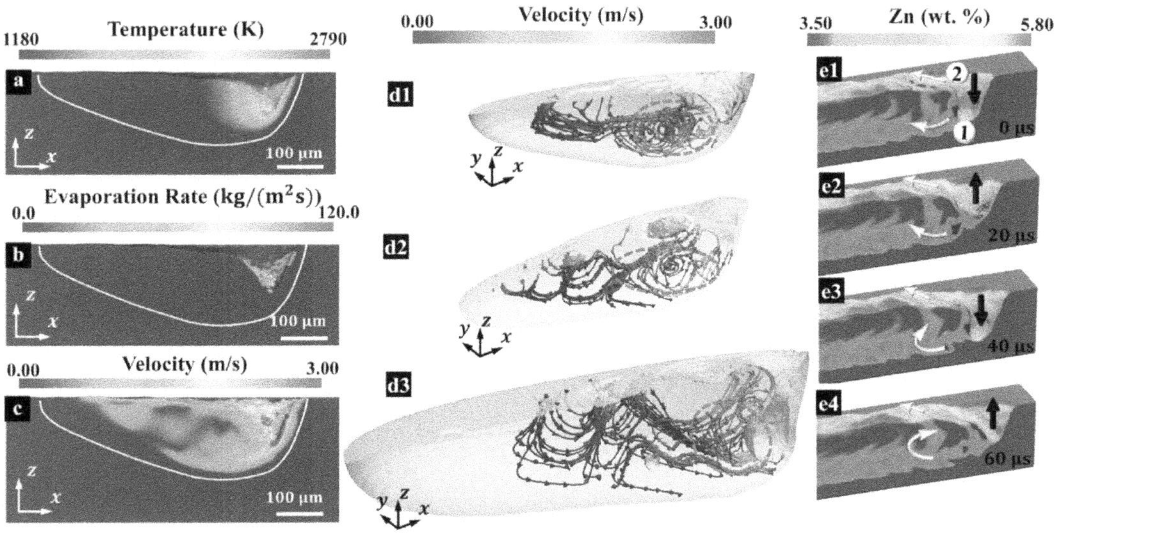

Figure 10.11 Mechanisms of evaporation-induced composition evolution. Simulation results of (a) the temperature, (b) evaporation rate on the molten pool surface, (c) the velocity magnitudes, (d) streamlines, and (e) snapshots of Zn evolution in the molten pool during the laser scanning. The white solid lines in (a–c) and contours in (d) are the boundaries of the molten pool. The black arrow in (e) indicates the keyhole bottom fluctuation direction, and white arrows 1 and 2 indicate the flow direction of low-Zn-concentration fluid in the upper and lower regions of molten pool [38].

mechanical deformation. For PBF, a range of numerical techniques has been employed to simulate process dynamics at multiple length scales. As illustrated in **Figure 10.12**, these methods can be broadly categorised into macroscopic, mesoscopic, and microscopic models, each addressing distinct aspects of the process while often requiring multi-scale integration to achieve high-fidelity predictions [8].

At the macroscopic scale, numerical models focus on the thermal and mechanical behaviour of the printed component, particularly residual stress development, part distortion, and thermal gradients induced by cyclic heating and cooling. These effects are particularly critical in PBF, where localised energy input, repeated layer-wise thermal cycling, and uneven cooling rates may result in defects such as warping, cracking, and delamination. The FEM has been widely adopted for thermo-mechanical analysis in PBF, offering a robust framework for solving heat conduction, elasticity, and plastic deformation equations. By discretising the computational domain into finite elements, FEM enables detailed predictions of thermal stress evolution and mechanical distortion. However, its reliance on static mesh structures makes it less effective in tracking free-surface evolution, such as the dynamic behaviour of the melt pool. The Finite Volume Method (FVM), on the other hand, is particularly effective for heat transfer and thermal history modelling, enforcing strict mass and energy conservation across control volumes. When two-phase interactions, such as gas–metal interactions and evaporation, become significant, FVM is often coupled with additional interface-capturing techniques, such as the Volume of Fluid method or the Level Set method, to improve accuracy. While both FEM and FVM provide valuable insights into large-scale process behaviour, they often require integration with mesoscopic models to capture finer details of melt pool evolution and powder-bed dynamics.

At the mesoscopic scale, numerical models shift their focus towards powder spreading, melt pool dynamics, and fluid–structure interactions, which are critical in defining the final

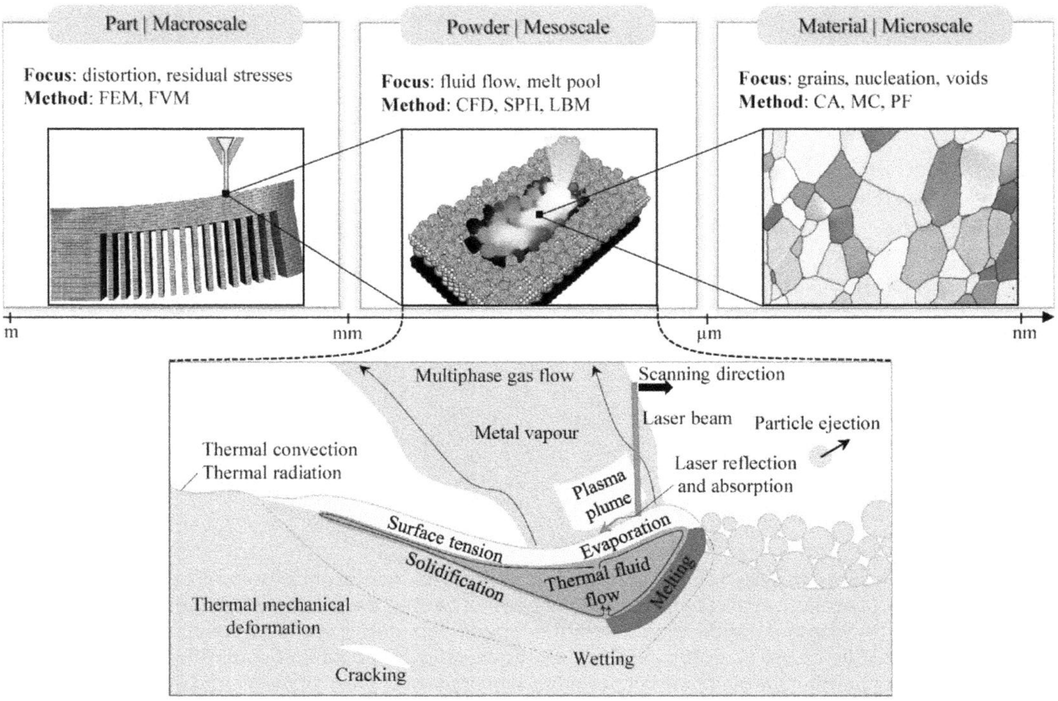

Figure 10.12 Schematic of modelling scales in MAM (top) and key physical phenomena near the melt pool at powder scale (bottom) [8].

microstructure and mechanical properties of PBF components. The complex interaction between the heat source and the powder bed leads to highly dynamic melt-pool behaviour, governed by convection, surface tension effects, phase transitions, evaporation, recoil pressure, and free-surface instability. CFD methods, typically implemented using the FVM, are widely used to solve the Navier–Stokes equations, enabling the simulation of melt pool convection, energy transport, and keyhole dynamics. However, the highly transient nature of PBF presents significant computational challenges, leading to the increasing adoption of alternative numerical approaches, such as the LBM and Smoothed Particle Hydrodynamics (SPH).

The LBM has gained prominence for modelling melt pool dynamics, Marangoni convection, capillary effects, and free-surface evolution in PBF. Unlike traditional CFD methods, which solve the Navier–Stokes equations directly, LBM operates at a mesoscopic scale by solving particle distribution functions, enabling high-fidelity simulations of fluid flow and interfacial behaviour. Its ability to handle complex boundary interactions makes it particularly advantageous for PBF, where irregular powder-bed morphology, dynamic phase transitions, and moving free surfaces are common. However, LBM remains computationally expensive and is less commonly integrated with stress analysis models, limiting its standalone applicability in predicting mechanical performance.

SPH, a mesh-free method based on the Lagrangian framework, has been widely applied to simulate powder spreading, melt pool evolution, and molten metal flow in PBF. Unlike traditional grid-based methods, SPH represents fluid and solid materials as discrete particles, making it particularly well-suited for free-surface flow modelling and powder-bed interactions. This method has been used to capture key effects, including recoater blade interaction, powder packing density, particle rearrangement, and melt pool instability. However, SPH simulations require significant computational resources, and maintaining numerical stability can be challenging in high-resolution applications, particularly when extending simulations to larger PBF domains.

At the microscopic scale, numerical modelling focuses on microstructural evolution, grain nucleation, and solidification behaviour, which dictate the mechanical properties and performance of PBF-fabricated components. Due to the rapid cooling rates and steep thermal gradients inherent to PBF, microstructural evolution is highly sensitive to process parameters, necessitating specialised modelling techniques such as Cellular Automaton (CA), Monte Carlo methods, and Phase-Field models. CA-based models provide a computationally efficient approach for predicting grain growth and competitive nucleation, allowing for the simulation of solidification textures under different PBF processing conditions. MC methods have been widely used to model recrystallisation and phase transformations, particularly in studies of post-processing and heat treatment effects on microstructure refinement. Phase-Field models, which describe solidification kinetics and dendritic growth dynamics based on thermodynamic and kinetic principles, offer high-fidelity insights into microstructure formation during PBF. While these methods provide critical insights into grain morphology, phase transformations, and defect formation, they often rely on precomputed thermal histories and require coupling with macroscopic thermal and mechanical models to enhance predictive accuracy.

Despite significant advancements in numerical modelling, no single method can comprehensively capture all the physical phenomena occurring in PBF. Macroscopic models, such as FEM and FVM, are effective for analysing thermal stress evolution and mechanical distortion, but they lack the resolution needed to capture melt pool instabilities and phase change phenomena. Mesoscopic approaches, including CFD, LBM, and SPH, provide high-fidelity simulations of fluid flow, energy transport, and powder dynamics, yet they remain computationally intensive and difficult to couple with large-scale stress models. Microscopic techniques, such as CA, MC, and PF models, offer valuable insights into solidification and grain structure evolution, but their predictive power is limited without integration with higher-scale thermal and fluid dynamics models.

To address these challenges, multi-scale modelling frameworks have been developed by integrating FEM-based stress analysis, CFD-based melt-pool simulations, and CA-based grain-growth models. These hybrid approaches enable a more comprehensive understanding of PBF process–structure–property relationships, facilitating defect prediction, process optimisation, and control of material properties. The continued refinement of these models, together with machine-learning-assisted surrogate modelling and data-driven process optimisation, has the potential to significantly enhance computational efficiency and predictive accuracy. Additionally, experimental validation through high-speed imaging, X-ray computed tomography, and in-situ process monitoring remains essential for refining these numerical frameworks and ensuring their applicability to practical PBF manufacturing scenarios.

10.3.2 Overview of Commonly Used Software

Numerical simulation has become integral to advancing AM technologies, providing critical insights into complex multi-physics phenomena. Several specialised software packages have emerged as essential tools within AM simulations, particularly in PBF processes. **Table 10.1** outlines the most commonly used software, highlighting their strengths and limitations:

Table 10.1 Overview of Commonly Used Numerical Simulation Software in AM

Numerical Method	*Software*	*Advantages*	*Limitations*
FEM	ABAQUS	Robust structural and thermal-stress analysis; Effective distortion and residual stress prediction; Strong industry recognition.	Limited fluid dynamics capabilities; Complex interface requiring significant user expertise.
	ANSYS	Comprehensive thermo-mechanical simulations; Extensive material database; Intuitive workflows tailored to AM.	High computational resources required; Costly licensing.
	COMSOL Multiphysics	Strong in multi-physics coupling (thermal, fluid, structural); User-friendly graphical interface; High flexibility for custom physics.	Computationally intensive; Relatively slower for large-scale simulations.
FVM	OpenFOAM	Highly customisable; Excellent for detailed CFD and thermal simulations; Extensive community support and resources.	Steep learning curve; Lacks direct professional technical support.
	FLOW-3D	Effective for complex free-surface flow and melt pool dynamics; User-friendly GUI; Established industry validation.	High computational cost for fine-scale simulations; Limited customisation compared to open-source tools.
DEM	LIGGGHTS	Open-source flexibility; Excellent for powder spreading, granular flows, particle interactions; Free and customisable.	Computationally demanding; Limited graphical interface.
	EDEM	Powerful powder flowability and compaction analysis; Strong integration capability with other commercial tools (ANSYS, ABAQUS); Good industry adoption.	High licensing costs; Resource-intensive simulations.
SPH	DualSPHysics	Open-source; Good scalability for GPU parallel computing; Widely used for highly dynamic free-surface flows.	Limited documentation; requires a solid understanding of the SPH method; Basic GUI.
	GPUSPH	Open-source; GPU-accelerated simulations, effective for free-surface flows; Good parallel performance.	Limited graphical interface and documentation; Moderate learning curve.
	nanoFluidX	Strong GPU acceleration for large-scale lubrication and multiphase flows; User-friendly interface; Industrially proven, excellent for automotive simulations.	High software licensing costs; Limited flexibility for custom physics and user modifications.
LBM	XFlow	No meshing required; Strong GUI, highly automated workflows; Industrially validated, especially automotive/aerospace aerodynamics.	Expensive commercial licensing; Limited flexibility in customisation.
	PowerFLOW	Highly efficient for large-scale aerodynamic simulations; Strong parallel computing performance; Widely used in the automotive industry.	High software licensing costs; Tailored primarily to specific industry applications, limiting general usability.

10.4 CONCLUSIONS

This chapter has presented the recent progress in numerical modelling approaches for PBF, with a focus on its underlying physical mechanisms and simulation strategies. The modelling of powder spreading and melt pool behaviour was discussed in detail, incorporating both discrete and continuum frameworks. Key physical phenomena, such as particle packing dynamics, heat transfer, phase transitions, and recoil pressure, were addressed using high-fidelity multi-physics formulations. These models provide essential tools for understanding defect formation and optimising processing parameters.

Looking ahead, process modelling in additive manufacturing is expected to advance towards greater fidelity, efficiency, and integration. Future research will continue to develop multi-physics and multiscale models that capture the complex interplay of thermal, mechanical, and microstructural phenomena. Meshfree and particle-based methods offer a promising direction, particularly for simulating the highly dynamic and discontinuous nature of melt pool evolution and powder behaviour. To meet the demands of real-time monitoring and control, streamlined numerical strategies that enable rapid simulation or reduced-order approximation will be essential.

At the same time, data-driven approaches, including machine learning, are increasingly coupled with physics-based models to improve predictive accuracy, accelerate parameter optimisation, and support model calibration when physical data are sparse [62, 63]. These hybrid frameworks are not only valuable for understanding process-structure-property relationships but also support the development of intelligent systems capable of adaptive decision-making. As modelling capabilities grow, numerical simulations are expected to play a central role in predicting build quality, identifying defects, and reducing the need for experimental iterations. Ultimately, process models will form a foundational pillar of digital twin architectures, integrated with in-situ sensing and data analytics to enable closed-loop control and autonomous optimisation in metal additive manufacturing.

As demonstrated in this chapter, combining physically informed models with emerging computational strategies will remain a key enabler for the development of reliable, scalable, and intelligent AM systems. This convergence is expected to bridge the gap between numerical modelling and practical deployment, accelerating the transition from process understanding to robust industrial implementation. In the longer term, such integration also lays the foundation for data-driven innovation in materials design and sustainable additive manufacturing.

REFERENCES

[1] S. Chowdhury et al., 'Laser powder bed fusion: A state-of-the-art review of the technology, materials, properties & defects, and numerical modelling', *J. Mater. Res. Technol.,* vol. 20, pp. 2109–2172, 2022, https://doi.org/10.1016/j.jmrt.2022.07.121.

[2] I. Gibson, D. Rosen, B. Stucker, and M. Khorasani, *Additive Manufacturing Technologies*. Cham: Springer International Publishing, 2021. https://doi.org/10.1007/978-3-030-56127-7.

[3] L. Jin et al., 'Big data, machine learning, and digital twin assisted additive manufacturing: A review', *Mater. Des.*, vol. 244, p. 113086, 2024, https://doi.org/10.1016/j.matdes.2024.113086.

[4] R. Singh et al., 'Powder bed fusion process in additive manufacturing: An overview', *Mater. Today Proc.*, vol. 26, pp. 3058–3070, 2020, https://doi.org/10.1016/j.matpr.2020.02.635.

[5] H. Chen, Q. Wei, Y. Zhang, F. Chen, Y. Shi, and W. Yan, 'Powder-spreading mechanisms in powder-bed-based additive manufacturing: Experiments and computational modeling', *Acta Mater.*, vol. 179, pp. 158–171, 2019, https://doi.org/10.1016/j.actamat.2019.08.030.

[6] H. Chen, Y. Chen, Y. Liu, Q. Wei, Y. Shi, and W. Yan, 'Packing quality of powder layer during counter-rolling-type powder spreading process in additive manufacturing', *Int. J. Mach. Tools Manuf.*, vol. 153, p. 103553, 2020, https://doi.org/10.1016/j.ijmachtools.2020.103553.

[7] W. J. Sames, F. A. List, S. Pannala, R. R. Dehoff, and S. S. Babu, 'The metallurgy and processing science of metal additive manufacturing', *Int. Mater. Rev.*, vol. 61, no. 5, pp. 315–360, 2016, https://doi.org/10.1080/09506608.2015.1116649.

[8] M. Afrasiabi and M. Bambach, 'Modelling and simulation of metal additive manufacturing processes with particle methods: A review', *Virtual Phys. Prototyp.*, vol. 18, no. 1, p. e2274494, 2023, https://doi.org/10.1080/17452759.2023.2274494.

[9] Y. Shanjani and E. Toyserkani, 'Material spreading and compaction in powder-based solid freeform fabrication methods: Mathematical modeling', in *Proceedings of the 19th Annual International Solid Freeform Fabrication Symposium*, SFF 2008, 2008, pp. 399–410.

[10] L. C. Capozzi, A. Sivo, and E. Bassini, 'Powder spreading and spreadability in the additive manufacturing of metallic materials: A critical review', *J. Mater. Process. Technol.*, vol. 308, p. 117706, 2022, https://doi.org/10.1016/j.jmatprotec.2022.117706.

[11] V. P. Katashinskii and M. B. Shtern, 'Stress-strain state of powder being rolled in the densification zone: II. Distribution of density, longitudinal strain, and contact stress in the densification zone', *Sov. Powder Metall. Met. Ceram.*, vol. 22, no. 12, pp. 972–976, 1983, https://doi.org/10.1007/BF00802421.

[12] P. A. Cundall and O. D. L. Strack, 'A discrete numerical model for granular assemblies', *Géotechnique*, vol. 29, no. 1, pp. 47–65, 1979, https://doi.org/10.1680/geot.1979.29.1.47.

[13] L. Wang, E. L. Li, H. Shen, R. P. Zou, A. B. Yu, and Z. Y. Zhou, 'Adhesion effects on spreading of metal powders in selective laser melting', *Powder Technol.*, vol. 363, pp. 602–610, 2020, https://doi.org/10.1016/j.powtec.2019.12.048.

[14] B. Blais, D. Vidal, F. Bertrand, G. S. Patience, and J. Chaouki, 'Experimental methods in chemical engineering: Discrete element method—DEM', *Can. J. Chem. Eng.*, vol. 97, no. 7, pp. 1964–1973, 2019, https://doi.org/10.1002/cjce.23501.

[15] P. S. Cook and A. B. Murphy, 'Simulation of melt pool behaviour during additive manufacturing: Underlying physics and progress', *Addit. Manuf.*, vol. 31, p. 100909, 2020, https://doi.org/10.1016/j.addma.2019.100909.

[16] M. Markl and C. Körner, 'Multiscale modeling of powder bed–based additive manufacturing', *Annu. Rev. Mater. Res.*, vol. 46, no. 1, pp. 93–123, 2016, https://doi.org/10.1146/annurev-matsci-070115-032158.

[17] D. Rosenthal, 'The theory of moving sources of heat and its application to metal treatments', *J. Fluids Eng.*, vol. 68, no. 8, pp. 849–865, 1946, https://doi.org/10.1115/1.4018624.

[18] P. Promoppatum, S.-C. Yao, P. C. Pistorius, and A. D. Rollett, 'A comprehensive comparison of the analytical and numerical prediction of the thermal history and solidification microstructure of inconel 718 products made by laser powder-bed fusion', *Engineering*, vol. 3, no. 5, pp. 685–694, 2017, https://doi.org/10.1016/J.ENG.2017.05.023.

[19] C. Körner, E. Attar, and P. Heinl, 'Mesoscopic simulation of selective beam melting processes', *J. Mater. Process. Technol.*, vol. 211, no. 6, pp. 978–987, 2011, https://doi.org/10.1016/j.jmatprotec.2010.12.016.

[20] J. Goldak, A. Chakravarti, and M. Bibby, 'A new finite element model for welding heat sources', *Metall. Trans. B*, vol. 15, no. 2, pp. 299–305, 1984, https://doi.org/10.1007/BF02667333.

[21] J. Cheng, Y. Huo, P. Fernandez-Zelaia, X. Hu, M. Li, and X. Sun, 'A Gaussian process-based extended goldak heat source model for finite element simulation of laser powder bed fusion additive manufacturing process', *Comput. Mater. Sci.*, vol. 244, p. 113185, 2024, https://doi.org/10.1016/j.commatsci.2024.113185.

[22] Z. Zhang et al., '3-Dimensional heat transfer modeling for laser powder-bed fusion additive manufacturing with volumetric heat sources based on varied thermal conductivity and absorptivity', *Opt. Laser Technol.*, vol. 109, pp. 297–312, 2019, https://doi.org/10.1016/j.optlastec.2018.08.012.

[23] M. A. Ilani and Y. M. Banad, 'Modeling melt pool geometry in metal additive manufacturing using Goldak's semi-ellipsoidal heat source: A data-driven computational approach'. arXiv:2404.08834, 2024. https://arxiv.org/abs/2404.08834.

[24] A. K. Mishra and A. Kumar, 'Numerical and experimental analysis of the effect of volumetric energy absorption in powder layer on thermal-fluidic transport in selective laser melting of Ti6Al4V', *Opt. Laser Technol.*, vol. 111, pp. 227–239, 2019, https://doi.org/10.1016/j.optlastec.2018.09.054.
[25] J. Yin et al., 'A finite element model of thermal evolution in laser micro sintering', *Int. J. Adv. Manuf. Technol.*, vol. 83, no. 9–12, pp. 1847–1859, 2016, https://doi.org/10.1007/s00170-015-7609-x.
[26] L. Ladani, J. Romano, W. Brindley, and S. Burlatsky, 'Effective liquid conductivity for improved simulation of thermal transport in laser beam melting powder bed technology', *Addit. Manuf.*, vol. 14, pp. 13–23, 2017, https://doi.org/10.1016/j.addma.2016.12.004.
[27] A. Foroozmehr, M. Badrossamay, E. Foroozmehr, and S. Golabi, 'Finite element simulation of selective laser melting process considering optical penetration depth of laser in powder bed', *Mater. Des.*, vol. 89, pp. 255–263, 2016, https://doi.org/10.1016/j.matdes.2015.10.002.
[28] M. Bayat, S. Mohanty, and J. H. Hattel, 'A systematic investigation of the effects of process parameters on heat and fluid flow and metallurgical conditions during laser-based powder bed fusion of Ti6Al4V alloy', *Int. J. Heat Mass Transf.*, vol. 139, pp. 213–230, 2019, https://doi.org/10.1016/j.ijheatmasstransfer.2019.05.017.
[29] A. V. Gusarov, I. Yadroitsev, Ph. Bertrand, and I. Smurov, 'Model of radiation and heat transfer in laser-powder interaction zone at selective laser melting', *J. Heat Transf.*, vol. 131, no. 7, p. 072101, 2009, https://doi.org/10.1115/1.3109245.
[30] C. Argento and D. Bouvard, 'A ray tracing method for evaluating the radiative heat transfer in porous media', *Int. J. Heat Mass Transf.*, vol. 39, no. 15, pp. 3175–3180, 1996, https://doi.org/10.1016/0017-9310(95)00403-3.
[31] H.-C. Tran and Y.-L. Lo, 'Heat transfer simulations of selective laser melting process based on volumetric heat source with powder size consideration', *J. Mater. Process. Technol.*, vol. 255, pp. 411–425, 2018, https://doi.org/10.1016/j.jmatprotec.2017.12.024.
[32] S.-W. Han, J. Ahn, and S.-J. Na, 'A study on ray tracing method for CFD simulations of laser keyhole welding: Progressive search method', *Weld. World*, vol. 60, no. 2, pp. 247–258, 2016, https://doi.org/10.1007/s40194-015-0289-1.
[33] C. Tang, L. Yao, and H. Du, 'Computational framework for the simulation of multi material laser powder bed fusion', *Int. J. Heat Mass Transf.*, vol. 191, p. 122855, 2022, https://doi.org/10.1016/j.ijheatmasstransfer.2022.122855.
[34] L. Wang, Y. Zhang, H. Y. Chia, and W. Yan, 'Mechanism of keyhole pore formation in metal additive manufacturing', *NPJ Comput. Mater.*, vol. 8, no. 1, p. 22, 2022, https://doi.org/10.1038/s41524-022-00699-6.
[35] Y. Yang, Z. Rao, T. Zhou, and P. Zhou, 'Ray tracing heat source model for polymer powder bed fusion considering laser energy absorption and multiple reflection', *Addit. Manuf.*, vol. 91, p. 104363, 2024, https://doi.org/10.1016/j.addma.2024.104363.
[36] W. Yan, J. Smith, W. Ge, F. Lin, and W. K. Liu, 'Multiscale modeling of electron beam and substrate interaction: A new heat source model', *Comput. Mech.*, vol. 56, no. 2, pp. 265–276, 2015, https://doi.org/10.1007/s00466-015-1170-1.
[37] A. Zakirov et al., 'Predictive modeling of laser and electron beam powder bed fusion additive manufacturing of metals at the mesoscale', *Addit. Manuf.*, vol. 35, p. 101236, 2020, https://doi.org/10.1016/j.addma.2020.101236.
[38] L. Wang, Z. Guo, G. Peng, S. Wu, Y. Zhang, and W. Yan, 'Evaporation-induced composition evolution in metal additive manufacturing', *Adv. Funct. Mater.*, p. 2412071, 2024, https://doi.org/10.1002/adfm.202412071.
[39] F. Verhaeghe, T. Craeghs, J. Heulens, and L. Pandelaers, 'A pragmatic model for selective laser melting with evaporation', *Acta Mater.*, vol. 57, no. 20, pp. 6006–6012, 2009, https://doi.org/10.1016/j.actamat.2009.08.027.
[40] V. Juechter, T. Scharowsky, R. F. Singer, and C. Körner, 'Processing window and evaporation phenomena for Ti–6Al–4V produced by selective electron beam melting', *Acta Mater.*, vol. 76, pp. 252–258, 2014, https://doi.org/10.1016/j.actamat.2014.05.037.
[41] J. Yin et al., 'Vaporization of alloying elements and explosion behavior during laser powder bed fusion of Cu–10Zn alloy', *Int. J. Mach. Tools Manuf.*, vol. 161, p. 103686, 2021, https://doi.org/10.1016/j.ijmachtools.2020.103686.

[42] L. Zhuo, B. Song, R. Li, Q. Wei, C. Yan, and Y. Shi, 'Effect of element evaporation on the microstructure and properties of CuZnAl shape memory alloys prepared by selective laser melting', *Opt. Laser Technol.*, vol. 127, p. 106164, 2020, https://doi.org/10.1016/j.optlastec.2020.106164.

[43] W. J. Lee, J. Kim, and H. W. Park, 'Improved corrosion resistance of Mg alloy AZ31B induced by selective evaporation of Mg using large pulsed electron beam irradiation', *J. Mater. Sci. Technol.*, vol. 35, no. 5, pp. 891–901, 2019, https://doi.org/10.1016/j.jmst.2018.12.004.

[44] W. E. King et al., 'Observation of keyhole-mode laser melting in laser powder-bed fusion additive manufacturing', *J. Mater. Process. Technol.*, vol. 214, no. 12, pp. 2915–2925, 2014, https://doi.org/10.1016/j.jmatprotec.2014.06.005.

[45] A. Ladewig, G. Schlick, M. Fisser, V. Schulze, and U. Glatzel, 'Influence of the shielding gas flow on the removal of process by-products in the selective laser melting process', *Addit. Manuf.*, vol. 10, pp. 1–9, 2016, https://doi.org/10.1016/j.addma.2016.01.004.

[46] H. Chen and W. Yan, 'Spattering and denudation in laser powder bed fusion process: Multiphase flow modelling', *Acta Mater.*, vol. 196, pp. 154–167, 2020, https://doi.org/10.1016/j.actamat.2020.06.033.

[47] L. Wang, Y. Zhang, and W. Yan, 'Evaporation model for keyhole dynamics during additive manufacturing of metal', *Phys. Rev. Appl.*, vol. 14, no. 6, p. 064039, 2020, https://doi.org/10.1103/PhysRevApplied.14.064039.

[48] S. A. Khairallah, A. T. Anderson, A. Rubenchik, and W. E. King, 'Laser powder-bed fusion additive manufacturing: Physics of complex melt flow and formation mechanisms of pores, spatter, and denudation zones', *Acta Mater.*, vol. 108, pp. 36–45, 2016, https://doi.org/10.1016/j.actamat.2016.02.014.

[49] C. Knight, 'Theoretical modeling of rapid surface vaporization with back-pressure', in *Proceedings of the11th Fluid and Plasma Dynamics Conference*, 1978, https://doi.org/10.2514/6.1978-1220.

[50] X. He, T. DebRoy, and P. W. Fuerschbach, 'Alloying element vaporization during laser spot welding of stainless steel', *J. Phys. Appl. Phys.*, vol. 36, no. 23, pp. 3079–3088, 2003, https://doi.org/10.1088/0022-3727/36/23/033.

[51] V. Semak and A. Matsunawa, 'The role of recoil pressure in energy balance during laser materials processing', *J. Phys. Appl. Phys.*, vol. 30, no. 18, pp. 2541–2552, 1997, https://doi.org/10.1088/0022-3727/30/18/008.

[52] T. Ytrehus and S. Østmo, 'Kinetic theory approach to interphase processes', *Int. J. Multiph. Flow*, vol. 22, no. 1, pp. 133–155, 1996, https://doi.org/10.1016/0301-9322(95)00056-9.

[53] M. Dal and R. Fabbro, '[INVITED] An overview of the state of art in laser welding simulation', *Opt. Laser Technol.*, vol. 78, pp. 2–14, 2016, https://doi.org/10.1016/j.optlastec.2015.09.015.

[54] T. DebRoy et al., 'Additive manufacturing of metallic components—Process, structure and properties', *Prog. Mater. Sci.*, vol. 92, pp. 112–224, 2018, https://doi.org/10.1016/j.pmatsci.2017.10.001.

[55] H. Wang and Y. Zou, 'Microscale interaction between laser and metal powder in powder-bed additive manufacturing: Conduction mode versus keyhole mode', *Int. J. Heat Mass Transf.*, vol. 142, p. 118473, 2019, https://doi.org/10.1016/j.ijheatmasstransfer.2019.118473.

[56] N. Kouraytem et al., 'Effect of laser-matter interaction on molten pool flow and keyhole dynamics', *Phys. Rev. Appl.*, vol. 11, no. 6, p. 064054, 2019, https://doi.org/10.1103/PhysRevApplied.11.064054.

[57] C. Zhao et al., 'Critical instability at moving keyhole tip generates porosity in laser melting', *Science*, vol. 370, no. 6520, pp. 1080–1086, 2020, https://doi.org/10.1126/science.abd1587.

[58] S. A. Khairallah et al., 'Controlling interdependent meso-nanosecond dynamics and defect generation in metal 3D printing', *Science*, vol. 368, no. 6491, pp. 660–665, 2020, https://doi.org/10.1126/science.aay7830.

[59] Y. Zhang, S. Wu, Z. Guo, G. Peng, L. Wang, and W. Yan, 'Defects caused by powder spattering and entrainment in laser powder bed fusion process: High-fidelity modeling of gas, melt pool and powder dynamics', *Acta Mater.*, vol. 288, p. 120816, 2025, https://doi.org/10.1016/j.actamat.2025.120816.

[60] A. Klassen, T. Scharowsky, and C. Körner, 'Evaporation model for beam based additive manufacturing using free surface lattice Boltzmann methods', *J. Phys. Appl. Phys.*, vol. 47, no. 27, p. 275303, 2014, https://doi.org/10.1088/0022-3727/47/27/275303.

[61] Z. Gan, Y. Lian, S. E. Lin, K. K. Jones, W. K. Liu, and G. J. Wagner, 'Benchmark study of thermal behavior, surface topography, and dendritic microstructure in selective laser melting of inconel 625', *Integrating Mater. Manuf. Innov.*, vol. 8, no. 2, pp. 178–193, 2019, https://doi.org/10.1007/s40192-019-00130-x.
[62] W. Yan et al., 'Data-driven multi-scale multi-physics models to derive process–structure–property relationships for additive manufacturing', *Comput. Mech.*, vol. 61, no. 5, pp. 521–541, 2018, https://doi.org/10.1007/s00466-018-1539-z.
[63] C. Wang, X. P. Tan, S. B. Tor, and C. S. Lim, 'Machine learning in additive manufacturing: State-of-the-art and perspectives', *Addit. Manuf.*, vol. 36, p. 101538, 2020, https://doi.org/10.1016/j.addma.2020.101538.

Chapter 11

Recent Developments in 3D Printing for Sustainable Construction

Biswajoy Bhattacharjee and Prakash Kumar Sahu

11.1 INTRODUCTION

The construction industry, which has long needed modernisation, is now on the verge of a transformative change, with the adoption of additive manufacturing (AM) technology (Žujović et al. 2022), which will completely redefine how we build structures (Zairul and Zaremohzzabieh 2023). The integration of AM into the construction sector has shown great promise, focusing on efficiency, productivity, cost-effectiveness, and sustainability (Maskuriy et al. 2019). This is further supported by the surge of research publications in this field, as shown in **Figure 11.1**, indicating growing interest in using AM for building applications. This analysis examines the developments and uses of AM technology in the building and construction sector, as well as its future potential.

Traditionally, the building and construction industry has been resistant towards change (Xue et al. 2014). However, global challenges such as urbanisation, population growth, resource scarcity, and sustainability concerns necessitate innovative solutions (Gambatese and Hallowell 2011). In this context, AM technology has emerged as a powerful catalyst for transformation (De Schutter et al. 2018). AM involves the layered deposition of materials to develop 3D objects from CAD models, offering a high level of design flexibility. Although there are different types of AM methods, it has been observed that material extrusion is best suited for construction applications. Printing concrete objects involves creating a digital model, slicing it into layers for the printer, selecting the appropriate concrete mix, setting up the printer, and starting the printing process. Once the concrete object is printed, it is then allowed to cure and, if required, refined by post-processing as shown in **Figure 11.2**. In the early days of its discovery, 3D printing was limited to producing small-scale prototypes and consumer goods.

Now, its possibilities are being expanded, and practical large-scale applications are being evaluated, offering the promise of increased efficiency, reduced waste, and unprecedented design freedom (Hager et al. 2016; Siddika et al. 2019). Over the last half-decade, numerous studies have evaluated the viability of integrating AM into the construction industry. Recent studies are now focused on improving material properties specific to the application to achieve sustainability and affordability (Hager et al. 2016).

11.2 REVIEW METHODOLOGY

The review methodology was designed to capture and analyse the literature on sustainable applications and developments in 3D concrete printing (3DCP), with a focus on advancing a cleaner future. The following search string was adopted for this review across the Scopus database: "(TITLE-ABS-KEY (3d AND printing) OR TITLE-ABS-KEY (additive AND manufacturing) OR TITLE-ABS-KEY (3d AND concrete AND printing) OR TITLEABS-KEY

DOI: 10.1201/9781003590774-11

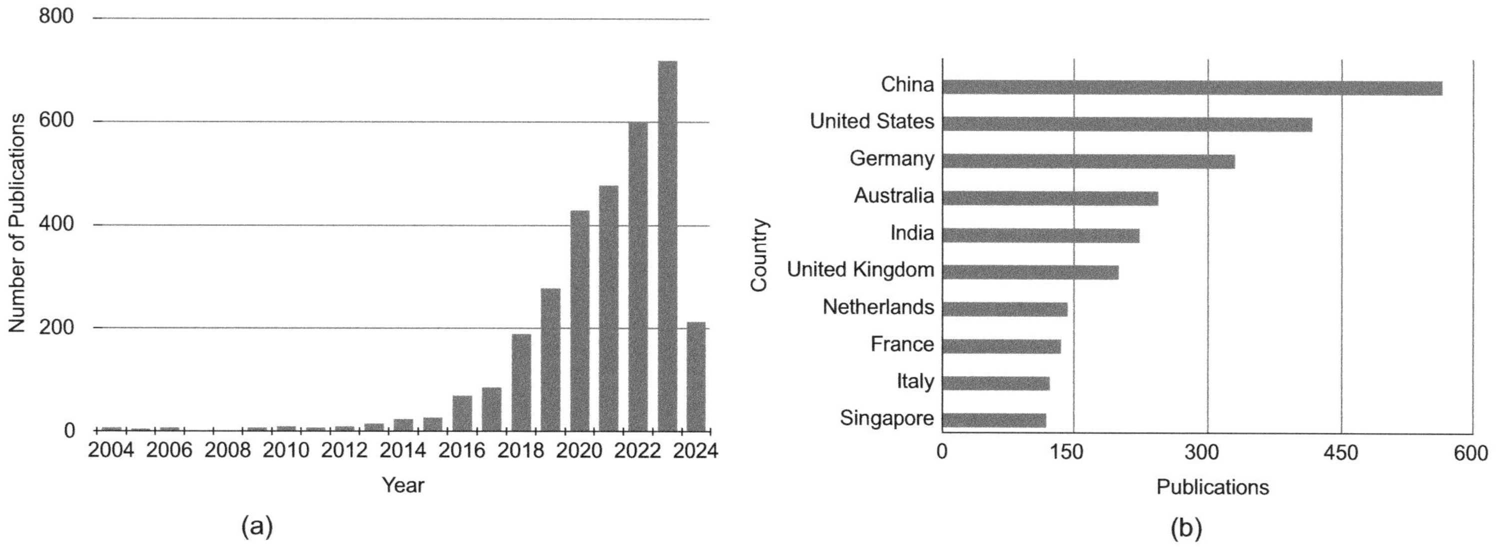

Figure 11.1 Breakdown of publications in 3DCP in Scopus database from 2004 to 2024, (a) Year-wise breakdown of publications in 3DCP, (b) Country-wise breakdown of publications in 3DCP.

Figure 11.2 Process flow of concrete 3D printing.

(concrete AND printing) AND TITLE-ABS-KEY (sustainability) OR TITLE-ABS-KEY (recycling)) AND PUBYEAR >2003 AND PUBYEAR <2025." A period of two decades, from 2004 to 2024, was selected for this search. The initial search retrieved 3515 publications as visualised in **Figure 11.3**. Two layers of exclusion criteria were applied to refine the selection process. The first excluded articles that did not directly address the sustainable aspects of 3DCP, while the second focused on recentness. To ensure relevancy, the screening process included a thorough assessment of titles, abstracts, and keywords. **Figure 11.4** shows a schematic diagram of the selection process.

11.3 LITERATURE REVIEW

11.3.1 Recycling in 3D Concrete Printing

11.3.1.1 Waste Tyres

Discarded tyres (D.T.) pose a significant environmental challenge, necessitating efficient reuse strategies. Abdulfattah et al. (2022) incorporated waste tyre-derived pyrolysed carbon black (PCB) into concrete mixes. Slump tests indicated reduced workability with increasing PCB content due to higher viscosity (Elyamany et al. 2014), although additions up to 5.5% remained within acceptable limits (**Figure 11.5a**). Optimal performance was observed at 4% PCB, where compressive strength peaked at approximately 39 MPa after 28 days, alongside reduced water absorption and improved abrasion resistance (**Figure 11.5b–e**). These trends are consistent with earlier studies on carbon-based additives (Rezania et al. 2019; Monteiro et al. 2017).

In a related study, Valente et al. (2023) also observed some difficulty in workability during the initial printing, due to increased viscosity, thereby reducing workability (see **Figure 11.6**). Flexural strength, thermo-mechanical behaviour, and the ink's heat and high-frequency acoustic insulation were found to be superior. In addition, incorporating carbon black has been acknowledged to enhance the strength of interlayer bonds (Hosseini et al. 2019). This environmentally friendly method, utilising waste tyre byproducts, offers a sustainable solution to the global problem of waste tyres while simultaneously enhancing concrete's properties.

11.3.1.2 Recycled Aggregates

Recycled glass, polymers, and aggregates influence rheology, strength, and interlayer behaviour in 3DCP. Increased incorporation of recycled materials generally reduces static yield stress while increasing dynamic yield stress and viscosity, affecting printability and bond

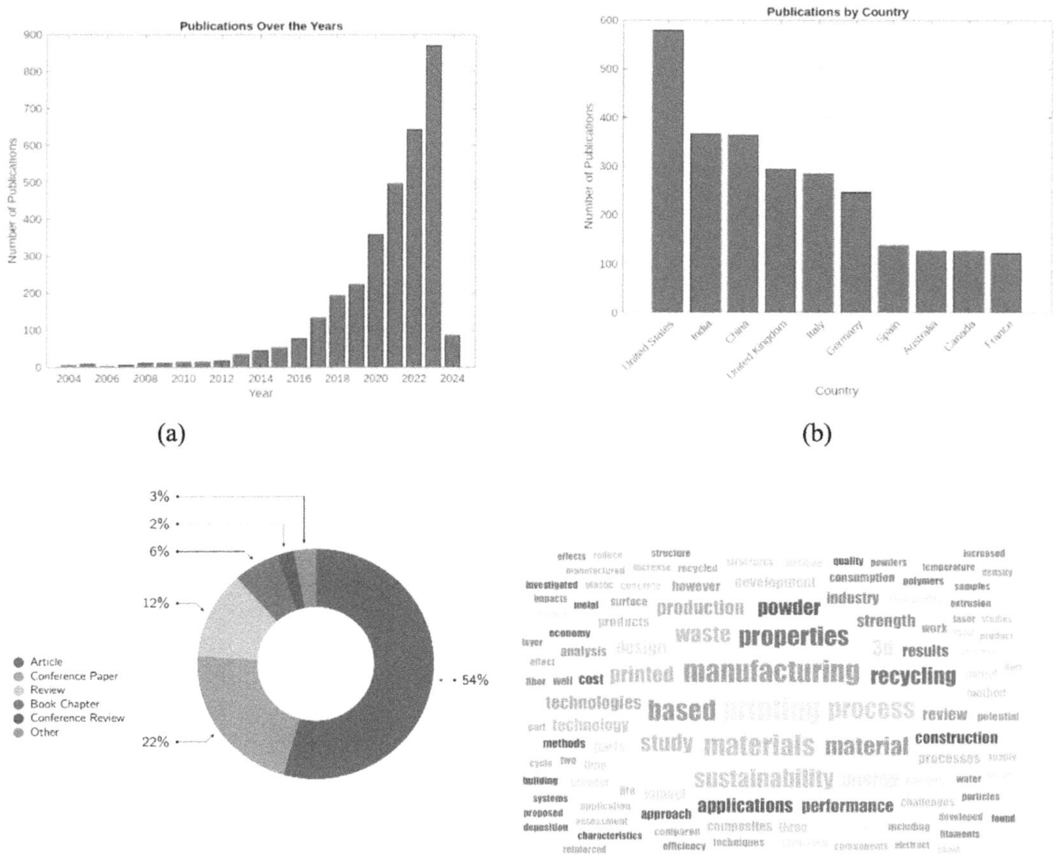

Figure 11.3 Breakdown of publications in the date range 2004–2024, (a) Year-wise breakdown of publications, (b) Country-wise breakdown of publications, (c) Category-wise breakdown of publications, (d) Word-cloud of publications.

strength (Ting et al. 2021; Zhou et al. 2024; Bai et al. 2021). For recycled glass cullets, this reduction in static yield stress and increase in viscosity lead to reduced compressive and inter-layer bond strength, as shown in **Figure 11.7**. Partial replacement, however, can maintain performance, as RPC mixes with 10 µm ultrafine glass powder showed a 27.54% increase in slump at 25% replacement with comparable strength at 20% replacement (Zhou et al. 2024). Recycled plastic aggregates further highlight the trade-off between flowability and structural stability (Oosthuizen et al. 2023; Bai et al. 2021). Increased replacement improves flow but reduces yield stress and buildability due to weaker matrix bonding and higher porosity. For example, replacing fine aggregate with Resin8 increased slump flow to 152–162 mm from 145 mm, while porosity rose from 5.8% to 16.8% at 15% replacement, reducing buildability though maintaining structural viability (**Figure 11.8**).

Enhancement of recycled aggregates through carbonation and nano-SiO_2 presoaking improves microstructure and mechanical performance despite reductions associated with higher replacement levels (Tong et al. 2023). Increasing RCA content reduces strength in all

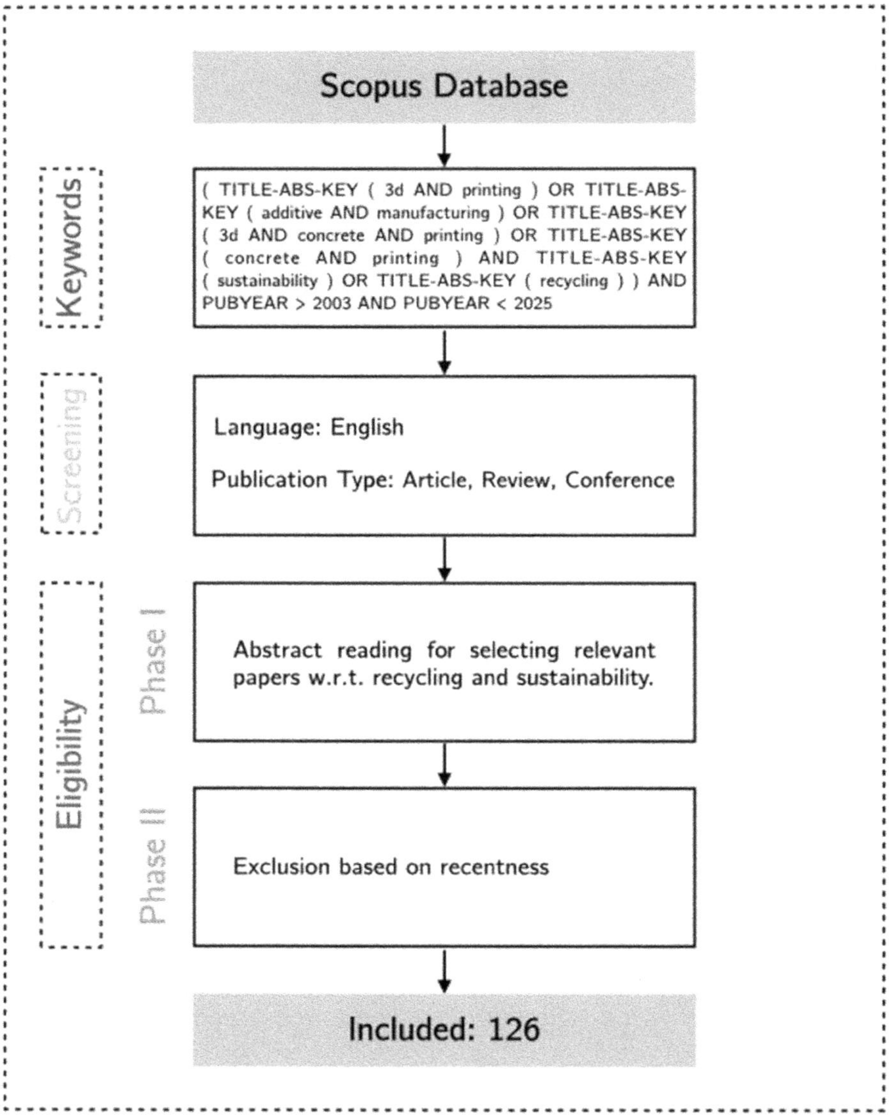

Figure 11.4 Schematic of the selection of literature.

directions (**Figure 11.9**), but carbonation at 100% RCA improved compressive, flexural, and splitting tensile strengths by up to 24.71%, 12.95%, and 18.27%, respectively, while reducing pore defects.

Reinforcement strategies and composite approaches address structural limitations in recycled systems (Liu et al. 2023; De Vlieger et al. 2023; Huawei Liu et al. 2022). Integration of engineered cementitious composites improves flexural behaviour, with ECC heights of 30–45 mm achieving load capacities comparable to reinforced concrete with 0.2% longitudinal reinforcement, and interfacial effects analysed using finite element modelling (**Figure 11.10**).

Recycled sand incorporation and interlayer bonding behaviour further demonstrate the coupling between rheology and structural performance (De Vlieger et al. 2023; Huawei Liu

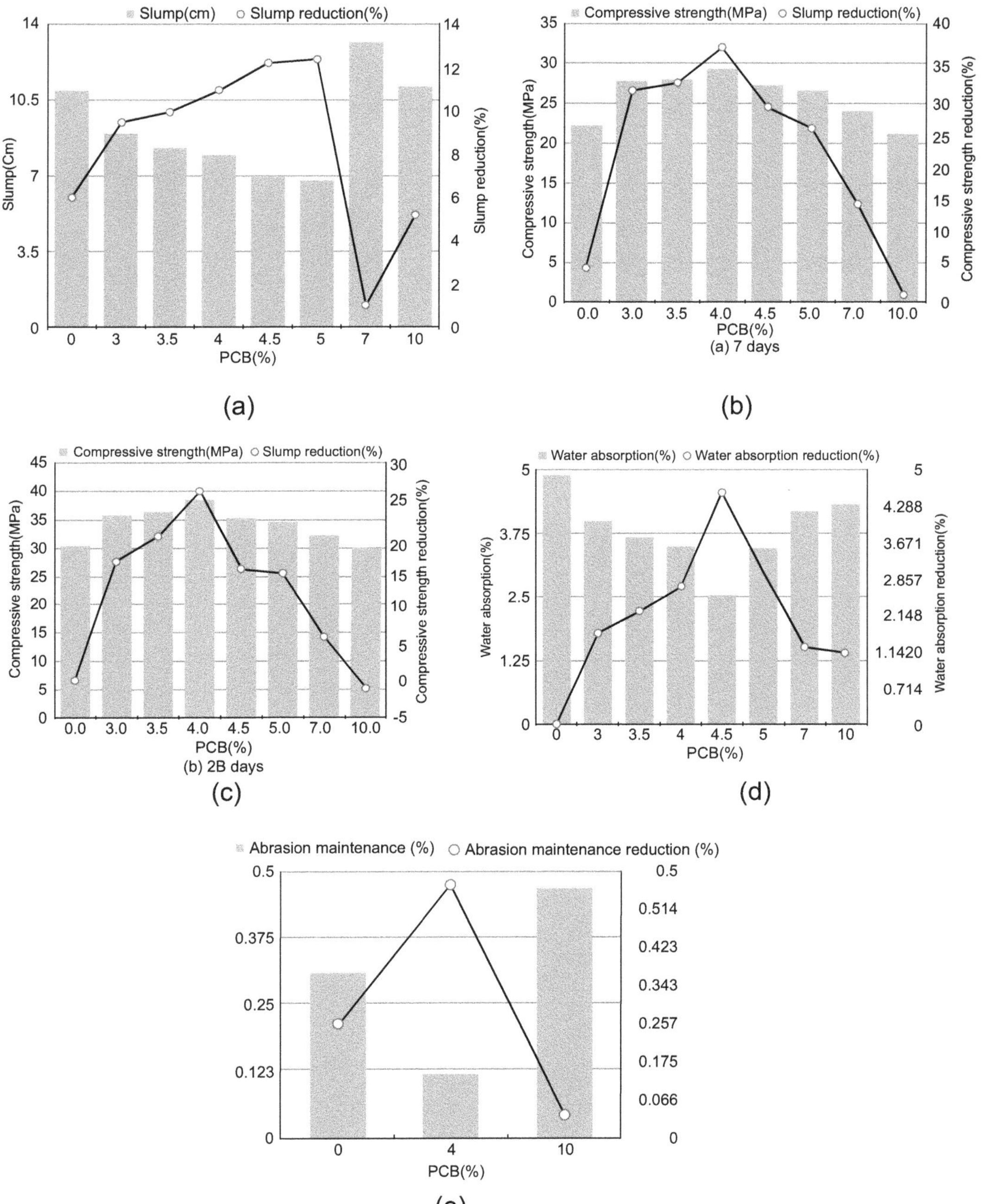

Figure 11.5 Various test results for concrete samples with different proportions of PCB, (a) Slump measurements for different proportions of PCB/c ratios (Abdulfattah et al. 2022), (b) Compressive strength outcomes assessed at different PCB percentages after 7 days (Abdulfattah et al. 2022), (c) Compressive strength outcomes assessed at different concrete samples at different proportions of PCB (Abdulfattah et al. 2022), (d) Mean water absorption test results obtained for PCB percentages after 28 days (Abdulfattah et al. 2022), (e) Average abrasion resistance across all PCB percentages (Abdulfattah et al. 2022).

Figure 11.6 Printing issues during preliminary printing test (Valente et al. 2023).

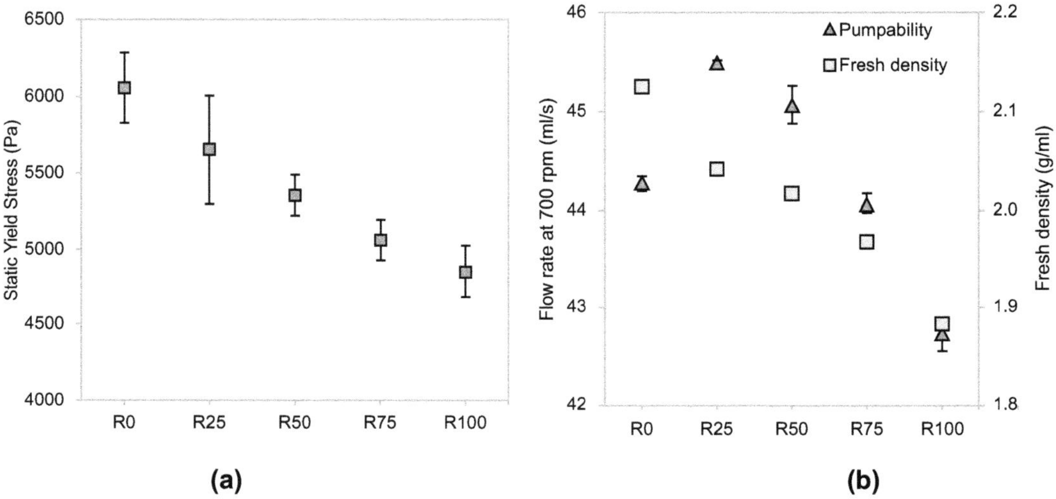

Figure 11.7 Effects of recycled glass content on concrete properties, (a) Static yield stress decrease with increased recycled glass content and Ting, Tay, and Tan (2021), (b) Pumpability and fresh density of fresh concrete mixture (Ting et al. 2021).

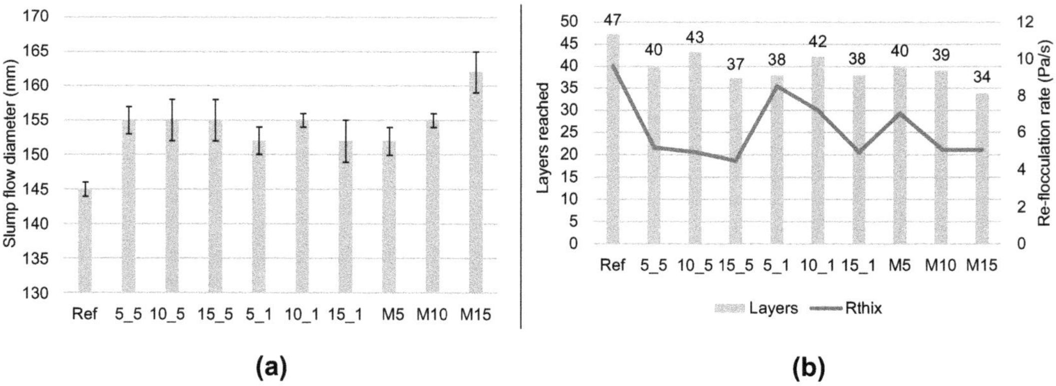

Figure 11.8 Effects of recycled plastic waste on concrete properties: (a) Slump values for different mixes (Oosthuizen et al. 2023), (b) Buildability (Oosthuizen et al. 2023).

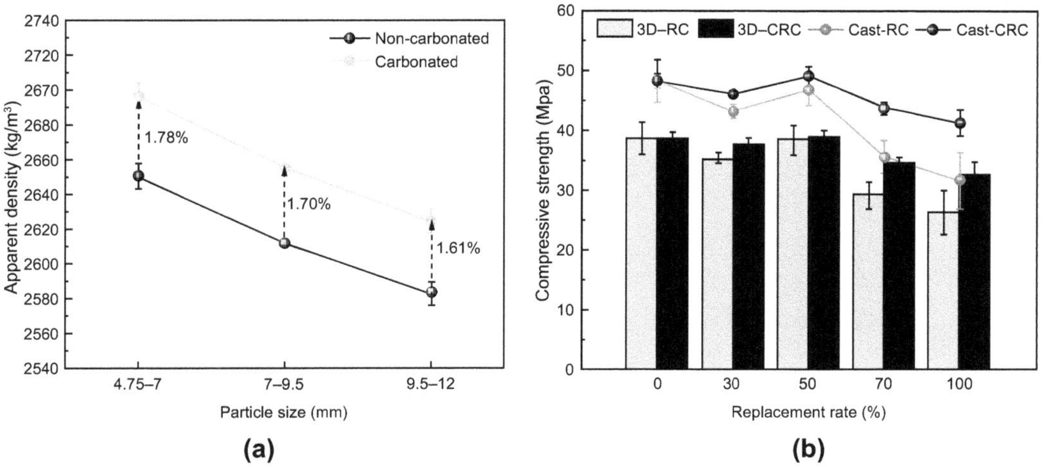

Figure 11.9 Properties of non-carbonated and carbonated RCA: (a) Apparent density (Tong et al. 2023), (b) Compressive strength comparison with cast and printed specimens (Tong et al. 2023).

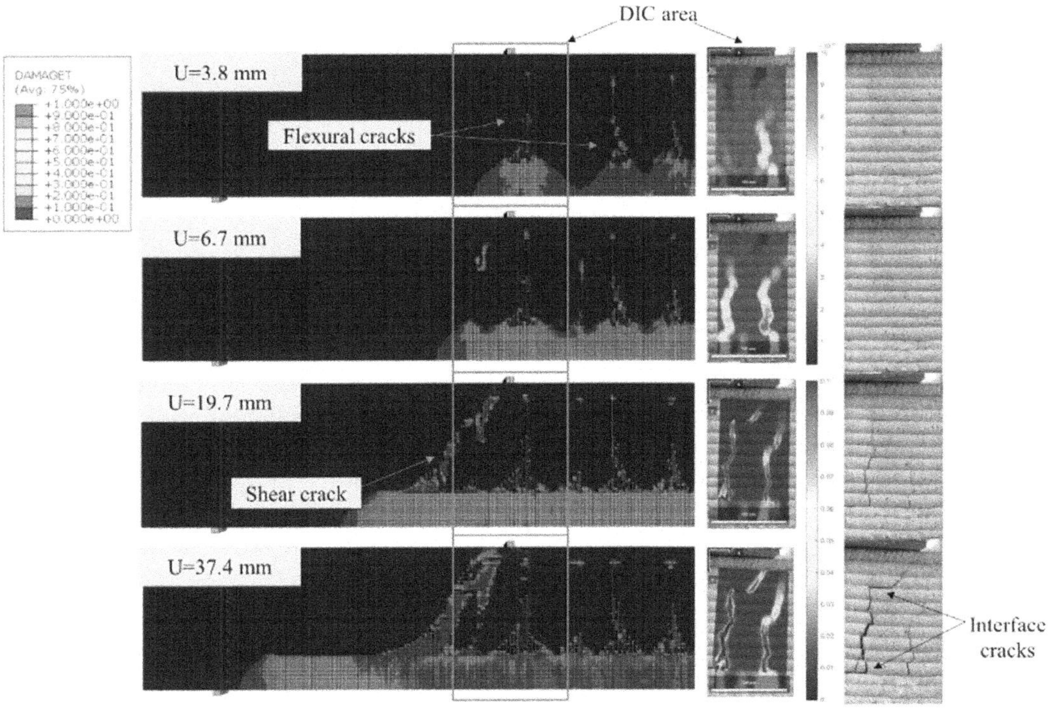

Figure 11.10 Comparison of test and FE results for the development of cracks.

Source: Liu et al. 2023.

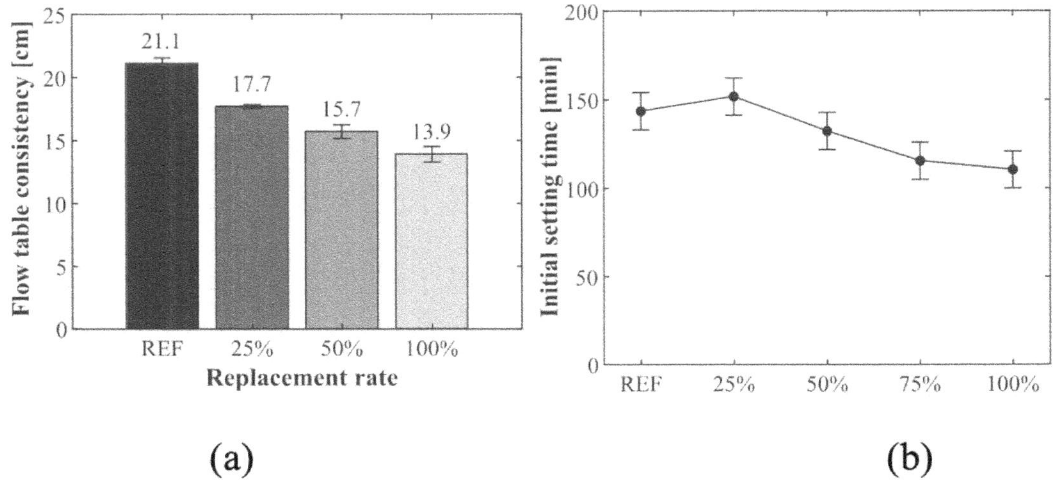

Figure 11.11 Effects of integrating fine recycled concrete aggregates for 3DCP, (a) Flow values for different replacement rates (De Vlieger et al. 2023), (b) Initial setting time (De Vlieger et al. 2023).

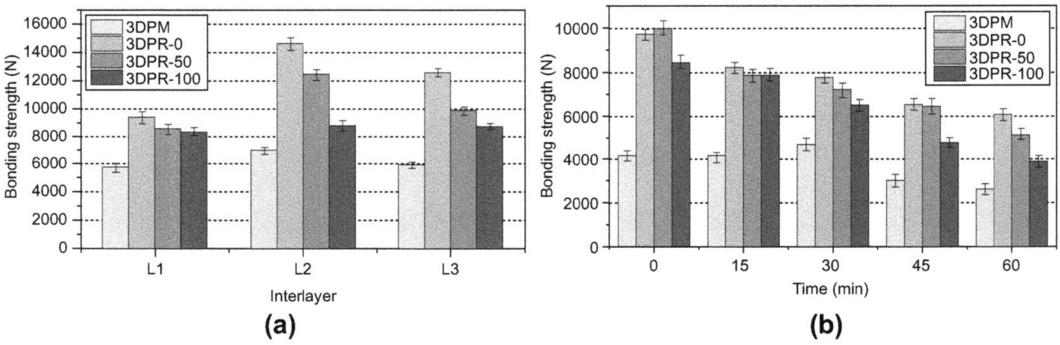

Figure 11.12 Interlayer bonding properties: (a) Variation between the interlayer height and the bond strength (Huawei Liu et al. 2022), (b) The variation in bond strength over the time of printing (Huawei Liu et al. 2022).

et al. 2022). Recycled sand reduces flowability and accelerates setting (**Figure 11.11**) but improves buildability by 33% and 83% at 25% and 50% replacement. Interlayer bond strength decreases with increasing RCA content and printing time interval, while layer height shows no clear influence, as shown in **Figure 11.12**, although RCA-based systems exhibit higher bond strength than mortar.

It is thus seen that the performance of recycled systems in 3DCP is governed by the balance between flowability, structural build-up, and interfacial bonding, rather than by material substitution alone.

11.3.2 Sustainable Incorporation of Supplementary Cementitious Materials

Mining tailings, gradation control, and alternative binders significantly influence rheology and mechanical performance in 3DCP systems. Partial replacement of natural sand with mining tailings improves buildability while maintaining acceptable strength, although printed specimens exhibit lower flexural performance compared to cast counterparts (Ma et al. 2017). Continuous gradation further enhances rheological stability, enabling improved shape retention during printing (Weng et al. 2018). Continuously graded mixtures achieved compressive strength of 49.7 MPa and flexural strength of 3.73 MPa after 28 days (**Figure 11.13a–c**), indicating that gradation models such as Fuller Thompson and Marson-Percy can guide rheological design in 3DCP.

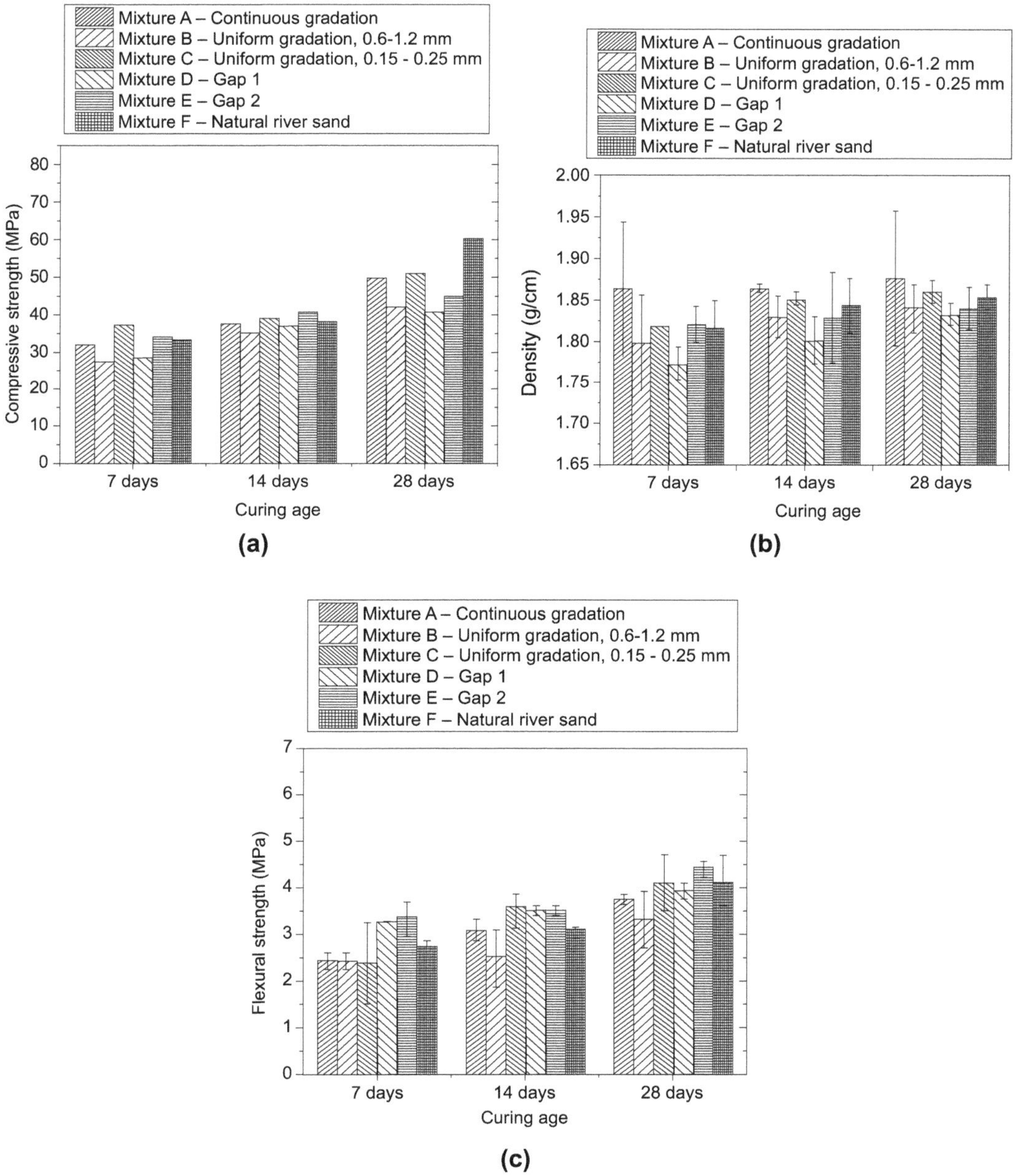

Figure 11.13 Impact of gradations on cementitious material rheology. (a) Density of each combination. (Weng et al. 2018), (b) Compressive strength of each combination (Weng et al. 2018), (c) Flexural strength of each combination (Weng et al. 2018).

Alternative binder systems, particularly magnesium potassium phosphate cement (MKPC) and calcium sulfoaluminate cement, demonstrate improved early-age performance and buildability (Weng et al. 2019; Van Der Putten et al. 2019). Incorporating 60 wt% fly ash and 10 wt% silica fume in MKPC increased working time to 22.9 minutes, enhanced yield stress and shape retention, and achieved early compressive strength of 5.36 MPa at 90 minutes, while silica fume reduced porosity by more than 70%. Calcium sulfoaluminate systems improve printability and compressive strength but may introduce voids due to their expansive behaviour, thereby affecting the microstructure. Further modifications using metakaolin and calcined clays improve thixotropy and structural build-up, although often reducing workability and open time (M. Chen et al. 2020; Yu Chen et al. 2021; Yu Chen et al. 2019; Yu Chen et al. 2020). High-grade calcined clay increased early-age strength by 38–40% and improved buildability but reduced workability time by 43% and flow spread by 10–15%.

Sustainability assessments indicate that binder selection significantly influences environmental impact (Bhattacherjee et al. 2021; Abu-Ennab et al. 2022; **Figure 11.14**). The use of SCMs, geopolymers, and calcium sulfoaluminate cement reduces carbon emissions and energy consumption, with ternary binder systems showing strong potential. Substituting masonry walls with 3D-printed kaolinite-based calcium sulfoaluminate cement concrete reduced energy consumption, global warming potential, and eutrophication potential by about 12%, 55%, and 4% per unit volume, and by 55%, 77%, and 53% per unit compressive strength. However, acidification potential increased by about 30% per unit volume. Geopolymer systems and fibre reinforcement strategies further enhance performance by improving flow behaviour, structural stability, and mechanical properties (Panda et al. 2019; Panda et al. 2017; Chen et al. 2023). One-part geopolymer systems address viscosity challenges, while short fibre reinforcement improves directional strength and structural robustness. Additives such as Na_2CO_3 and nano-silica extend printable duration and improve extrudability, buildability, and interlayer strength (**Figure 11.15**).

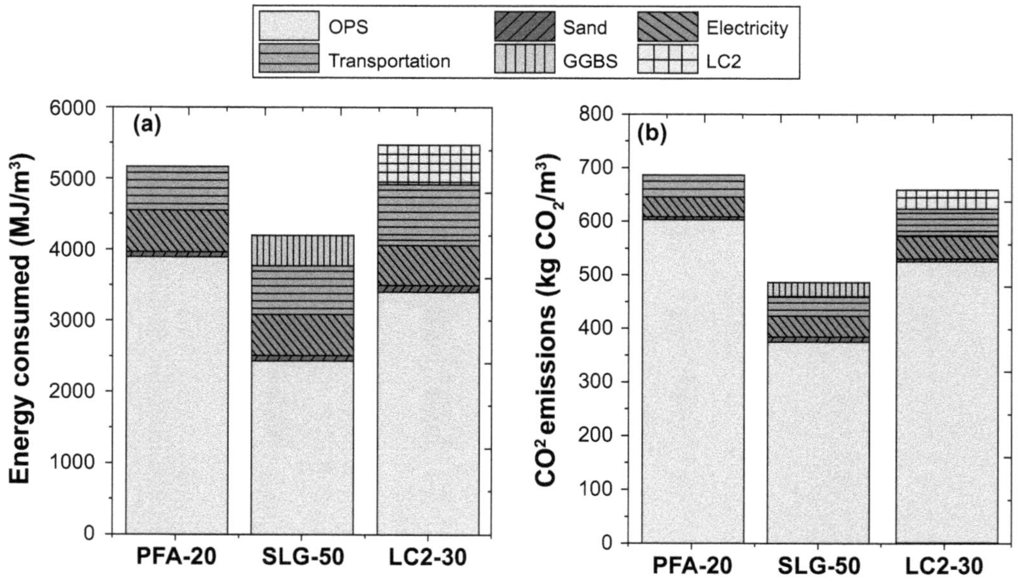

Figure 11.14 3D-printed concrete mixtures have environmental effects, all with an initial yield stress of 1.6 kPa (Bhattacherjee et al. 2021).

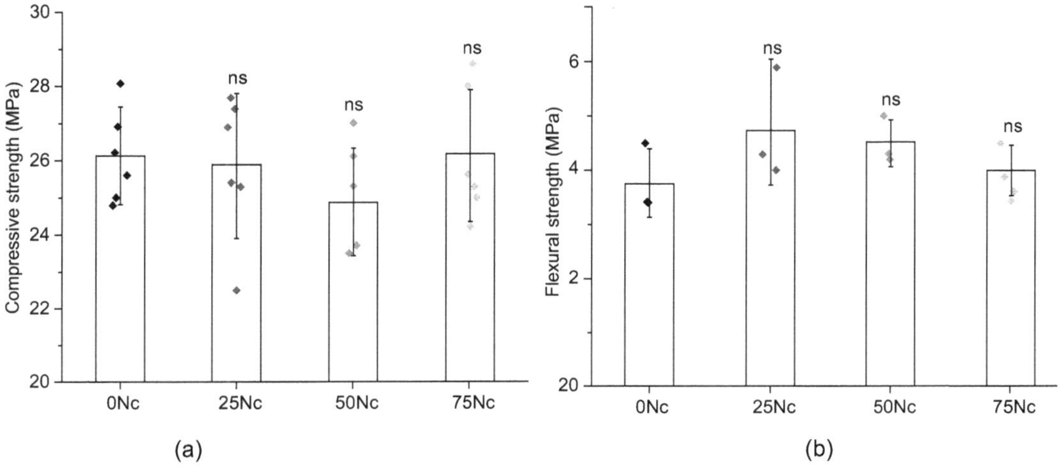

Figure 11.15 Mechanical properties of cast samples following a typical 3-day curing period with varying Nc replacement dosages; (a) Compressive strength. (Chen et al. 2023), (b) Flexural strength (Chen et al. 2023).

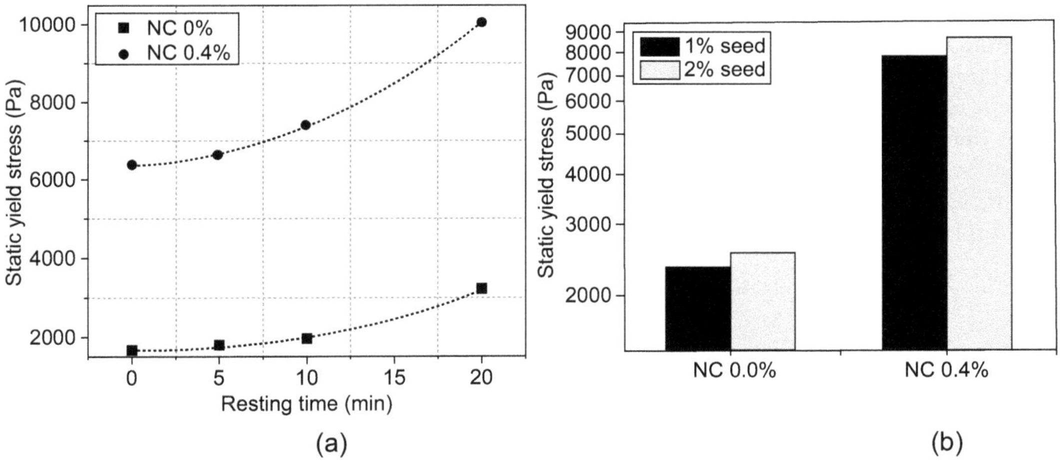

Figure 11.16 Influence of NC and nucleation seed on static yield stress; (a) Static yield stress evolution in AAS mortars with and without 0.4% NC (Panda et al. 2018), (b) The impact of nucleation seeds on the yield stress of AAS mixes with 0.4% NC determined immediately after mixing (Panda et al. 2018).

11.3.2.1 Slag-Based Geopolymers

Panda et al. (2018) explored the properties of alkali-activated slag mixes incorporating nanoclay and nucleation seeds to enhance thixotropy, resulting in improved extrudability and buildability. Microstructural characterisation revealed the formation of various hydration products, providing valuable insights into material behaviour (see **Figure 11.16**).

Bayat and Kashani (2023) investigated the use of silica fume, blast furnace slag, and hydrated lime in 3DCP mortars. They found that 6% silica fume with lime increased initial yield stress and cohesiveness, improving flocculation, thixotropy, and buildability, whereas replacing up to 20% of the cement with blast furnace slag had minimal effect. Lu et al. (2020)

examined a cement-free material for spray-based 3DCP using ground granulated blast furnace slag (GGBS) and evaluated magnesium oxide (MgO) and fly ash cenosphere (FAC). MgO significantly reduced setting time, with up to 78% reduction at 40% replacement, and FTIR analysis confirmed slag activation and improved water retention. FAC reduced pumping pressure and viscosity while increasing the static yield stress, thereby improving printing performance. The optimal mix (20%/40% FAC) reduced pumping pressure by 29%/31% and increased the critical ratio by 78%/68% compared to MgO-activated slag, successfully printing stable vertical filaments and profiles.

11.3.2.2 Fly-Ash-Based Geopolymers

Panda and Tan (2018a), Biranchi Panda et al. (2019), and Panda et al. (2018) collectively examined fly ash-based geopolymer mortars for 3DCP, focusing on rheology, buildability, and early-age performance. These studies consistently report that incorporating nanoclay and microfibres enhances yield stress, viscosity recovery, and structural build-up, improving extrudability and buildability. For instance, the addition of 0.1–0.5% nanoclay increased structural build-up, while clay and microfibre combinations enabled stable printing of complex geometries, including a life-sized freeform structure (Panda and Tan 2018a). Guo et al. (2020) further analysed fly ash-based geopolymer systems with slag powder, silica fume, and thixotropic additives, showing that rheological behaviour is better described by the Herschel-Bulkley model than the Bingham model. They observed that apparent viscosity, yield stress, and thixotropy varied non-linearly with composition, with optimal rheological performance at 10% slag powder and 10% silica fume.

In large-scale applications, Panda et al. (2017a) demonstrated that fly ash-based geopolymer cement can be used to fabricate structural components via robotic 3D printing. The results show that mechanical properties depend on loading direction (**Figure 11.18**), confirming anisotropic behaviour inherent to layer-wise deposition. Panda and Tan (2018b) further reported that increasing fly ash content reduces yield stress and viscosity compared to Portland cement-based mixes, while viscosity recovery remains a critical parameter governing printability.

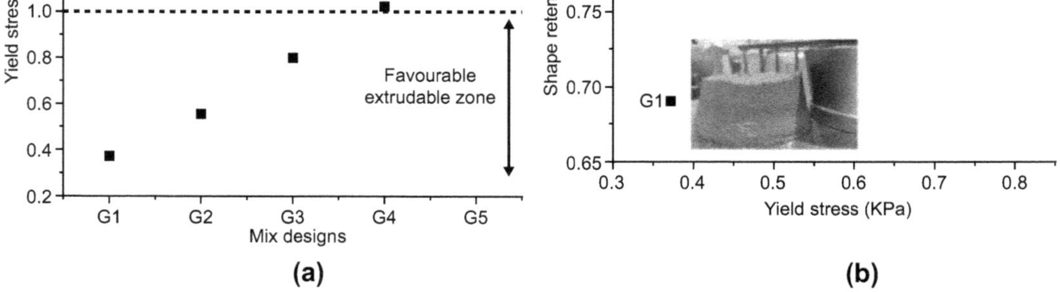

Figure 11.17 Properties of a geopolymer mortar derived from fly ash: (a) The geopolymer mixes' static yield stresses (Panda and Tan 2018a), (b) Relationship between the extrudable mix's yield stress and shape retention factor (SRF) (Panda and Tan 2018a).

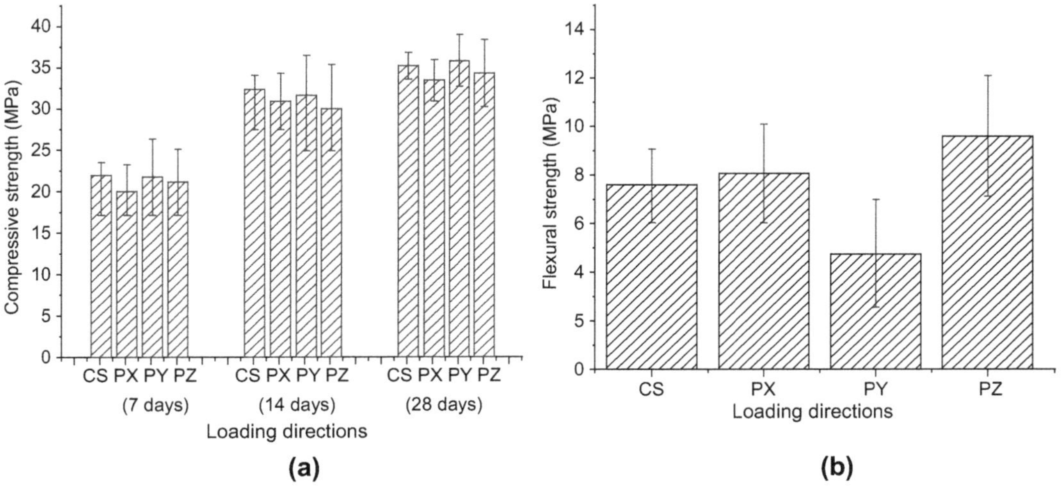

Figure 11.18 Influence of printing direction on mechanical properties: (a) Compression values (Panda et al. 2017a), (b) Flexural strength (Panda et al. 2017a).

11.3.3 Design and Structural Factors

Interlayer bonding, buildability, and structural performance in 3DCP are governed by material rheology and printing parameters. Studies consistently highlight the influence of thixotropy, yield stress, printing velocity, and interface conditions on mechanical response and failure behaviour (Panda et al. 2017b; Prem et al. 2024; Paul et al. 2017). These factors determine load-bearing capacity, failure modes, and construction efficiency. For example, Paul et al. (2017) reported acceptable workability with slump values of 1–3 mm and open times of 30–60 minutes, with flexural strength varying by up to 12% depending on printing direction. Process parameters such as nozzle geometry, printing speed, and material flow rate are critical for ensuring printability and structural integrity (L. Yang et al. 2022; Khan et al. 2024). These studies emphasise the need for optimised nozzle dimensions, controlled deposition rates, and alignment between volumetric flow and nozzle movement to avoid defects and instability. For instance, L. Yang et al. (2022) recommend a nozzle movement speed of two to three times the volumetric flow rate and a nozzle-to-aggregate size ratio of at least four to prevent blockage. Material modification and reinforcement strategies have been widely explored to enhance strength and anisotropic behaviour in 3DCP (Panda et al. 2017c; Panda et al. 2018; Lim et al. 2018; Liu and Chun 2022). These approaches improve structural robustness and directional performance through fibre addition and hybrid reinforcement systems. For example, Lim et al. (2018) demonstrated a 290% increase in flexural strength by embedding a continuous steel cable within the printed matrix. Structural design approaches for 3DCP have also evolved to address load-bearing requirements and failure mechanisms (Panda et al. 2017b). These studies integrate reinforcement strategies, interface considerations, and predictive modelling to improve bond strength and structural reliability. Sustainable material design and optimisation frameworks have been proposed to enhance performance and reduce environmental impact (Li et al. 2023; Alyami et al. 2023). These approaches combine fibre reinforcement, SCM utilisation, and data-driven optimisation techniques to improve mix design and predict mechanical properties. For instance, Li et al. (2023) developed a stepwise optimisation approach incorporating steel and PVA fibres to achieve printable and structurally efficient mixtures.

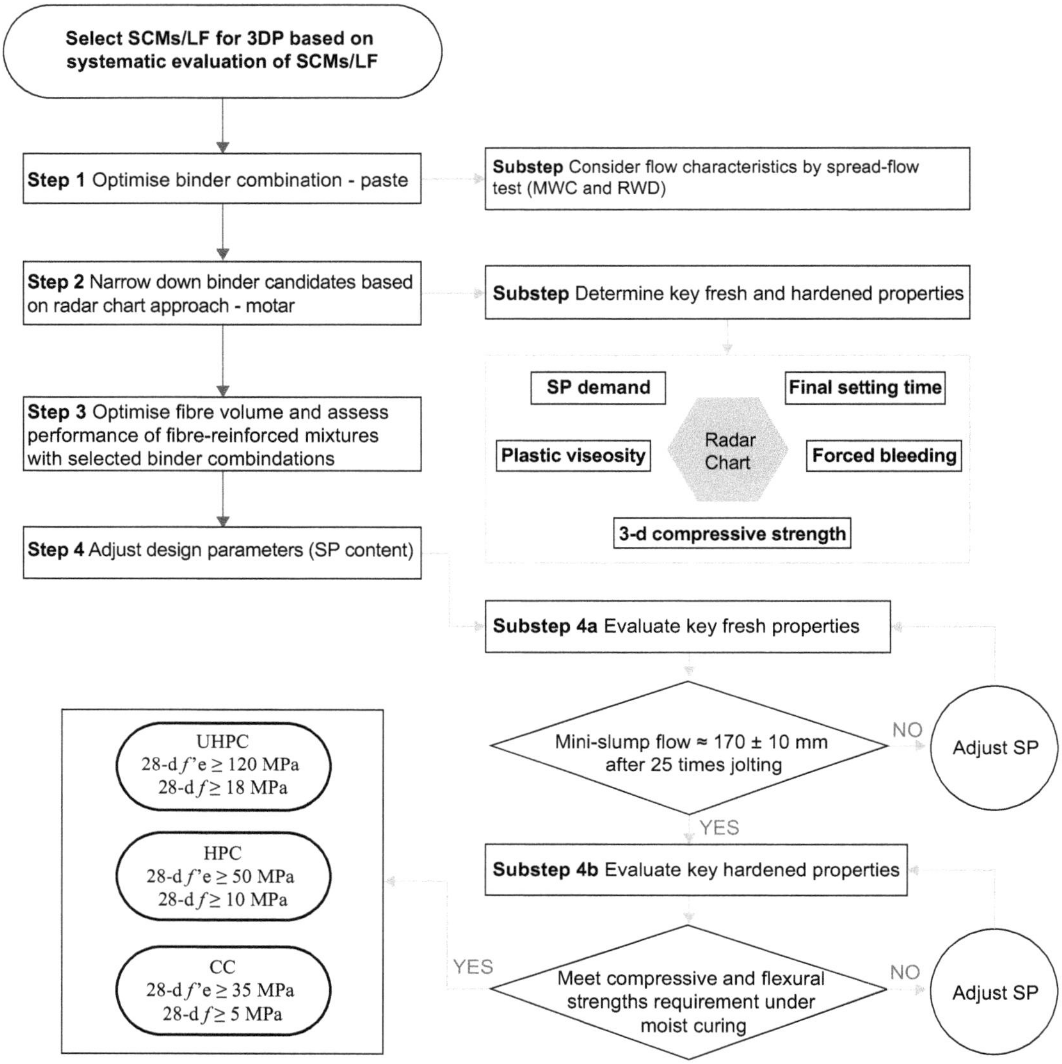

Figure 11.19 Optimisation design flowchart for fibre-reinforced mixtures suitable for printing (Li et al. 2023).

Process modelling and performance evaluation methods further support the efficient implementation of 3DCP (Foteinopoulos et al. 2019; Lu et al. 2021; Inaty et al. 2022). These studies focus on system-level optimisation, analytical modelling, and durability assessment under aggressive environments. For example, Lu et al. (2021) developed an analytical model for spray-based 3DCP, achieving average relative errors of 15.29% in width and 9.92% in thickness prediction, while Inaty et al. (2022) demonstrated improved early-stage durability of printed concrete under sulphuric acid exposure due to a denser matrix.

11.3.4 Applications of 3DCP

West et al. (2021) propose using 3DCP to build machine tool bases as an alternative to existing fabrication methods. According to the study, concrete bases made with 3DCP showed excellent stiffness and damping, with a damping ratio up to 4 times higher than that of cast-iron construction. This made them a competitive alternative to conventional cast-iron machine tool bases. Pajonk et al. (2021) investigated the feasibility of combining multiple materials or material properties within a unified additive manufacturing method to create multi-material products. They reviewed the numerous benefits and obstacles associated with this strategy. Mata-Falcón et al. (2022) investigated the environmental impact of concrete floor slabs, highlighting that traditional solid slabs contribute significantly to greenhouse gas emissions due to their high concrete content. They argue that ribbed slabs, which were abandoned in favour of faster and cheaper construction methods, offer a more sustainable alternative by requiring less concrete. The study then explores using digital fabrication techniques, such as 3D printing and polymer extrusion, to create complex formwork for these ribbed slabs, demonstrating a possible 50% reduction in concrete use compared to solid slabs. However, they acknowledge the current limitations of these digital methods, including inefficiency and labour-intensive nature, and call for further development towards sustainable, competitive digital fabrication processes for eco-friendly ribbed slab construction. Duarte et al. (2023) suggested that 3DCP could address the global shortage of affordable housing, especially in remote areas with labour and material constraints. Their research focused on printing entire structures, including enclosing surfaces, using 3DCP technology. They designed a modular housing system inspired by traditional vaults, optimised for structural, thermal, and printing considerations, and validated by printing a scaled-down version that confirmed the design's feasibility.

11.4 CHALLENGES AND FUTURE OUTLOOK

The future of 3DCP demands further investigations in key areas. There have been significant efforts to utilise technology to improve the sustainability of the construction industry, such as incorporating recycled materials, waste aggregates, and even waste-based binders (Alhumayani et al. 2020). However, optimising mixtures and addressing potential trade-offs in durability and performance remain crucial questions (Panda et al. 2018; Al-Tamimi et al. 2023). Machine Learning could simulate and achieve optimal mix designs (Ali et al. 2023). SCMs such as fly ash and slag offer exciting opportunities to reduce reliance on Portland cement, but their printability, buildability, and mechanical properties require closer examination (Chang et al. 2021; Khan and Koç 2023; Joh et al. 2020). Alternative binders like geopolymers hold promise for a lower carbon footprint, but long-term field validation, focusing on rheology, extrudability, and strength development, is essential (Souza et al. 2021; Sanjayan et al. 2020; Souza et al. 2020; Basha et al. 2023). Mechanical properties and anisotropy pose challenges, but innovations in fibre-reinforced engineered cementitious composites (ECC) and embedded reinforcement show promise. Further research is needed to adapt these approaches specifically for 3D printing. Beyond materials, advanced modelling and simulation are key to unlocking the potential of complex geometries in 3DCP, particularly for structural and seismic applications (Vantyghem et al. 2020; Ko 2021; Bos et al. 2019). However, translating these advancements into practical tools and ensuring code compliance remain hurdles. Finally, quality control and standardisation are crucial for industry confidence and widespread adoption. Robust solutions for consistency, interlayer bonding, surface finish, tolerances, and durability metrics are essential (Atta Ur Rehman et al. 2023; Lao et al. 2020; S. P. Liu et al. 2022; Y. Yang et al. 2023). Additionally, multi-material printing has significant potential, as it can integrate functionalities such as electrical wiring and plumbing directly into structural builds. However, material compatibility and process complexity must be addressed before broad use. 3DCP has the potential to revolutionise the construction sector and move it towards a more efficient, flexible, and sustainable future by addressing these issues.

11.5 CONCLUSION

This review highlights the potential of 3DCP to transform construction towards an environmentally friendly, automated, and versatile future. The review suggests five key areas that demand further investigation:

- Sustainable Materials: Utilising recycled materials offers attractive environmental benefits, but optimising mixtures to ensure optimal durability and mechanical performance remains a critical challenge.
- Alternative Binders: Supplementary cementitious materials and novel binders like geopolymers present exciting avenues for reducing reliance on traditional cement. However, further research is crucial to investigate the printability, buildability, and long-term field validation data for these alternatives.
- Material and Reinforcement Innovations: Promising approaches like fibres and embedded reinforcement hold the potential to address limitations such as anisotropy and tensile strength. Nevertheless, adaptation and optimisation specifically for 3DCP printing are essential.
- Advanced Modelling and Design: Rapid advancements in modelling and simulation methods pave the way for complex structure design. However, bridging the gap to develop practical structural tools aligned with building codes and standards remains crucial.
- Quality Control and Standardisation: Establishing robust quality control procedures focused on consistency, interlayer bonding, surface finish, and dimensional tolerances is vital for gaining industry confidence in 3DCP technology.

By systematically tackling these challenges, 3DCP can progressively evolve into an efficient, sustainable, resilient, and cost-effective mainstream construction approach. The promising results obtained thus far strongly warrant continued cross-domain investments in AM technologies tailored for the construction sector.

DECLARATION OF COMPETING INTEREST

The authors declare that none of the work described in this study appears to have been influenced by any known conflicts of interest or personal relationships.

ACKNOWLEDGEMENT

We acknowledge the researchers, writers, and organisations whose input has formed this review. Their efforts have been crucial in progressing the field, and we value their commitment to innovation and sustainability.

REFERENCES

Abdulfattah, Osayd, Ihab H. Alsurakji, Amjad El-Qanni, Mohammad Samaaneh, Mohammad Najjar, Ramez Abdallah, and Iyad Assaf. 2022. "Experimental Evaluation of Using Pyrolysed Carbon Black Derived from Waste Tires as Additive Towards Sustainable Concrete." *Case Studies in Construction Materials* 16 (February): e00938. https://doi.org/10.1016/j.cscm.2022.e00938.

Abu-Ennab, L., M.K. Dixit, B. Birgisson, and P. Pradeep Kumar. 2022. "Comparative Life Cycle Assessment of Large-scale 3D Printing Utilising Kaolinite-based Calcium Sulfoaluminate Cement Concrete and Conventional Construction." *Cleaner Environmental Systems* 5 (April): 100078. https://doi.org/10.1016/j.cesys.2022.100078.

Alhumayani, Hashem, Mohamed Gomaa, Veronica Soebarto, and Wassim Jabi. 2020. "Environmental Assessment of Large-scale 3D Printing in Construction: A Comparative Study Between Cob and Concrete." *Journal of Cleaner Production* 270 (June): 122463. https://doi.org/10.1016/j.jclepro.2020.122463.

Ali, Ammar, Raja Dilawar Riaz, Umair Jalil Malik, Syed Baqar Abbas, Muhammad Usman, Mati Ullah Shah, In-Ho Kim, Asad Hanif, and Muhammad Faizan. 2023. "Machine Learning-Based Predictive Model for Tensile and Flexural Strength of 3D-Printed Concrete." *Materials* 16 (11): 4149. https://doi.org/10.3390/ma16114149.

Al-Tamimi, Adil K., Habib H. Alqamish, Ahlam Khaldoune, Haidar Alhaidary, and Kamyar Shirvani-moghaddam. 2023. "Framework of 3D Concrete Printing Potential and Challenges." *Buildings* 13 (3): 827. https://doi.org/10.3390/buildings13030827.

Alyami, Mana, Majid Khan, Muhammad Faisal Javed, Mujahid Ali, Hisham Alabduljabbar, Taoufik Najeh, and Yaser Gamil. 2023. "Application of Metaheuristic Optimisation Algorithms in Predicting the Compressive Strength of 3D-printed Fibre-reinforced Concrete." *Developments in the Built Environment* 17 (December): 100307. https://doi.org/10.1016/j.dibe.2023.100307.

Bai, Gang, Li Wang, Guowei Ma, Jay Sanjayan, and Mingke Bai. 2021. "3D Printing Eco-friendly Concrete Containing Under-utilised and Waste Solids as Aggregates." *Cement and Concrete Composites* 120 (March): 104037. https://doi.org/10.1016/j.cemconcomp.2021.104037.

Basha, Shaik Inayath, Atta Ur Rehman, Md. Abdul Aziz, and Jung-Hoon Kim. 2023. "Cement Composites with Carbon-Based Nanomaterials for 3D Concrete Printing Applications—a Review." *The Chemical Record* 23 (4). https://doi.org/10.1002/tcr.202200293.

Bayat, Hamid, and Alireza Kashani. 2023. "Analysis of Rheological Properties and Printability of a 3D-printing Mortar Containing Silica Fume, Hydrated Lime, and Blast Furnace Slag." *Materials Today Communications* 37 (September): 107128. https://doi.org/10.1016/j.mtcomm.2023.107128.

Bhattacherjee, Shantanu, Anusha S. Basavaraj, A.V. Rahul, Manu Santhanam, Ravindra Gettu, Biranchi Panda, Erik Schlangen, et al. 2021. "Sustainable Materials for 3D Concrete Printing." *Cement and Concrete Composites* 122 (June): 104156. https://doi.org/10.1016/j.cemconcomp.2021.104156.

Bos, Freek, Zeeshan Ahmed, Claudia Romero Rodriguez, and Stefan Chaves Figueiredo. 2019. "3D Concrete Printing for Structural Applications." *DOAJ (DOAJ: Directory of Open Access Journals)*, December. https://doaj.org/article/7c80e168258749aaa4a39faf84425c16.

Chang, Ze, Yading Xu, Yu Chen, Yidong Gan, Erik Schlangen, and Branko Šavija. 2021. "A Discrete Lattice Model for Assessment of Buildability Performance of 3D-printed Concrete." *Computer-Aided Civil and Infrastructure Engineering* 36 (5): 638–655. https://doi.org/10.1111/mice.12700.

Chen, Mingxu, Lei Yang, Yan Zheng, Yongbo Huang, Laibo Li, Piqi Zhao, Shoude Wang, Lingchao Lu, and Xin Cheng. 2020. "Yield Stress and Thixotropy Control of 3D-printed Calcium Sulfoaluminate Cement Composites With Metakaolin Related to Structural Build-up." *Construction and Building Materials* 252 (April): 119090. https://doi.org/10.1016/j.conbuildmat.2020.119090.

Chen, Yu, Shan He, Yu Zhang, Zhi Wan, Oğuzhan Çopuroğlu, and Erik Schlangen. 2021. "3D Printing of Calcined Clay-limestone-based Cementitious Materials." *Cement and Concrete Research* 149 (August): 106553. https://doi.org/10.1016/j.cemconres.2021.106553.

Chen, Yu, Zhenming Li, Stefan Chaves Figueiredo, Oğuzhan Çopuroğlu, Fred Veer, and Erik Schlangen. 2019. "Limestone and Calcined Clay-Based Sustainable Cementitious Materials for 3D Concrete Printing: A Fundamental Study of Extrudability and Early-Age Strength Development." *Applied Sciences* 9 (9): 1809. https://doi.org/10.3390/app9091809.

Chen, Yu, Claudia Romero Rodriguez, Zhenming Li, Boyu Chen, Oğuzhan Çopuroğlu, and Erik Schlangen. 2020. "Effect of Different Grade Levels of Calcined Clays on Fresh and Hardened Properties of Ternary-blended Cementitious Materials for 3D Printing." *Cement and Concrete Composites* 114 (July): 103708. https://doi.org/10.1016/j.cemconcomp.2020.103708.

Chen, Yuning, Kailun Xia, Zijian Jia, Yueyi Gao, Zuhua Zhang, and Yamei Zhang. 2023. "Extending Applicability of 3D-printable Geopolymer to Large-scale Printing Scenario via Combination of

Sodium Carbonate and Nano-silica." *Cement and Concrete Composites* 145 (October): 105322. https://doi.org/10.1016/j.cemconcomp.2023.105322.

De Schutter, Geert, Karel Lesage, Viktor Mechtcherine, Venkatesh Naidu Nerella, Guillaume Habert, and Isolda Agusti-Juan. 2018. "Vision of 3D Printing With Concrete—Technical, Economic and Environmental Potentials." *Cement and Concrete Research* 112 (August): 25–36. https://doi.org/10.1016/j.cemconres.2018.06.001.

De Vlieger, Jentel, Luc Boehme, Jan Blaakmeer, and Jiabin Li. 2023. "Buildability Assessment of Mortar With Fine Recycled Aggregates for 3D Printing." *Construction and Building Materials* 367 (January): 130313. https://doi.org/10.1016/j.conbuildmat.2023.130313.

Duarte, Jose Pinto, Gonçalo Duarte, Nathan Brown, Shadi Nazarian, Ali Memari, Sven Bilén, and Aleksandra Radlinska. 2023. "Designing a Habitat for 3D Concrete Printing in Permafrost Regions." *Joelho Revista De Cultura Arquitectonica*, no. 14 (July): 49–70. https://doi.org/10.14195/1647-8681_14_3.

Elyamany, Hafez E., Abd Elmoaty M. Abd Elmoaty, and Basma Mohamed. 2014. "Effect of Filler Types on Physical, Mechanical and Microstructure of Self Compacting Concrete and Flow-able Concrete." *Alexandria Engineering Journal* 53 (2): 295–307. https://doi.org/10.1016/j.aej.2014.03.010.

Foteinopoulos, Panagis, Alexios Papacharalampopoulos, and Panagiotis Stavropoulos. 2019. "Block-based Analytical Hierarchy Process Applied for the Evaluation of Construction Sector Additive Manufacturing." *Procedia CIRP* 81 (January): 950–955. https://doi.org/10.1016/j.procir.2019.03.233.

Gambatese, John A., and Matthew Hallowell. 2011. "Factors That Influence the Development and Diffusion of Technical Innovations in the Construction Industry." *Construction Management and Economics* 29 (5): 507–517. https://doi.org/10.1080/01446193.2011.570355.

Guo, Xiaolu, Junyi Yang, and Guiyan Xiong. 2020. "Influence of Supplementary Cementitious Materials on Rheological Properties of 3D Printed Fly Ash Based Geopolymer." *Cement and Concrete Composites* 114 (September): 103820. https://doi.org/10.1016/j.cemconcomp.2020.103820.

Hager, Izabela, Anna Golonka, and Roman Putanowicz. 2016. "3D Printing of Buildings and Building Components as the Future of Sustainable Construction?" *Procedia Engineering* 151 (January): 292–299. https://doi.org/10.1016/j.proeng.2016.07.357.

Hosseini, Ehsan, Mohammad Zakertabrizi, Asghar Habibnejad Korayem, and Guanzhong Xu. 2019. "A Novel Method to Enhance the Interlayer Bonding of 3D Printing Concrete: An Experimental and Computational Investigation." *Cement and Concrete Composites* 99 (March): 112–119. https://doi.org/10.1016/j.cemconcomp.2019.03.008.

Inaty, François El, Bilal Baz, and Georges Aouad. 2022. "Long-term Durability Assessment of 3D Printed Concrete." *Journal of Adhesion Science and Technology* 37 (12): 1921–1936. https://doi.org/10.1080/01694243.2022.2102717.

Joh, Changbin, Jungwoo Lee, Quang Bui, Jihun Park, and In-Hwan Yang. 2020. "Buildability and Mechanical Properties of 3D Printed Concrete." *Materials* 13 (21): 4919. https://doi.org/10.3390/ma13214919.

Khan, Shoukat Alim, Hüseyin Ilcan, Ramsha Imran, Ehsan Aminipour, Oğuzhan Şahin, Ans Al Rashid, Mustafa Şahmaran, and Muammer Koç. 2024. "The Impact of Nozzle Diameter and Printing Speed on Geopolymer-based 3D-Printed Concrete Structures: Numerical Modeling and Experimental Validation." *Results in Engineering* 21 (January): 101864. https://doi.org/10.1016/j.rineng.2024.101864.

Khan, Shoukat Alim, and Muammer Koç. 2023. "Buildability Analysis of 3D Concrete Printing Process: A Parametric Study Using Design of Experiment Approach." *Processes* 11 (3): 782. https://doi.org/10.3390/pr11030782.

Ko, Chien-Ho. 2021. "Constraints and Limitations of Concrete 3D Printing in Architecture." *Journal of Engineering Design and Technology* 20 (5): 1334–1348. https://doi.org/10.1108/jedt-11-2020-0456.

Lao, Wenxin, Mingyang Li, Teck Neng Wong, Ming Jen Tan, and Tegoeh Tjahjowidodo. 2020. "Improving Surface Finish Quality in Extrusion-based 3D Concrete Printing Using Machine Learning-based Extrudate Geometry Control." *Virtual and Physical Prototyping* 15 (2): 178–193. https://doi.org/10.1080/17452759.2020.1713580.

Li, Haodao, Alfred Addai-Nimoh, Eric Kreiger, and Kamal H. Khayat. 2023. "Methodology to Design Eco-friendly Fiber-reinforced Concrete for 3D Printing." *Cement and Concrete Composites* 147 (December): 105415. https://doi.org/10.1016/j.cemconcomp.2023.105415.

Lim, Jian Hui, Biranchi Panda, and Quang-Cuong Pham. 2018. "Improving Flexural Characteristics of 3D Printed Geopolymer Composites with In-process Steel Cable Reinforcement." *Construction and Building Materials* 178 (May): 32–41. https://doi.org/10.1016/j.conbuildmat.2018.05.010.

Liu, Haoran, Jianzhuang Xiao, and Tao Ding. 2023. "Flexural Performance of 3D-printed Composite Beams With ECC and Recycled Fine Aggregate Concrete: Experimental and Numerical Analysis." *Engineering Structures* 283 (March): 115865. https://doi.org/10.1016/j.engstruct.2023.115865.

Liu, Huawei, Chao Liu, Yiwen Wu, Guoliang Bai, Chunhui He, Yizhou Yao, Rongfei Zhang, and Youqiang Wang. 2022. "3D Printing Concrete With Recycled Coarse Aggregates: The Influence of Pore Structure on Interlayer Adhesion." *Cement and Concrete Composites* 134 (September): 104742. https://doi.org/10.1016/j.cemconcomp.2022.104742.

Liu, Jie, and Chun Lv. 2022. "Properties of 3D-Printed Polymer Fiber-Reinforced Mortars: A Review." *Polymers* 14 (7): 1315. https://doi.org/10.3390/polym14071315.

Lu, Bing, Mingyang Li, Teck Neng Wong, and Shunzhi Qian. 2021. "Effect of Printing Parameters on Material Distribution in Spray-based 3D Concrete Printing (S-3DCP)." *Automation in Construction* 124 (February): 103570. https://doi.org/10.1016/j.autcon.2021.103570.

Lu, Bing, Weiping Zhu, Yiwei Weng, Zhixin Liu, En-Hua Yang, Kah Fai Leong, Ming Jen Tan, Teck Neng Wong, and Shunzhi Qian. 2020. "Study of MgO-activated Slag as a Cementless Material for Sustainable Spray-based 3D Printing." *Journal of Cleaner Production* 258 (February): 120671. https://doi.org/10.1016/j.jclepro.2020.120671.

Ma, Guowei, Zhijian Li, and Li Wang. 2017. "Printable Properties of Cementitious Material Containing Copper Tailings for Extrusion Based 3D Printing." *Construction and Building Materials* 162 (December): 613–627. https://doi.org/10.1016/j.conbuildmat.2017.12.051.

Maskuriy, Raihan, Ali Selamat, Petra Maresova, Ondrej Krejcar, and Oladipo Olalekan David. 2019. "Industry 4.0 for the Construction Industry: Review of Management Perspective." *Economies* 7 (3): 68. https://doi.org/10.3390/economies7030068.

Mata-Falcón, Jaime, Patrick Bischof, Tobias Huber, Ana Anton, Joris Burger, Francesco Ranaudo, Andrei Jipa, et al. 2022. "Digitally Fabricated Ribbed Concrete Floor Slabs: A Sustainable Solution for Construction." *RILEM Technical Letters* 7 (September): 68–78. https://doi.org/10.21809/rilemtechlett.2022.161.

Monteiro, André O., Paulo B. Cachim, and Pedro M.F.J. Costa. 2017. "Self-sensing Piezoresistive Cement Composite Loaded With Carbon Black Particles." *Cement and Concrete Composites* 81 (April): 59–65. https://doi.org/10.1016/j.cemconcomp.2017.04.009.

Oosthuizen, Jandré Daniel, Adewumi John Babafemi, and Richard Shaun Walls. 2023. "3D-printed Recycled Plastic Eco-aggregate (Resin8) Concrete." *Construction and Building Materials* 408 (October): 133712. https://doi.org/10.1016/j.conbuildmat.2023.133712.

Pajonk, Adam, Alejandro Prieto, Ulrich Blum, and Ulrich Knaack. 2021. "Multi-material Additive Manufacturing in Architecture and Construction: A Review." *Journal of Building Engineering* 45 (November): 103603. https://doi.org/10.1016/j.jobe.2021.103603.

Panda, Biranchi, Suvash Chandra Paul, Lim Jian Hui, Yi Wei Daniel Tay, and Ming Jen Tan. 2017a. "Additive Manufacturing of Geopolymer for Sustainable Built Environment." *Journal of Cleaner Production* 167 (August): 281–288. https://doi.org/10.1016/j.jclepro.2017.08.165.

Panda, Biranchi, Suvash Chandra Paul, Nisar Ahamed Noor Mohamed, Yi Wei Daniel Tay, and Ming Jen Tan. 2017b. "Measurement of Tensile Bond Strength of 3D Printed Geopolymer Mortar." *Measurement* 113 (September): 108–116. https://doi.org/10.1016/j.measurement.2017.08.051.

Panda, Biranchi, Suvash Chandra Paul, and Ming Jen Tan. 2017c. "Anisotropic Mechanical Performance of 3D Printed Fiber Reinforced Sustainable Construction Material." *Materials Letters* 209 (July): 146–149. https://doi.org/10.1016/j.matlet.2017.07.123.

Panda, Biranchi, Shaoqin Ruan, Cise Unluer, and Ming Jen Tan. 2018. "Improving the 3D Printability of High Volume Fly Ash Mixtures via the Use of Nano Attapulgite Clay." *Composites Part B Engineering* 165 (November): 75–83. https://doi.org/10.1016/j.compositesb.2018.11.109.

Panda, Biranchi, Gvp Bhagath Singh, Cise Unluer, and Ming Jen Tan. 2019. "Synthesis and Characterization of One-part Geopolymers for Extrusion Based 3D Concrete Printing." *Journal of Cleaner Production* 220 (February): 610–619. https://doi.org/10.1016/j.jclepro.2019.02.185.

Panda, Biranchi, and Ming Jen Tan. 2018a. "Experimental Study on Mix Proportion and Fresh Properties of Fly Ash Based Geopolymer for 3D Concrete Printing." *Ceramics International* 44 (9): 10258–1065. https://doi.org/10.1016/j.ceramint.2018.03.031.

Panda, Biranchi, and M. J. Tan. 2018b. Rheological behavior of high volume fly ash mixtures containing micro silica for digital construction application. *Materials Letters*, *237*, 348–351. https://doi.org/10.1016/j.matlet.2018.11.131

Panda, Biranchi, Y.W.D. Tay, S.C. Paul, and M.J. Tan. 2018. "Current Challenges and Future Potential of 3D Concrete Printing." *Materialwissenschaft Und Werkstofftechnik* 49 (5): 666–673. https://doi.org/10.1002/mawe.201700279.

Panda, Biranchi, Cise Unluer, and Ming Jen Tan. 2019. "Extrusion and Rheology Characterization of Geopolymer Nanocomposites Used in 3D Printing." *Composites Part B Engineering* 176 (August): 107290. https://doi.org/10.1016/j.compositesb.2019.107290.

Paul, Suvash Chandra, Yi Wei Daniel Tay, Biranchi Panda, and Ming Jen Tan. 2017. "Fresh and Hardened Properties of 3D Printable Cementitious Materials for Building and Construction." *Archives of Civil and Mechanical Engineering* 18 (1): 311–319. https://doi.org/10.1016/j.acme.2017.02.008.

Prem, Prabhat Ranjan, P. S. Ambily, Shankar Kumar, and Swapnil Balasaheb Ghodke. 2024. "A Theoretical Model to Predict the Structural Buildability of 3D Printable Concrete." *Mechanics of Time-Dependent Materials*, January. https://doi.org/10.1007/s11043-024-09666-8.

Rehman, Atta Ur, Arnaud Perrot, Bizu Melesse Birru, and Jung-Hoon Kim. 2023. "Recommendations for Quality Control in Industrial 3D Concrete Printing Construction With Mono-component Concrete: A Critical Evaluation of Ten Test Methods and the Introduction of the Performance Index." *Developments in the Built Environment* 16 (September): 100232. https://doi.org/10.1016/j.dibe.2023.100232.

Rezania, M., M. Panahandeh, N. Razavi, and F. Berto. 2019. "Experimental Study of the Simultaneous Effect of Nano-silica and Nano-carbon Black on Permeability and Mechanical Properties of the Concrete." *Theoretical and Applied Fracture Mechanics* 104 (October): 102391. https://doi.org/10.1016/j.tafmec.2019.102391.

Sanjayan, J. G., R. Jayathilakage, and P. Rajeev. 2020. "Vibration Induced Active Rheology Control for 3D Concrete Printing." *Cement and Concrete Research* 140 (November): 106293. https://doi.org/10.1016/j.cemconres.2020.106293.

Siddika, Ayesha, Md. Abdullah Al Mamun, Wahid Ferdous, Ashish Kumer Saha, and Rayed Alyousef. 2019. "3D-printed Concrete: Applications, Performance, and Challenges." *Journal of Sustainable Cement-Based Materials* 9 (3): 127–164. https://doi.org/10.1080/21650373.2019.1705199.

Souza, Marcelo Tramontin, Igor Maia Ferreira, Elisângela Guzi De Moraes, Luciano Senff, Sabrina Arcaro, José Renato Castro Pessôa, Manuel J. Ribeiro, and Antonio Pedro Novaes De Oliveira. 2021. "Role of Chemical Admixtures on 3D Printed Portland Cement: Assessing Rheology and Buildability." *Construction and Building Materials* 314 (November): 125666. https://doi.org/10.1016/j.conbuildmat.2021.125666.

Souza, Marcelo Tramontin, Igor Maia Ferreira, Elisângela Guzi De Moraes, Luciano Senff, and Antonio Pedro Novaes De Oliveira. 2020. "3D Printed Concrete for Large-scale Buildings: An Overview of Rheology, Printing Parameters, Chemical Admixtures, Reinforcements, and Economic and Environmental Prospects." *Journal of Building Engineering* 32 (September): 101833. https://doi.org/10.1016/j.jobe.2020.101833.

Ting, Guan Heng Andrew, Yi Wei Daniel Tay, and Ming Jen Tan. 2021. "Experimental Measurement on the Effects of Recycled Glass Cullets as Aggregates for Construction 3D Printing." *Journal of Cleaner Production* 300 (April): 126919. https://doi.org/10.1016/j.jclepro.2021.126919.

Tong, Jiaqi, Yahong Ding, Xiuwen Lv, and Wei Ning. 2023. "Effect of Carbonated Recycled Coarse Aggregates on the Mechanical Properties of 3D Printed Recycled Concrete." *Journal of Building Engineering* 80 (October): 107959. https://doi.org/10.1016/j.jobe.2023.107959.

Valente, Marco, Matteo Sambucci, Mehdi Chougan, and Seyed Hamidreza Ghaffar. 2023. "Composite Alkali-activated Materials with Waste Tire Rubber Designed for Additive Manufacturing: An Eco-sustainable and Energy Saving Approach." *Journal of Materials Research and Technology* 24 (April): 3098–3117. https://doi.org/10.1016/j.jmrt.2023.03.213.

Van Der Putten, Jolien, Mahzad Azima, Karel Lesage, Geert De Schutter, and Kim Van Tittelboom. 2019. *Combination of Portland Cement and Calcium-Sulfoaluminate Cement: The Highroad for Durable 3D Printing? RheoCon2 & SCC9*. https://biblio.ugent.be/publication/8631266.

Vantyghem, Gieljan, Ticho Ooms, and Wouter De Corte. 2020. "FEM Modelling Techniques for Simulation of 3D Concrete Printing." *arXiv (Cornell University)*, January. https://doi.org/10.48550/arxiv.2009.06907.

Weng, Yiwei, Mingyang Li, Ming Jen Tan, and Shunzhi Qian. 2018. "Design 3D Printing Cementitious Materials via Fuller Thompson Theory and Marson-Percy Model." *Construction and Building Materials* 163 (January): 600–610. https://doi.org/10.1016/j.conbuildmat.2017.12.112.

Weng, Yiwei, Shaoqin Ruan, Mingyang Li, Liwu Mo, Cise Unluer, Ming Jen Tan, and Shunzhi Qian. 2019. "Feasibility Study on Sustainable Magnesium Potassium Phosphate Cement Paste for 3D Printing." *Construction and Building Materials* 221 (June): 595–603. https://doi.org/10.1016/j.conbuildmat.2019.05.053.

West, Justin L., Emma D. Betters, Tony L. Schmitz, Scott Smith, Alex Roschli, David Nuttall, John Lindahl, and Lonnie Love. 2021. "Rethinking Production of Machine Tool Bases: Polymer Additive Manufacturing and Concrete." *Manufacturing Letters* 31 (June): 33–35. https://doi.org/10.1016/j.mfglet.2021.05.008.

Xue, Xiaolong, Ruixue Zhang, Rebecca Yang, and Jason Dai. 2014. "Innovation in Construction: A Critical Review and Future Research." *International Journal of Innovation Science* 6 (2): 111–126. https://doi.org/10.1260/1757-2223.6.2.111.

Yang, Liming, Samad M. E. Sepasgozar, Sara Shirowzhan, Alireza Kashani, and David Edwards. 2022. "Nozzle Criteria for Enhancing Extrudability, Buildability and Interlayer Bonding in 3D Printing Concrete." *Automation in Construction* 146 (December): 104671. https://doi.org/10.1016/j.autcon.2022.104671.

Yang, Yan, Hangzi Wu, Lifang Han, Qingling Huang, Shuo Yang, Jie Bai, and Mingyang Feng. 2023. "Investigation on Geometric and Surface Finish Quality of 3D Concrete Printed Walls with Hollow Section." *Advances in Structural Engineering* 27 (3): 386–401. https://doi.org/10.1177/13694332231222088.

Zairul, Mohd, and Zeinab Zaremohzzabieh. 2023. "Thematic Trends in Industry 4.0 Revolution Potential Towards Sustainability in the Construction Industry." *Sustainability* 15 (9): 7720. https://doi.org/10.3390/su15097720.

Zhou, Yiyi, Haoran Luo, Kamal Anand, Amardeep Singh, and Yi Min Xie. 2024. "Sustainable Use of Ultrafine Recycled Glass in Additive Manufactured (3D Printed) Reactive Powder Concrete." *Construction and Building Materials* 419 (February): 135556. https://doi.org/10.1016/j.conbuildmat.2024.135556.

Žujović, Maša, Radojko Obradović, Ivana Rakonjac, and Jelena Milošević. 2022. "3D Printing Technologies in Architectural Design and Construction: A Systematic Literature Review." *Buildings* 12 (9): 1319. https://doi.org/10.3390/buildings12091319.

Chapter 12

Implementation of NDT in Additive Manufacturing

Vikas Diwakar, Mohd Zaheer Khan Yusufzai, and Meghanshu Vashista

12.1 INTRODUCTION

AM has evolved significantly, enabling the fabrication of complex metal components with applications spanning aerospace, nuclear, energy, and automotive industries. Defined by ISO/ASTM 52900:2021 as a layer-by-layer manufacturing process that constructs objects from 3D model data, AM contrasts with traditional subtractive methods (Elkaseer et al. 2022). The AM workflow includes pre-processing, where CAD models are prepared; manufacturing; and post-processing, which may involve surface finishing and heat treatments (hot isostatic pressing) to enhance mechanical properties. AM offers advantages, including reduced material waste, rapid customisation, and design flexibility. However, the complexity of AM processes introduces various defects that affect the quality, mechanical properties, and microstructure of manufactured parts.

The consistency and structural integrity of AM parts depend on effective defect detection and quality assessment. NDT has become essential for identifying flaws in AM parts without compromising their functionality. Unlike conventionally manufactured components, AM parts exhibit unique defect types influenced by process parameters, material properties, and build orientation. The intricate geometries of AM components further complicate inspection, necessitating optimal selection of NDT techniques or a combination of methods for thorough evaluation. Several NDT techniques have been employed in AM, both during and after fabrication, to detect and monitor defects. "Methods such as X-ray computed tomography (CT), acoustic emission (AE), ultrasonic testing (UT), thermography, computer vision, structured light scanning, and laser line scanning have demonstrated efficacy in defect identification (Rao et al. 2023; Segovia Ramírez et al. 2023)." Researchers have explored the use of NDT for instantaneous process inspection, quality control, and in-service inspection to enhance the reliability of AM parts. Efforts are also being made to establish standardised testing protocols to ensure consistency across different AM processes and materials.

12.2 FUNDAMENTALS OF NDT

NDT is a critical approach for evaluating materials and components without causing damage, ensuring the detection of defects that could compromise performance and structural integrity. In the context of AM, where complex geometries and layered deposition techniques are employed, NDT plays a crucial role in quality assurance, both during the manufacturing process and in-service inspections. Defects in AM components can originate at various stages, including powder-bed inconsistencies, thermal stresses, and residual porosity, necessitating

 DOI: 10.1201/9781003590774-12

advanced NDT methods to ensure reliability. The fundamental principle of NDT is the use of physical interactions, such as acoustic, thermal, electromagnetic, or radiographic techniques, to identify internal imperfections. The responses to these interactions are observed as signals or images, which are then analysed to characterise material integrity. However, it is essential to distinguish between the detected indication and the actual defect, as signal interpretation depends on the sensitivity and resolution of the chosen NDT technique. NDT is not limited to the manufacturing phase; it is equally important for in-service monitoring to assess long-term performance and detect service-induced degradation, such as fatigue cracks and corrosion. The effectiveness of NDT techniques varies according to their detection capabilities, precision in defect characterisation, and applicability across different materials and component geometries.

According to ISO 9712, non-destructive testing (NDT) techniques are organised into distinct categories based on the physical principles they employ. Similarly, ASTM F2792 classifies additive manufacturing (AM) technologies into seven main processes, as illustrated in **Figure 12.1**: directed energy deposition (DED), material extrusion (ME), powder bed fusion (PBF), material jetting (MJ), binder jetting (BJ), vat photopolymerisation (VP), and sheet lamination (SL). Each of the NDT techniques has unique advantages and limitations based on the type of defects they can detect, material compatibility, resolution, and inspection conditions. The growing adoption of AM in industries necessitates continuous advancements in NDT to meet stringent quality and safety standards. Despite significant progress, challenges remain in implementing NDT for AM components. The variability in AM processes, high-temperature gradients, and layer-by-layer fabrication introduce complexities that traditional NDT methods may struggle to address. The development of in-situ monitoring techniques, improved standards for defect characterisation, and AI-driven data analysis are pivotal to enhancing NDT efficiency in AM.

Modern NDT approaches integrate machine learning (ML) and artificial intelligence (AI) algorithms to efficiently process large datasets, enabling real-time defect detection and predictive maintenance. However, while AI-driven techniques offer improved defect classification and automation, ensuring the reliability of these algorithms remains a key research focus. Statistical validation and hybrid data analysis approaches are required to improve accuracy and minimise false positives in defect identification.

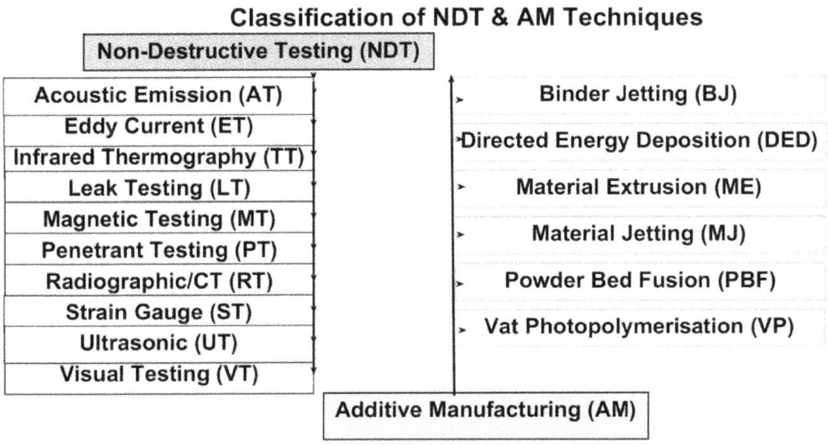

Figure 12.1 NDT and AM classification.

12.3 STUDIES OF NDT IMPLEMENTATION IN AM

NDT plays an integral part in confirming the quality and reliability of components manufactured through AM. As AM processes are increasingly utilised for their ability to produce intricate geometries and reduce material waste, ensuring the integrity of the manufactured parts becomes paramount. The application of NDT methods in AM addresses critical challenges in defect detection, microstructural analysis, and surface characterisation. This section explores the implementation of NDT in metal and polymer AM parts, and visual inspection in large-scale AM. AM processes are likely to exhibit various imperfections that can impact the mechanical performance, surface quality, and structural integrity of the fabricated components. These defects can be broadly classified into geometrical defects, metallurgical defects, mechanical defects, process-induced defects, and surface defects, as shown in **Figure 12.2**, each arising from specific process attributes and material interactions. Considering these defects is crucial for improving the reliability and quality of AM parts. Geometrical defects such as dimensional inaccuracy, warping, and overhang issues stem from improper thermal management, insufficient support structures, or layer misalignment during printing. These defects lead to deviations from the intended design, reducing precision and functionality. Meanwhile, metallurgical defects, including porosity, cracks, and inhomogeneous microstructures, arise from improper melting, rapid solidification, or inconsistent powder distribution, creating structural weaknesses that compromise mechanical strength. Mechanical defects, such as residual stresses, delamination, and fatigue failure, arise from thermal gradients and cyclic loading, often leading to internal stress buildup and eventual material failure under repeated use. Similarly, process-induced defects, including lack of fusion, unmelted powder, and keyhole formation, result from incorrect laser or electron beam energy settings, leading to incomplete melting and weak interlayer bonding. These defects significantly affect part density and mechanical performance. Surface defects, such as rough surface texture, oxidation, and layer shifting, often occur due to inconsistent powder deposition, improper recoating mechanisms, or exposure to high-temperature environments, affecting surface finish, wear resistance, and overall aesthetics.

12.3.1 Defect Detection in Metal AM Parts

Defects such as porosity, cracks, and lack of fusion in metal AM parts can significantly affect mechanical performance and long-term durability. Advanced NDT methods such as

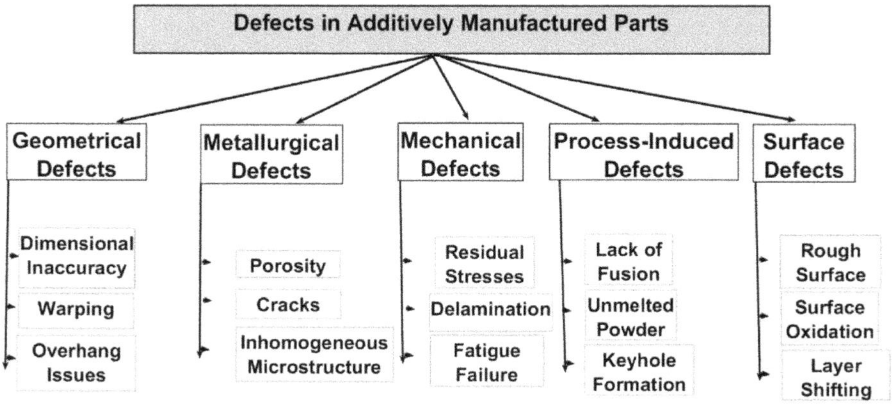

Figure 12.2 Defects in the AM process.

CT scans and ultrasonic testing (UT) have proven effective in identifying and characterising these defects (Zhang et al. 2022; Khosravani and Reinicke 2020). CT scans, in particular, are invaluable for providing high-resolution, three-dimensional imaging of internal structures. This method allows detailed inspection of the complex geometries inherent to AM parts, enabling the detection of subsurface defects that are otherwise inaccessible with traditional inspection methods (Shah et al. 2016). By using X-ray CT, porosity distributions and their morphologies can be quantified, which is critical for applications where fatigue performance and load-bearing capacity are essential. A creep test was conducted on an AM Al-Cu-Mn-Zr alloy specimen, machined to a gauge diameter of 6.35 mm and a gauge length of 31.75 mm. The specimen's loading axis was aligned with the build direction (BD), as shown in **Figure 12.3**. Creep testing was performed under tensile loading at 300°C and an applied stress of 35 MPa until fracture. "**Figure 12.3 (a)** provides a visual comparison of the specimen before and after the test, alongside the XCT-scanned region of the crept sample (Rahman et al. 2024)."

Ultrasonic inspection serves as a quick, cost-effective alternative to computed tomography for detecting defects. In this approach, high-frequency acoustic pulses are sent through the component, and the returning echoes are examined to reveal any irregularities. **Figure 12.4** depicts a study that employs phased-array ultrasonics to assess residual stresses in wire-arc additively manufactured Ti-6Al-4V alloy parts. To validate the findings, the contour method

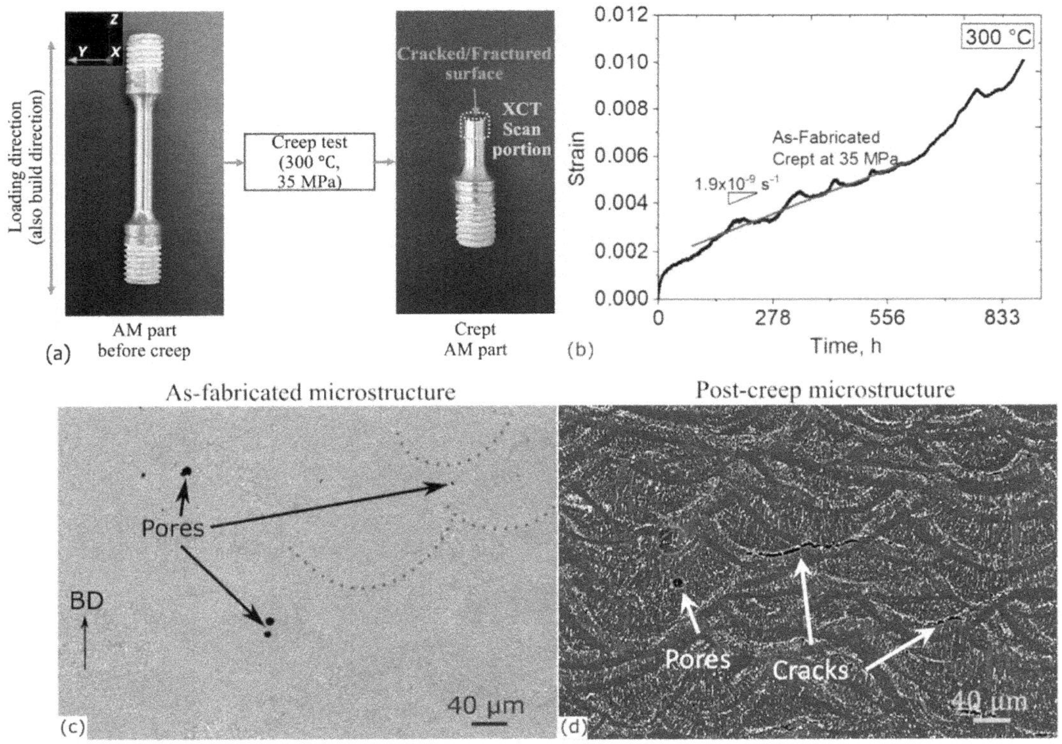

Figure 12.3 (a) AM Al-Cu-Mn-Zr alloy specimen before and after creep testing, with the XCT-scanned region marked; (b) Temporal strain evolution curve for the alloy subjected to creep at 300°C and 35 MPa; (c) SEM micrograph illustrating pores and faint MPBs (dotted lines) in the as-fabricated condition; (d) SEM image near the fracture zone, revealing pores, cracks, and distinct MPBs (bright arcs) captured using dual X-ray CT (Rahman et al. 2024).

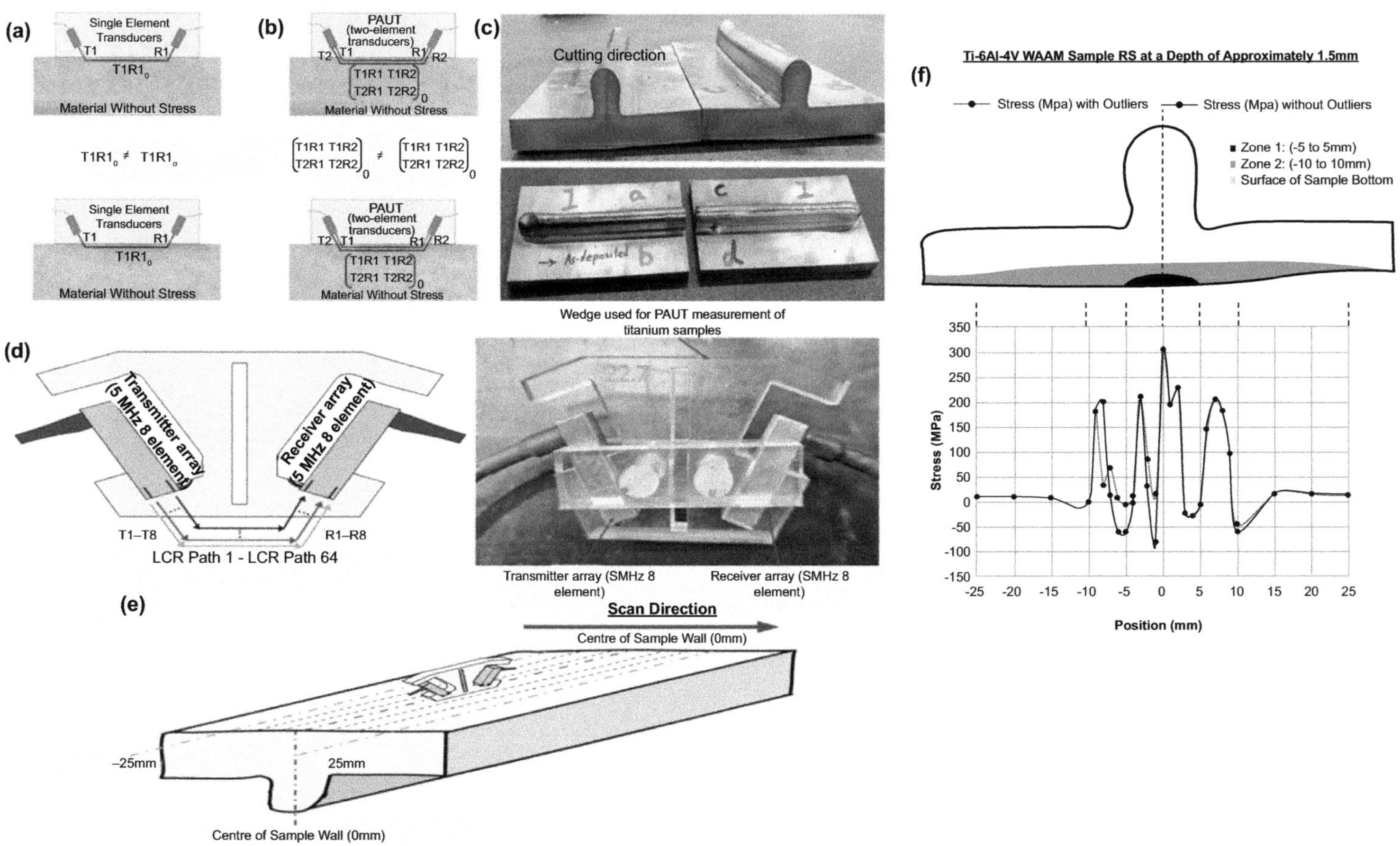

Figure 12.4 Schematic representation of "residual stress measurement: (a) Using single-element transducers; (b) Application of phased array ultrasonics; (c) Cutting direction for Ti-6Al-4V sample preparation for the contour method alongside the resulting cut sample"; (d) PAUT-LCR residual stress measurement setup; (e) Scan path and movement direction during PAUT on a Ti-6Al-4V sample; (f) Residual stress measurement results obtained via the phased array ultrasonics method (Walker et al. 2024).

(CM) was employed as a reference, enabling a comparative analysis of both methods to evaluate residual stress distribution in the samples (Walker et al. 2024). The experimental results demonstrated strong alignment between the phased-array ultrasonic measurements and the CM data, confirming similar stress-distribution trends across the tested specimens. Notably, phased-array ultrasonics provided improved accuracy when analysing individual elements, enabling the detection and mitigation of anomalies through averaging results. This capability highlights a key advantage of the phased-array ultrasonic technique: it allows for a more reliable assessment of residual stress than traditional methods.

12.3.2 Residual Stress Analysis in LPBF

Microstructural characteristics such as grain size, orientation, and residual stresses have a more significant role in defining the "mechanical properties" of LPBF components. The Barkhausen Noise Analyser (BNA) has emerged as a valuable tool for analysing residual stresses and material properties in AM parts (Diwakar et al. 2024). This technique measures magnetic noise generated by domain-wall motion in ferromagnetic materials under an applied magnetic field. Changes in the noise signal correspond to variations in residual stress and microstructural features. Recent technological advancements have accelerated the industrial adoption of LPBF. However, challenges such as residual stress formation remain significant due to the process's inherent complexity. It also explores the feasibility of using Barkhausen Noise Analysis (BNA) as an "in-situ monitoring tool for residual stress" in LPBF-manufactured "Maraging steel 300 (18Ni-300/1.2709)," as shown in **Figure 12.5**. Two experimental setups were conducted to validate BNA's applicability. The first employed an external load to verify the correlation between Barkhausen noise signals and induced stress variations in the LPBF-processed material. The second involved on-plate samples with varying residual stress states to analyse signal responses. Results indicated microstructural stability, an essential factor for reliable BNA measurements. Furthermore, the external load experiment confirmed that BNA signal variations align with induced stress states. Similarly, on-plate measurements exhibited signal patterns consistent with literature findings for residual stress changes.

12.3.3 NDT in Polymer AM Parts

The use of polymers in AM has grown significantly due to their versatility, lightweight properties, and cost-effectiveness. However, polymer AM parts are also susceptible to defects such as delamination, voids, and surface irregularities, which can compromise their structural integrity. Thermographic methods have emerged as effective NDT tools for identifying delamination and other defects in polymer components. Thermography uses infrared imaging to detect surface temperature variations on a part. These variations indicate subsurface defects, as heat conduction is disrupted in areas with voids or delaminations. Active thermography, which involves applying an external heat source to the part, enhances defect detection by creating a thermal contrast between defective and defect-free regions. For instance, effective temperature management is essential for achieving high-quality parts in Fused Filament Fabrication (FFF) (Ferraris, Zhang, and Van Hooreweder 2019). An analysis is conducted, which includes an infrared-based setup for monitoring 3-D and progressive temperature distribution in FFF printing, as shown in **Figure 12.6**. The system employs an Optris PI640 infrared camera with a wavelength range of 7.5–13 μm, a spatial resolution of 30.2 μm/pixel, and a field of view (FOV) of 21.7 × 14.9 mm^2. The camera, mounted on a Prusa MK3 printer, captures temperature changes in the x–z plane, enabling the observation of primary and secondary reheating effects during printing. To reduce thermal reflection, the build platform

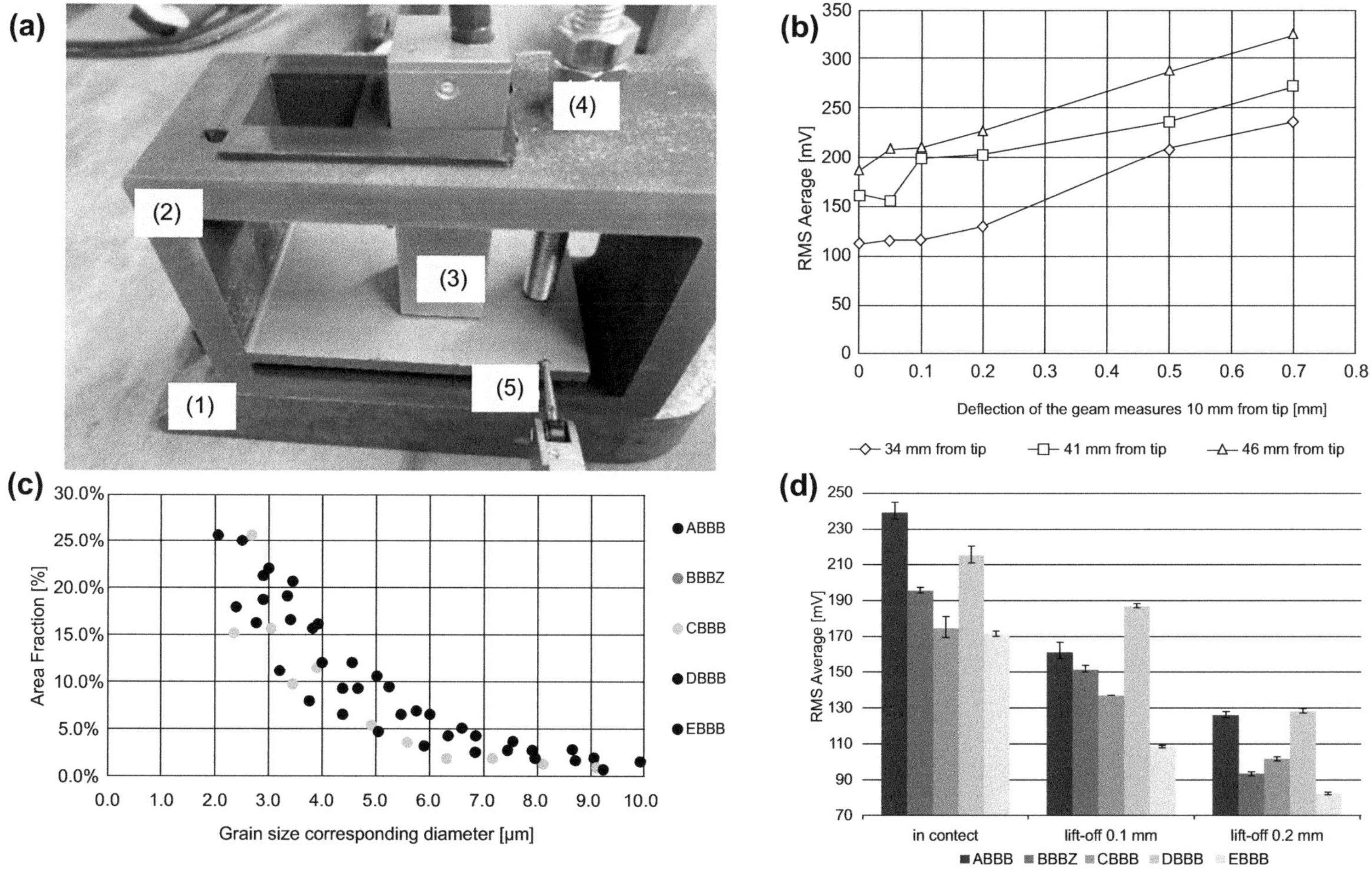

Figure 12.5 (a) Bending beam method setup: (1) Supporting plate, (2) Setup fabricated using LPBF, (3) Flat BNA sensor, (4) Screw, (5) Deflection measurement gauge. (b) Barkhausen noise measurement using a flat sensor along the deflection beam at varying distances from the beam tip. (c) "Grain size distribution across all samples, including a power fit function (, R^2=0.879)." (d) "RMS average measurements for different LPBF samples, comparing in-contact measurements and lift-off measurements at 0.1 mm and 0.2 mm, with the magnetic field aligned parallel to the final layer scanning direction (Staub et al. 2022)."

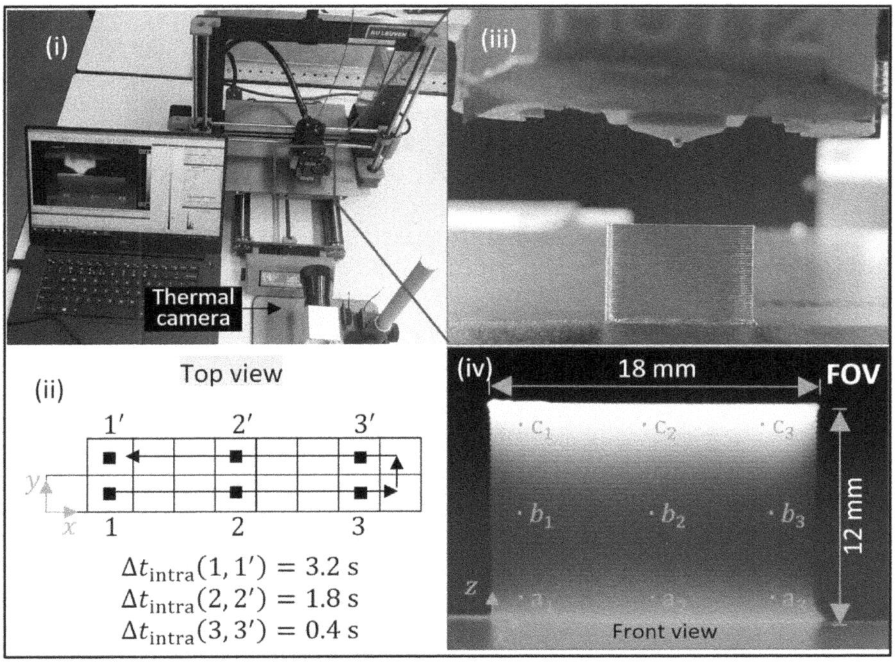

Figure 12.6 (i) Thermal monitoring setup for "Fused Filament Fabrication (FFF) during the printing process"; (ii) "Sample geometry, deposition sequence, and intra-layer time (⟦Δt⟧_intra)"; (iii) Fabricated sample; (iv) Thermographic image highlighting x–z locations (Ferraris, Zhang, and Van Hooreweder 2019).

was coated with low-reflectivity paint, and the nozzle was shielded. The captured temperature profiles were experimentally validated and compared with predictions from a custom-developed FFF numerical model. While initial results demonstrate that temperature data can help correlate with bond lengths, challenges remain, including emissivity calibration, signal variations due to the additive process, and improving the numerical model to better account for radiation and convection effects.

12.3.4 Topography and Surface Characterisation of Cladded Layers

In AM processes such as laser cladding, where a "high-energy laser is applied to fuse powder or wire feedstock on top of a workpiece," the quality of the cladded layers is important for ensuring functionality and durability. The topography and surface characteristics of cladded layers can significantly influence wear, corrosion, and thermal performance. NDT methods such as laser scanning and visual inspection have proven effective in assessing the topography and surface integrity of cladded layers. Laser scanning systems use high-resolution laser beams to map the surface profile of clad layers. By generating detailed 3D surface models, this technique provides valuable information on surface roughness, waviness, and contour accuracy. Variations in surface topography can indicate process inconsistencies, such as uneven powder deposition or fluctuations in laser energy. For example, multi-clad structures deposited through coaxial laser cladding can exhibit height variations between adjacent tracks due to differences in overlapping ratios. Laser scanning allows for precise measurement of these variations, enabling process optimisation to achieve uniform topography. Laser cladding technology melts desired metal powder by applying a laser as a heat source to the substrate, forming a cladding layer that enhances surface properties. However, defects such as cracks can occur, affecting cladding quality. An approach is proposed to monitor defects using AE signals for real-time state identification and crack detection during laser cladding (Li et al. 2021). AE signals are preprocessed to extract key features, such as peak amplitude and energy, which are then analysed using a deep learning neural network for cladding state recognition and crack detection. The experimental setup includes a coaxial laser cladding system with an LDF 4000–100 semiconductor laser, a KUKA KR30HA 6-DOF robot, and a PrecitecYC52 laser head delivering up to 6 kW power, as shown in **Figure 12.7.** A PAC USB AE Node system acquires AE signals for analysis, capturing frequency ranges from 1.0 kHz to 1.0 MHz. Results show that crack signals are significantly amplified relative to base-cladding signals, and that shunting these signals improves the neural network's accuracy. This method outperforms traditional models and demonstrates the potential of AE detection for operational monitoring and flaw detection in laser cladding.

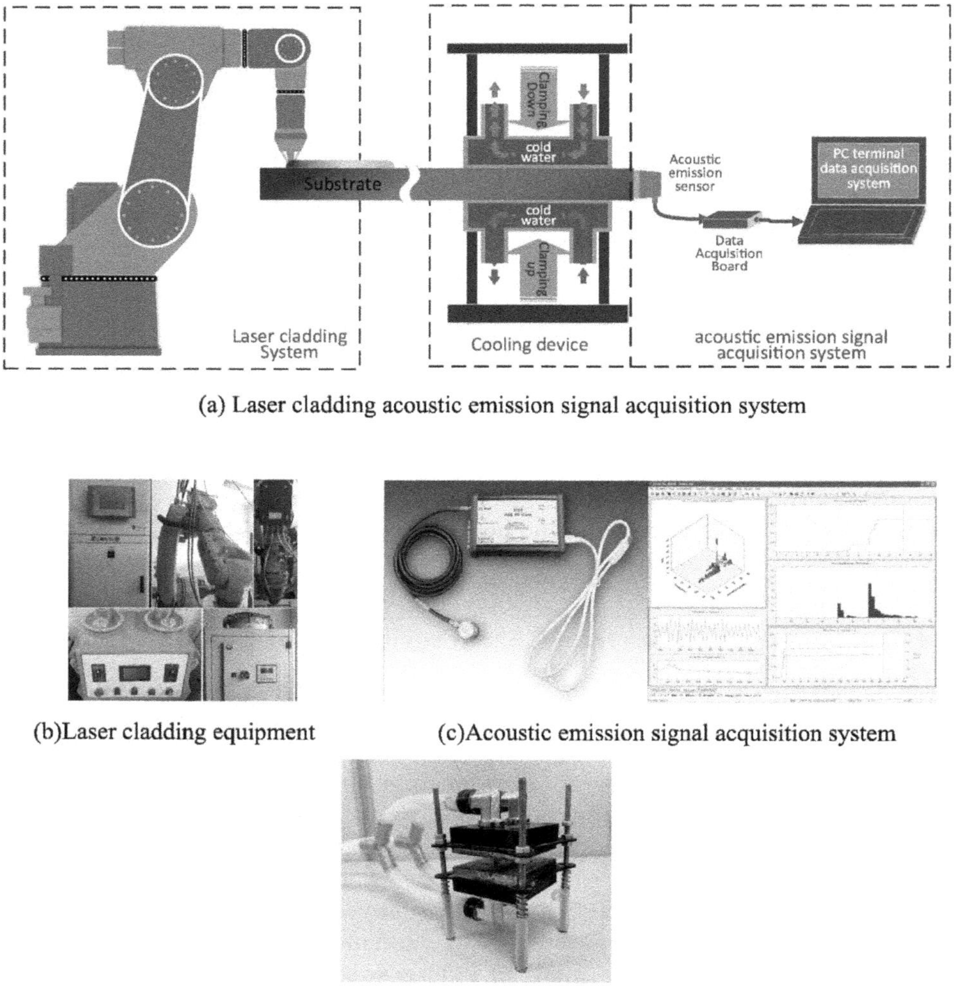

(a) Laser cladding acoustic emission signal acquisition system

(b)Laser cladding equipment

(c)Acoustic emission signal acquisition system

(d) Cooling device

Figure 12.7 Schematic diagram of experimental equipment for AE system (Li et al. 2021).

Visual inspection, while less advanced than laser scanning, remains a practical and cost-effective method for detecting surface defects such as cracks, pores, and discolouration. High-resolution optical cameras and microscopes are often used to enhance the accuracy of visual inspections. In some cases, automated systems equipped with machine vision technology are used to streamline inspection processes and reduce human error. By combining visual inspection with other NDT methods, manufacturers can achieve a comprehensive evaluation of cladded layers. The integration of NDT techniques into the AM workflow for topography and surface characterisation extends beyond defect detection. These methods also contribute to process validation and quality control by providing feedback on process stability and repeatability. For instance, consistent surface roughness measurements can indicate stable process conditions, while deviations may signal the need to adjust process parameters. This iterative feedback loop is essential for maintaining the high-quality standards required in aerospace and medical devices.

12.4 INTEGRATION OF NDT IN THE AM WORKFLOW

NDT has become an essential component of AM workflow, ensuring quality control at every stage of the process. From material preparation to in-process monitoring and final inspection. The integration of NDT into the AM workflow can be broadly divided into three stages: pre-process inspection, in-situ NDT techniques, and post-process inspection. The incorporation of automation and artificial intelligence (AI) further enhances the efficacy and efficiency of NDT in AM.

12.4.1 Pre-Process Inspection

Before the AM process begins, the pre-process inspection focuses on ensuring the quality of raw materials, including powders, wires, and filaments. The properties of feedstock materials directly influence the mechanical, thermal, and structural characteristics of the final part. NDT methods are used extensively at this stage for material quality control and feedstock characterisation. Powder-based AM processes, such as LPBF or DED, rely on powders with specific particle sizes, morphologies, and chemical compositions. Techniques such as XRD and SEM are typically used to analyse powder morphology, but NDT methods, such as laser-based particle size analysis and optical imaging systems, offer a non-invasive approach for rapid evaluation. These methods ensure uniformity in particle size distribution, which is crucial for consistent layer deposition and melt pool dynamics. For metal powders, NDT techniques such as acoustic resonance testing and eddy current testing can detect contaminants, inclusions, and irregularities in bulk powder lots. Similarly, thermographic methods can assess the thermal conductivity of polymer filaments to identify variations that might affect melting and extrusion during AM processes. By implementing these NDT techniques during pre-process inspection, manufacturers can significantly reduce the risk of defects originating from substandard materials, leading to enhanced reliability and reduced waste.

12.4.2 In-Situ NDT Techniques

The ability to monitor and control the build process in real time is one of the most promising advancements in AM. In-situ NDT techniques are applied to record the data during the manufacturing process, enabling the detection of defects as they form and providing opportunities for immediate corrective action (Shevchik et al. 2019). These techniques employ sensors and advanced analytical tools to monitor thermal, acoustic, and other parameters during part fabrication. Thermal imaging and infrared cameras are widely used to monitor heat distribution across the building. Variations in thermal profiles can indicate issues such as insufficient fusion, overmelting, or uneven energy input. Studies have shown that thermal data collected during laser cladding processes can be analysed to identify critical defects such as porosity or incomplete fusion zones. Also, thermal cameras integrated into AM systems can provide feedback for dynamic adjustments to laser power or scanning speed, improving the overall build quality. Acoustic sensors capture sound waves generated during material deposition, which are indicative of phenomena such as crack formation, delamination, or improper layer bonding. AE techniques have proven effective for monitoring deposition processes in DED and wire-arc AM.

Laser metal deposition (LMD) is an advanced AM process used to build or repair metal parts layer by layer for various applications. Defects in the deposited material can alter mechanical properties and potentially cause failure, increasing machining time and the costs of corrective actions. To address this, a novel real-time defect monitoring system was developed using "an AE sensor combined with unsupervised pattern recognition analysis (K-means clustering) and principal component analysis (PCA)" (Gaja and Liou 2017). Time and frequency domain descriptors were employed to enhance defect characterisation, effectively distinguishing between cracks and porosities. Cracks were identified as generating high-energy AE events with shorter decay times and lower amplitudes, while pores exhibited distinct acoustic signatures. Signal energy proved to be the key feature for defect source identification, while frequency had minimal impact.

Real-time laser scanning systems monitor surface roughness and track deviations in layer height during deposition (Lin et al. 2019). Optical cameras with high-resolution imaging can capture surface anomalies and irregularities, enabling automated detection of build inconsistencies. These systems are particularly useful in processes like LPBF, where layer thicknesses are on the micrometre scale. A novel in-line quality control approach employing laser scanning has been developed for defect detection in material extrusion AM. In this technique, surface point clouds captured by a 2D laser scanner are first preprocessed and then compared against an idealised point cloud derived from the pre-sliced STL model. Both datasets are transformed into depth maps, which are analysed using advanced image-processing routines to pinpoint overfill and underfill anomalies and delineate their exact geometry. The detected flaws are reconstructed into 3D representations, providing the control system with precise quantitative feedback. This information enables the printer to halt automatically when critical defects arise, thereby conserving both material and production time. **Figure 12.8** (**a**) and **Figure 12.8** (**b**) illustrate the real-time integration of laser scanning, defect identification, and feedback control within the extrusion workflow.

12.4.3 Post-Process Inspection

After the AM process is completed, post-process inspection serves as the final verification step to confirm that the fabricated part meets the essential quality standards. NDT methods at this stage focus on identifying internal defects, measuring mechanical properties, and characterising the part's surface and subsurface features. CT scans are among the most effective NDT techniques for post-process inspection, offering high-resolution 3D imaging of the internal structure of AM parts. CT scans are particularly useful for detecting porosity, cracks, or inclusions within the material. For metal AM parts, CT imaging has revealed defect clusters and pore distributions that are often linked to variations in process parameters. UT is another widely used post-process inspection technique. Phased Array Ultrasonic Testing (PAUT) enables detailed imaging of subsurface defects, such as cracks or delaminations, by utilising multiple transducers arranged in an array.

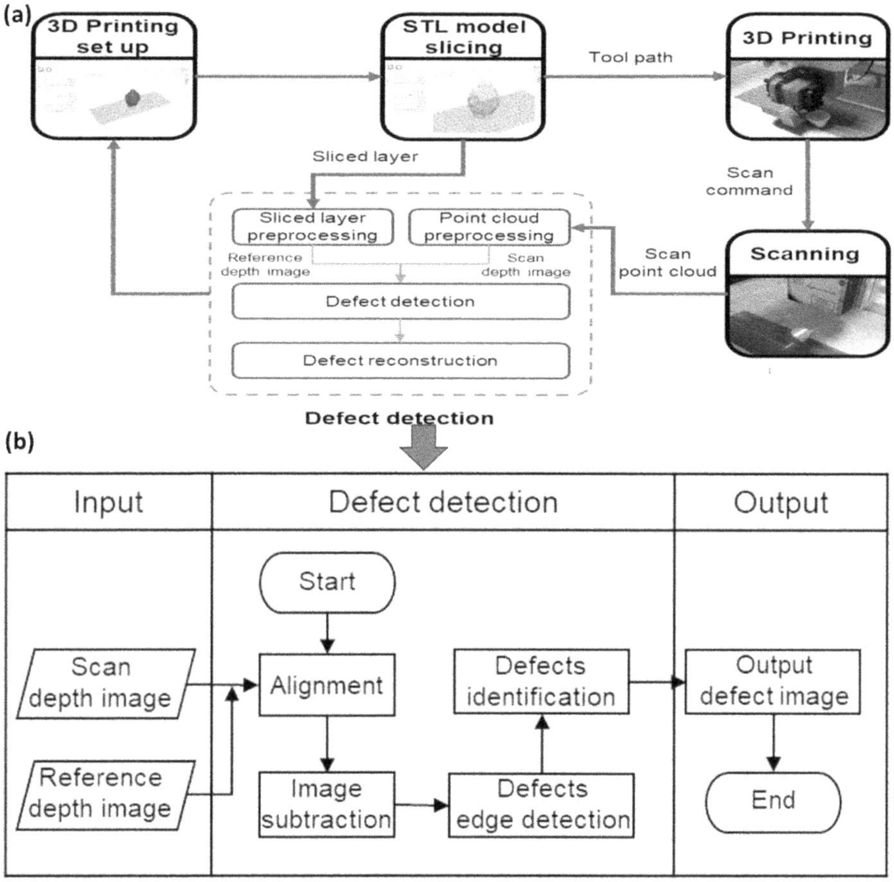

Figure 12.8 "Methodology based on laser scanning technology for defect detection and feedback control in the material extrusion process" (a) Flowchart depicting the research methodology; (b) Flowchart illustrating the defect detection process (Lin et al. 2019).

Visual inspection methods combined with automated image analysis tools are effective for detecting surface-level defects such as cracks, undercuts, and incomplete fusion zones. For applications requiring tight tolerances, laser profilometry and 3D surface scanning provide precise measurements of surface topography and dimensional accuracy. NDT techniques like microhardness mapping and Barkhausen Noise Analysis (BNA) are used to assess residual stresses and hardness variations across the part (Diwakar et al. 2023). These methods provide insights into the mechanical integrity and performance of AM parts, helping to ensure compliance with industry standards and specifications.

12.4.4 Automation and AI in NDT for AM

The incorporation of automation and artificial intelligence (AI) into NDT workflows has revolutionised quality assurance in AM. These technologies enable faster data analysis, more accurate defect detection, and enhanced decision-making capabilities. Machine learning (ML) algorithms have been developed to predict defects based on real-time data collected during the AM process. By analysing thermal images, acoustic signals, or laser scanning data,

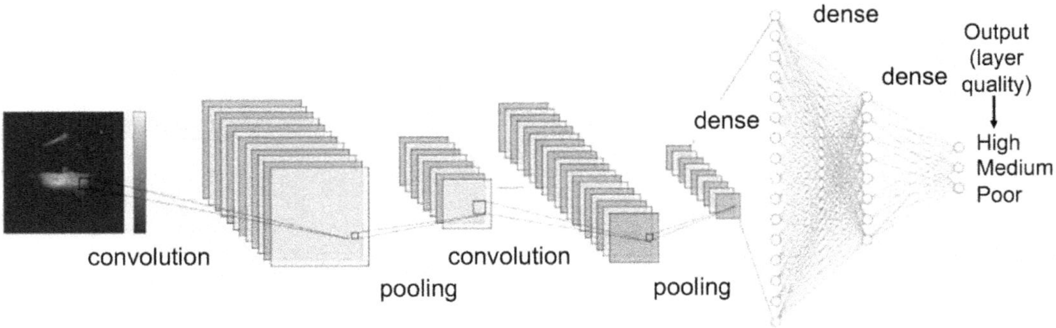

Figure 12.9 Overview of the Convolutional Neural Network (CNN) architecture utilising infrared thermal imaging for defect detection in directed energy deposition AM (Valizadeh and Wolff 2022).

ML models can identify patterns indicative of defect formation. For instance, convolutional neural networks (CNNs) have been used to process CT scan images of AM parts (Valizadeh and Wolff 2022), achieving high accuracy in detecting porosity and voids, as mentioned in **Figure 12.9**. Predictive models trained on historical data can also suggest optimal process parameters to minimise the likelihood of defects. Automated image processing tools integrated with NDT systems streamline defect detection by reducing manual inspection efforts (Bakas et al. 2022). Advanced software algorithms can analyse high-resolution optical or thermal images to identify cracks, inclusions, or surface irregularities. For example, automated image analysis tools have been used to evaluate the surface quality of multi-clad layers in laser cladding, identifying regions with insufficient overlap or excessive roughness. The use of robotic arms equipped with NDT sensors has enabled precise and consistent inspection of AM parts (Shevchik et al. 2019). Robotic systems can navigate complex geometries and perform inspections with high repeatability, making them ideal for large-scale production environments. Coupled with AI, these systems can adapt their inspection strategies in real time, further improving efficiency. The concept of digital twins involves creating a virtual replica of the AM process and the manufactured part. NDT data collected during the process is integrated into the digital twin, allowing for real-time simulation and defect prediction. This approach enhances the ability to foresee potential issues and implement corrective actions before they manifest in the physical part.

12.5 ADVANCES IN NDT FOR AM

The latest advancements in NDT for AM highlight improvements in imaging, in-situ monitoring, material-specific methods, and automation. Imaging remains one of the most crucial aspects of NDT, and recent developments have significantly enhanced its capabilities for AM. High-resolution CT has become a gold standard for inspecting AM parts, particularly for identifying internal defects such as porosity, inclusions, and cracks. Modern CT systems now boast improved scanning speeds and resolutions, allowing for faster inspections without compromising accuracy. These advancements enable the inspection of intricate geometries and lattice structures that are common in AM. Ultrasonic techniques have also advanced significantly. Phased Array Ultrasonic Testing (PAUT), for instance, allows for precise defect localisation and characterisation. When combined with time-of-flight diffraction (TOFD), these methods provide complementary information about defect size and orientation

(Sanchez Duo et al. 2024). These developments are particularly useful for inspecting dense metallic parts or hybrid materials produced through AM.

One of the most transformative advancements in NDT for AM is the adoption of in-situ monitoring systems. Real-time monitoring during the build process has become essential for defect prevention and process optimisation. Thermal monitoring, using infrared cameras or pyrometers, provides continuous feedback on the build's temperature profile. Deviations from expected profiles can indicate issues such as thermal distortion, incomplete melting, or excessive heat input. AE sensors represent another critical in-situ tool (Gaja and Liou 2017). These sensors capture high-frequency sound waves emitted during material deposition, which can reveal anomalies such as cracking, delamination, or insufficient bonding between layers. AE systems are particularly valuable in high-energy processes like LPBF and DED, where thermal gradients are significant. Optical monitoring systems (Zenzinger et al. 2015) have also evolved, incorporating high-speed cameras and laser-based scanners (Yi et al. 2022). These systems detect surface irregularities, layer height deviations, and insufficient overlaps. The integration of advanced machine vision algorithms has further improved their ability to identify defects in real-time, reducing the need for post-process inspection (Lu and Wong 2017).

The diversity of materials used in AM necessitates tailored NDT approaches for metals, polymers, and composites. For metal AM parts, magnetic methods like Eddy Current Testing (ECT) and Barkhausen Noise Analysis (BNA) are being increasingly applied. BNA, in particular, provides valuable information on residual stress and microstructural changes by analysing magnetic signals generated under cyclic magnetisation (Spurek et al. 2022; Gel'atko et al. 2024; He et al. 2019). These methods are highly effective for ferromagnetic materials, offering a non-invasive way to assess both surface and subsurface properties. In polymer AM, thermographic methods have emerged as a robust tool for detecting delamination and voids. By analysing heat flow across the material, thermography identifies temperature anomalies associated with internal defects. Ultrasonic testing has also been adapted for polymer applications, using transducers designed to account for the material's acoustic impedance. For composite AM parts, advanced ultrasonic methods like C-scan imaging are widely used (Wu et al. 2024). These techniques provide detailed insights into layer bonding and internal structure, which are critical for ensuring the mechanical integrity of complex composite designs. Laser shearography is another method gaining popularity for composites, as it effectively detects surface deformations caused by internal defects (Francis, Tatam, and Groves 2010).

12.6 CHALLENGES AND LIMITATIONS

The integration of NDT into AM workflows has advanced significantly, but several challenges and limitations still hinder its widespread adoption. These issues, stemming from technical, standardisation, and workforce-related barriers, must be addressed to fully unlock the potential of NDT in ensuring the quality and reliability of AM-produced parts. Technical limitations remain a foremost challenge in applying NDT to AM. Equipment constraints, such as the resolution of imaging systems and the sensitivity of sensors, often restrict the ability to detect fine defects in complex geometries. Additively manufactured parts frequently feature internal cavities, lattice structures, and overhangs that traditional NDT methods struggle to inspect with precision. For example, while advanced techniques such as high-resolution CT scans and phased array ultrasonic testing have made strides in defect detection, their inspection speed and cost remain prohibitive for high-throughput production environments (Salarian and Toyserkani 2018). Furthermore, resolution issues are particularly problematic for small-scale defects, such as micro-cracks or subsurface porosity, which can have a disproportionate impact on the mechanical properties and fatigue life of AM parts. Methods like thermography and acoustic emission, while effective for certain applications, face challenges in adapting to the high thermal gradients and noise levels inherent to AM processes. As a result, the development of new equipment and methodologies tailored specifically for AM is crucial to overcoming these barriers. A major obstacle in the adoption of NDT in AM is the lack of standardised protocols across industries (Wu et al. 2024). Unlike traditional manufacturing processes, where well-established standards guide quality assurance, AM presents unique challenges due to its wide range of materials, processes, and applications. For instance, the same NDT method may require different calibration or interpretation when applied to LPBF parts versus DED parts. The absence of unified standards complicates both the validation of NDT techniques and the comparison of results across different facilities or industries. This fragmentation not only slows the adoption of NDT in AM but also undermines confidence in the reliability of the inspection process. Industry-wide efforts, led by organisations such as ASTM International and ISO, are underway to address this gap. However, the dynamic nature of AM technology, with new materials and processes continuously emerging, makes it difficult to establish universal guidelines.

Moreover, standardisation issues extend to the interpretation of inspection data. For instance, the criteria for acceptable levels of porosity or residual stress may vary by application, making it difficult to define clear thresholds. Establishing consistent, application-specific standards will require close collaboration between researchers, manufacturers, and regulatory bodies. The effective implementation of NDT in AM requires a workforce skilled in both domains, but the demand for such expertise often outpaces supply. AM and NDT are both highly specialised fields, and their intersection creates a niche skill set that includes understanding AM processes, material behaviours, and advanced NDT techniques. Without proper training, inspectors may struggle to accurately interpret the complex data generated during AM inspections, such as CT scan reconstructions or Barkhausen noise signals. The multidisciplinary nature of NDT in AM also poses challenges for training programmes. Professionals must be well-versed not only in conventional NDT methods but also in emerging techniques such as in-situ monitoring and AI-driven analysis. This requires significant investment in education and training, as well as access to cutting-edge equipment and software.

12.7 FUTURE DIRECTIONS

The future of NDT in AM lies in advancing its capabilities, integrating emerging technologies, and addressing critical challenges to meet the needs of modern manufacturing. The evolving landscape of AM demands innovative solutions to enhance quality assurance while maintaining efficiency and sustainability. The development of hybrid NDT methods offers a significant opportunity to improve defect detection and material characterisation in AM. By combining multiple NDT techniques, hybrid approaches can leverage the strengths of individual methods while mitigating their limitations. For instance, integrating ultrasonic testing with thermography enables simultaneous detection of surface and subsurface defects, providing a more comprehensive analysis of AM parts. Similarly, combining X-ray CT with AE can enable real-time monitoring of internal defect propagation during mechanical testing (Khosravani and Reinicke 2020). Hybrid methods are particularly valuable in AM due to the complex geometries and heterogeneous material properties of additively manufactured components. For example, in lattice structures or multi-material parts, a single technique may not provide sufficient resolution or coverage. Hybrid systems can offer a more robust solution, enhancing both accuracy and reliability. As research in this area progresses, the development of automated hybrid NDT systems tailored to specific AM processes and materials is expected to play a crucial role in advancing quality assurance.

The integration of NDT with Industry 4.0 technologies, such as the Internet of Things (IoT), cloud computing, and data analytics, is poised to revolutionise quality control in AM. Smart manufacturing environments can benefit from IoT-enabled sensors embedded in AM equipment, which provide real-time data on process parameters and part quality. For instance, thermal sensors and acoustic monitoring devices can continuously capture data during the build process, enabling immediate detection of anomalies and reducing the risk of defects. Cloud computing further enhances NDT capabilities by enabling the storage, sharing, and analysis of large datasets. Advanced algorithms, powered by artificial intelligence and machine learning, can analyse this data to identify patterns, predict defects, and optimise inspection strategies. Such systems can also provide actionable insights for process optimisation, reducing the likelihood of defects in future builds. Moreover, digital twins, virtual replicas of physical parts and processes, can incorporate real-time NDT data to simulate and predict part performance under various conditions. This integration not only enhances the efficiency of quality assurance but also aligns with Industry 4.0's broader goals by enabling data-driven decision-making and continuous improvement. As sustainability becomes a critical consideration in manufacturing, efforts to minimise the environmental impact of NDT in AM are gaining traction. Traditional inspection methods often involve significant energy consumption and material waste, particularly when destructive testing is required. In contrast, advanced NDT techniques can reduce waste by preserving the integrity of inspected parts while ensuring their quality.

One approach to enhancing sustainability is the development of energy-efficient inspection systems. For instance, low-power ultrasonic transducers and optimised scanning algorithms can reduce the energy footprint of NDT processes. Additionally, AI-driven analysis can streamline inspections, minimising time and resources required. Another aspect of sustainability is the potential to integrate NDT with AM processes to enable in-situ monitoring and defect correction. By identifying and addressing defects during the build process, manufacturers can reduce material waste and avoid the need for costly post-process repairs. These advancements not only support environmental goals but also contribute to the economic viability of AM. Establishing AM-specific NDT standards is essential to ensuring consistent, reliable quality assurance across industries. Organisations such as ASTM International and ISO are actively developing guidelines to address the unique challenges of inspecting AM parts. These efforts focus on defining standardised procedures for defect detection, material characterisation, and process validation, tailored to the diverse range of AM technologies and materials. For example, ASTM's Committee F42 on AM Technologies has introduced standards for evaluating the mechanical properties and dimensional accuracy of AM parts, which include recommendations for NDT methods (Kawalkar, Dubey, and Lokhande 2021; Spierings et al. 2015). Similarly, ISO's Technical Committee 261 on AM is collaborating with industry stakeholders to create a comprehensive framework for quality assurance in AM (Pei and Seidel 2023). The development of such standards will provide manufacturers with clear quality benchmarks, facilitating the adoption of NDT in AM workflows. Moreover, standardised protocols will enable better comparability of inspection results, fostering collaboration and innovation across the industry. As AM continues to evolve, these standards will need to be updated to address emerging technologies and applications, ensuring their relevance and applicability.

12.8 CASE STUDY

Coaxial laser cladding is a widely used AM technique for surface enhancement and component repair. It involves depositing a metallic powder or wire onto a substrate using a high-powered laser to form a metallurgically bonded clad layer. Multi-clad layer deposition is essential for achieving sufficient coating thickness, wear resistance, and corrosion protection. However, the formation of these layers introduces microstructural variations, including dendritic growth, porosity, residual stresses, dilution effects, and elemental segregation, which can significantly impact the mechanical performance and service life of the coated component. In this case study, BNA are employed to evaluate microstructural variations in multi-clad layers produced via coaxial laser cladding. These techniques enable in-depth assessment without damaging the clad parts, ensuring high-quality manufacturing standards. The multi-clad layers were deposited using a coaxial laser cladding system, where a controlled feed of metal powder was melted and fused onto a steel substrate using a 1500W fibre laser. The process parameters, including laser power, scanning speed, powder feed rate, and overlapping ratio, were optimised to achieve homogeneous and defect-free coatings. A 50% overlapping ratio was maintained between successive clads to ensure uniform coverage and minimal dilution. The final obtained single-layer multi-track clad and its corresponding microstructure are shown in **Figure 12.10**.

BNA and hysteresis loop (HL) techniques are utilised as a non-destructive tool to characterise the magnetic and mechanical aspects of MLCs' HEA cladding. The reason is effective and faster assessment of the thermal damage region and irregularities in in-line production and inspection. It measures the noise generated by abrupt changes in the magnetisation of the LC

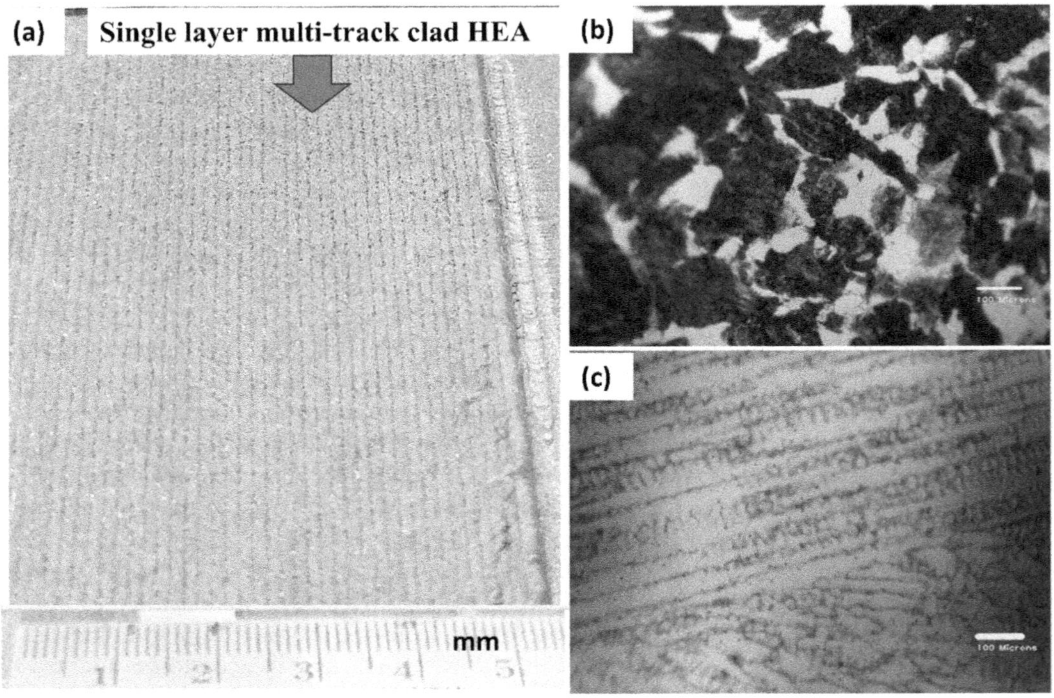

Figure 12.10 (a) Single-layer multi-track clads of HEA, (b) Microstructure of steel substrate, (c) Microstructure of HEA clad.

samples when subjected to a varying magnetic frequency (MF) and magnetic field intensity (MFI). This noise, known as Barkhausen noise, is sensitive to microstructural features such as grain size, residual stress, microhardness, and dislocation density. This section elaborates on the microstructural and microhardness changes observed in the cladded region compared to the base material, emphasising the confirmability and applicability of these enhancements as measured by the Barkhausen noise analyser. **Figure 12.11** shows the BNA and HL envelopes of the MLCs HEA cladding, which clearly shows that the excellent BNA response in terms of magnetic envelope broadening is observed in base substrate compared to the cladded portion. Furthermore, BNA and HL responses, that is, higher RMS (43.1), average permeability (286.87), coercivity (100.63 Oe), and remanence (12652 Gauss) were observed under base metal. The reason is ferritic structure along with coarse grain, which helps in magnetic domain wall rotation and movement along with the magnetic field intensity direction.

On the other hand, poor BNA response, characterised by narrow and small envelopes and a lower RMS (22.9), was observed in the clad region, as shown in **Figure 12.11 (a, c)**. At the same time, the HL envelope has a lower average permeability (13.87), coercivity (49.86 Oe), and remanence (388 Gauss) in the case of cladding, as followed by the base region, as shown in **Figure 12.11 (b, d)**. These BNA and HL envelopes and response outcomes can be correlated with the microstructure refinement and microhardness. A poor magnetic response was observed in the cladding region, due to a uniform fine-grain columnar dendritic structure that restricts magnetic domain wall movement in the direction of the magnetic field.

A fine-grained structure having higher hardness according to the Hall-Petch relationship. The observed reduction in Barkhausen Noise Amplitude response, characterised by narrow envelopes and a lower RMS value (22.9) in the cladded region, directly correlates with the microstructural refinement induced during the cladding process. The cladding technique, such as laser or thermal spraying, promotes rapid solidification, leading to finer grains and increased grain boundaries or secondary phases. These microstructural features act as pinning sites, obstructing the movement of magnetic domain walls. Consequently, the restricted domain wall motion suppresses the intensity of Barkhausen noise jumps, lowering the RMS value. This phenomenon aligns with the Hall-Petch relationship, in which finer grains enhance microhardness by impeding dislocation motion, further reinforcing the interplay between mechanical and magnetic properties. Similarly, the HL parameters lower average permeability (13.87), coercivity (49.86 Oe), and remanence (388 Gauss) reflect the cladded region's magnetic "softening" despite its hardened microstructure. The reduced permeability indicates greater difficulty in magnetising the material due to microstructural obstacles (e.g., dislocations, precipitates) that hinder domain alignment under an external field. While grain refinement typically increases coercivity by increasing pinning sites, the unexpected lower coercivity in the cladded region may stem from residual compressive stresses or from phase composition (e.g., the dominance of a softer magnetic phase) overriding the grain boundary effects. The diminished remanence, indicating weaker retained magnetisation, further underscores the role of microstructural barriers in limiting domain reorientation. Overall, the interplay between magnetic responses (BNA and HL) and mechanical properties (microhardness) is governed by the cladded region's refined microstructure, where enhanced hardness and magnetic property degradation coexist due to competing effects of defect density, stress states, and phase distribution.

12.9 CONCLUSION

The integration of NDT in AM represents an essential step in ensuring the quality, reliability, and performance of AM-produced components. NDT has demonstrated its capability

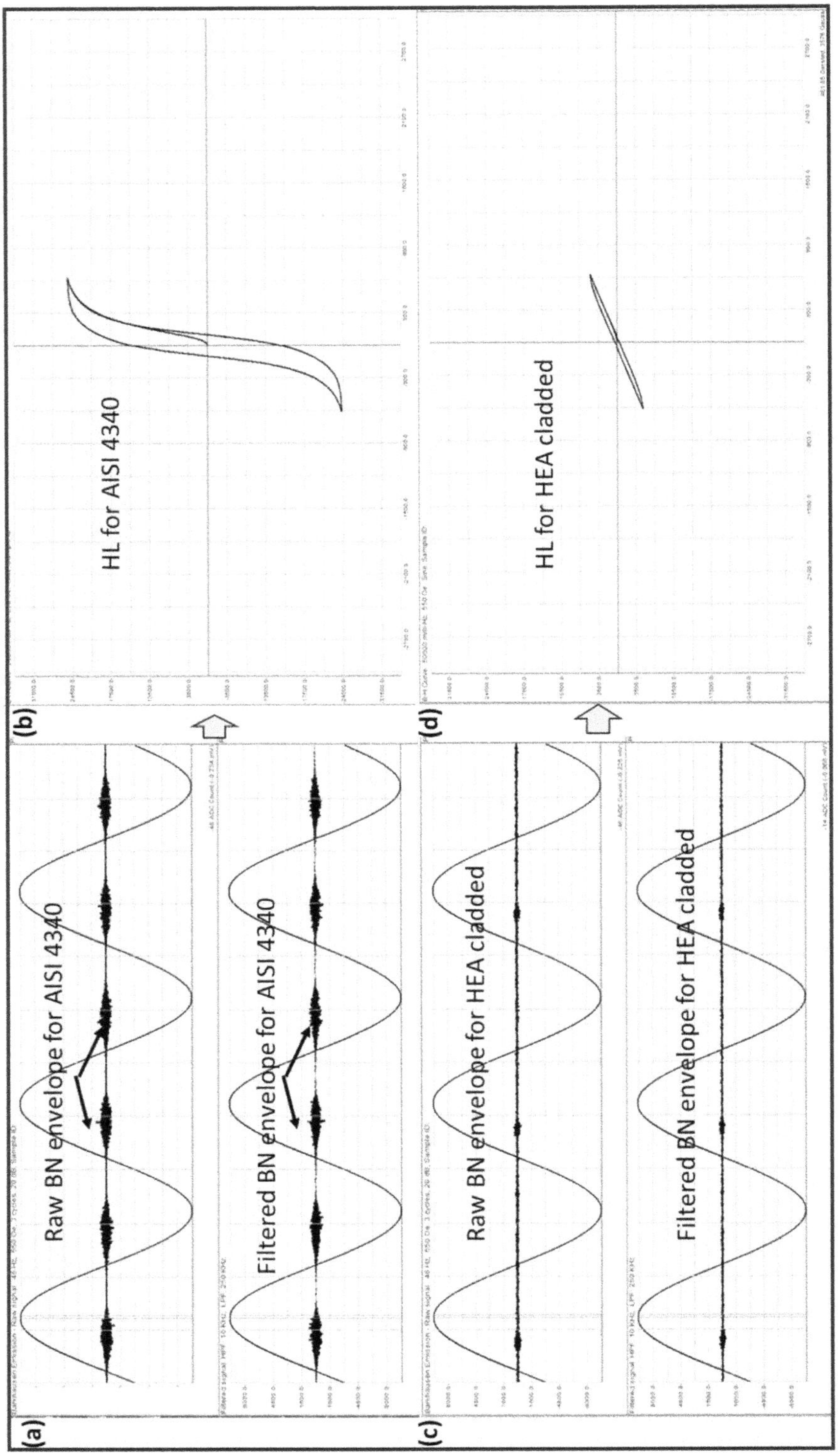

Figure 12.11 BN and HL output of (a–b) Steel substrate, (c–d) HEA clads.

to address critical challenges in AM, including defect detection, microstructure evaluation, residual stress measurement, and surface characterisation, while preserving the integrity of the part. By applying advanced techniques such as X-ray computed tomography, ultrasonic testing, thermography, and Barkhausen noise analysis, manufacturers can achieve a comprehensive understanding of part quality across various AM processes and materials. These methods not only enhance the consistency and repeatability of AM production but also support the industry's transition towards high-value applications in aerospace, healthcare, and automotive sectors.

A key takeaway is the growing importance of integrating NDT with emerging technologies, such as artificial intelligence, machine learning, and IoT, to enable real-time monitoring and predictive analytics for defect prevention. The development of hybrid NDT methods and AM-specific standards further highlights the need for continued innovation and collaboration across academia, industry, and standardisation bodies. Recommendations for the future include prioritising research into energy-efficient and sustainable NDT methods, expanding training programmes to build a skilled workforce, and fostering global standardisation to ensure uniformity in quality assurance practices. Also, BNA is used as NDT to check the capability of the clad region. The cladded region's magnetic responses, characterised by reduced BNA RMS and HL parameters (permeability, coercivity, and remanence), are strongly influenced by its refined microstructure. The cladding process induces rapid solidification, leading to finer grains, increased grain boundaries, and secondary phases, which enhance microhardness but hinder magnetic domain wall motion.

REFERENCES

Bakas, Georgios, Spyridon Dimitriadis, Stavros Deligiannis, Leonidas Gargalis, Ioannis Skaltsas, Kyriaki Bei, Evangelia Karaxi, and Elias P. Koumoulos. 2022. "A Tool for Rapid Analysis Using Image Processing and Artificial Intelligence: Automated Interoperable Characterization Data of Metal Powder for Additive Manufacturing with SEM Case." *Metals* 12 (11). https://doi.org/10.3390/met12111816.

Diwakar, Vikas, Ashwani Sharma, Mohd Zaheer Khan Yusufzai, and Meghanshu Vashista. 2023. "Measurement and Analysis of Magnetic Properties in IS 2062 Steel and AISI D2 Tool Steel Via Non-Destructive Testing." *Mapan—Journal of Metrology Society of India* 38 (2): 489–498. https://doi.org/10.1007/s12647-023-00637-x.

Diwakar, Vikas, Ashwani Sharma, Mohd Zaheer Khan Yusufzai, and Meghanshu Vashista. 2024. "Machine Learning-Based Prediction of Single Clad Characteristics and Non-Destructive Characterization of Multi-Layer Deposited FeCoNiCrMo HEA on EN24 via Laser Cladding." *Materials Today Communications* 41 (October): 110839. https://doi.org/10.1016/j.mtcomm.2024.110839.

Elkaseer, Ahmed, Karin J. Chen, Jan C. Janhsen, Oliver Refle, Veit Hagenmeyer, and Steffen G. Scholz. 2022. "Material Jetting for Advanced Applications: A State-of-the-Art Review, Gaps and Future Directions." *Additive Manufacturing* 60 (PA): 103270. https://doi.org/10.1016/j.addma.2022.103270.

Ferraris, E., J. Zhang, and B. Van Hooreweder. 2019. "Thermography Based In-Process Monitoring of Fused Filament Fabrication of Polymeric Parts." *CIRP Annals* 68 (1): 213–216. https://doi.org/10.1016/j.cirp.2019.04.123.

Francis, D., R. P. Tatam, and R. M. Groves. 2010. "Shearography Technology and Applications: A Review." *Measurement Science and Technology* 21 (10). https://doi.org/10.1088/0957-0233/21/10/102001.

Gaja, Haythem, and Frank Liou. 2017. "Defects Monitoring of Laser Metal Deposition Using Acoustic Emission Sensor." *International Journal of Advanced Manufacturing Technology* 90 (1–4): 561–574. https://doi.org/10.1007/s00170-016-9366-x.

Geľatko, Matúš, Michal Hatala, František Botko, Radoslav Vandžura, and Jiří Hajnyš. 2024. "Study of Eddy Current Testing Ability on SLM Aluminium Alloy." *Materials* 17 (14). https://doi.org/10.3390/ma17143568.

He, Dongfeng, Zhi Wang, Masahiro Kusano, Satoshi Kishimoto, and Makoto Watanabe. 2019. "Evaluation of 3D-Printed Titanium Alloy Using Eddy Current Testing with High-Sensitivity Magnetic Sensor." *NDT and E International* 102 (November 2018): 90–95. https://doi.org/10.1016/j.ndteint.2018.11.007.

Kawalkar, Rajat, Harrsh Kumar Dubey, and Satish P. Lokhande. 2021. "A Review for Advancements in Standardization for Additive Manufacturing." *Materials Today: Proceedings* 50: 1983–1990. https://doi.org/10.1016/j.matpr.2021.09.333.

Khosravani, Mohammad Reza, and Tamara Reinicke. 2020. "On the Use of X-Ray Computed Tomography in Assessment of 3D-Printed Components." *Journal of Nondestructive Evaluation* 39 (4). https://doi.org/10.1007/s10921-020-00721-1.

Li, Kaiqiang, Tao Li, Min Ma, Dong Wang, Weiwei Deng, and Huitian Lu. 2021. "Laser Cladding State Recognition and Crack Defect Diagnosis by Acoustic Emission Signal and Neural Network." *Optics and Laser Technology* 142 (November 2020): 107161. https://doi.org/10.1016/j.optlastec.2021.107161.

Lin, W., Hongyao Shen, Jianzhong Fu, and Senyang Wu. 2019. "Online Quality Monitoring in Material Extrusion Additive Manufacturing Processes Based on Laser Scanning Technology." *Precision Engineering* 60 (June): 76–84. https://doi.org/10.1016/j.precisioneng.2019.06.004.

Lu, Q. Y., and C. H. Wong. 2017. "Applications of Non-Destructive Testing Techniques for Post-Process Control of Additively Manufactured Parts." *Virtual and Physical Prototyping* 12 (4): 301–321. https://doi.org/10.1080/17452759.2017.1357319.

Pei, Eujin, and Christian Seidel. 2023. "Standardisation Efforts of ISO/TC 261 'Additive Manufacturing' 21st Plenary Meeting of ISO/TC 261 'Additive Manufacturing'." *Progress in Additive Manufacturing* 8 (6): 1143–1144. https://doi.org/10.1007/s40964-023-00448-1.

Rahman, Obaidullah, Jovid Rakhmonov, Sumit Bahl, Curtis Frederick, Amit Shyam, Ryan Dehoff, Alex Plotkowski, and Amirkoushyar Ziabari. 2024. "Dual X-Ray Computed Tomography-Aided Classification of Melt Pool Boundaries and Flaws in Crept Additively Manufactured Parts." *Materials Characterization* 217 (May): 114317. https://doi.org/10.1016/j.matchar.2024.114317.

Rao, Jing, Swee Leong Sing, Peipei Liu, Jilai Wang, and Hoon Sohn. 2023. "Non-Destructive Testing of Metal-Based Additively Manufactured Parts and Processes: A Review." *Virtual and Physical Prototyping* 18 (1). https://doi.org/10.1080/17452759.2023.2266658.

Salarian, Mehrnaz, and Ehsan Toyserkani. 2018. "The Use of Nano-Computed Tomography (Nano-CT) in Non-Destructive Testing of Metallic Parts Made by Laser Powder-Bed Fusion Additive Manufacturing." *International Journal of Advanced Manufacturing Technology* 98 (9–12): 3147–3153. https://doi.org/10.1007/s00170-018-2421-z.

Sanchez Duo, Irati, Jose Luis Lanzagorta, Iratxe Aizpurua Maestre, and Lander Galdos. 2024. "Enhancing Time-of-Flight Diffraction (TOFD) Inspection through an Innovative Curved-Sole Probe Design." *Sensors* 24 (19). https://doi.org/10.3390/s24196360.

Segovia Ramírez, Isaac, Fausto Pedro García Márquez, and Mayorkinos Papaelias. 2023. "Review on Additive Manufacturing and Non-Destructive Testing." *Journal of Manufacturing Systems* 66 (January): 260–286. https://doi.org/10.1016/j.jmsy.2022.12.005.

Shah, Paras, Radu Racasan, and Paul Bills. 2016. "Comparison of Different Additive Manufacturing Methods Using Computed Tomography." *Case Studies in Nondestructive Testing and Evaluation* 6: 69–78. https://doi.org/10.1016/j.csndt.2016.05.008.

Shevchik, Sergey A., Giulio Masinelli, Christoph Kenel, Christian Leinenbach, and Kilian Wasmer. 2019. "Deep Learning for in Situ and Real-Time Quality Monitoring in Additive Manufacturing Using Acoustic Emission." *IEEE Transactions on Industrial Informatics* 15 (9): 5194–5203. https://doi.org/10.1109/TII.2019.2910524.

Spierings, A.B., M. Voegtlin, T. Bauer, and K. Wegener. 2015. "Materials Testing Standards for Additive Manufacturing of Polymer Materials." *Prog Addit Manuf* 1: 9–20.

Spurek, Marvin A., Adriaan B. Spierings, Marc Lany, Bernard Revaz, Gilles Santi, Jonatan Wicht, and Konrad Wegener. 2022. "In-Situ Monitoring of Powder Bed Fusion of Metals Using Eddy Current Testing." *Additive Manufacturing* 60 (PB): 103259. https://doi.org/10.1016/j.addma.2022.103259.

Staub, Alexandre, Muriel Scherer, Pascal Zehnder, Adriaan Bernardus Spierings, and Konrad Wegener. 2022. "Residual Stresses Measurements in Laser Powder Bed Fusion Using Barkhausen Noise Analysis." *Materials* 15 (7). https://doi.org/10.3390/ma15072676.

Valizadeh, Mahsa, and Sarah Jeannette Wolff. 2022. "Convolutional Neural Network Applications in Additive Manufacturing: A Review." *Advances in Industrial and Manufacturing Engineering* 4 (February). https://doi.org/10.1016/j.aime.2022.100072.

Walker, Joseph, Brandon Mills, Yashar Javadi, Charles MacLeod, Yongle Sun, Pradeeptta Kumar Taraphdar, Bilal Ahmad, Sundar Gurumurthy, Jialuo Ding, and Fiona Sillars. 2024. "Study of Residual Stress Using Phased Array Ultrasonics in Ti-6AL-4V Wire-Arc Additively Manufactured Components." *Sensors* 24 (19): 6372. https://doi.org/10.3390/s24196372.

Wu, Di, Wenhan Qu, Yintang Wen, Yuyan Zhang, and Bo Liang. 2024. "The Application, Challenge, and Developing Trends of Non-Destructive Testing Technique for Large-Scale and Complex Engineering Components Fabricated by Metal Additive Manufacturing Technology in Aerospace." *Journal of Nondestructive Evaluation* 43 (3). https://doi.org/10.1007/s10921-024-01107-3.

Yi, Li, Alborz Shokrani, Rachele Bertolini, Unai Mutilba, Maria Grazia Guerra, Evripides G. Loukaides, and Peer Woizeschke. 2022. "Optical Sensor-Based Process Monitoring in Additive Manufacturing." *Procedia CIRP* 115 (March): 107–112. https://doi.org/10.1016/j.procir.2022.10.058.

Zenzinger, Guenter, Joachim Bamberg, Alexander Ladewig, Thomas Hess, Benjamin Henkel, and Wilhelm Satzger. 2015. "Process Monitoring of Additive Manufacturing by Using Optical Tomography." *AIP Conference Proceedings* 1650 (March): 164–170. https://doi.org/10.1063/1.4914606.

Zhang, Jun, Xin Zhao, Bing Yang, Jingyu Li, Yan Liu, Guanbing Ma, Shuxian Yuan, and Jinfeng Wu. 2022. "Nondestructive Evaluation of Porosity in Additive Manufacturing by Laser Ultrasonic Surface Wave." *Measurement: Journal of the International Measurement Confederation* 193 (May 2021): 110944. https://doi.org/10.1016/j.measurement.2022.110944.

Chapter 13

In-Space Wire Robotics for Large-Scale Extra-Terrestrial Manufacturing (LSETM)

Yash Gopal Mittal, Vedant Agarwal, Pushkar Kamble, Gopal Gote, Yogesh Patil, Avinash Kumar Mehta, Dixita Arvind Yadav, Paras Mandal, and K. P. Karunakaran

13.1 INTRODUCTION

As emphasised by Ehrenfreund et al., space exploration fosters scientific discovery, technological innovation, and international collaboration, paving the way for humanity's current and future endeavours beyond Earth [1]. Malshe et al. highlighted the necessity of *factories-in-space* (FiS) as crucial components for servicing, assembly, and manufacturing of specialised components tailor-made for space applications [2]. This self-reliance is essential for enabling long-duration space missions, facilitating extraterrestrial habitation, and sustenance in space.

Additive manufacturing (AM) provides geometric and material (and its form) flexibility, making it an ideal choice for the space industry where low-volume and highly customised production is needed [3]. AM is a promising option for revolutionising space exploration, as it enables the utilisation of local resources, i.e., space regolith [4], and on-site fabrication. This concept of using locally available materials is referred to as *in-situ resource utilisation* (ISRU).

In this research, a laboratory-scale wire-robot prototype is proposed, with the potential to be appropriately scaled up for large-scale space prototyping and the formation of human habitats.

13.2 BACKGROUND AND LITERATURE REVIEW

Conventional manufacturing processes such as *electron beam welding* (EBW) [5] and *electron beam machining* (EBM) [6] can give better results when operated in space. Welding in space is very effective due to the low-vacuum conditions. It is also essential for space construction, maintenance, and repair [7]. Processes such as powder metallurgy, vapour droplet deposition, and AM can also be employed for space applications [8].

AM is a revolutionary method of combining materials to form models, prototypes, patterns, tools, and production parts [1]. ASTM F42 has categorised AM into seven standard techniques, as shown in **Figure 13.1**.

Sacco and Moon emphasised that AM techniques have begun to be employed in space, primarily due to substantial mass (& cost) savings of up to 90% [3]. Space agencies like the *National Aeronautics and Space Administration* (NASA) and *European Space Agency* (ESA) are working on integrating AM for space-related applications such as printing of propulsion systems [9] and ceramics [10], to using *laser engineered net shaping* (LENS) [11] with space regolith [3, 12, 13], to even potentially printing an entire orbital spacecraft [14].

Novel feedstock materials have to be developed for space applications. NASA proposed ice as a candidate material for creating extraterrestrial human-habitable structures [15, 16]. Ice structures can be formed directly via jetting-based AM techniques [17, 18] or indirectly used as patterned materials for *investment casting* (IC) [19, 20].

 DOI: 10.1201/9781003590774-13

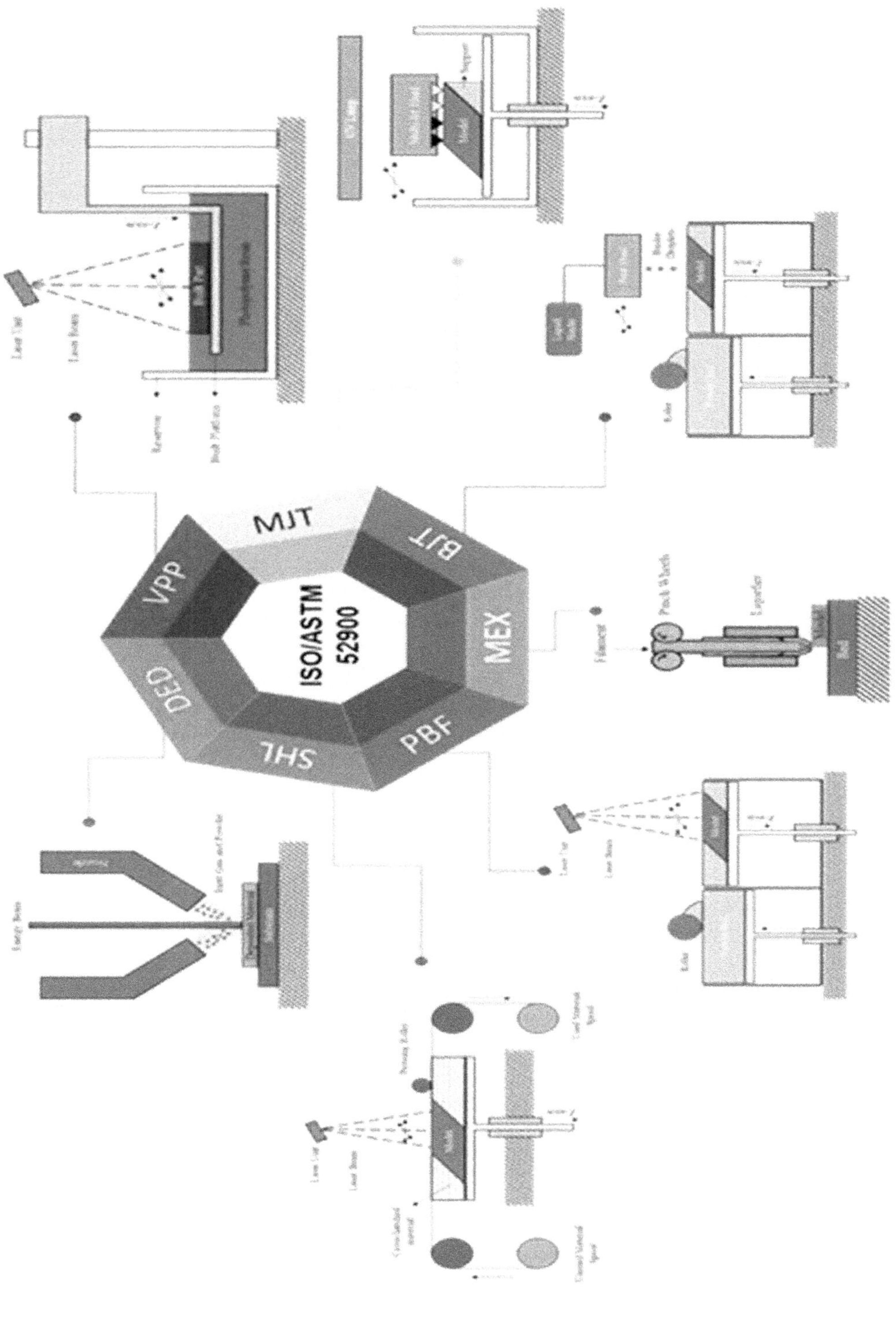

Figure 13.1 ASTM classification of AM techniques. Figure by the author.

13.2.1 Extrusion-Based Additive Manufacturing (EAM) for Space Applications

Extrusion-based AM (EAM) is one of the most extensively used AM techniques and is categorised into three major types: pinch-wheel extrusion, piston/plunger extrusion, and screw extrusion [21]. Pinch-wheel extrusion forms the basis of the *fused filament fabrication* (FFF) technique (popularised by Stratasys as fused deposition modelling (FDM)). FDM is the earliest and the most extensively tested AM technique in microgravity conditions [22]. The idea of using AM is to enable astronauts to print instruments on demand rather than bringing them pre-made from Earth [23]. Low-gravity conditions can also facilitate printing without support structures. FDM is also a popular choice for unmanned space missions [3]. Piston extrusion uses a reciprocating arrangement to extrude material. It is often employed in small-batch productions such as bioprinting [24]. *Screw-extrusion-based AM* (SEAM) uses a rotating screw to continuously deposit material and offers a high throughput (~10x of FDM [25]). Mitchell et al. reviewed the potential of *4D printing* (4DP) in space applications [26], emphasising the use of this shape-shifting technology in the highly adaptive and changing environments of outer space. 4DP can be used to efficiently realise complex multi-material geometries with self-assembling and self-repair properties.

13.2.2 Wire Robots as Motion Systems for In-Space Manufacturing

Wire robots, also known as *cable-driven parallel robots* (CDPRs), are a class of parallel manipulator-based motion systems that utilise flexible kinematic elements such as cables for actuation. Unlike serial manipulators, which use independent drives for each axis, the flexible kinematic linkages in a parallel motion system are bound together to move in synchronous motion. **Table 13.1** summarises the key differences between serial and parallel kinematics.

Various wire robots have been designed for a range of engineering and real-world applications, including material handling [27], large-scale painting [28–30], sports events [31], rehabilitation facilities [32], and rescue operations [33]. All these applications benefit from the inherent advantages of wire robots, such as flexible installation [34], unconstrained boundaries [35], a large workspace, low inertia, and a high payload-to-weight ratio [36]. Wire robots offer a range of motion capabilities, from a planar system [37, 38] to a 6-DoF 8-wire arrangement [39] with a specialised end-effector [40]. Although these motion systems are advantageous and established, they are limited by their workspaces, which are defined by their arm length [41]. However, wire robots can overcome these limitations by operating at much larger scales, being easily transportable and scalable, and not requiring additional scaffolding [42]. Khoshnevis et al. of USC (*University of Southern California*) proposed a layered fabrication technique, called *contour crafting* (CC), for the automated construction of civil infrastructures. CC is a hybrid method that combines the extrusion of cementitious material

Table 13.1 Serial vs. Parallel Kinematics

Aspect	*Serial Motion Systems*	*Parallel Motion Systems*
Definition	Parts move sequentially along a single path	Components move simultaneously along many paths
Examples	CNC and Articulated Robots	Cable Robots and Delta Robots
Control Complexity	Usually easier because every part moves one after the other	More complicated, as it involves synchronising several axes
Accuracy	Generally poorer because of the sequential nature	Higher Accuracy due to synchronous movement

with surface-forming using transverse trowels to create integrated structures from lunar or Martian regolith [43]. On the other hand, a cable-driven system would be simpler to transport and set up [44]. A research group from Ohio University further explored CC and proposed a translation-only cable-suspended wire robot for contour crafting [45, 46].

13.3 LABORATORY-LEVEL CASE STUDY

13.3.1 6-Axis Wire Robot with Screw Extruder

This chapter describes a novel (laboratory-scale) wire robot, equipped with a custom-built screw extruder, which can be scaled up for possible in-space extraterrestrial AM applications.

13.3.1.1 Design Iterations

The setup uses a stationary screw extruder, continuously depositing material on a synchronised moving/build platform controlled in 5 degrees via six cable wires. This arrangement ensures a multi-axis motion configuration required for any advanced AM setup. It facilitates supportless fabrication via tilting the built platform and conformal non-planar deposition. Wire robots follow a direct relation (Eq. (13.1)) between the minimum number of cable wires (m) required for a given DoF (n) based on the cable redundancy and inability to bear compressive loads.

$$m = n + 1 \tag{13.1}$$

Hence, for a six-DoF system, at least seven cables are required. To achieve symmetry in the system, an initial redundant design with eight motors is chosen, as shown in **Figure 13.2 (I)**. In the second iteration (**Figure 13.2 (II)**), the number of motors (wires) is reduced to six to save

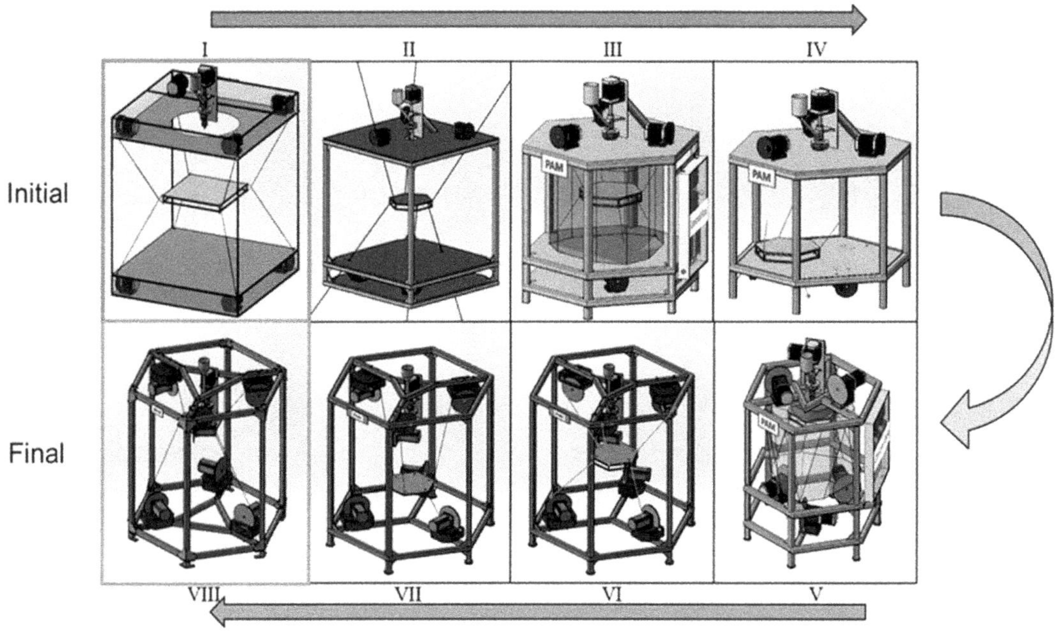

Figure 13.2 Design iterations for the wire robotics motion setup. Figure by the author.

on the overall setup cost while maintaining a multi-axis motion but with five DoF (3 translations and 2 rotations). To further employ symmetry in the design, the cuboidal machine frame is changed to a hexagonal one (**Figure 13.2 (III)**). This design uses three hexagonal plates to mount the six drive motors, each in an upright fashion. In the next design (**Figure 13.2 (IV)**), a hexagonal plate is reduced by flipping the orientation of the bottom motor. These plates are removed by mounting the motors directly to the frame, as shown in **Figure 13.2 (V)**.

Still, in this arrangement, the wire has three contact points (pulley, anchor point, and moving platform) and 1 (constant) rubbing surface, i.e., the anchor point. A novel swivel pulley design is incorporated, in which the entire drive assembly is mounted on a bracket with passive swivelling motion (**Figure 13.2 (VI)**). This arrangement reduced the number of wire contact points to two and removed any rubbing surface. In the next iteration (**Figure 13.2 (VII)**), the thickness of the moving platform is reduced to achieve a higher tilt (explained later). The overall system is made entirely symmetric by keeping the top and the bottom hexagonal frames identical. The final design setup is shown in **Figure 13.2 (VIII)**.

13.3.1.2 Components and Parts

Frame: It is the central part of the motion system and is made from hollow stainless steel (AISI 104) extrusions with a 40 mm × 40 mm cross-section and a 2 mm wall thickness, providing the necessary stiffness and rigidity to the setup (**Figure 13.3**).

Swivelling Assembly: A swivelling mechanism is introduced into the design to reduce the number of wire contact points by eliminating rubbing surfaces. The swivelling arrangement has two perpendicular rotation axes: one for the pulley rotation (shown in **Figure 13.4**), which is coaxial with the motor shaft; the other is the swivelling axis (shown in **Figure 13.4**) that adjusts the pulley orientation according to the moving platform's position.

Motor Drive: Wire robotics requires exceptionally high torques to balance the tensions generated by the weight of the moving platform, the added weight of the depositing material

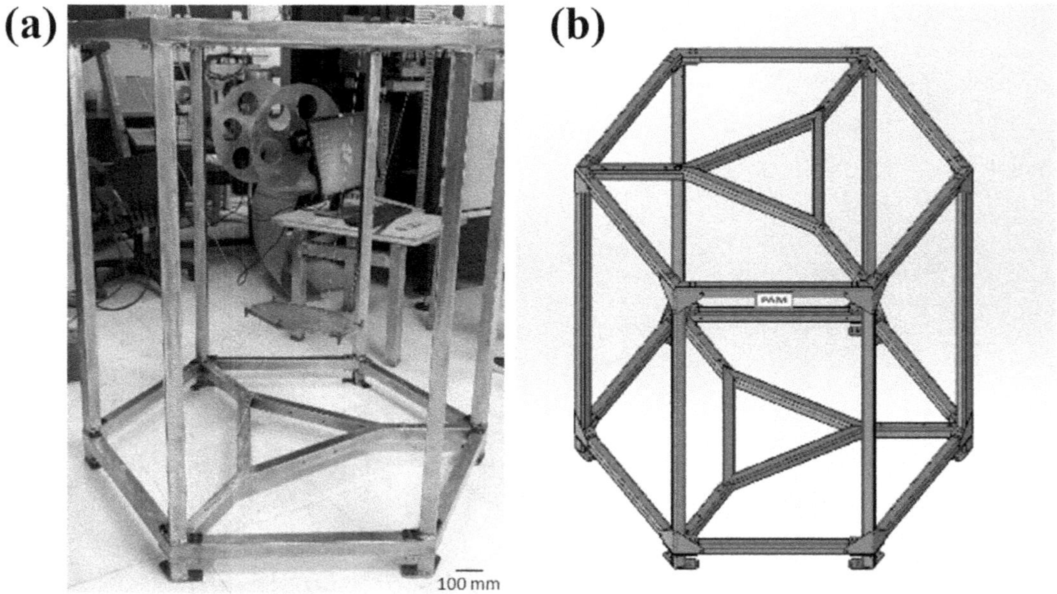

Figure 13.3 Frame Design: (a) Conceptual CAD, (b) Actual Setup. Figure by the author.

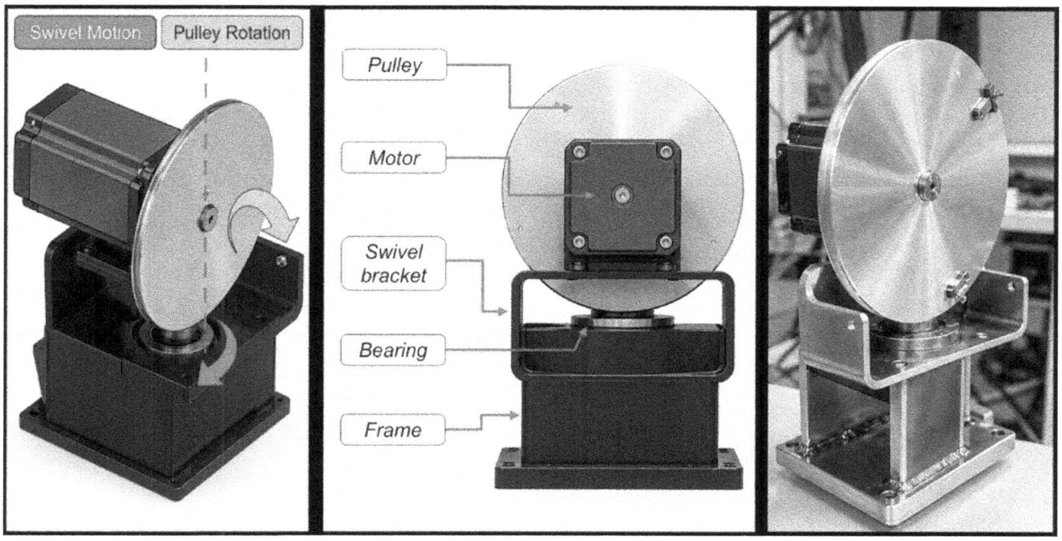

Figure 13.4 Swivelling assembly design and actual setup (far right). Figure by the author.

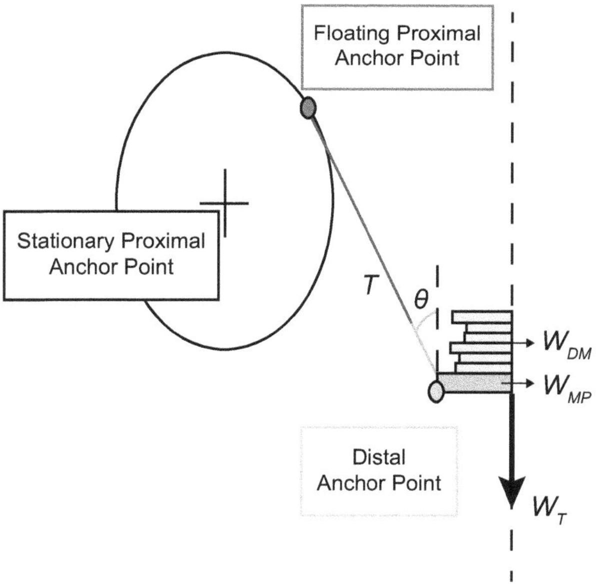

Figure 13.5 Tension balancing in cable wires. Figure by the author.

(in the case of AM), and the torque at high overhangs at the wire anchor points [36]. The working volume/*wrench feasible workspace* (WFW) is heavily affected by the torque capacity of the drive motor [47].

For the given design (**Figure 13.2 (VIII)**), the torque requirement is calculated by determining the maximum wire tension. For a safe-side analysis, only the top three cable wires are considered for torque analysis, as they support most of the load in the static condition [7]. The fundamental force balance equation, as shown in **Figure 13.5**, is next (Eq. (13.2)).

$$n.T.coscos\theta = W_T \tag{13.2}$$

In the aforementioned relation, 'n' is the number of wires involved, i.e., 3, 'T' is the wire tension (in Newtons), 'θ' is the inclination of the cable wires from the vertical, and 'W_T' is the total load that is balanced by the cable wires, including the weight of the moving platform 'W_{MP}' and the deposited material 'W_{DM}'. The maximum tension is experienced at the maximum inclination (θ), i.e., 27°, for the given setup. The peak load (W_T) is set at 300 N, corresponding to a maximum permissible tension of 112 N. The pulley's effective wire-wrapping radius is 100 mm (explained in the next section). Using these two properties (wire tension and pulley radius), the torque capacity of the motor drive is calculated to be 11.2 N-m (112 kg-cm). The NEMA34 geared stepper motor with a rated torque of 122 kg-cm is selected for the given torque rating.

Pulley: It is a crucial element in the design of a CDPR motion system. The proposed AM system accommodates parts up to 300 mm. Therefore, a safe side wire length of 600 mm is chosen to accommodate such high geometries. It corresponds to a radius of ~95.5 mm (600/2π). As a precautionary measure, the effective pulley radius is 100 mm.

The initial pulley design consisted of a groove of height just enough for the cable wire (Ø1.2 mm), as shown in **Figure 13.6 (a and b)**. In the next iteration, the groove height is made almost eight times the wire diameter to arrest that (**Figure 13.6 (c and d)**). Weight optimisation via slot cutting is done. The ridge thickness (t) is determined per the bending criteria given in Eq. (13.3).

$$\sigma_b \leq \frac{\frac{\frac{T}{R}(R-r)}{n} \times \left(\frac{t}{2}\right)}{\frac{bt^3}{12}} \tag{13.3},$$

σ_b is the bending strength of the material (SS 304), i.e., ~621 MPa, τ is the motor torque capacity (122 kg-cm), R (77.5 mm), and r (20 mm) are the radii, from the pulley centre, at which the ridge is defined, n is the number of ridges (6 here), and b is the pulley width (8 mm). Eq. (13.3) is solved, and a safe design value of t = 5 mm is selected. The various pulley dimensions and the final fabricated pulley design are shown in **Figure 13.6** (**e**).

Cable Wires: These are the primary kinematic linkages that characterise the motion system. High-strength, stainless steel (SS) multi-strand cable wires are used for the current setup. Each wire is made using 10 stainless-steel filaments, wound together for increased strength and durability. The total width of the cable wire is 1.2 mm. Six such wires need to be mounted on each motor assembly. The pulley groove size and length are designed as per the wire size. In the current arrangement, these are designed such that a single pulley rotation is enough to cover the entire wire length.

Moving Platform: The moving element and the wire robot's end-effector. The platform uses standard 10 mm stainless-steel extrusions with a 0.5 mm thickness (**Figure 13.7**). These extrusions provide sufficient strength while maintaining a lightweight structure. The cable wires are attached according to a cable crossing layout [6]. The arrangement provides precise positioning of the cable wires as a milli-metric length adjustment, as described in **Figure 13.8**. With this arrangement, the wire lengths, which are in the order of 700 mm, can be precisely calibrated up to 1 mm resolution.

13.3.1.3 Screw-Based EAM System

Conventional extruder screws have three discrete zones with different pitch and diameter [48], which help develop the necessary pressure, up to 10,000 psi [49]. Applications like AM don't require high pressure and often require repeated start-and-stop. This can place significant stress on the drive motor [50]. The current design significantly reduces the load on the

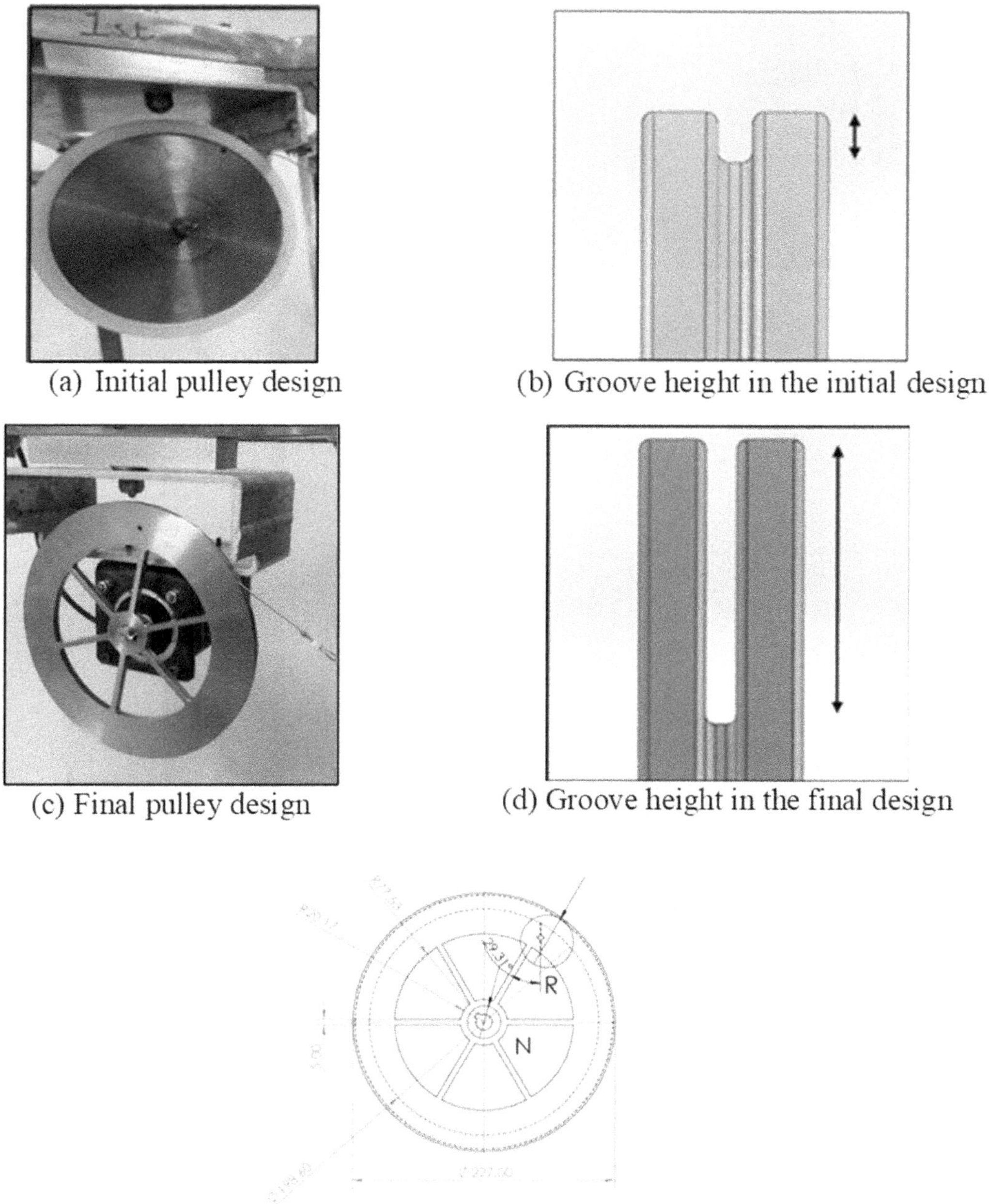

(a) Initial pulley design (b) Groove height in the initial design

(c) Final pulley design (d) Groove height in the final design

(e) Dimensions of the final pulley design

Figure 13.6 Pulley design iterations. Figure by the author.

screw while optimising the system's flow and space requirements. This arrangement provides phenomenal deposition rates of up to 200 cc/hr, compared with conventional FDM systems, which offer 10 cc/hr [51]. **Figure 13.9** describes the presented extruder screw setup and compares it with the conventional injection moulding screw. The geometric design parameters of the fabricated extruder screw are given in **Table 13.2.**

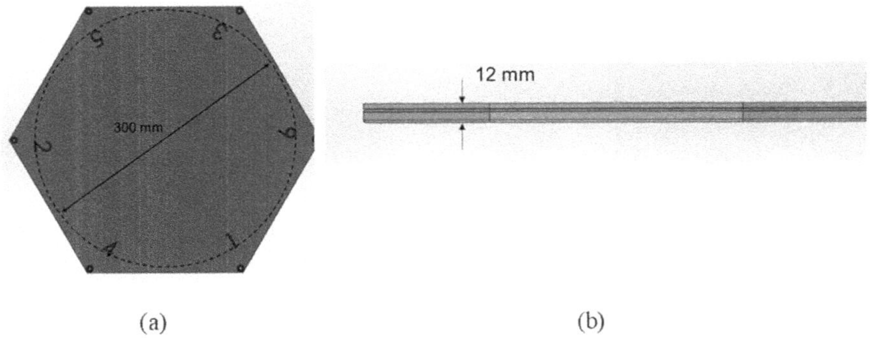

Figure 13.7 Design of the moving platform: (a) Top view, (b) Front view. Figure by the author.

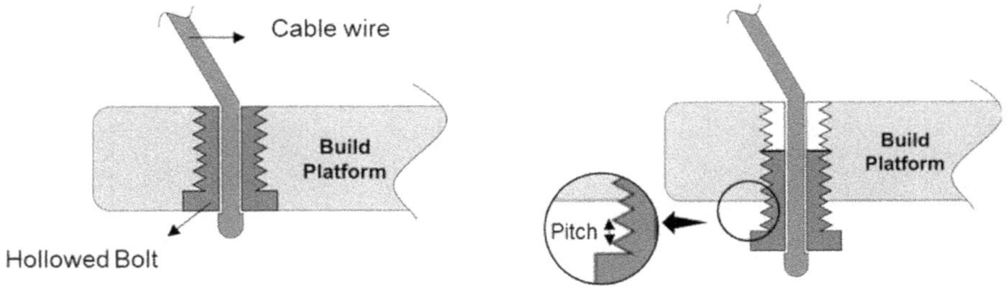

Figure 13.8 Milli-metric wire length adjustment system. Figure by the author.

13.3.2 Inverse Kinematics and Feasible Workspace Analysis

In the present setup, the wire robot comprises a moving platform connected to a stationary frame by m cables (6 here). The posture/pose of the moving/build platform, b, is in terms of the Cartesian coordinates of the platform centre-of-mass (COM), $\overrightarrow{r^F}$, and the platform's orientation, as defined by the roll (∅), pitch (θ) and yaw (φ) angles. The posture is described with respect to the absolute reference frame, FF, with origin OF, at the nozzle tip and reference axes x_F, y_F, and z_F [36]. The Cartesian coordinate vectors a_i^F, $i = 1, 2, 3, \ldots, m$, describe the position of the stationary proximal anchor points, A_i^F, $i = 1, 2, 3, \ldots, m$, relative to F_F. This anchor point does not lie on the cable wire but instead at the centre of the pulley and remains inconsequential to the swivelling movements, as the wrapping plane of the pulley is set to coincide with the swivelling plane. The vector set, a_i^s, $i = 1, 2, 3, \ldots, m$, contains the floating proximal anchor points relative to the swivel frame, F_s, which is attached at the centre of each pulley, with an origin, O_s, and axes, x_s, y_s, and z_s. The Cartesian coordinate vectors b_i^b, $i = 1, 2, 3, \ldots, m$ represent the positions of the platform-wire connection points, B_i^b, $i = 1, 2, 3, \ldots, m$, about the local frame, F_b, attached to the moving platform, with an origin, O_b, and axes, x_b, y_b, and z_b. The overall inverse kinematic chain is shown in **Figure 13.10**. The geometric model based on the chain closure equations of the cables w.r.t. the F_F is used to calculate the active lengths, l_i^F, $i = 1, 2, 3, \ldots, m$, as described in Eq. (13.4). R_x^y denotes the orientation matrix between the x (s, b) and y (F) frames. While R_b^F can be easily calculated using Eq. (13.5).

$$\overrightarrow{l_i^F} = \overrightarrow{a_i^F} - \overrightarrow{r^F} - \overrightarrow{b_i^b} \times R_b^F + \overrightarrow{a_i^s} \times R_s^F \tag{13.4}$$

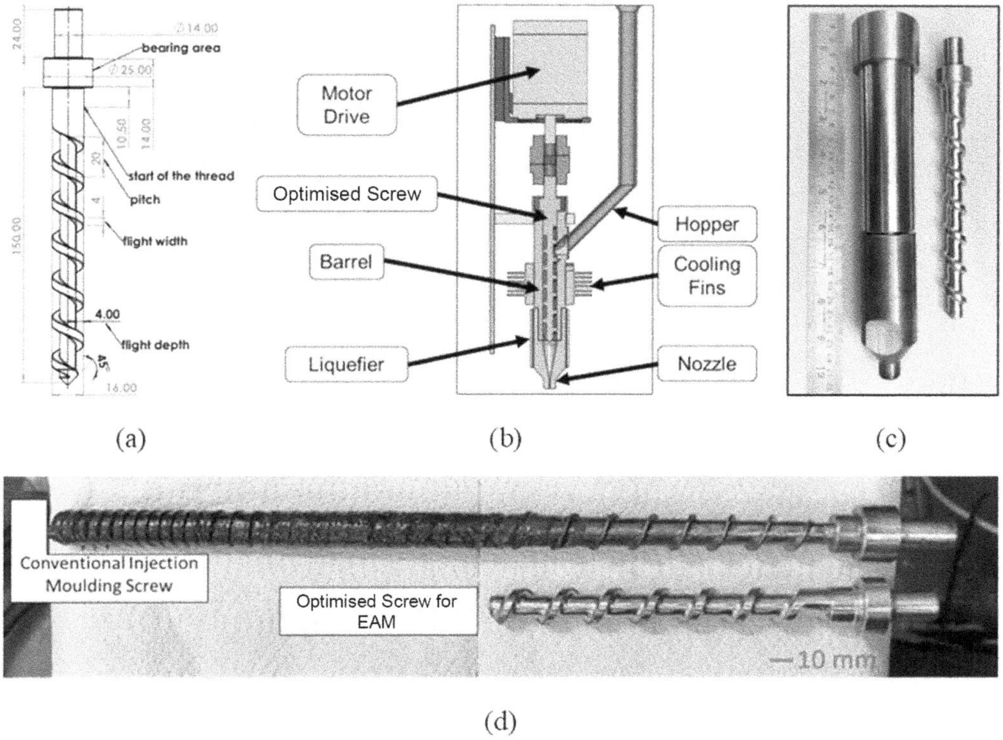

Figure 13.9 EAM setup: (a) Screw geometry, (b) CAD model, (c) Fabricated parts, and (d) comparison of the injection moulding screw. Figure by the author.

Table 13.2 Geometric Design Parameters of the Extruder Screw Setup

Geometric Parameter	*Values*
Screw Length	150 mm
Screw Diameter	16 mm
Helix Angle	20.84°
Channel Depth	4 mm
Metering Zone	70 mm
Nozzle Length	10 mm
Nozzle Diameter	1 mm

$$R_b^F = R_z(\varnothing)\times R_y(\theta)\times R_x(\varphi)$$
$$= \begin{bmatrix} c\varnothing.c\theta & c\varnothing.s\theta.s\varphi - s\varnothing.c\varphi & c\varnothing.s\theta.s\varphi + s\varnothing.s\varphi \\ s\varnothing.c\theta & s\varnothing.s\theta.s\varphi + c\varnothing.c\varphi & s\varnothing.s\theta.c\varphi - c\varnothing.s\varphi \\ -s\theta & s\varphi.c\theta & c\varphi.c\theta \end{bmatrix} \tag{13.5}$$

To avoid R_s^F, only the magnitude of the active length ($|l_i^F|$) is calculated, as described in Eq. (13.6). L_i^F is the vector in frame F_F, connecting the stationary proximal anchor point (A_i^F)

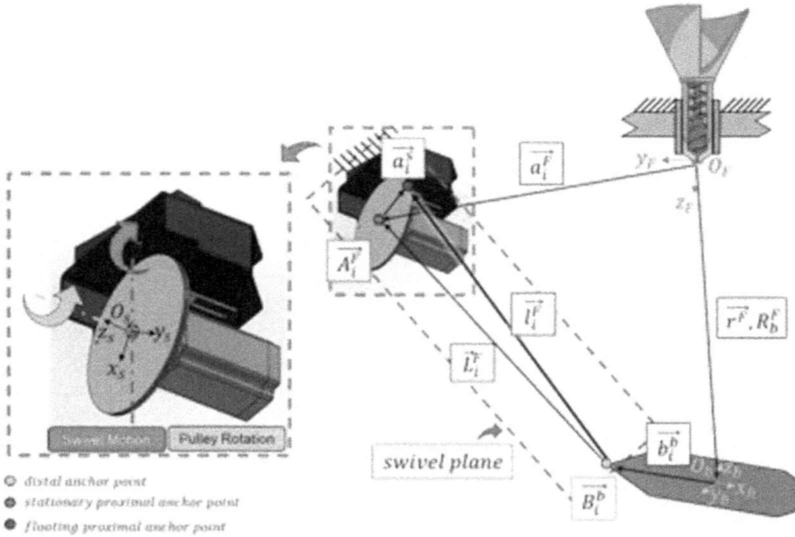

Figure 13.10 Inverse kinematics with the swivelling arrangement. Figure by the author.

from the distal anchor point in F_F frame (B_i^F), as given in Eq. (13.7). Also, the magnitude of a_i^s, i.e.,$|a_i^s|$ is nothing but the pulley radius, R_p.

$$|l_i^F| = \sqrt{\left(\left|L_i^F\right|\right)^2 - \left(\left|a_i^s\right|\right)^2} \tag{13.6}$$

$$\overrightarrow{L_i^F} = \overrightarrow{a_i^F} - \overrightarrow{r^F} - \overrightarrow{b_i^b} \times R_b^F \tag{13.7}$$

After determining the wire length for a given pose, the change in the active length for a given cable wire, $|\Delta l_i^F|$, is calculated (Eq. (13.8)), when the moving platform is shifting from one given pose to another ($\overrightarrow{r_1^F}$ to $\overrightarrow{r_2^F}$). This change in length has to be accommodated by the pulley rotations, $N|_{12}$ (Eq. (13.9)).

$$|\Delta l_i^F|_{12} = \left|l_i^F\right| |_{r_2^F} - \left|l_i^F\right| |_{r_1^F} \tag{13.8}$$

$$N|_{12} = \frac{|\Delta l_i^F|_{12}}{2\pi R_p} \tag{13.9}$$

13.3.2.1 Feasible Workspace

The first constraint concerns the build platform's location relative to the machine frame. Because the build platform is contained within the polygon formed by joining all the proximal anchor points of the setup, the work volume is constrained to this volume. Second, the fundamental property of cable wires is examined: they cannot sustain a compressive load.

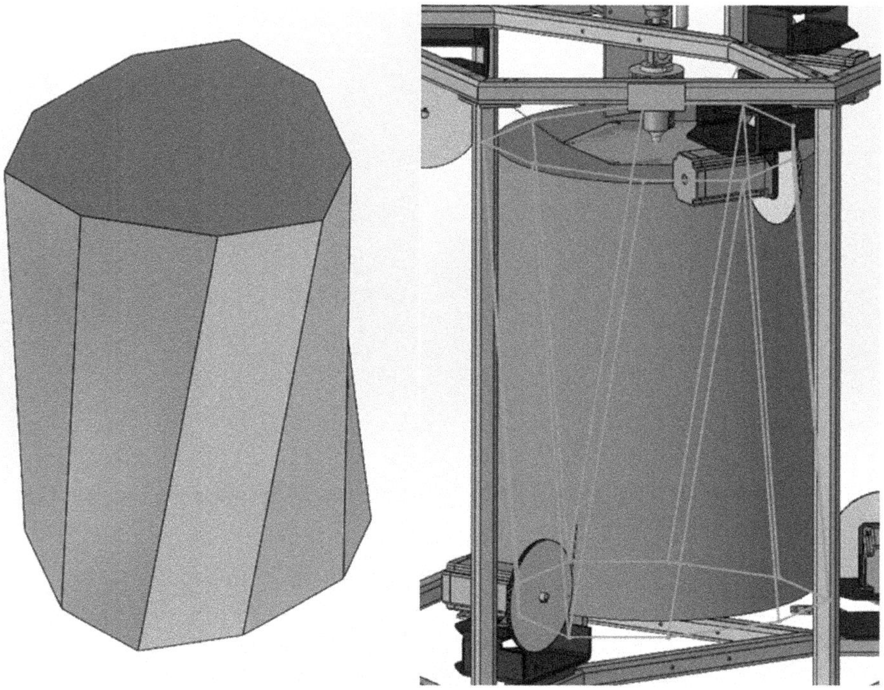

Figure 13.11 The actual work volume (left) and the smooth approximation as a cylinder (right) inside the setup. Figure by the author.

Hence, under no circumstances can the cable wire inclination (refer to **Figure 13.5**) become negative. The presence of the screw extruder and the total available wire length (600 mm) are also considered, which set the upper and lower limits of the moving platform's position. Based on these constraints, a complex feasible workspace volume is obtained, as shown in **Figure 13.11** (**a**). For simple depictions, the feasible workspace is approximated as a cylinder (**Figure 13.11** (**b**)) with a diameter of ~565 mm, a length of ~580 mm, and a total work volume of ~145,000 cc. This workspace is suitable for large-scale prototyping.

13.3.2.2 Angular Limits

It is assumed that the maximum platform tilt, in either direction, is reached when the set of diagonally opposite cable wires (1–5, 2–6, 3–4) becomes collinear (**Figure 13.12** (**a**)). After this point, the motor drive cannot provide sufficient tension on the cable wires to further tilt the platform, as it reaches a singularity. Because of the hexagonal nature of the machine frame and build platform, the angular limits for platform tilting differ in the x and y directions (**Figure 13.12** (**b**) & (**c**)). The expressions for the platform tilts are given in Eqs. (13.10) and (13.11).

$$\varphi = \left(\frac{H}{A}\right) - \left(\frac{b}{2a}\right) \tag{13.10}$$

$$\theta = \left(\frac{H}{\sqrt{3}A}\right) - \left(\frac{b}{\sqrt{3}a}\right) \tag{13.11}$$

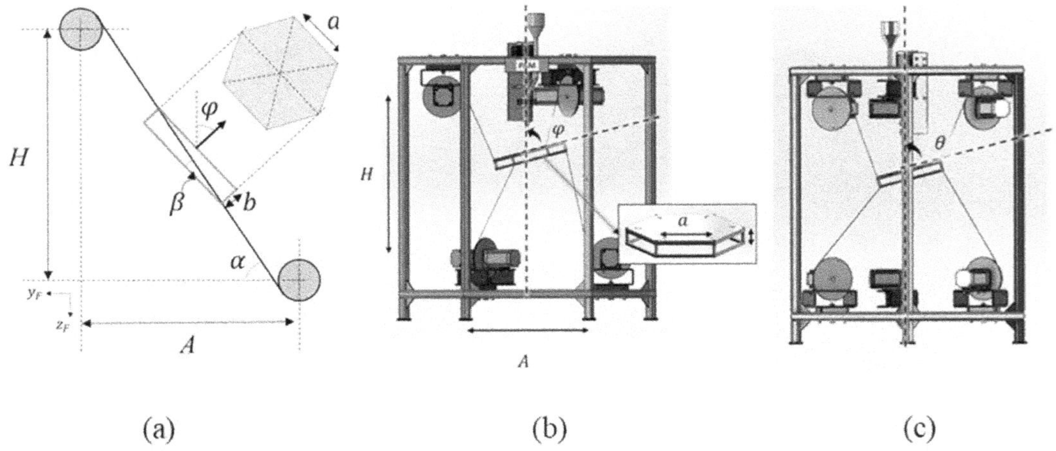

Figure 13.12 (a) Tilt limit reaching by the collinearity of wire set, (b) Tilt along the x_F-axis, (c) Tilt along the y_F-axis. Figure by the author.

'*H*' is a measure of the height of the setup. It is the vertical centre-to-centre distance between the upper and lower pulleys, set at 787.5 mm in this case. '*A*' is the measure of the spread/width of the setup. It is the side length of the hexagonal prism formed by joining all the stationary proximal anchor points. In this case, its value is 403.7 mm. '*a*' and '*b*' are the side length and thickness of the platform and are kept at 173.4 mm and 12 mm, respectively. For these given values, the angular limits of the platform tilts came out to be $\varphi = 60.87^\circ$ and $\theta = 46.11^\circ$. Both of these angles are above the angular threshold required for support generation in conventional EAM, i.e., 45°. Hence, the current setup can be used as a multi-axis motion system.

13.3.2.3 Tension Measurement

The wire tension also affects the overall wrench of the system [36] and the feasible work-volume limits [47]. Conventional techniques are not feasible enough for the reasons explained in **Table 13.3**. A unique in-situ spring-based tension measurement approach is explored.

13.3.2.4 Spring Design

The design procedure follows [52]. As the springs are in a serial arrangement with the cable wire, the maximum spring force is the same as the maximum tension experienced by the cable wire [53]. A safe-side value of 120 N is selected for the maximum spring force (P). The workspace analysis and inverse kinematics show that the minimum possible active length of any cable wire is ~57.2 mm. The spring deflection (δ) is set at 50 mm to have a conservative design. Spring Steel Grade-2 is a suitable spring material with an ultimate tensile strength (S_{ut}) of 1,873 MPa [21]. The permissible shear stress (τ) is calculated as 50% of the ultimate tensile strength, i.e., $\tau = 0.5S_{ut}$. The spring index (C) of 8 is chosen for the spring design. The spring index is defined as the coil diameter (D) divided by the wire diameter (d). Subsequently, the Wahl factor (K) is calculated using Eq. 13.12 [54].

$$K = \frac{4C-1}{4C-4} + \frac{0.615}{C} \tag{13.12}$$

Table 13.3 Various Approaches to Tension Measurement and Their Limitations When Measuring Tension Near the Platform's Wire Mounting Point

#No.	*Method & Description*	*Schematics*	*Limitation*
1	*Direct loading using known weights* A known weight is attached to the wire at the equilibrium position. The weight force is the wire tension.	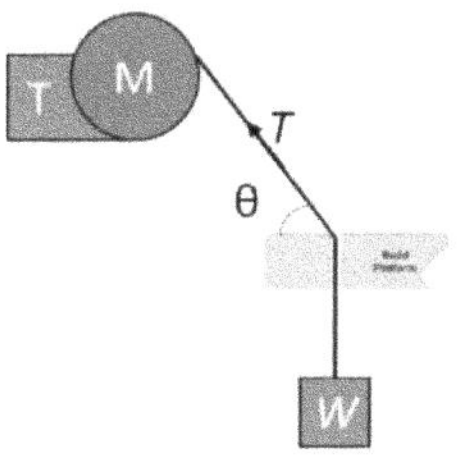	Not feasible for in-situ measurement.
2	*Measurement using force sensors* Electronic sensors, such as load cells and force-sensitive resistors, can be attached at the anchor point to read the tension values directly.	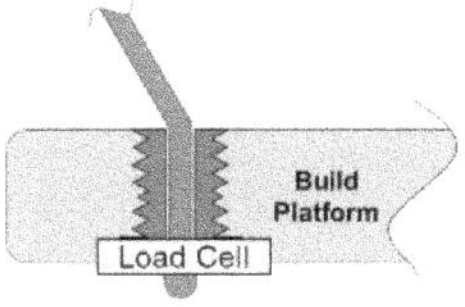	Extra electronic components & controls.
3	*Correlating tension with vibrations* The wire tension and the associated vibration frequency correlate for a given length and wire mass-per-unit length.	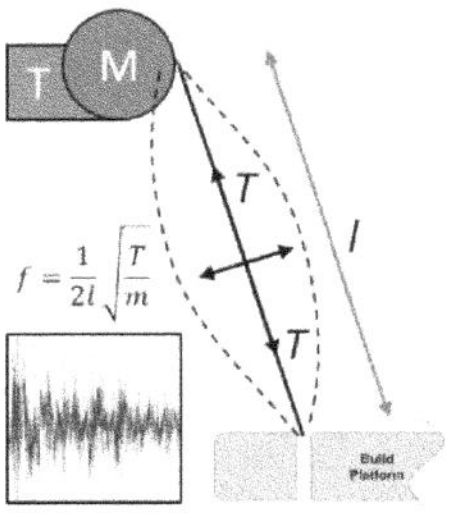	It needs to be more accurate & more accessible to implement in situ.
4	*Using a handheld Tension-Meter* Commercially available handheld spoke tension measuring devices based on 3-point bending.	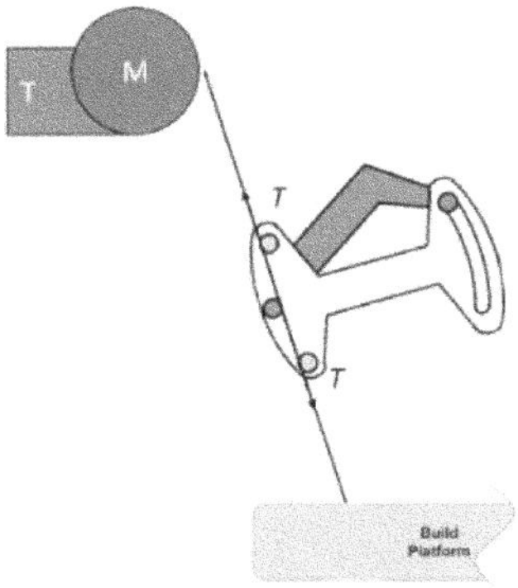	Such devices are not suitable for cable wires.

The wire diameter is calculated by equating the maximum permissible shear stress with the induced shear stress in the wire (Eq. 13.13).

$$\tau = K\left(\frac{8PC}{\pi d^2}\right) \tag{13.13}$$

The mean coil diameter is then determined using the spring index (D=$C.d$). The number of active coils (N_a) is then calculated by equating the overall spring deflection with the torsional twist in the wires (Eq. 13.14).

$$\delta = \frac{8PD^3 N_a}{Gd^4} \tag{13.14}$$

Table 13.4 Various Spring Dimensions and Properties

Dimension/Property	*Value*
Wire Diameter, d	2 mm
Coil Diameter, D	16 mm
Number of Coils, N_a	17
Spring Stiffness, k_s	2.4 kN/m

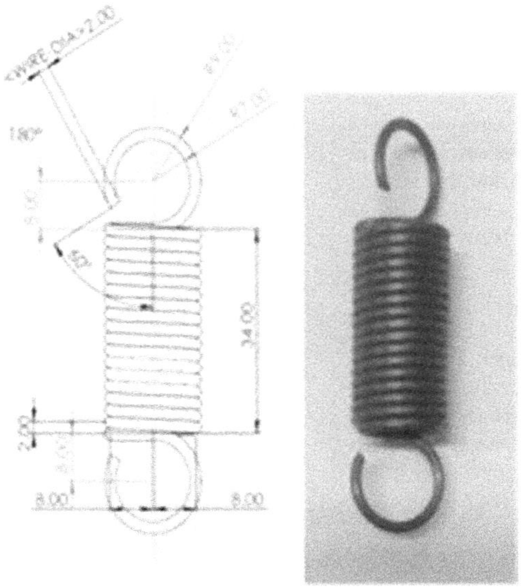

Figure 13.13 Spring design and fabricated geometry. Figure by the author.

G is the modulus of rigidity of the spring wires and is taken as 83.56 GPa for the spring material. The following relation can also be used to calculate the spring stiffness (k_s) (Eq. 13.15).

$$k_s = \frac{P}{\delta} = \frac{Gd^4}{8D^3N_a} \tag{13.15}$$

For the given setup, the spring dimensions and stiffness are mentioned in **Table 13.4**. The spring ends are squared and given a full hook for easy mounting, as shown in **Figure 13.13**.

Physical testing of the designed spring under uniaxial tension is done on the Tinius Olsen H25KS UTM machine (**Figure 13.14**). The experimental results provided a stiffness value of 2.22 kN/m, within a 10% error bound. The *F-d* characteristics are also shown in **Figure 13.14**.

13.3.2.5 Vibrational Analysis

The wire tension (T_i) in any given cable (i=1, 2, 3, . . ., m) at any given position inside the workspace can be directly calculated, in situ, by measuring the final length of the spring (l_{si}). The associated wire tension is then calculated using Eq. (13.16).

$$T_i = k\left(l_{si} - N_a d\right) \tag{13.16}$$

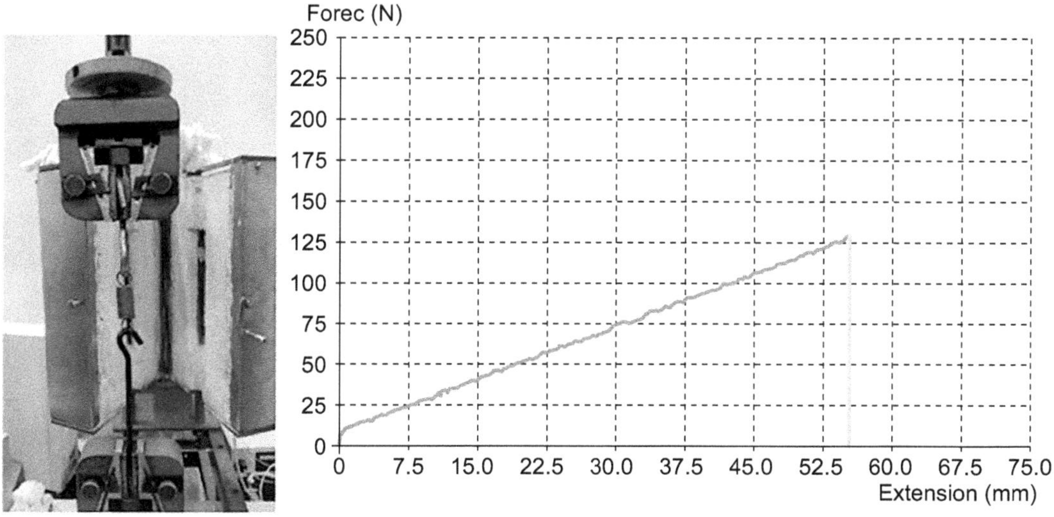

Figure 13.14 Experimental trial under uniaxial tension (left) and the fabricated springs' Force-Displacement curve (right). Figure by the author.

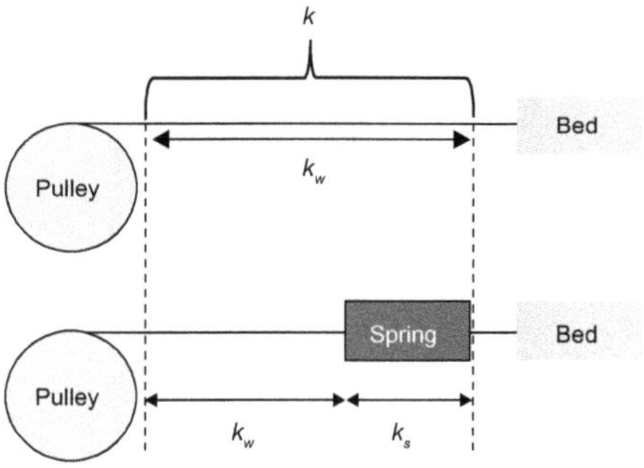

Figure 13.15 Spring arrangement and overall stiffness of the spring-wire system. Figure by the author.

The presented setup has a considerable working volume for large-scale prototyping. This enormous workspace offers a wide range of active wire lengths, varying from 50 mm to 600 mm. The corresponding change in the cable wire stiffness (k_w), calculated using Eq. (13.17), can range from 396 to 4,750 kN/m.

$$k_w = \frac{E_w . A_c}{l_w} \tag{13.17}$$

E_w is Young's modulus of the wire (210 GPa), A_c is the cross-sectional area of the 1.2 mm diameter cable wire, and l_w is the active length of the cable wire. After adding the spring in a series arrangement with the cable wire, the overall stiffness of the spring-wire system (k) can be calculated using Eq. (13.18), as depicted in **Figure 13.15**.

$$k = \left(k_w^{-1} + k_s^{-1}\right)^{-1} = \left(\left(\frac{E_w . A_c}{l_w - l_s}\right)^{-1} + k_s^{-1}\right)^{-1} \tag{13.18}$$

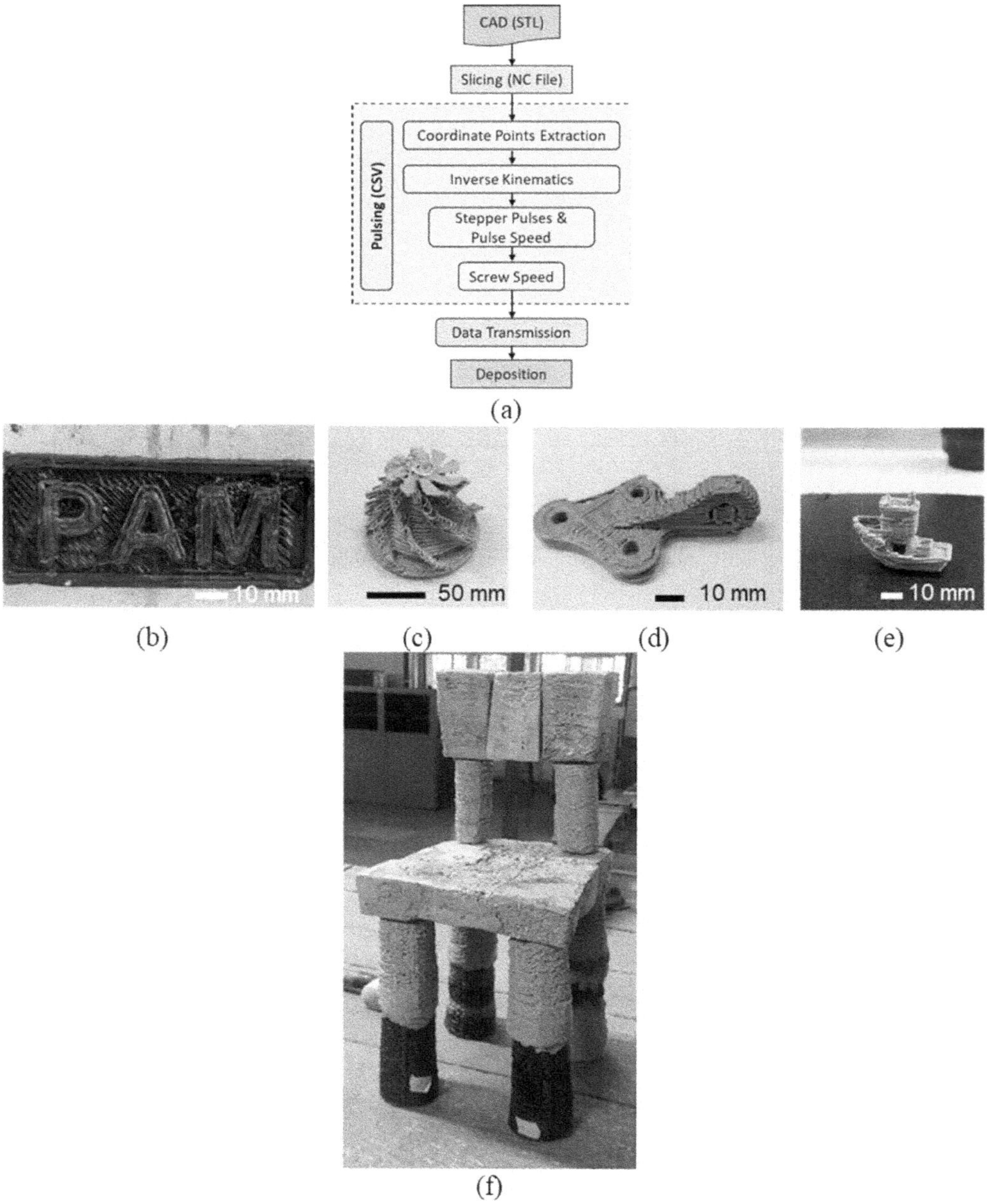

Figure 13.16 (a) Process flow chart and the various fabricated geometries using the presented setup: (b) Name tag, (c) Impeller, (d) Bracket, (e) Benchy, (f) Mini chair model (height ~400 mm). Figure by the author.

It is seen that after the addition of the spring, the overall stiffness variation is brought down to 2.38–2.4 kN/m. Zhang et al. also analysed a similar design configuration in a CDPR. The authors reported that the addition of passive tension springs enabled the CDPR to achieve a higher acceleration while avoiding redundancy [55].

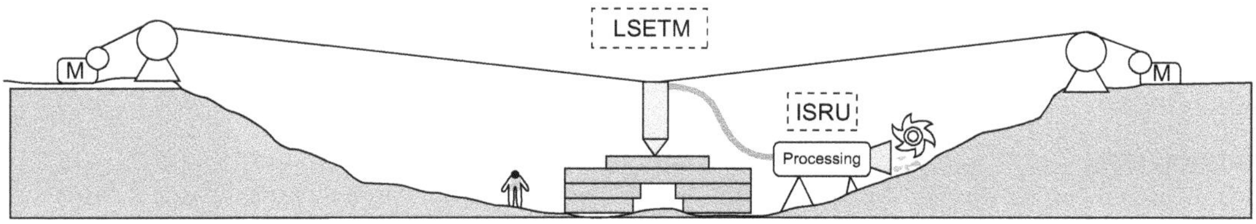

Figure 13.17 A schematic vision for LSETM. Figure by the author.

13.3.2.6 Fabricated Geometries

The overall process flowchart, along with the fabricated geometries generated using these heuristics, is shown in **Figure 13.16.** The technology has been proven using thermoplastic but can easily work with lunar or martian regolith as well. The system can achieve layer heights from 0.2 mm to 2 mm.

13.4 SCALING UP FOR EXTRA-TERRESTRIAL MANUFACTURING

The presented arrangement contributes to the development of wire-based parallel robots and also holds the potential to revolutionise additive manufacturing motion systems for LSETM. An existing crater can also serve as the frame for the wire robot, similar to the technique used for the Lunar Crater Radio Telescope (LCRT) [56]. The screw extruder can be controlled, suspended, and actively controlled via motor-driven cable wires. A vision of such an arrangement for LSETM is given in **Figure 13.17.**

This vision of wire robots building large-scale, human-habitable structures on crater surfaces represents a paradigm shift in space exploration. It presents a future in which humanity expands its presence beyond Earth, opening new frontiers and unlocking the vast potential of space towards a truly interplanetary civilisation.

13.5 CONCLUSION AND FUTURE DIRECTIONS

Outer space is a potential area for scientific discovery and technological advancement. Additive manufacturing (AM) is a disruptive technology suited for space-based manufacturing, offering extensive geometrical and material flexibility. The technique of screw extrusion-AM (EAM) is highly versatile and can easily create 3D objects from space regolith. Cable-driven parallel robots (CDPRs) or wire robots demonstrate considerable potential as multi-axis motion systems, offering up to 6 degrees of freedom and efficient work volumes within a compact footprint. Their simple, cost-effective, and scalable design is highly favourable for in-space servicing, assembly, and manufacturing (ISAM) tasks, including the construction of extraterrestrial habitats. This research has presented a laboratory-level case study that can be extrapolated for space environments, facilitating large-scale extra-terrestrial manufacturing (LSETM). The setup integrates a 6-axis wire robot with a tailor-made screw extruder for multi-axis EAM applications, featuring an innovative swivelling mechanism to minimise wear and enhance process longevity and reliability. Special consideration must be given to the pulley design, particularly the groove geometry, to achieve efficient swivelling motion. The inverse kinematics and the feasible workspace of such a motion system are also presented, suggesting that the setup can provide 5-axis deposition beyond 45°. A vision is presented in which a wire robot navigates the crater surfaces with a specialised multi-purpose end-effector to realise expansive human-habitable structures.

FUNDING

The authors have no funding to disclose.

CONFLICT OF INTEREST

There are no conflicts of interest.

REFERENCES

[1] Ehrenfreund, P., McKay, C., Rummel, J. D., Foing, B. H., Neal, C. R., Masson-Zwaan, T., Ansdell, M., Peter, N., Zarnecki, J., Mackwell, S., and Perino, M. A., 2012, "Toward a global space exploration program: A stepping stone approach," *Adv. Space Res.*, 49(1), pp. 2–48. https://doi.org/10.1016/J.ASR.2011.09.014

[2] Malshe, H., Bapat, S., Vickers, J., and Malshe, A. P., 2023, "Factories-in-space for servicing, assembly, & manufacturing," *Manuf. Lett.*, 38, pp. 24–28. https://doi.org/10.1016/J.MFGLET.2023.08.142 Sacco, E., and Moon, S. K., 2019, "Additive manufacturing for space: Status and promises," *Int. J. Adv. Manuf. Technol.*, 105, pp. 4123–4146. https://doi.org/10.1007/S00170-019-03786-Z

[3] Leach, N., 2014, "3D printing in space," *Architect. Des.*, 84(6), pp. 108–113. https://doi.org/10.1002/AD.1840

[4] Benfield, M. P., and Landrum, B. D., 1998, "International joint ventures in space: The international space welding experiment," *J. Technol. Transfer*, 23(3), pp. 25–38. https://doi.org/10.1007/BF02509573

[5] Ma, Y., Zhao, J., Yang, W., Ma, T., Gu, S., He, J., Yu, B., and Tang, X., 2023, "Future application and prospect of space electron beam processing technology," *J. Phys.: Conf. Ser.*, 2658(1), p. 012044. https://doi.org/10.1088/1742-6596/2658/1/012044

[6] Kovalchuk, D., Melnyk, V., and Melnyk, I., 2022, "A coaxial wire-feed additive manufacturing of metal components using a profile electron beam in space application," *J. Mater. Eng. Perform.*, 31(8), pp. 6069–6082. https://doi.org/10.1007/S11665-022-06994-Z

[7] Avchare, K. R., Tarwani, A., Jain, D., Saini, U., and Purohit, R., 2014, "Space manufacturing techniques: A review," *Int. J. Sci. Res. Publ.*, 4(4), pp. 1–10. ISSN 2250–3153

[8] Marshall, W. M., Stegeman, J. D., Zemba, M., MacDonald, E., Shemelya, C., Wicker, R., Kwas, A., and Kief, C., 2015, "Using additive manufacturing to print a cubesat propulsion system," *Proceedings of the 51st AIAA/SAE/ASEE Joint Propulsion Conference*, Orlando, FL, July 27–29, p. 4184. https://doi.org/10.2514/6.2015-4184

[9] Halbig, M.C. and Singh, M., 2015, "Additive Manufacturing of SiC Based Ceramics and Ceramic Matrix Composites," In *Proceedings of the 11th International Conference on Ceramic Materials and Components for Energy and Environmental Applications*, Vancouver, BC, Canada, 14–19 June, 2015, NASA Paper No. GRC-E-DAA-TN24157.

[10] Frazier, W.E., 2014, "Metal Additive Manufacturing: A Review," *J. Mater. Eng. Perform.*, 23, pp. 1917–1928. https://doi.org/10.1007/S11665-014-0958-Z

[11] Balla, V. K., Roberson, L. B., O'Connor, G. W., Trigwell, S., Bose, S., and Bandyopadhyay, A., 2012, "First demonstration on direct laser fabrication of lunar regolith parts," *Rapid Prototyp. J.*, 18(6), pp. 451–457. https://doi.org/10.1108/13552541211271992

[12] Mueller, R. P., Townsend, I. I., Tamasy, G. J., Evers, C. J., Sibille, L. J., Edmunson, J. E., Fiske, M. R., Fikes, J. C., and Case, M. P., 2018, "Additive construction with mobile emplacement (ACME)/automated construction of expeditionary structures (ACES) materials delivery system (MDS),"

Proceedings of the ASCE Earth and Space Conference 2018, Cleveland, OH. NASA Paper No. KSC-E-DAA-TN48594. https://doi.org/10.1061/9780784481899.020
[13] National Research Council, 2014, *3D Printing in Space*, The National Academies Press, Washington, DC. https://doi.org/10.17226/18871
[14] Morris, M., Ciardullo, C., Lents, K., Montes, J., Rudakevych, O., Sono, M., Sono, Y., and Yashar, M., 2016, "Mars ice house: Using the physics of phase change in 3D printing a habitat with H2O," In *AIAA Space*, 13–16 July, 2016, pp. 673–681. https://doi.org/10.2514/6.2016-5528
[15] Kamble, P., Hodgir, R., Gote, G., Mittal Y., and Karunakaran K. P., 2022, "Sub-zero additive manufacturing: A review of peculiarities and applications of additive manufacturing at temperatures below 0 °C," *Prog. Addit. Manuf.* 7, pp. 993–1008. https://doi.org/10.1007/S40964-022-00273-Y
[16] Kamble, P. P., Chavan, S., Hodgir, R., Gote, G., and Karunakaran, K. P., 2022, "Multi-jet ice 3D printing," *Rapid Prototyp. J.*, 28(6), pp. 989–1004. https://doi.org/10.1108/RPJ-03-2021-0065
[17] Kamble, P., Mittal, Y., Gote, G., Patil, M., and Karunakaran, K. P., 2024, "A mathematical surface roughness model for objects made by material jetting," *Prog. Addit. Manuf.*, pp. 1–12. https://doi.org/10.1007/S40964-024-00573-5
[18] Hodgir, R., Mittal, Y. G., Kamble, P., Gote, G., Patil, Y., Patel, A. K., Bernard, A., and Karunakaran, K. P., 2022, "Mathematical modelling of pattern sublimation in rapid ice investment casting," *Inter Metalcast*, 16(2), pp. 1002–1009. https://doi.org/10.1007/S40962-021-00665-W
[19] Hodgir, R., Mittal, Y. G., Kamble, P., Gote, G., Patil, Y., Patel, A. K., and Karunakaran, K. P., 2023. "Comparative study of rapid ice investment casting vs rapid casting processes of aluminium alloy," *Int. J. Precis. Eng. Manuf.*, 24, pp. 853–866. https://doi.org/10.1007/S12541-023-00788-6
[20] Gonzalez-Gutierrez, J., Cano, S., Schuschnigg, S., Kukla, C., Sapkota, J., and Holzer, C., 2018, "Additive manufacturing of metallic and ceramic components by the material extrusion of highly-filled polymers: A review and future perspectives," *Materials*, 11(5), p. 840. https://doi.org/10.3390/MA11050840
[21] Johnston, M. M., Werkheiser, M. J., Cooper, K. G., Snyder, M. P., and Edmunson, J. E., 2014, "3d printing in zero-g ISS technology demonstration," In: *AIAA SPACE 2014 Conference and Exposition*, San Diego, CA, 4–7 August, p. 4470. https://doi.org/10.2514/6.2014-4470
[22] Wong J. Y., and Pfahnl A. C., 2014, "3D printing of surgical instruments for long-duration space missions," *Aviat. Space Environ. Med.*, 85(7), pp. 758–763. https://doi.org/10.3357/ASEM.3898.2014
[23] Daly, A. C., Prendergast, M. E., Hughes, A. J., and Burdick, J. A., 2021, "Bioprinting for the biologist," *Cell*, 184(1), pp. 18–32. https://doi.org/10.1016/J.CELL.2020.12.002
[24] Pappas, J. M., Thakur, A. R., Leu, M. C., and Dong, X., 2021, "A comparative study of pellet-based extrusion deposition of short, long, and continuous carbon fiber-reinforced polymer composites for large-scale additive manufacturing," *ASME J. Manuf. Sci. Eng.*, 143(7), p. 071012. https://doi.org/10.1115/1.4049646
[25] Mitchell, A., Lafont, U., Hołyńska, M., and Semprimoschnig, C. J. A. M., 2018, "Additive manufacturing—A review of 4D printing and future applications," *Addit. Manuf.*, 24, pp. 606–626. https://doi.org/10.1016/J.ADDMA.2018.10.038
[26] Pott, A., Meyer, C., and Verl, A., 2010, "Large-scale assembly of solar power plants with parallel cable robots," In *41st International Symposium on Robotics, and ROBOTIK 2010 (6th German Conference on Robotics)*, Munich, Germany, June, pp. 1–6.
[27] Albus, J., Bostelman, R., and Dagalakis, N., 1992, "The NIST spider: A robot crane," *J. Res. Natl. Inst. Stand. Technol.*, 97(3), pp. 373–385. https://doi.org/10.6028/JRES.097.016
[28] Gagliardini, L., Caro, S., Gouttefarde, M., Wenger, P., and Girin, A., 2014, "Optimal design of cable-driven parallel robots for large industrial structures," In *2014 IEEE International Conference on Robotics and Automation (ICRA)*, Hong Kong, China, pp. 5744–5749. https://doi.org/10.1109/ICRA.2014.6907703
[29] Gagliardini, L., Caro, S., Gouttefarde, M., Wenger, P., and Girin, A., 2015, "A reconfigurable cable-driven parallel robot for sandblasting and painting of large structures," In *Proceedings of the 2nd International Conference on Cable-driven Parallel Robots*, Pott A. et al., eds., 32, pp. 275–291. https://doi.org/10.1007/978-3-319-09489-2_20
[30] Lukas, Mike. "The NFL SkyCam: Ultimate guide to the floating camera." *WSN*, January 27, 2022, accessed October 5, 2023. https://www.wsn.com/nfl/skycam-ultimate-guide.

[31] Rosati, G., Andreolli, M., Biondi, A., and Gallina, P., 2007, "Performance of cable suspended robots for upper limb rehabilitation," In *2007 IEEE 10th International Conference on Rehabilitation Robotics*, Noordwijk, Netherlands, pp. 385–392. https://doi.org/10.1109/ICORR.2007.4428454
[32] Merlet, J.,-P., and Daney, D., "A portable, modular parallel wire crane for rescue operations," In *2010 IEEE International Conference on Robotics and Automation*, Anchorage, AK, pp. 2834–2839. https://doi.org/10.1109/ROBOT.2010.5509299
[33] Boschetti, G., Passarini, C., and Trevisani, A., 2017, "A strategy for moving cable driven robots safely in case of cable failure," *Proceedings of the Advances in Italian Mechanism Science, in Advances in Italian Mechanism Science: Proceedings of the First International Conference of IFToMM Italy*, G. Boschetti et al., eds., Springer, Cham, Switzerland, 47, pp. 203–211. https://doi.org/10.1007/978-3-319-48375-7_22
[34] Werner, K., 2016, "Force Control of Cable-Driven Parallel Robots," Ph.D. dissertation, University of Stuttgart, Germany. https://doi.org/10.18419/OPUS-6899
[35] Gagliardini, L., Caro, S., and Gouttefarde, M., 2015, "Dimensioning of cable-driven parallel robot actuators, gearboxes and winches according to the twist feasible workspace," *Proceedings of the 2015 IEEE International Conference on Automation Science and Engineering (CASE)*, Gothenburg, Sweden, pp. 99–105. https://doi.org/10.1109/COASE.2015.7294046
[36] Jin, XueJun, Dae Ik Jun, Andreas Pott, Sukho Park, Jong-Oh Park, and Seong Young Ko. "Four-cable-driven parallel robot." *13th International Conference on Control*, Automation and Systems (ICCAS 2013), Gwangju, Korea, October 20–23, pp. 879–883.
[37] Rushton, M., and Khajepour, A., 2021, "Planar variable structure cable-driven parallel robots for circumventing obstacles," *ASME. J. Mech. Rob.*, 13(2), p.021011. https://doi.org/10.1115/1.4049440
[38] Tho, T. P., and Thinh, N. T., 2021, "Using a cable-driven parallel robot with applications in 3D concrete printing." *Appl. Sci.*, 11(2), p. 563. https://doi.org/10.3390/app11020563
[39] Métillon, M., Cardou, P., Subrin, K., Charron, C., and Caro, S., 2021, "A cable-driven parallel robot with full-circle end-effector rotations," *ASME. J. Mech. Rob.*, 13(3), p. 031017. https://doi.org/10.1115/1.4049631
[40] Brell-Cokcan, S., and Braumann J., 2012, *Rob|Arch 2012: Robotic Fabrication in Architecture, Art and Design*, Springer, Vienna, Austria. https://doi.org/10.1007/978-3-7091-1465-0
[41] Hahlbrock, D., Braun, M., Heidel, R., Lemmen, P., Boumann, R., Bruckmann, T., Schramm, D., Helm, V., and Willmann, J., 2022, "Cable Robotic 3D-printing: Additive manufacturing on the construction site," *Constr. Robotics*, 6(3), pp. 305–318. https://doi.org/10.1007/s41693-022-00082-3
[42] Khoshnevis, B., Bodiford, M., Burks, K., Ethridge, E., Tucker, D., Kim, W., Toutanji, H., and Fiske, M., 2005, "Lunar Contour Crafting-A Novel Technique for ISRU-Based Habitat Development," In *43rd AIAA Aerospace Sciences Meeting and Exhibit*, Reno, NV, January 10–13, p. 538. https://doi.org/10.2514/6.2005-538
[43] Bhardwaj, A., Jones, S. Z., Kalantar, N., Pei, Z., Vickers, J., Wangler, T., Zavattieri, P., and Zou, N., 2019, "Additive manufacturing processes for infrastructure construction: A review." *ASME. J. Manuf. Sci. Eng.*, 141(9), p. 091010. https://doi.org/10.1115/1.4044106
[44] Bosscher, P., Williams II, R. L., Bryson, L. S., and Castro-Lacouture, D., 2007, "Cable-suspended robotic contour crafting system," *Autom. Constr.*, 17(1), pp. 45–55. https://doi.org/10.1016/J.AUTCON.2007.02.011
[45] Williams, R. L., Xin, M., and Bosscher, P., 2008, "Contour-crafting-cartesian-cable robot system concepts: Workspace and stiffness comparisons," *Proceedings of the ASME 2008 International Design Engineering Technical Conferences & Computers and Information in Engineering Conference*, Brooklyn, USA, August 3–6, 43260, pp. 31–38. https://doi.org/10.1115/DETC2008-49478
[46] Bruckmann, T., Mikelsons, L., Brandt, T., Hiller, M., and Schramm, D., 2008, "Wire robots part I kinematics, analysis & design," *Parallel Manipulators*, New Developments, I-Tech Education and Publishing, Vienna, Austria, pp. 109–132.
[47] Avila-Alfaro, J. A., Grünschloß, E., Epple, S., and Bonten, C., 2015, "Analysis of a single screw extruder with a grooved plasticating barrel—Part I: The melting model," *Int. Polym. Proc.*, 30(2), pp. 284–296. https://doi.org/10.3139/217.3021

[48] Rosato, D. V., Rosato, D. V., and Rosato, M. V., 2004, *Plastic Product Material and Process Selection Handbook*, Elsevier, New York, pp. 1–618. https://doi.org/10.1016/B978-1-85617-431-2.X5000-2
[49] Whyman, S., Arif, K. M., and Potgieter, J., 2018, "Design and development of an extrusion system for 3D printing biopolymer pellets." *Int. J. Adv. Manuf. Technol.*, 96, pp. 3417–3428. https://doi.org/10.1007/s00170-018-1843-y
[50] Go, J., Schiffres, S. N., Stevens, A. G., and Hart, A. J., 2017, "Rate limits of additive manufacturing by fused filament fabrication and guidelines for high-throughput system design," *Addit. Manuf.*, 16, pp. 1–11. https://doi.org/10.1016/J.ADDMA.2017.03.007
[51] Bhandari, V. B., 2020, *Design of Machine Elements*, 5th ed., Tata McGraw-Hill, Noida, India.
[52] An, H., Zhang, Y., Yuan, H., Xu, W., and Wang, X., 2021, "Design control and performance of a cable-driving module with external encoder and force sensor for cable-driven parallel robots," *ASME. J. Mech. Rob.*, 14(1), p. 014502. https://doi.org/10.1115/1.4051608
[53] Wahl, A. M., and Bisshopp, K. E., 1964 "Mechanical springs," *J. Appl. Mech.*, 31(1), p. 159. https://doi.org/10.1115/1.3629561
[54] Cuvillon, L., Weber, X., and Gangloff, J., 2020, "Modal control for active vibration damping of cable-driven parallel robots," *ASME. J. Mech. Rob.*, 12(5), p. 051004. https://doi.org/10.1115/1.4046434
[55] Zhang, Z., Shao, Z., Peng, F., Li, H., and Wang, L., 2020, "Workspace analysis and optimal design of a translational cable-driven parallel robot with passive springs," *ASME. J. Mech. Rob.*, 12(5), p. 051005. https://doi.org/10.1115/1.4046030
[56] Bandyopadhyay, S., Mcgarey, P., Goel, A., Rafizadeh, R., Delapierre, M., Arya, M., Lazio, J., et al., 2021, "Conceptual design of the Lunar Crater Radio telescope (LCRT) on the far side of the moon," *Proceedings of the 2021 IEEE Aerospace Conference (50100)*, Big Sky, MT, pp. 1–25. https://doi.org/10.1109/AERO50100.2021.9438165

Chapter 14

Microstructural Refinement through Heat Treatment for Enhancing the Mechanical Characteristics of LPBF-Processed TI-6Al-4V Alloy

Mumtaz Rizwee and Deepak Kumar

14.1 INTRODUCTION

The inherent cost of raw materials and the challenges associated with traditional manufacturing (welding and machining) make Ti-6Al-4V particularly well-suited for 3D printing (AM) [Rizwee et al. 2024b; Choi et al. 2020]. Furthermore, additive-manufactured Ti-6Al-4V presents reduced labour costs and material waste, minimised processing, shortened manufacturing times, and most importantly, customised characteristics in the end parts. The AM demonstrates distinctly different cooling rates and temperature gradients compared to traditional manufacturing techniques [Bian et al. 2015; Rizwee & Kumar 2024]. Although it offers various advantages, the AM Ti-6Al-4V exhibits problematic microstructures characterised by extensive needle β-Ti throughout the builds, resulting in anisotropy [Shamsaei et al. 2015; Donoghue 2017].

The LPBF is an effective AM technique for fabricating Ti-6Al-4V. The L-PBF (DMLS) method employs a laser to melt and consolidate metallic powder [Bremen et al. 2012]. The unique microstructure and thermal gradient resulting from the L-PBF method require specific treatments to achieve optimal mechanical properties [Pegues et al. 2018; Rizwee et al. 2024a]. The microstructure of the Ti-6Al-4V produced via the L-PBF generally features a fine acicular martensite referred to as the α′ [Murr et al. 2009; Thijs et al. 2010; Rizwee & Kumar 2025]. While L-PBF Ti-6Al-4V components generally demonstrate tensile strength (σ_u) and yield stress (σ_y), they possess relatively low ductility, typically below 10% [Vrancken et al. 2012]. To achieve desired mechanical attributes for specific applications of the LPBF-processed components, suitable post-treatment processes must be developed. Furthermore, these HTs reduce thermal stresses accumulated during AM processes, thereby improving part performance.

Numerous studies [Kumar & Ramamurty 2019; Zhang et al. 2021; Ding et al. 2022; Rizwee et al. 2025] have reported on the influence of HT on microstructure and micromechanical attributes of traditionally fabricated Ti. Typically, the HT of conventionally produced Ti-6Al-4V alloy results in increased hardness at higher cooling rates, attributed to the formation of martensite during rapid cooling [Zhang et al. 2006; Callister et al. 2020]. While there is a wealth of information on conventionally produced Ti-6Al-4V, the existing literature to date is scarce in addressing the influence of microstructure and mechanical properties of additively fabricated Ti-6Al-4V [Zhang et al. 2017].

Vrancken et al. [2012] examined the varied responses of the LPBF-processed Ti-6Al-4V alloy to commonly utilised HTs, elucidating the effects of temperature and cooling rate. Their findings indicated that heat treating at intermediate to high temperatures, just below the β-transus, followed by FC, was the most effective approach for the comprehensive enhancement of tensile properties. It was asserted that the unique process conditions and distinct microstructure of L-PBF-produced components result in standard HTs yielding atypical

 DOI: 10.1201/9781003590774-14

outcomes, necessitating a different approach than for bulk alloy parts. Gai et al. [2018] investigated that the passive film on electron beam-melted (ET) Ti-6Al-4V alloy is thinner, denser, and forms more rapidly than that on the wrought type (WT). This is due to ET's fine-grain structure and high grain-boundary density, which promote 3D progressive nucleation. As a result, ET exhibits superior corrosion resistance in phosphate-buffered saline at 37°C. Yan et al. [2018] fabricated Ti64 samples via SLM and post-processed them by vacuum HT and hot isostatic pressing (HIP). The as-built microstructure showed columnar grains and acicular α′ martensite, with high strength but low ductility and fatigue resistance. HT modified the grain structure, improved ductility and fatigue performance, and reduced strength. HIP further enhanced ductility and fatigue life by eliminating internal defects, such as pores and cracks. Sercombe et al. [2008] optimised the mechanical properties of L-PBF Ti-6Al-4V alloy through HT below and above the β-transus. HT below β-transus enhanced β-grain boundary visibility, while HT above β-transus led to grain growth, forming semi-equiaxed β grains. The study highlights the impact of HT on microstructural evolution and mechanical behaviour.

This work seeks to elucidate the influence of annealing on the surface morphology of L-PBF-processed Ti-6Al-4V and its effect on mechanical characteristics. The HTs used in this study were conducted at 700°C, followed by furnace cooling (FC). Quantitative microstructural investigations, namely SEM and EDAX, were conducted to find the surface morphology and microstructure.

14.2 EXPERIMENTATION

The Ti-6Al-4V (Grade 23) was used to fabricate the test coupons. The constituent analysis of the Ti64 powder was executed using EDAX analysis, as seen in **Figure 14.1**. The powder has the major constituents Ti, Al and V, which is confirmed by EDAX analysis (**Figure 14.1**). The specimens were produced using the L-PBF method (**Figure 14.2**). The parameters used for the L-PBF process to manufacture components are summarised in **Table 14.1**. These parameters are the suggested performance specifications specified by the equipment manufacturer. The process parameters may substantially influence the resultant mechanical characteristics. This research examines the impact of annealing treatment on mechanical characteristics. While parameter optimisation is not the primary focus here, it is essential that the beginning conditions for all samples are consistent. The uniformity of the fabrication procedure across all specimens indicates that any notable changes in mechanical characteristics are directly

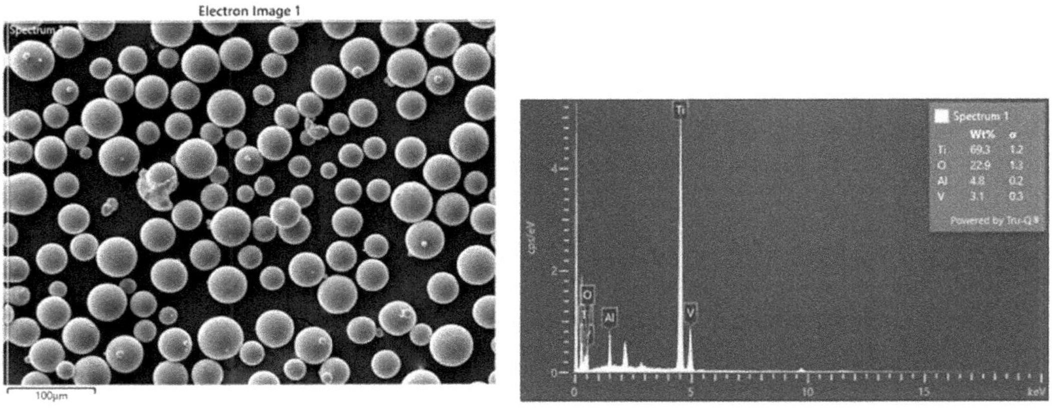

Figure 14.1 Elemental analysis of Ti64 powder.

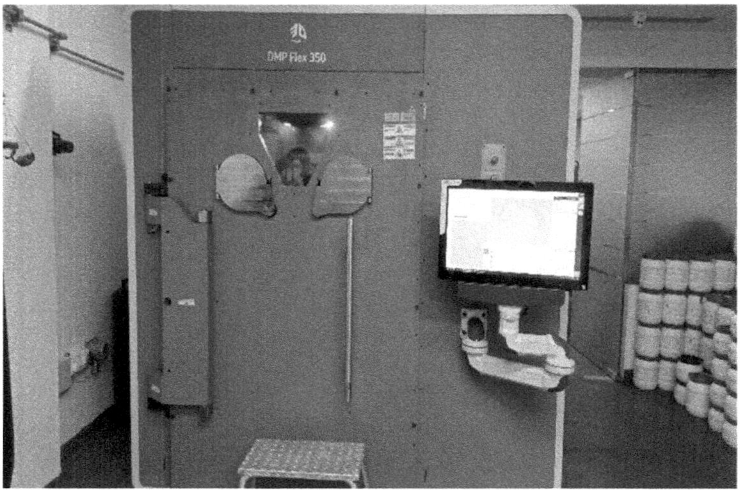

Figure 14.2 DMP Flex 350 LPBF Printer.

Table 14.1 Printing Setup

Process Variables	*Value*	*Process Variables*	*Value*
Laser power (W)	500	Layer thickness	60
Print speed (mm/s)	1250	Shielding gas	Argon
Substrate material	Ti	Laser spot diameter	65
Mean powder diameter (μm)	45–60	Infill rate (%)	100

attributable to the HT. The samples underwent HT in a vacuum furnace with a ±1°C consistency. The specimens were heated to 700°C at 20°C/min and held for 5 hours. Following the annealing procedure, furnace cooling (FC) was performed. The furnace was cooled by deactivating the heating system.

The cubic coupons, measuring 8 × 8 × 8 mm³, were produced to analyse the microstructure of both the as-printed (AP) and HT samples. Microstructural characterisation of the HT coupons utilised EDAX and SEM on samples that underwent mirror polishing and etching with Kroll's reagent. The mechanical behaviour of L-PBF coupons was assessed by evaluating tensile stress-strain responses. The tensile tests were conducted on specimens shaped like dog bones, featuring a gauge length of 25 mm and a cross-section measuring 6 mm × 2 mm. Tensile tests were performed at a constant rate of 0.01 mm/s, in accordance with the ASTM [Rizwee et al. 2025]. **Figure 14.3** presents the tensile samples processed using L-PBF.

14.3 RESULTS AND DISCUSSION

Figure 14.4 (**a**) **and** (**b**) display the SEM photographs of the AP and annealed samples, respectively. The AP component exhibits acicular and needle-like α'. A coarsened microstructure is observed after the annealing treatment. The morphology of the oxide black spots on the surface of the HT coupon is indicated in **Figure 14.4** (**b**). The annealing procedure partially converts the α′ martensite into α and β titanium. Particulate and rod-shaped β Ti were observed on the annealed surface. In addition to microstructural alterations in AP components, another

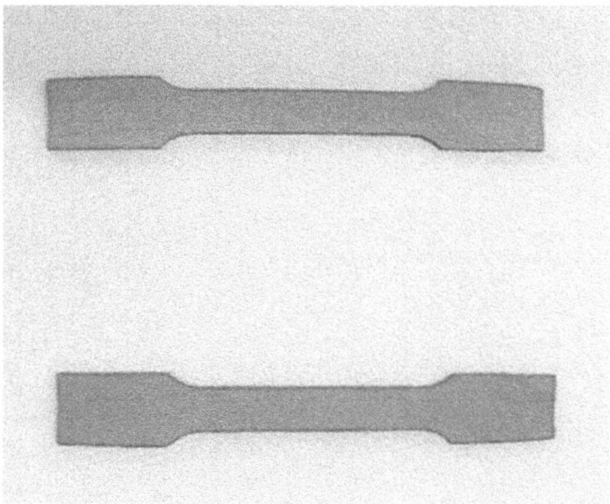

Figure 14.3 Fabricated tensile sample.

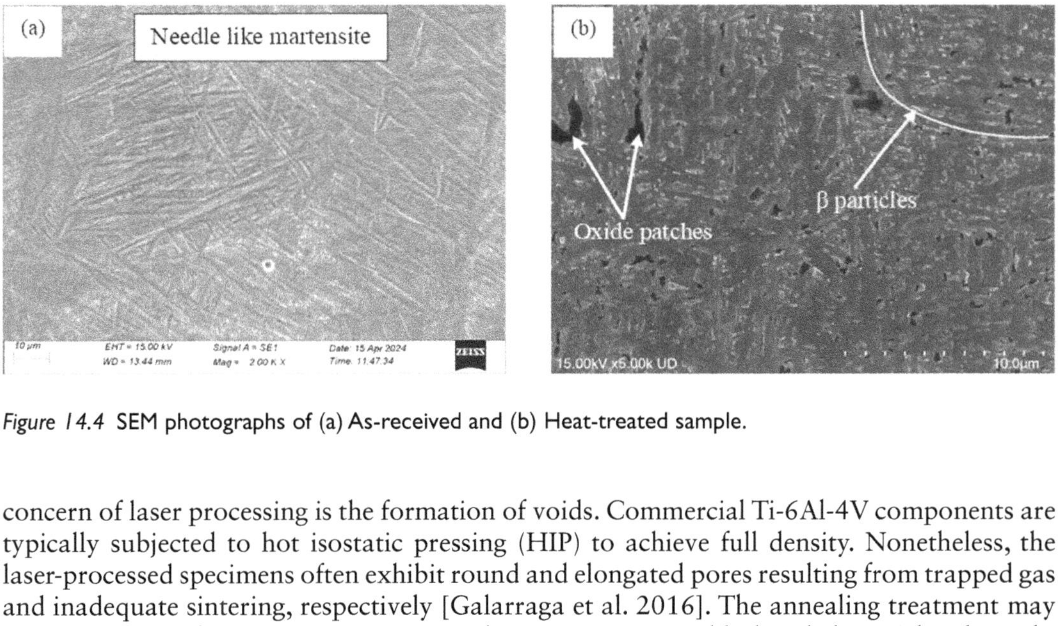

Figure 14.4 SEM photographs of (a) As-received and (b) Heat-treated sample.

concern of laser processing is the formation of voids. Commercial Ti-6Al-4V components are typically subjected to hot isostatic pressing (HIP) to achieve full density. Nonetheless, the laser-processed specimens often exhibit round and elongated pores resulting from trapped gas and inadequate sintering, respectively [Galarraga et al. 2016]. The annealing treatment may reduce porosity from 0.29% to 0.074% by re-sintering weakly bonded particles throughout the process. Other researchers asserted that porosity significantly influences mechanical properties, particularly tensile and fatigue characteristics, as pores can serve as sites for fracture initiation, leading to failure [Leuders et al. 2013; Galarraga et al. 2016]. Coarsened α laths and agglomerated precipitates are observable at annealing temperatures (**Figure 14.4b**).

As shown in **Figure 14.5**, stress-strain responses were recorded on various coupons both before and after HT. **Table 14.2** summarises tensile characteristics obtained from these plots, including strain to failure (e sub f), yield stress (sigma sub y), and ultimate tensile strength (sigma sub u). The percentage change in these attributes compared to the as-LPBF sample measurements. **Table 14.2** presents the specific significant findings derived from the data. (i) As anticipated, after HT, σ_y rises (3.82%) but σ_u falls (8.5%). (ii) The yield strength (σ_y) is 1005 MPa, whereas the ultimate tensile strength (σ_u) is 1131 MPa. These strength values remain better than those of conventionally produced alloys [Lütjering 1998; Rice et al. 2003; Singh et al. 2012]. (iii) The level of strain hardening observed in the heat-treated sample is comparatively small, as indicated by the difference between σ_u and σ_y (126 MPa). (iv) Upon

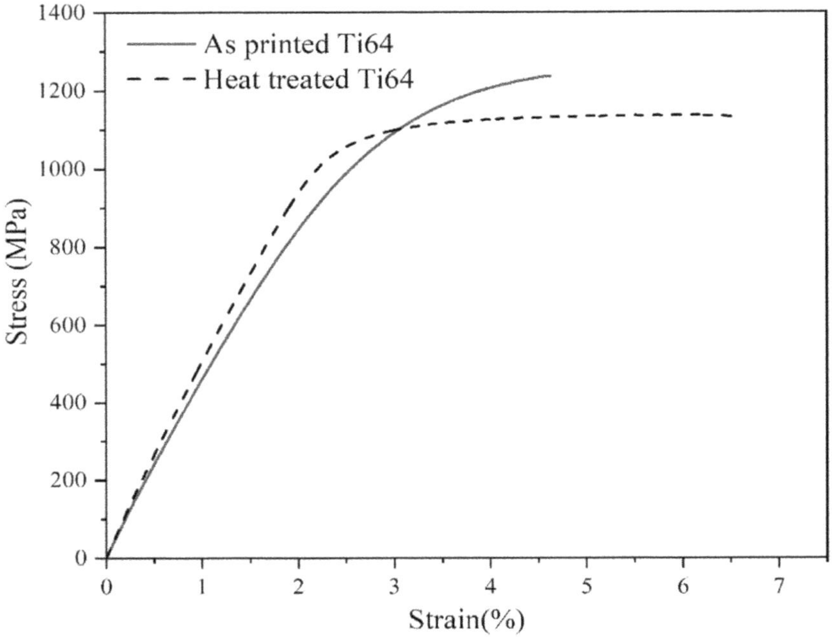

Figure 14.5 Experimental results of tensile test.

Table 14.2 Mechanical Characteristics

Sample	σ_y *(MPa)*	σ_u *(MPa)*	e_f *(%)*
As printed	968	1236	4.63
Heat treated	1005	1131	5.45

comparison of σ_y and σ_u in both AP and heat-treated samples, it is observed that both samples exhibit near-isotropic characteristics. (v) The strength drops observed during annealing are sufficiently compensated by improvements in ductility, as represented by e_f (18%). In certain instances, the enhancements exceed 50%.

14.4 SUMMARY AND CONCLUSIONS

The surface morphology and mechanical attributes of the AP and annealed LPBF-processed Ti-6Al-4V coupons were investigated in this article. The post-annealing treatment effectively sinters loosely bound particles and helps minimise porosity. Reducing the pores responsible for crack nucleation can enhance the uniaxial tensile properties of annealed parts. Annealing treatment improved the σ_y and ductility while decreasing the σ_u of the specimen. The σ_y and ductility were improved by 3.82% and 18%, respectively, while a reduction in σ_u was 8.5%. The LPBF-processed specimen has superior mechanical properties as compared to conventional manufactured Ti64 alloy.

Furthermore, the HP technique can be used after annealing treatment to improve the mechanical properties. The optimisation technique can also be used to improve mechanical characteristics by optimising printing parameters.

ACKNOWLEDGEMENT

The experimental work was supported by IIT (ISM), Dhanbad, and CSIR-NML, Jamshedpur, Jharkhand.

CONFLICT OF INTERESTS

None

REFERENCES

Bian, Linkan, Scott M. Thompson, and Nima Shamsaei. "Mechanical properties and microstructural features of direct laser-deposited Ti-6Al-4V." *Jom* 67, no. 3 (2015): 629–638. https://doi.org/10.1007/s11837-015-1308-9

Bremen, Sebastian, Wilhelm Meiners, and Andrei Diatlov. "Selective laser melting: A manufacturing technology for the future?" *Laser Technik Journal* 9, no. 2 (2012): 33–38. https://doi.org/10.1002/latj.201290018

Callister Jr, William D., and David G. Rethwisch. *Materials Science and Engineering: An Introduction.* John Wiley & Sons, 2020.

Choi, Gwanghyo, Won Seok Choi, Jeongho Han, and Pyuck-Pa Choi. "Additive manufacturing of titanium-base alloys with equiaxed microstructures using powder blends." *Additive Manufacturing* 36 (2020): 101467. https://doi.org/10.1016/j.addma.2020.101467

Ding, Xueping, Honglin Ma, Qi Zhang, Jie Yang, Daoyuan Li, and Shuqian Fan. "Effect of annealing heat treatment on microstructure and corrosion behavior of Ti6Al4V alloy fabricated by multi-laser beam wire-feed additive manufacturing in vacuum environment." *Journal of Alloys and Compounds* 914 (2022): 165363. https://doi.org/10.1016/j.jallcom.2022.165363

Donoghue, Jack Marsden. *Hybrid Additive Manufacture and Deformation Processing for Large Scale Near-Net Shape Manufacture of Titanium Aerospace Components.* The University of Manchester (United Kingdom), 2017.

Gai, Xin, Yun Bai, Ji Li, Shujun Li, Wentao Hou, Yulin Hao, Xing Zhang, Rui Yang, and R. D. K. Misra. "Electrochemical behaviour of passive film formed on the surface of Ti-6Al-4V alloys fabricated by electron beam melting." *Corrosion Science* 145 (2018): 80–89. https://doi.org/10.1016/j.corsci.2018.09.010

Galarraga, Haize, Diana A. Lados, Ryan R. Dehoff, Michael M. Kirka, and Peeyush Nandwana. "Effects of the microstructure and porosity on properties of Ti-6Al-4V ELI alloy fabricated by electron beam melting (EBM)." *Additive Manufacturing* 10 (2016): 47–57. https://doi.org/10.1016/j.addma.2016.02.003

Kumar, Punit, and Upadrasta Ramamurty. "Microstructural optimization through heat treatment for enhancing the fracture toughness and fatigue crack growth resistance of selective laser melted Ti6Al4V alloy." *Acta Materialia* 169 (2019): 45–59. https://doi.org/10.1016/j.actamat.2019.03.003

Leuders, Stefan, M. Thöne, Andre Riemer, T. Niendorf, T. Tröster, H. A. A. Richard, and H. J. Maier. "On the mechanical behaviour of titanium alloy TiAl6V4 manufactured by selective laser melting: Fatigue resistance and crack growth performance." *International journal of fatigue* 48 (2013): 300–307. https://doi.org/10.1016/j.ijfatigue.2012.11.011

Lütjering, G. E. R. D. "Influence of processing on microstructure and mechanical properties of (α+ β) titanium alloys." *Materials Science and Engineering: A* 243, no. 1–2 (1998): 32–45. https://doi.org/10.1016/S0921-5093(97)00778-8

Murr, L. E., S. A. Quinones, S. M. Gaytan, M. I. Lopez, A. Rodela, E. Y. Martinez, D. H. Hernandez, E. Martinez, F. Medina, and R. B. Wicker. "Microstructure and mechanical behavior of Ti–6Al–4V produced by rapid-layer manufacturing, for biomedical applications." *Journal of the Mechanical Behavior of Biomedical Materials* 2, no. 1 (2009): 20–32. https://doi.org/10.1016/j.jmbbm.2008.05.004

Pegues, Jonathan, Michael Roach, R. Scott Williamson, and Nima Shamsaei. "Surface roughness effects on the fatigue strength of additively manufactured Ti-6Al-4V." *International Journal of Fatigue* 116 (2018): 543–552. https://doi.org/10.1016/j.ijfatigue.2018.07.013

Rice, R. C., J. L. Jackson, J. Bakuckas, and S. Thompson. "Metallic materials properties development and standardization (MMPDS) Handbook." *NTIS Virginia Scientific Report* (2003). National Technical Information Service (NTIS).

Rizwee, Mumtaz, and Deepak Kumar. "Evaluating residual stresses in metal additive manufacturing: A comprehensive review of detection methods, impact, and mitigation strategies." *International Journal of Material Forming* 17, no. 6 (2024): 53. https://doi.org/10.1007/s12289-024-01855-7

Rizwee, Mumtaz, and Deepak Kumar. "Microstructural and tribological characterisation of an 3D printed Ti-6Al-4V alloy under annealing treatment for prosthetic implantation." *Welding International* (2025): 1–9.https://doi.org/10.1080/09507116.2025.2458042

Rizwee, Mumtaz, Deepak Kumar, and Md Murtuja Husain. "Enhancing the tribological characteristics of the biocompatible Ti-6Al-4 V alloy via heat treatment: An experimental analysis." *Tribology International* 199 (2024a): 110052. https://doi.org/10.1016/j.triboint.2024.110052

Rizwee, Mumtaz, Deepak Kumar, and Md Murtuja Husain. "Impact of heat treatment on microstructure and mechanical characteristics of laser powder bed fused Ti-6Al-4V alloy: A comprehensive investigation." *Journal of Materials Engineering and Performance* (2025): 1–17. https://doi.org/10.1007/s11665-025-10663-2

Rizwee, Mumtaz, Deepak Kumar, and Rahul Kumar. "Computational analysis of the mechanical compression of additive manufactured lattice structures: A numerical study." In *Advances in Industrial Engineering in the Industry 4.0 Era*, pp. 147–156. CRC Press, 2024b. http://doi.org/10.1201/9781003486244-12

Sercombe, Tim, Noel Jones, Rob Day, and Alan Kop. "Heat treatment of Ti-6Al-7Nb components produced by selective laser melting." *Rapid Prototyping Journal* 14, no. 5 (2008): 300–304. https://doi.org/10.1108/13552540810907974

Shamsaei, Nima, Aref Yadollahi, Linkan Bian, and Scott M. Thompson. "An overview of Direct Laser Deposition for additive manufacturing; Part II: Mechanical behavior, process parameter optimization and control." *Additive Manufacturing* 8 (2015): 12–35. https://doi.org/10.1016/j.addma.2015.07.002

Singh, Gaurav, Indrani Sen, K. Gopinath, and U. Ramamurty. "Influence of minor addition of boron on tensile and fatigue properties of wrought Ti–6Al–4V alloy." *Materials Science and Engineering: A* 540 (2012): 142–151. https://doi.org/10.1016/j.msea.2012.01.119

Thijs, Lore, Frederik Verhaeghe, Tom Craeghs, Jan Van Humbeeck, and Jean-Pierre Kruth. "A study of the microstructural evolution during selective laser melting of Ti–6Al–4V." *Acta Materialia* 58, no. 9 (2010): 3303–3312. https://doi.org/10.1016/j.actamat.2010.02.004

Vrancken, Bey, Lore Thijs, Jean-Pierre Kruth, and Jan Van Humbeeck. "Heat treatment of Ti6Al4V produced by Selective Laser Melting: Microstructure and mechanical properties." *Journal of alloys and Compounds* 541 (2012): 177–185. https://doi.org/10.1016/j.jallcom.2012.07.022

Yan, Xingchen, Shuo Yin, Chaoyue Chen, Chunjie Huang, Rodolphe Bolot, Rocco Lupoi, Min Kuang et al. "Effect of heat treatment on the phase transformation and mechanical properties of Ti6Al4V fabricated by selective laser melting." *Journal of Alloys and Compounds* 764 (2018): 1056–1071. https://doi.org/10.1016/j.jallcom.2018.06.076

Zhang, Bin, Kyungmin Ham, Shuai Shao, Nima Shamsaei, and Scott M. Thompson. "Effect of heat treatment and hot isostatic pressing on the morphology and size of pores in additive manufactured Ti-6Al-4V parts." (2017). https://hdl.handle.net/2152/89803

Zhang, W., B. Wu, W. S. Zhao, D. X. Li, and M. L. Sui. "Formation of novel β-Ti martensites in Ti–6Al–4V under an electric-current-pulse heat treatment." *Materials Science and Engineering: A* 438 (2006): 320–323. https://doi.org/10.1016/j.msea.2005.12.067

Zhang, Yifei, Limin Feng, Tao Zhang, Haoyuan Xu, and Jianzhong Li. "Heat treatment of additively manufactured Ti-6Al-4V alloy: Microstructure and electrochemical properties." *Journal of Alloys and Compounds* 888 (2021): 161602. https://doi.org/10.1016/j.jallcom.2021.161602

Chapter 15

Current Trends in 4D and 5D Printing

Shivashankarayya Hiremath, Anupama Rajendra Hiremath, Rahul Shukla, Mallikarjuna B, and Arunkumar Jayakumar

15.1 INTRODUCTION

Additive manufacturing has evolved beyond traditional 3D printing to encompass 4D and even 5D printing paradigms. 4D printing utilises adaptive materials that enable printed structures to change shape or properties over time in response to environmental stimuli, creating dynamic devices with self-assembly capabilities at the macroscale [1]. This technology employs multifunctional materials with shape-memory and self-healing properties, offering solutions for demanding environments where human intervention is not possible [2]. 5D printing enables printing along five axes, allowing layers to be deposited in various orientations and curves [3]. The new printing concept combines 4D and 5D processes, implying that 5 degrees of freedom will be used to create the final object. As a result, the finished object might have been intelligent enough to alter its appearance, form, and properties in response to external stimuli [4]. These advances enable stronger, more material-efficient structures with enhanced adaptive capabilities.

The primary impetus for 4D printing is the incorporation of time-dependent functionality into 3D-printed products using smart or stimuli-responsive materials. Unlike static 3D constructions, 4D-printed items can change shape, characteristics, or function in response to environmental stimuli such as temperature, humidity, pH, light, or magnetic fields [5]. This dynamic adaptability is especially important for biomedical applications like self-deploying stents, programmable tissue scaffolds, and drug delivery systems, where shape-morphing and responsiveness can reduce the need for invasive procedures and improve patient outcomes [6]. Materials research gains substantially from 4D printing because it encourages the creation of innovative programmable materials such as shape-memory polymers, hydrogels, and multi-material composites that can encode temporal behaviour [7]. This broadens design options and improves the performance of functional components across multiple domains. Furthermore, 4D printing eliminates or reduces the need for complex assemblies by enabling self-actuation, thereby increasing efficiency and lowering costs in industrial applications [8]. While 4D printing focuses on temporal adaptability, 5D printing adds an extra layer of information or intelligence incorporated into the manufactured structures. The goal here is to create higher-order interactivity with the environment by allowing printed things to not only morph or respond to stimuli but also carry, process, or release information in a programmed manner [9]. This opens up new possibilities for improved biomedical scaffolds that can morph and release biomolecules, as well as smart gadgets for aerospace, robotics, and environmental monitoring. Thus, the 5D printing aims to provide hierarchical control by combining physical geometry (3D), dynamic adaptation (4D), and embedded information (5D) to produce intelligent, multipurpose systems [10]. The concept originated with William Yerazunis at Mitsubishi Electric Research Laboratories and offers considerable advantages, such as 25% less material utilisation than 3D printing and finished items that are three to

DOI: 10.1201/9781003590774-15

four times stronger [11]. Despite its conceptual and embryonic stages, 5D printing has transformative promise for precision medicine, smart robotics, and environmental monitoring systems. The emerging technology, which can be combined with 4D and 5D printing, represents a paradigm shift in manufacturing: from producing static components to fabricating adaptable, programmable, and intelligent systems. Hence, the promise of transforming healthcare, robotics, aerospace, and other industries drives the integration of 4D and 5D printing, paving the way for a new era of adaptive, intelligent production.

This chapter discusses recent advancements in 4D and 5D printing technologies, with an emphasis on adoption pathways, geometric factors, material developments, and application fields. While 4D printing introduces the concept of time-dependent adaptation via smart and stimuli-responsive materials, 5D printing takes this paradigm a step further by incorporating intelligence and multifunctionality into printed structures. The discussion emphasises the importance of geometry in both technologies: in 4D printing, geometry is designed to allow shape-morphing and actuation in response to external stimuli, whereas in 5D printing, geometry and structural orientation are optimised to incorporate programmable behaviours and multi-directional reinforcement. Materials advancements include shape-memory polymers, hydrogels, and 4D composites, as well as hybrid multifunctional biomaterials and intelligent 5D composites. Applications include biomedicine, automotive, and food technology, where these technologies enable self-deploying devices, adaptive scaffolds, soft actuators, and intelligent structural components [12, 13]. This comprehensive overview provides future scholars with a thorough understanding of the technological, material, and application-based aspects of 4D and 5D printing. It also discusses the future outlook: research should focus on improving multi-material integration, scalable manufacturing techniques, enhanced computational design tools, and standardised testing protocols. These directions would be important for advancing 4D and 5D printing from emerging research concepts to mainstream manufacturing technologies capable of revolutionising the next generation of adaptive and intelligent systems.

15.2 EVOLUTION OF PRINTING TECHNOLOGIES: FROM 1D TO 5D

Figure 15.1 illustrates the progression of printing technologies, from the most basic one-dimensional (1D) printing to emerging paradigms such as four-dimensional (4D) and five-dimensional (5D) printing. This evolution is depicted not just by geometric shapes but also by the symbolic progression of the ongoing expansion of technology in tandem with the development of manufacturing technology.

The first level, 1D, is portrayed as a simple straight line, representing the most fundamental type of production. This shows linear material deposition, similar to basic extrusion or single-line printing. While not suitable for complicated constructions, 1D demonstrates the rudimentary origins of additive manufacturing techniques. The second stage depicts a flat plane composed of squares, representing the transition to two-dimensional (2D) printing. At this point, printing is limited to planar geometries that result in surfaces or sheet-like objects. 2D printing applications include conventional paper printing and thin-film manufacturing. This level laid the groundwork for extending designs into more complex fields. The third level depicts the cube, which represents the three-dimensional (3D) printing revolution. This step enabled the creation of volumetric structures with height, width, and depth, marking the true start of additive manufacturing as we know it today. 3D printing enables prototyping, personalised part creation, and the production of complex geometries that traditional manufacturing would find difficult to produce. It has numerous applications across aerospace, automotive, healthcare, and consumer products.

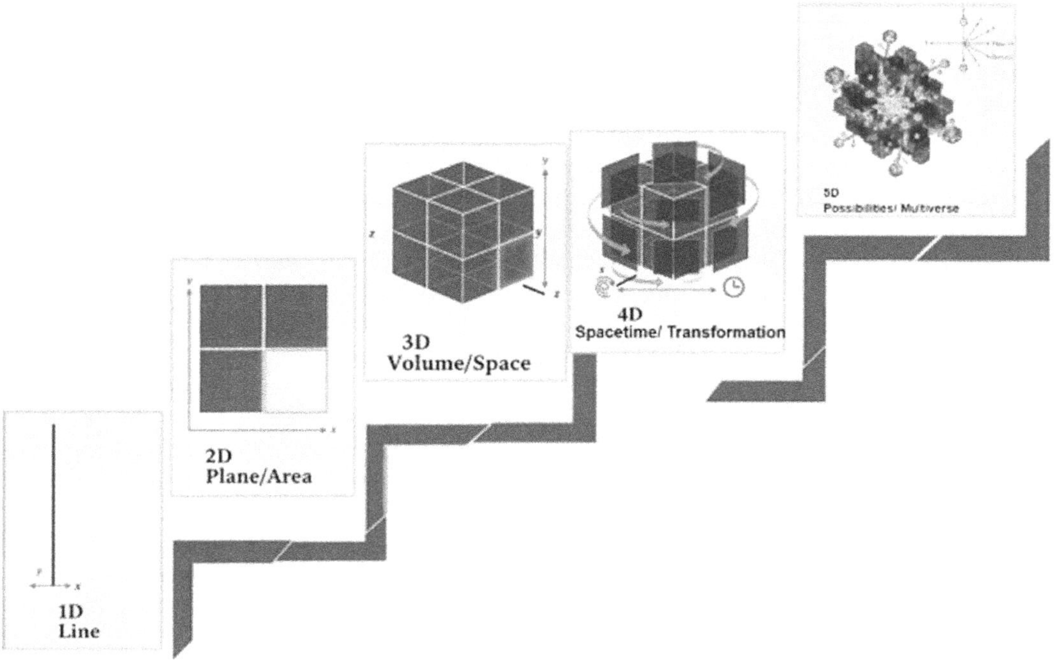

Figure 15.1 1D to the 5D in dimensional changes considerations in manufacturing.

The fourth level introduces the concept of time (the fourth dimension), represented by arrows that indicate motion and transformation. 4D printing creates objects from stimuli-responsive materials (such as shape-memory polymers, hydrogels, or composites) that alter shape, properties, or functionality over time when triggered by stimuli such as heat, light, pH, or magnetic fields. This makes structures adaptable, self-assembling, and changeable. Medical stents that stretch inside the body, as well as aircraft structures that adapt to environmental changes, are examples of 4D printing applications. The figure emphasises that 4D printing is 3D printing plus time-dependent behaviour. The most recent phase focuses on 5D printing, which incorporates not just flexibility but also intelligence, multifunctionality, and increased geometric optimisation. In this sense, the fifth dimension refers to enhanced geometric reinforcement and programmable intelligence, enabling structures to be built through curved layer deposition and multidirectional strength. This improves endurance, mechanical qualities, and the ability to embed data within the structure. Applications for 5D printing include biological scaffolds with bioactive properties, aerospace components with real-time structural monitoring, and intelligent robotics [14, 15].

15.3 FABRICATION OF 4D AND 5D OBJECTS

The fabrication of 4D printed structures follows a systematic pathway similar to 3D printing, but with the additional consideration of time-dependent functionality and stimuli-responsive materials, as illustrated in **Figure 15.2.** Shape morphing or actuation is enabled by incorporating geometric features into computer-aided design (CAD) models. To mimic structural changes under external stimuli, finite element modelling, or FEM, is used to analyse the part by simulation. The selection of materials is crucial and typically includes liquid crystal elastomers, hydrogels, shape-memory polymers, and multi-material composites. To achieve controlled deformation, hybrid constructions may incorporate both passive and active materials. Additive manufacturing methods, such as Fused Deposition Modelling (FDM), which is popular for printing thermoplastics with shape-memory properties, are employed for fabrication [16]. Stimulus-responsive hydrogels can be printed with high-resolution printers, such as stereolithography (SLA). For multi-material composites with programmable deformation, Direct Ink Writing (DIW) is helpful. For regulated behaviour, printing factors including temperature, speed, and layer orientation are tuned.

The printed object is programmed using post-processing techniques such as thermal treatment, UV curing, or mechanical pre-straining. External stimuli such as heat, humidity, light, pH, and magnetic/electric fields are subsequently used to cause the required alteration. The printed part's performance is tested by monitoring mechanical strength, recovery ratio, actuation speed, and stability under repeated stimulation. Finally, advanced imaging and sensing technologies are employed to verify the programmed behaviour.

Pei et al. [17] demonstrated various 4D-printed structures in their study, showing the capability of these systems to transform shape in response to external triggers. The shape-memory effect encompasses one-way, two-way, and three-way transformations in response to external stimuli. To achieve such shape-changing behaviours as self-folding, self-assembling, and self-disassembling, different homogeneous and heterogeneous composite materials have been

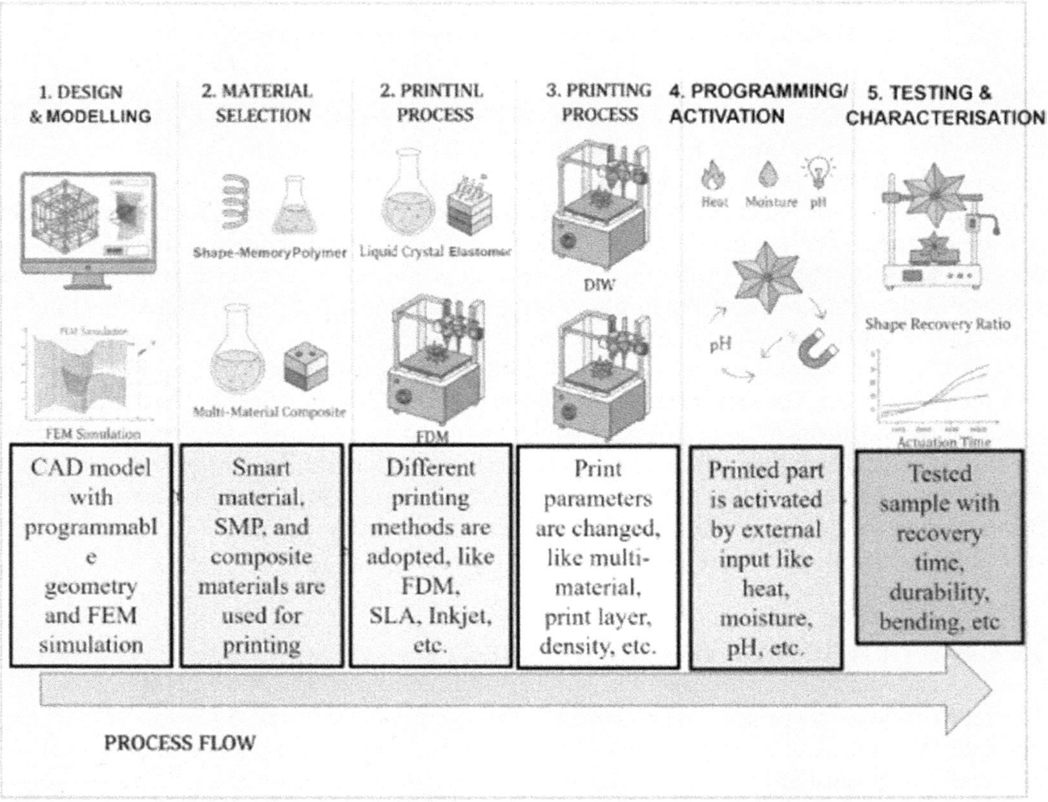

Figure 15.2 4D printing from CAD modelling, material selection, selection of a 3D printer, printing, activation of the printed part, and testing of the specimens.

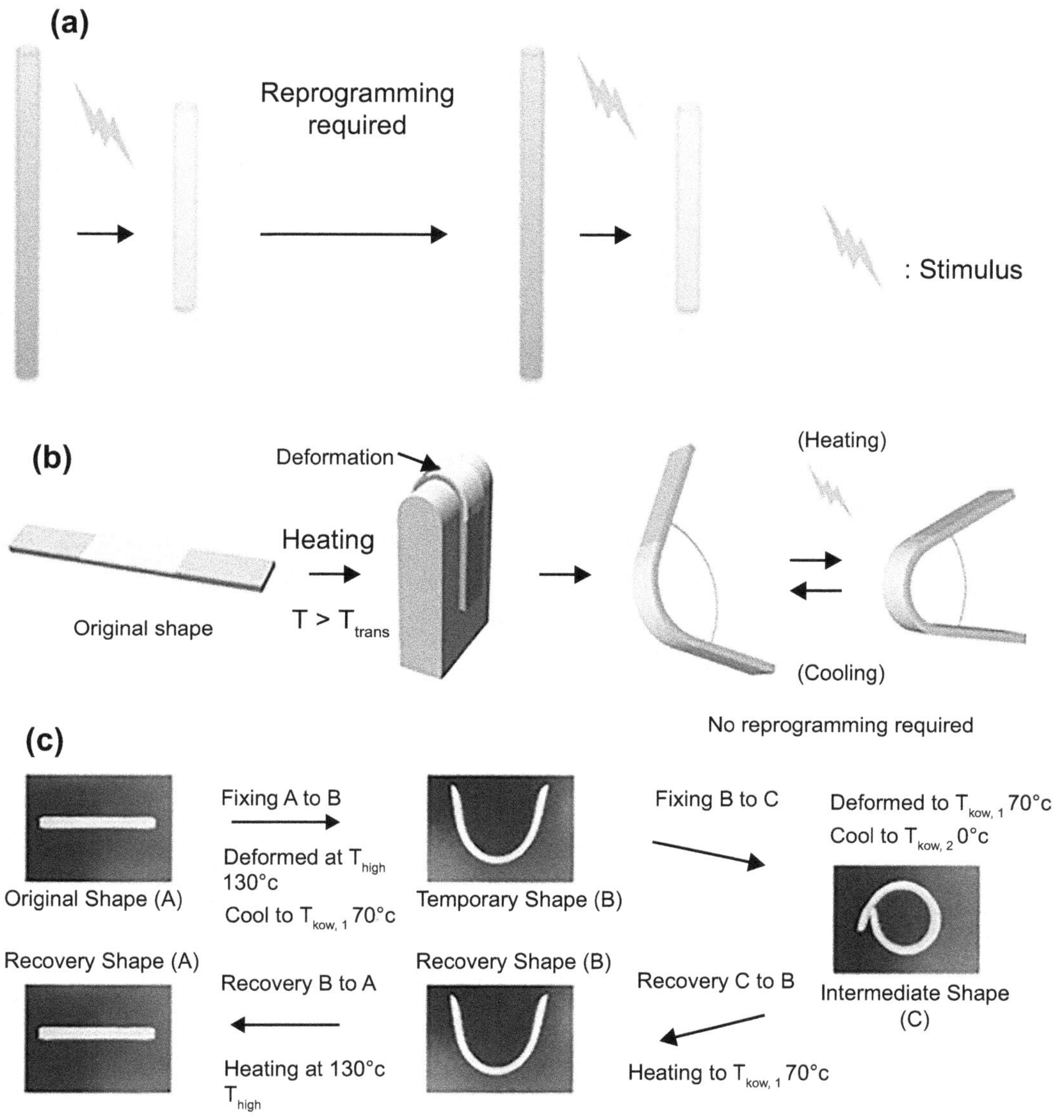

Figure 15.3 Illustration of the (a) One-way shape memory effect, (b) Two-way shape memory effect, (c) Three-way shape memory effect [17] (Original is from an open-access journal).

employed. Overall, 4D printing represents a transformative advancement towards intelligent and adaptive manufacturing.

Sydney Gladman, A et al. [18], inspired by botanical systems, developed a biomimetic 4D-printed structure capable of flower-like shape transformations, including opening and closing in response to external stimuli. The flower-inspired structure was fabricated using an ink-filament printing method, where the inter-filament arrangement enabled rapid actuation. These transformations were primarily triggered by water absorption and temperature variations. This demonstration highlights how printing pathways and complex geometries can be effectively utilised in 4D printing technology, as illustrated in **Figure 15.4.**

Figure 15.4 Flower-like structures were generated by biomimetic 4D printing [18] (Original is from an open-access journal).

Zhang, Hang, et al. [19] have introduced a novel 4D printing strategy for creating electrically responsive shape memory composites (SMCs). The printed structure features a sandwich configuration with a silver nanowire (AgNW) electrothermal layer embedded between multiwalled carbon nanotube–reinforced shape memory polyurethane acrylate (MWCNT–SMPUA) layers. Using digital light processing (DLP), the process temporarily halts UV curing to insert the AgNW layer, thereby ensuring strong interfacial bonding. Upon electrical stimulation, the structure exhibits fast, controllable, and reversible deformation, with the shape recovery speed tunable by the applied current. The approach demonstrates potential in developing deployable, biomimetic actuators, as shown in the bat-inspired soft actuator, where electrical heating drives wing-like motion and angle change up to 45° within 60 s, as illustrated in **Figures 15.5a** to **15.5e.**

Cui, H., et al. [20] have presented a nanomaterial-assisted 4D printing platform that enables spatiotemporally controlled, remote, and on-demand shape transformations. By incorporating near-infrared (NIR)–responsive nanocomposites into conventional stimuli-responsive polymers, the system allows printed structures to deform dynamically under localised photothermal stimulation. A proof-of-concept model demonstrates precise, reversible, and site-specific transformations induced by NIR light, highlighting potential applications in biomedical engineering and smart soft robotics. The approach provides a pathway for fabricating self-morphing, programmable structures capable of real-time control and biocompatible actuation, advancing the versatility and functionality of 4D printed systems, as illustrated in **Figure 15.6.**

Five-dimensional (5D) printing is an advanced form of additive manufacturing that surpasses the capabilities of regular 3D and 4D printing by enabling more degrees of freedom

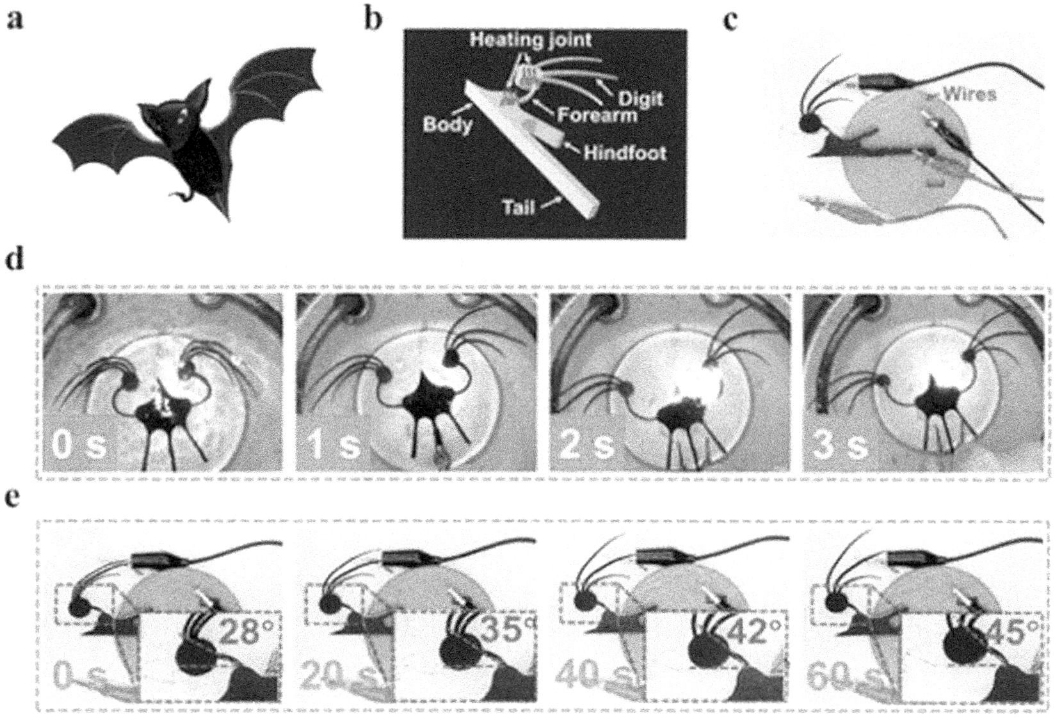

Figure 15.5 The 4D printing strategy is used to print electrically responsive, deployable, and extendible structures of the designed bat wing [19] (original is from an open-access journal).

in motion and material orientation during construction, as shown in **Figure 15.7**. Unlike traditional 3D printing, which moves the print head along fixed X-Y-Z axes, 5D printing uses rotational and translational axes to allow the substrate or print head to tilt and rotate, resulting in the deposition of curved layers that better conform to the natural geometry of the printed object [21, 22]. This multi-axis printing technique enhances the mechanical strength, durability, and surface polish of printed objects by aligning material filaments with stress trajectories, thereby reducing delamination and poor interlayer bonding.

Furthermore, 5D printing opens new avenues for creating sophisticated, bioinspired, load-bearing structures with optimal performance, lightweight design, and superior structural integrity, especially in biomedical and tissue engineering applications [23–25]. When combined with smart or stimuli-responsive materials, 5D printing can create multifunctional architectures that combine geometric versatility and active, self-adjusting features. As research advances, 5D printing is projected to play a transformative role in next-generation manufacturing by bridging the divide between digital fabrication, material science, and intelligent design.

One approach that shows promise for creating biological replacements that aid in tissue regeneration and repair is scaffold-based tissue engineering. The precise creation of complex, biocompatible, and biodegradable scaffolds that resemble the extracellular matrix (ECM) and provide transient structural support for cell development has been enabled by advancements in 3D printing. To improve anatomical conformance and integration, 4D printing introduces dynamic functionality, enabling structures to change their shape or features in response to physiological cues. Further developing, 5D printing combines functional programming and spatial design to produce multi-layered, cell-rich structures that can distribute

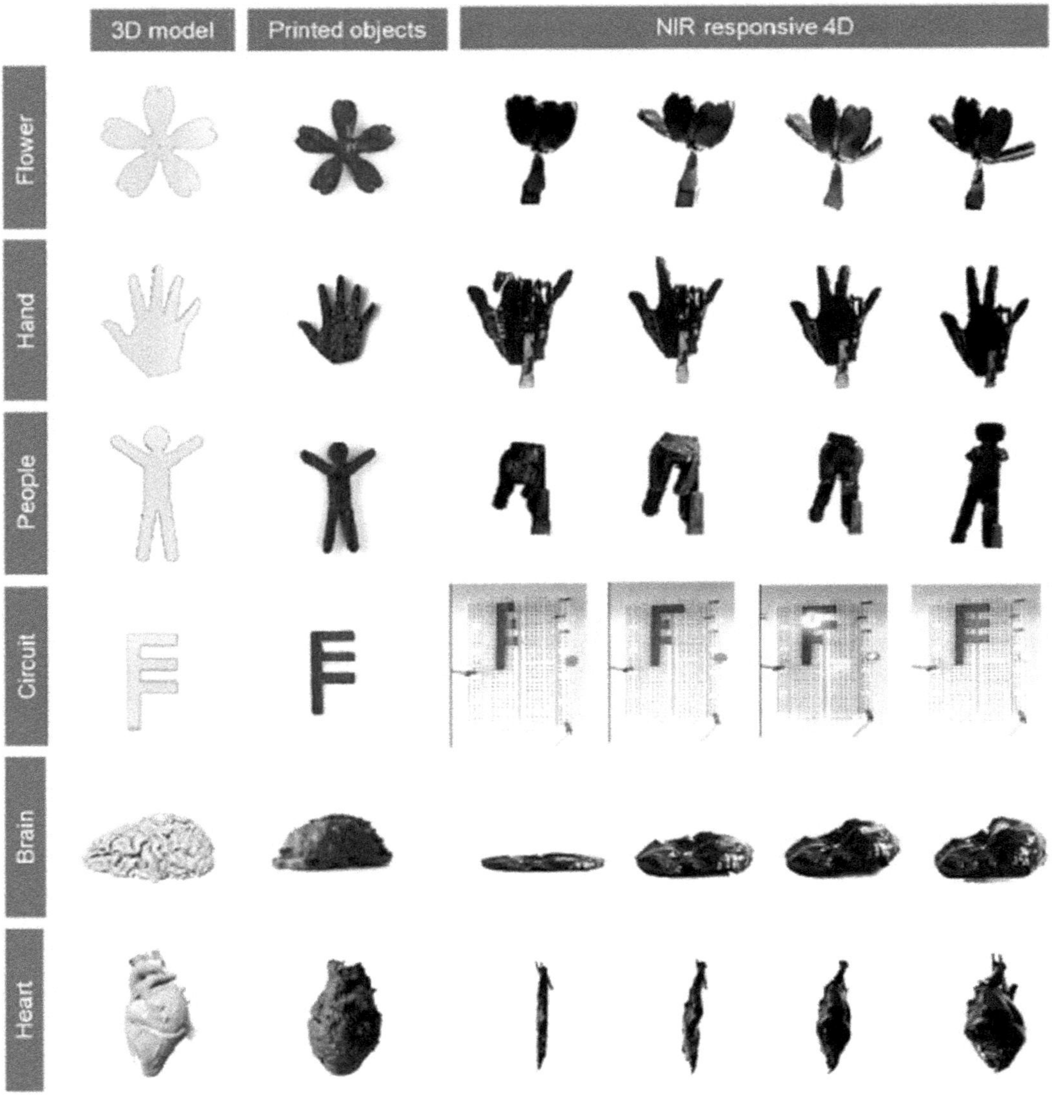

Figure 15.6 Illustration of a different 4D-printed model, including a blooming flower, a hand gesture, an exerciser, a controllable circuit switch, a folded brain, and a dilated heart [20] (original from an open-access journal).

biomolecules and change shape. The creation of tubular tissue structures (e.g., blood arteries) with a shape-memory polymer (PDLLA-co-TMC) as the morphing layer and a biomolecule layer loaded with rat bone marrow mesenchymal stem cells (rBMSC) for biological functionality is illustrated in **Figure 15. 8.** The glass transition temperature (~37°C) of the polymer allows for temperature-triggered shape recovery, which is in line with physiological conditions and advances the creation of adaptive, bio-functional scaffolds of the next generation for regenerative medicine.

The fundamental advantage of 5D printing technology is its ability to print along five distinct axes, enabling the creation of curved, multidirectional layers that significantly improve the mechanical strength and structural integrity of printed components. Unlike traditional

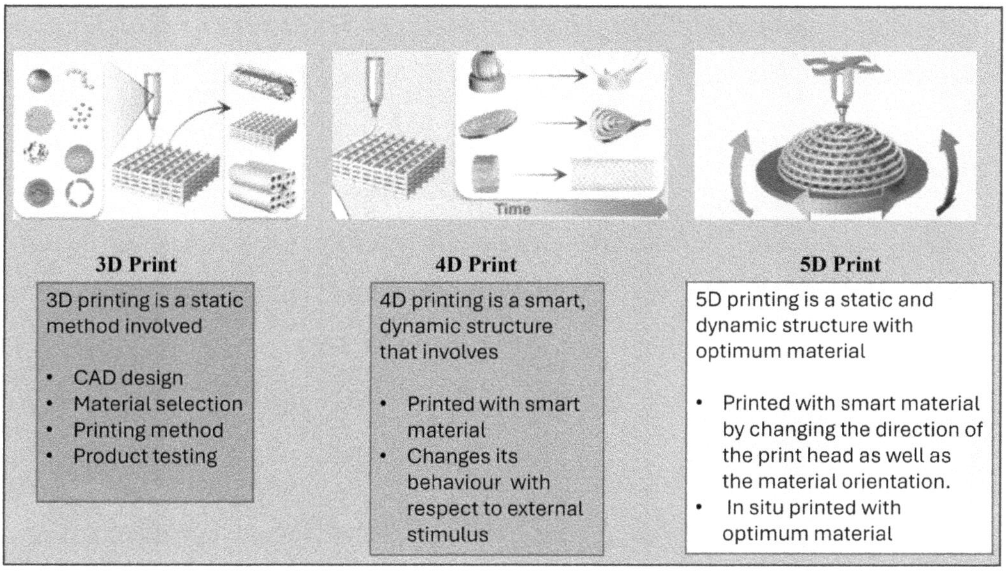

Figure 15.7 Schematic illustration of 3D to 5D printing.

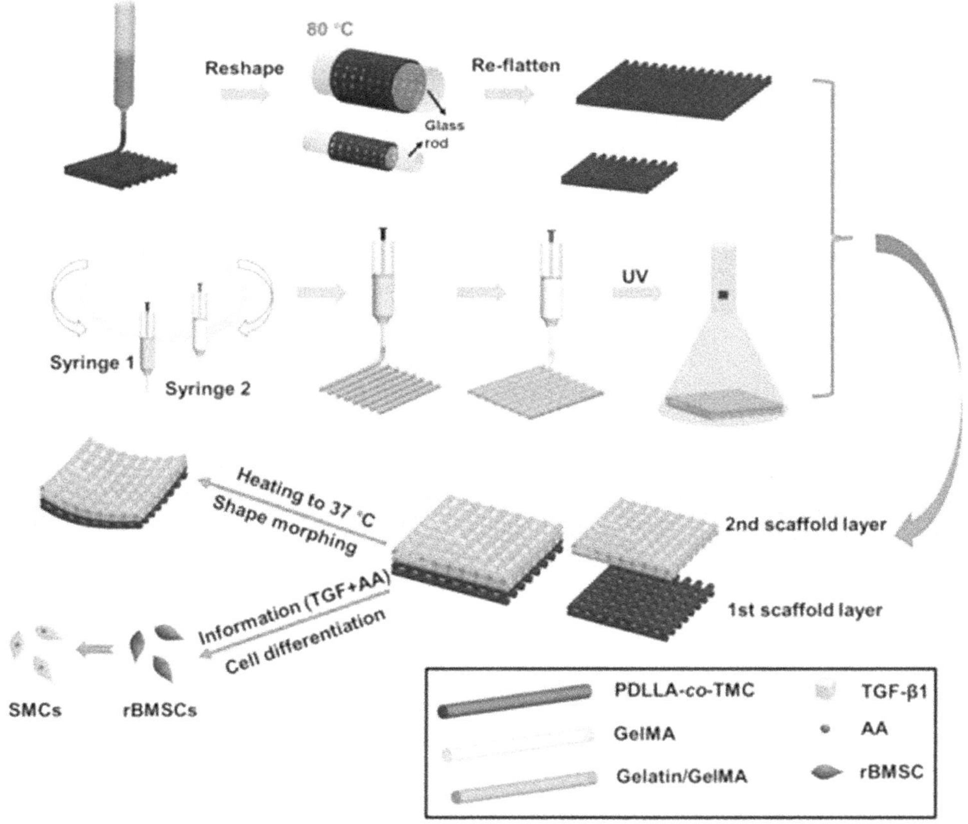

Figure 15.8 Schematic of the 5D bioprinting approach for bilayer cell–scaffold constructs with programmable shape transformation [26] (Original is from an open-access journal).

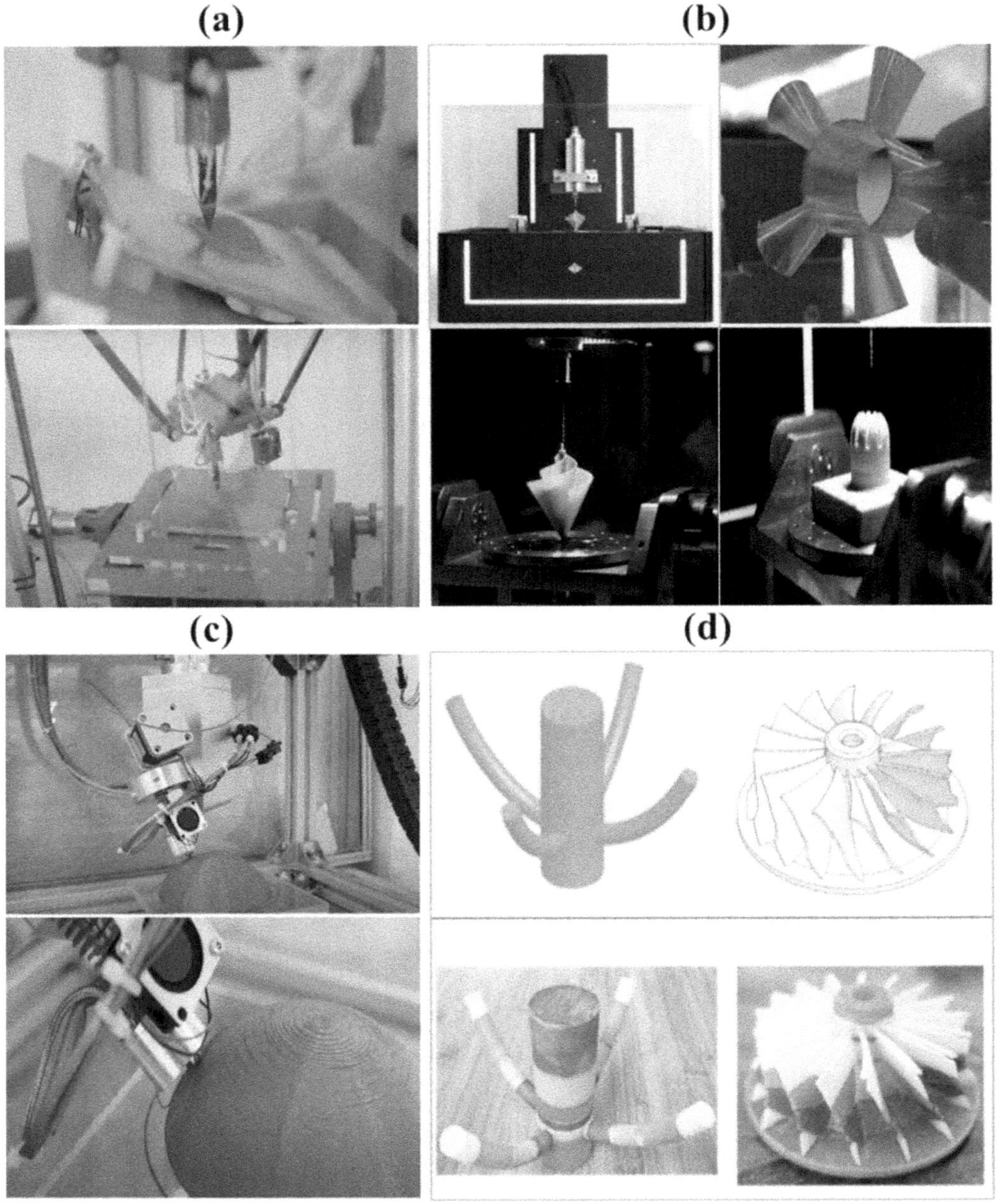

Figure 15.9 5D printed objects: (a) MERL has used a trunnion table as a print bed to create a five-axis 3D printer [29], (b) Ethereal Machines has built a subtractive and additive printing in a single machine [30], (c) 5D printing is for complex structures, (d) 3D Printing in five axes makes the world flat [31] (Original is from an open-access journal).

3D printing, which relies on horizontal layers and is susceptible to interlayer weakness, 5D printing permits deposition along complicated geometries and contours, resulting in more robust items without the need for extra support material, as illustrated in **Figure 15.9** [27, 28]. Experimental results from the MERL Laboratory showed that 5D-printed products are three to five times stronger than their 3D-printed equivalents, owing to superior layer bonding and stress distribution. Furthermore, the method improves material efficiency, cutting usage by about 25%. This multi-axis printing methodology marks a significant leap in additive manufacturing, enabling the creation of complex, high-strength, and lightweight structures that were previously impossible with typical 3D printing processes.

15.4 4D AND 5D PRINTING METHOD

Technologies for 4D printing, an emerging field of additive manufacturing, are fundamentally based on a core set of current 3D printing technologies. These have evolved into a variety of fabrication processes, including extrusion-based methods like fused deposition modelling (FDM) and direct ink writing (DIW), as well as light-assisted techniques like selective laser sintering (SLS) and digital light processing (DLP) [32], as shown in **Figure 15.10.**

15.4.1 Fused Deposition Modelling

Fused deposition modelling (FDM), also known as melt material extrusion (MME), is an additive manufacturing technique that uses a heated nozzle to extrude thermoplastic filaments, yielding layered 3D objects. PLA, ABS, and PC are popular materials with adjustable mechanical properties, including stiffness and elasticity. FDM is simple, affordable, and dependable, making it perfect for fabricating porous scaffolds and specialised medical devices for tissue engineering and medicine delivery [33]. However, FDM cannot be used with hydrogels or cell-containing bio-inks because the high heat it uses destroys them. Problems like clogged nozzles and reduced accuracy occur when FDM is used with polymers that contain fillers or melt at low temperatures. To fix these issues, people have tried using polymers that need high heat to melt, carefully sized nozzles, and mixed materials. Ongoing studies and cheaper printers are making FDM a strong, inexpensive method for making medical devices and shape-memory polymers [34].

15.4.2 Direct Ink Writing

Direct ink writing (DIW) is a printing method in which a computer guides a nozzle to extrude ink in layers, building intricate shapes. This process uses pressure to push the ink, typically with air, motors, and magnets. Because it uses tiny nozzles, DIW can print very small details, which is ideal for creating devices like tiny electronic components, radio-frequency devices, and self-sufficient gadgets [35]. DIW has many benefits, such as working with various materials, printing with multiple materials at once, being openly available, and being inexpensive. It can easily print materials that combine nanoparticles with other substances, as well as special inks that include materials that remember their shape, respond to magnets, are flexible like liquid crystal elastomer, and conductive filler [36]. These inks become thinner when squeezed but hold their form once printed until they are hardened by heat or light, making DIW a strong tool for building working objects that react to their surroundings.

15.4.3 Selective Laser Sintering

Selective laser sintering (SLS) is an additive manufacturing method that uses a high-energy laser to selectively fuse polymer powder particles into solid layers. The process constructs objects layer by layer without requiring external supports, as unsintered powder acts as a natural support medium. SLS is appreciated for its ability to produce mechanically strong, autoclavable, and cost-effective parts with high production throughput. The method is compatible with polymers such as nylon, polyurethane, and polyether-ether-ketone (PEEK), whose properties can be enhanced through mineral or clay reinforcement. Additional advantages

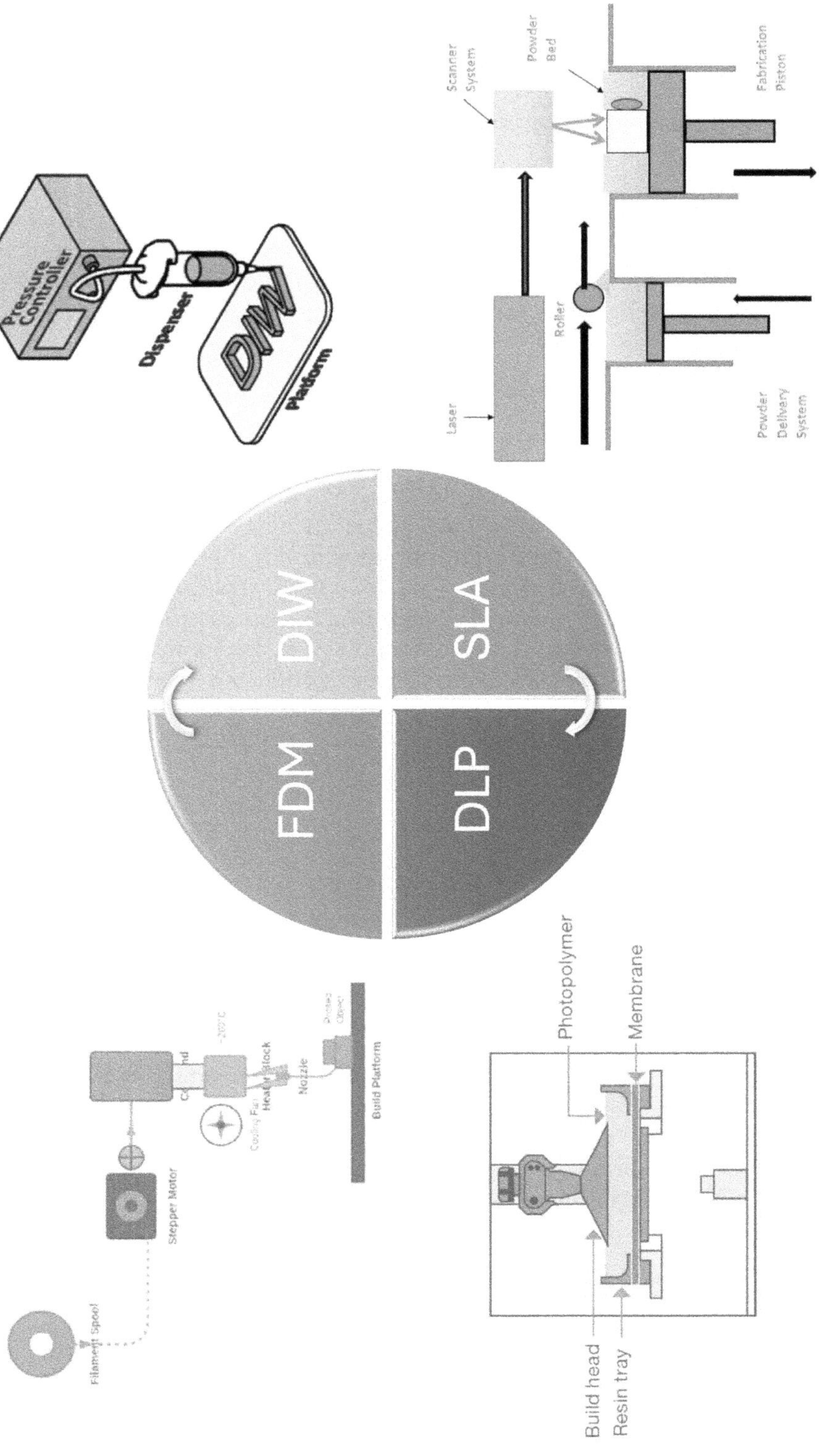

Figure 15.10 Various 4D printing techniques.

include the powder's recyclability and mechanical strength comparable to that of injection-moulded parts. However, high equipment costs, powder handling risks, and the requirement for a temperature-controlled chamber remain key limitations of this technology [36].

15.4.4 Digital Light Processing (DLP)

Digital light processing (DLP) is a photopolymerisation-based additive manufacturing technique similar to stereolithography (SLA), in which liquid photosensitive resin is selectively cured layer by layer using projected UV light patterns. The process employs a digital micromirror device (DMD) to project each layer's image onto the resin surface, enabling rapid and uniform solidification across the entire layer [37].

While both DLP and SLA rely on UV curing, they differ in their exposure mechanisms SLA scans point by point with a laser beam, whereas DLP cures entire layers simultaneously. As a result, DLP offers higher printing speed but lower resolution compared to SLA. The adjustable light intensity in DLP also provides greater control over curing depth and material properties, making it well-suited for rapid prototyping and large-scale component fabrication.

Five-dimensional (5D) printing is an advanced additive manufacturing technique that operates along five axes, combining the movements of both the print head and the build platform. This enables the deposition of curved and multi-directional layers, unlike conventional 3D printing, which builds flat, stacked layers. As a result, 5D-printed parts exhibit 3 to 4 times higher durability than those made by 3D or 4D printing, with reported strength improvements from 0.1 MPa to 3.7 MPa [38, 39].

The ability to fabricate complex curved geometries without additional supports makes 5D printing particularly valuable in aerospace, automotive, and biomedical applications. **Figure 15.11**, illustrating the synchronised rotational movement of the print head and translational motion of the print bed, highlights the enhanced geometric flexibility of this technology [40].

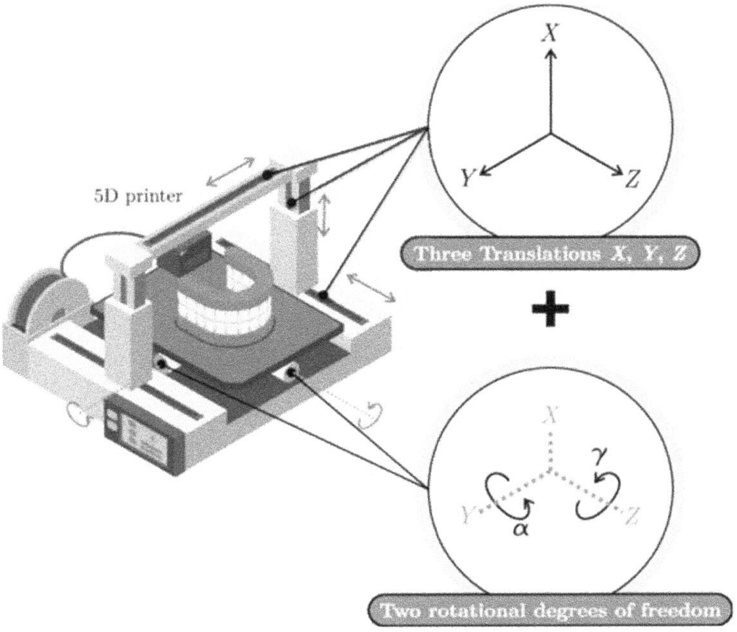

Figure 15.11 5D printing method [40] (Origin of the figure is from an open-access journal).

15.5 MATERIALS FOR 4D AND 5D PRINTING

The advancement of 4D and 5D printing technologies has greatly enlarged the range of materials utilised in additive manufacturing, allowing for the creation of adaptable, multifunctional, and mechanically durable structures. While 4D printing focuses on smart, stimuli-responsive materials that can evolve, 5D printing prioritises the usage of materials tailored for high mechanical strength and geometric complexity via multi-axis production.

15.5.1 Materials for 4D Printing

4D printing incorporates the dimension of time into traditional 3D printing by using stimulus-responsive smart materials that undergo controlled shape, property, or functional transformations in response to external stimuli such as heat, light, moisture, pH, magnetic fields, or electricity [41, 42]. The most commonly used materials in 4D printing include Shape Memory Polymers (SMPs), which are among the most prevalent materials for 4D printing and have a remarkable capacity to return to specified shapes when exposed to external stimuli [43]. Several investigations indicate their relevance, highlighting their distinctive properties: high recoverable strain (up to 800%), form recoverability in bending (~93%), and tensile (~87%) modes [44]. Polyurethane (PU), polylactic acid (PLA), and polycaprolactone (PCL) are constantly referenced [45]. These materials excel in biomedical and soft-robotic applications, enabling self-deploying structures through their stimulus-responsive nature, notably to heat, light, and other external triggers [46].

Hydrogels are novel, stimuli-responsive polymers with enormous potential for biological applications, including drug delivery and tissue engineering. Poly(N-isopropylacrylamide) (PNIPAM) represents these smart hydrogels, displaying significant temperature sensitivity around its lower critical solution temperature of 32°C [47]. These materials can absorb large amounts of water while maintaining structural integrity, making them excellent for regulated therapeutic operations [48]. Their unique capacity to inflate and deswell in response to external cues, such as temperature, allows for precise, tailored medicine release mechanisms. Critically, its excellent biocompatibility and customisable network structure allow for customisation across numerous biological applications, including wound healing and tissue engineering [49].

Liquid crystal elastomers (LCEs) are novel materials that combine rubber-like elasticity with liquid crystal molecular alignment, allowing for programmable deformation in response to external stimuli. Several investigations demonstrate that LCEs can bend, twist, and actuate when exposed to heat, light, electric, or magnetic fields [50]. These materials have shown intriguing uses in soft robotics, such as artificial muscles, sensors, and actuators [51]. LCEs' molecular structure allows them to convert modest amounts of external energy into significant mechanical motion, making them ideal for advanced technological applications.

Magnetoactive and electroactive composites are a cutting-edge family of smart materials that enable programmable, responsive constructions by strategically including conductive and magnetic fillers. These composites combine magnetic nanoparticles like Fe_3O_4 and conductive elements like carbon nanotubes to create materials with exceptional shape-changing properties. It is found that these materials can accomplish quick, remote-controlled actuation while being extremely versatile. It confirms the materials' revolutionary potential in the creation of adaptive, intelligent structures capable of responding dynamically to external inputs [52].

15.5.2 Materials for 5D Printing

Unlike 4D printing, 5D printing focuses on mechanical optimisation and structural strength through five-axis fabrication, with curved-layer deposition improving stress distribution within the material. As a result, materials utilised in 5D printing are often engineering-grade thermoplastics, composites, and metals with high strength-to-weight ratios and exceptional durability. 5D printing can increase the mechanical performance of thermoplastic polymers by deliberately arranging polymer chains along stress paths. The current additive manufacturing techniques have limited property ranges, leading to strong interest in high-performance engineering polymers. Polymers such as PEEK, polycarbonate, ABS, and nylon are particularly promising for 5D printing due to their exceptional mechanical and thermal properties.

Composites and fibre-reinforced materials, such as carbon fibre, glass fibre, and Kevlar, are commonly used in aerospace and automotive sectors to create lightweight yet strong components [53]. These reinforcements enhance dimensional stability and fatigue resistance under complex loading conditions. Metals and alloys: While 5D metal printing is still in development, metals such as titanium alloys, stainless steel, and aluminium have been examined using directed energy deposition (DED) and wire-arc additive manufacturing (WAAM) for multi-axis motion [54]. These materials are perfect for creating bent, load-bearing, corrosion-resistant components. 5D printing can create patient-specific implants and tubular scaffolds from biodegradable polymers like poly(L-lactide-co-ε-caprolactone) and PDLLA-co-TMC. These materials have the potential to change shape while remaining biocompatible, enabling cellular adhesion and tissue integration [55].

15.5.3 4D and 5D Materials and Response Mechanism

4D printing uses smart materials that can change shape, functionality, or qualities over time when subjected to external stimuli, including heat, light, moisture, pH, magnetic, or electric fields. Internal stress recovery, molecular reorientation, or swelling behaviour are common driving forces in the transformation mechanisms of 4D materials. In contrast, 5D printing does not rely on material transformation but rather improves the mechanical behaviour of printed structures through multi-axis construction. By depositing layers along curved routes, the microstructure aligns with stress trajectories, increasing load-bearing capacity, fatigue resistance, and isotropic strength. Thus, while 4D printing uses material intelligence for dynamic response, 5D printing uses geometric intelligence and material alignment to achieve higher structural performance, as summarised in **Table 15.1**.

Table 15.1 4D and 5D Print Material and Response Mechanisms

Material Type	*Stimuli*	*Response Mechanism*	*Printing Type*	*Reference*
Shape Memory Polymer (SMP)	Heat, Light	Thermal activation above Tg causes molecular chain recovery and shape reconfiguration	4D print	[56]
Hydrogel (e.g., PNIPAM, Alginate, PEGDA)	Moisture, pH, Temperature	Swelling/deswelling due to solvent diffusion and ionic interaction	4D print	[57]
Liquid Crystal Elastomer (LCE)	Heat, Light	Reversible anisotropic expansion/contraction from molecular alignment	4D print	[58]
Magnetoactive Polymer Composite	Magnetic Field	Magnetic particle reorientation induces actuation or bending	4D print	[59, 60]
Electroactive Polymer	Electric Field	Ion migration or electrostatic force leads to deformation	4D print	[61]

(Continued)

Table 15.1 (Continued) 4D and 5D Print Material and Response Mechanisms

Material Type	*Stimuli*	*Response Mechanism*	*Printing Type*	*Reference*
Fiber-Reinforced Polymer	Mechanical Design (Multi-axis Deposition)	Fibre alignment along stress trajectories enhances strength and stiffness	5D print	[62, 63]
Functionally Graded Material (FGM)	Geometric Control	Gradual variation in composition or structure improves stress distribution	5D print	[64]
Thermoplastic (e.g., PLA, PEEK)	Multi-axis Printing Strategy	Layer orientation and curvature improve isotropy and mechanical durability	5D print	[65, 66]
Metal/Composite Hybrid	Deposition Path and Tool Orientation	Curved-layer deposition increases fatigue resistance and load transfer efficiency	5D print	[67]

15.6 APPLICATIONS OF 4D AND 5D PRINTING IN DIVERSE INDUSTRIES

4D and 5D printing have transformative applications in healthcare, engineering, food processing, and sophisticated manufacturing, with a particularly high potential for biomedical and personalised product creation.

Healthcare includes tissue engineering, artificial organs, drug delivery systems, dental implants, and orthopaedic devices [68]. The food industry is developing sophisticated food structures, tailored confectionery, and packaging [69]. Engineering involves creating complex curved surfaces with increased strength, self-assembling materials, and adaptive structures [70]. The technologies provide considerable benefits, including 70–90% time reductions, the ability to print complex curved surfaces, and materials that can dynamically change shape in response to environmental stimuli. The key areas of application for 4D and 5D printing technologies are shown in **Figure 15.12**.

15.6.1 Biomedical Applications

4D and 5D printing have transformed biomedical engineering by enabling the creation of smart, adaptable, and biocompatible structures that respond to physiological signals. 4D printing is widely used in tissue engineering, where shape-memory polymers (SMPs) and hydrogels produce scaffolds that change shape to fit anatomical structures after implantation, promoting cell growth and blood vessel formation. Examples include self-expanding stents, drug delivery devices, and artificial organs with programmable mechanical and biological features [71]. Meanwhile, 5D printing enhances the mechanical strength of medical implants by enabling manufacturing along curved axes, resulting in parts that are three to five times stronger than typical 3D-printed parts [72]. It has been used to make patient-specific prostheses, orthopaedic devices, and dental implants with excellent fit and durability [73].

15.6.2 Manufacturing and Aerospace Applications

In the sophisticated industrial and aerospace sectors, both 4D and 5D printing help to create lightweight, adaptable, and functionally graded structures. 4D printing enables the creation

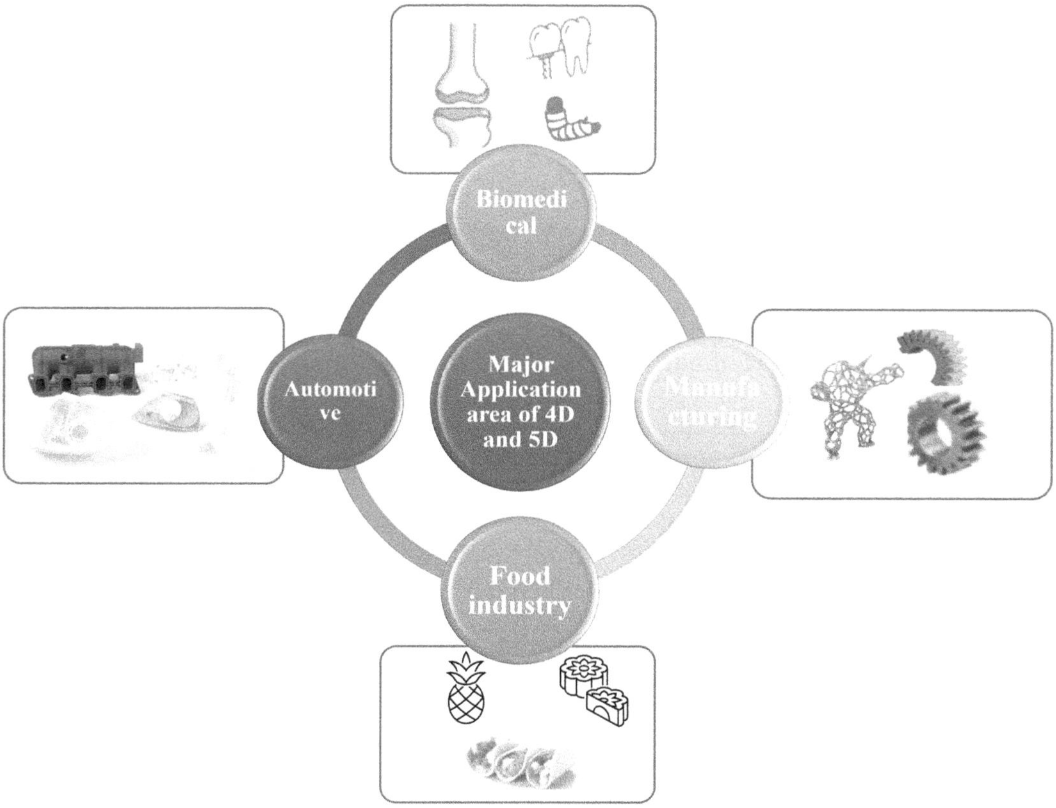

Figure 15.12 The major application areas of 4D and 5D printing.

of self-assembling components and morphing surfaces that adapt to environmental factors like temperature, light, and magnetic fields [74]. This allows for deployable satellite structures, self-repairing aircraft coverings, and reconfigurable turbine blades [75]. Multi-axis deposition in 5D printing enables the fabrication of curved, stress-aligned components with high load-bearing capacity and minimal material waste, which is especially useful in aerospace, defence, and industrial tooling [76].

15.6.3 Food Industry Applications

The food business has begun to use 4D printing to create personalised, responsive, and nutritionally adaptable food products. 4D-printed foods, which use temperature-, pH-, and moisture-sensitive materials, can change shape, texture, and flavours during manufacturing or consumption [77]. Examples include pasta that transforms while boiling and candy that changes colour or texture when hydrated. Although still in its early stages, 5D printing has the potential to create multi-layered food structures with varying densities and textures, thereby increasing both aesthetic presentation and mechanical integrity [78].

15.6.4 Automotive Applications

In the automobile industry, 4D printing enables the creation of self-healing materials, adaptive interiors, and smart aerodynamic components. Shape-memory alloys and electroactive polymers are used to create self-adjusting chairs, variable air vents, and morphing exteriors [79]. 5D printing adds to this by providing structurally designed, lightweight components that improve fuel efficiency and crash resistance. Complex geometries, such as curved panels, engine housings, and suspension systems, can be manufactured without supports, lowering production costs and material waste [80].

15.7 CASE STUDY: SOLAR PV-SMART AND ADAPTIVE MATERIALS

Researchers have explored 4D printing of solar concentrators inspired by flower shapes, specifically the ability of some flowers to track the sun. These concentrators, mimicking the shape-shifting of diurnal and nocturnal flowers, can switch between parabolic and hyperbolic shapes based on the sun's position. Fabricated using shape-memory polymers through 4D or 5D printing, these concentrators respond autonomously to environmental cues. This bio-inspired design aims to improve solar energy collection efficiency by optimising the concentrator's shape throughout the day, potentially enhancing optical efficiency by over 25% compared to traditional fixed-shape designs [81]. Thus, through 4D printing, it's technically feasible to create solar panels that automatically adjust their configuration to maximise energy capture based on light intensity and direction. For example, a black-coloured, elliptical solar tracker could shrink and tilt up to 30 degrees when exposed to sunlight, maintaining optimal exposure for the solar cell. In addition, 4D-printed materials can be designed with anisotropic thermal conductivity, meaning they can conduct heat differently in different directions. This can be used to create systems that automatically adjust heat flow to optimise energy conversion or prevent overheating in solar cells. 4D printing can create building facades with self-adjusting elements that regulate temperature by opening or closing in response to sunlight, reducing the need for artificial heating or cooling. By optimising the amount of sunlight entering the building, these facades can significantly reduce reliance on heating and cooling systems, leading to energy savings and, in turn, passive solar thermal conditioning [82]. Integrating SMPs with other materials like shape memory alloys (SMAs) or reinforcing fibres can enhance mechanical properties and deformation capabilities. A 4D-printed device with programmable anisotropic thermal conductivity (using shape-memory polymers and thermally conductive materials) can direct heat flow based on lighting conditions, potentially improving thermal management in buildings. Apparently, 4D printing can be used to create solar panels with self-healing capabilities that automatically repair minor damage. Thus, novel additive manufacturing techniques, such as four-dimensional (4D) printing, have emerged as next-generation processes for fabricating complex, multifunctional materials, especially for Solar PV applications. Previously, studies have used biocompatible polymers such as polyethene glycol (PEG) to create 4D-printed structures for biomedical applications, which can also be adapted for solar energy systems [83]. PEG's role in solar energy applications, primarily for thermal management and as an additive in solar cells due to its attributes of high latent heat, non-toxicity, and tunable phase transition temperatures, makes it suitable for thermal energy storage [84]. PEG can be incorporated into composite materials to enhance light absorption and thermal conductivity, improving the efficiency of solar thermal conversion and storage. The exploration of metal-based additive manufacturing also offers sustainable advantages through design freedom, reduced material use, and avoidance of hazardous chemicals [85]. Both the solar PV and solar thermal modules can have three main elements: a receiver (mandatory), a concentrator (optional), and a tracking mechanism (optional). The primary reason for using concentrators is to concentrate sunlight and achieve the same efficiency with fewer solar cell materials in solar PV applications (Parida et al. [86]) and to increase the receiver temperature in solar thermal applications (Tian & Zhao [87]). Currently, solar concentrators are designed with a fixed shape, and most existing concentrators are parabolic. Elliptic, V-trough, and hyperbolic concentrators have been studied in a few specific cases, and the parabolic concentrators have been recognised as a suitable form for general purposes and usual operations [88]. Advanced energy solutions include SMP-based passive solar trackers [89, 90] and 4D-printed thermal management devices leveraging anisotropic thermal conductivity and photothermal effects for net-zero energy buildings [91]. Hybrid systems integrating conventional PV panels with 3D-printed metasurface units fabricated via stereolithography (SLA) have demonstrated improved electrical performance [92]. The exploration of metal-based 3D printing for renewable energy applications further expands the possibilities in this domain [93]. Additive manufacturing also offers sustainable advantages through design freedom, reduced material use, and avoidance of hazardous chemicals [94].

15.7.1 Payback Analysis Economic of 4D Printing in Solar Photovoltaic Systems

This section evaluates the techno-economic potential of 4D-printed solar photovoltaic (PV) systems compared to conventional PV installations. Traditional PV modules cost approximately \$0.20–\$0.30/W, excluding tracking systems, but require additional expenditure and maintenance for mechanical trackers. In contrast, 4D-printed PV structures integrate adaptive geometry for sun-tracking, eliminating most moving parts. While 4D printing increases initial capital expenditure (CAPEX), operational expenditure (OPEX) is reduced due to lower maintenance requirements. Using Levelised Cost of Energy (LCOE) modelling, results indicate that a 10–20% improvement in energy yield, combined with a 50% reduction in OPEX, can lower LCOE by 10–25% compared to traditional systems, as shown in **Table 15.3**. In high-insulation regions, payback periods could be shortened to as little as two years. Photovoltaic technology costs have declined significantly in the past decade, reaching \$0.20–\$0.30/W for module fabrication, excluding trackers [95]. Mechanical sun-tracking systems can increase yield but add to both CAPEX and OPEX [96]. 4D printing, an evolution of additive manufacturing, enables smart, adaptive solar structures that can self-reorient using shape-memory materials [97]. These systems aim to enhance light capture without the need for mechanical actuators, thus reducing maintenance costs. This economic analysis compares the cost-effectiveness of 4D-printed PV systems with that of traditional fixed-tilt PV installation is shown in **Table 15.2** and **Figure 15.13**.

15.7.2 Cost Parameters

Table 15.2 Cost Parameters 4D Print Used in PV System

Parameter	*Traditional PV System*	*4D-Printed PV System*
CAPEX (\$/W)	\$0.25/W (mid-range estimate)	\$0.3125/W (+25% premium)
OPEX (\$/W-year)	\$0.015/W-year	\$0.0075/W-year (50% reduction)
System Lifetime	25 years	25 years
Discount Rate	6%	6%
Capacity Factor (CF)	20% (moderate insolation) 25% (high insolation)	Same assumptions
Energy Price	\$0.08/kWh (low) \$0.12/kWh (high)	Same assumptions

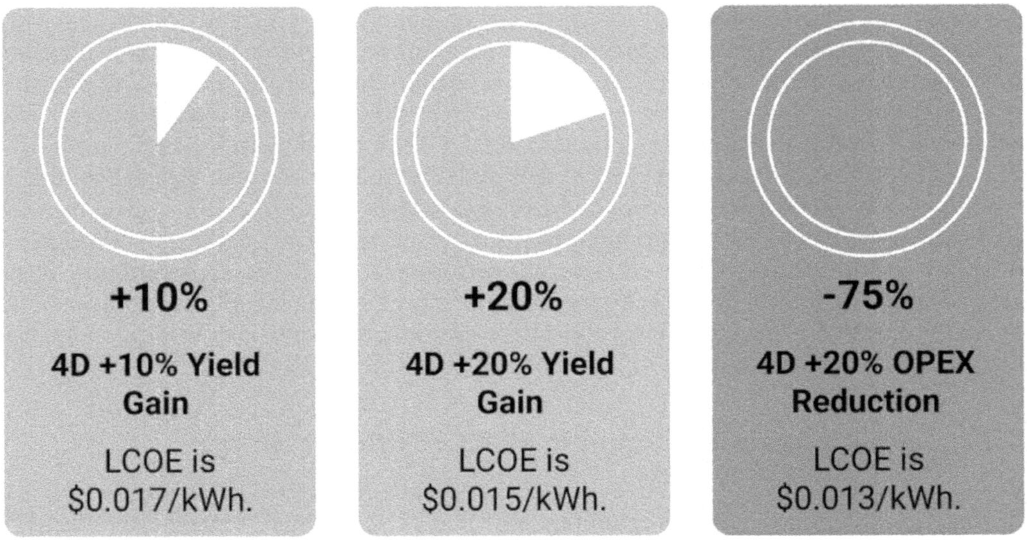

Figure 15.13 LCOE comparison for solar PV systems.

15.7.3 LCOE Model

The Levelised Cost of Energy (LCOE) is calculated as:

$$\text{LCOE} = \frac{\text{CAPEX*CRF} + \text{OPEX}}{\text{Annual energy output per Watt}} \quad (15.1),$$

where
CRF=Capital Recovery Factor

$$\text{CRF} = \frac{r}{1-(1+r)^{-n}} \quad (15.2),$$

where
r = discount rate
n = life time
Annual energy per Watt = CF *8.76kWh/W-year
Scenarios Considered

i. Base case: No yield improvement.
ii. Case A: +10% yield due to improved sun-tracking.
iii. Case B: +20% yield due to improved sun-tracking.
iv. Sensitivity: OPEX reduction of 75% combined with +20% yield.

15.7.4 LCOE Comparison

Table 15.3 Moderate Insolation (CF = 20%)

Case	*CAPEX ($/W)*	*OPEX ($/W-yr)*	*Yield Gain (%)*	*LCOE ($/kWh)*
Traditional	0.25	0.0150	0	0.020
4D Base	0.3125	0.0075	0	0.018
4D +10%	0.3125	0.0075	10	0.017
4D +20%	0.3125	0.0075	20	0.015
4D+20% (OPEX −75%)	0.3125	0.0038	20	0.013

Table 15.4 High Insolation (CF = 25%)

Case	*CAPEX ($/W)*	*OPEX ($/W-yr)*	*Yield Gain*	*LCOE ($/kWh)*
Traditional	0.25	0.0150	0%	0.016
4D Base	0.3125	0.0075	0%	0.015
4D +10%	0.3125	0.0075	10%	0.013
4D +20%	0.3125	0.0075	20%	0.012
4D +20% (OPEX −75%)	0.3125	0.0038	20%	0.011

15.7.5 High Insolation (Cf = 25%)

For CF = 20%, +20% yield, OPEX –50%:

i. Extra CAPEX per W = $0.0625.
ii. Annual benefit per W = Energy gain revenue + OPEX savings.
  a. Energy gain = 0.350 kWh/year × $0.08/kWh = $0.028/W-year.
  b. OPEX savings = $0.0075/W-year.
  c. Total = $0.0355/W-year.
iii. Payback ≈ 1.8 years.

For CF = 25%, +20% yield, OPEX –50%, price $0.12/kWh: Payback ≈ 1.0 year.

Table 15.4 shows that even with higher initial investment, 4D-printed PV systems can achieve lower LCOE due to enhanced energy yield and reduced maintenance [98, 99]. Yield improvement of 10–20% plays a decisive role; at only 5% gain, benefits diminish considerably. High-insulation sites further enhance economic performance, making them ideal for early deployment [100]. However, the assumed CAPEX premium of 25% is optimistic; higher costs or shorter lifetimes could negate the advantage. Material durability, manufacturing scalability, and real-world OPEX reduction remain critical uncertainties [101]. Thus, in an optimistic assumption, 4D-printed PV systems have the potential to outperform traditional PV in both LCOE and payback time, particularly in regions with high solar insolation.

15.8 ADVANCEMENTS IN 4D AND 5D PRINTING TECHNOLOGY

The subsequent 4D and 5D printing technologies would no longer be focused on materials that respond to a single stimulus but rather on multi-responsive, programmable smart materials capable of reacting to several environmental changes, such as temperature, pH, moisture, magnetic field, or light intensity [102]. Material science innovations are expected to lead to flexible mechanical structures with electrical, thermal, or optical characteristics made of hybrid composites and nanomaterial-based systems. Just to illustrate, liquid crystal elastomers (LCEs) and magnetoactive soft materials have been the case in the past to exhibit precise shape alteration and also self-sensing features [103]. The next scientific investigations will be committed to merging nature-inspired hierarchical structures with digitally designed metamaterials to fabricate 4D autonomous and adaptable systems endowed with memory and self-learning abilities [104].

A major next step in the future of 4D and 5D printing would be the integration of AI-powered modelling, real-time sensing, and digital twin architectures. Various machine learning algorithms may be implemented to predict material deformation, optimise toolpaths, and even adjust printing parameters for local, real-time control of precision [105]. The use of AI-driven kinematic optimisation and sensor feedback to control the 5D printing motion, which is complicated by the interplay of many rotational and translational axes, could be the next breakthrough. Digital twins' virtual replicas of the printed objects would allow the prediction of maintenance, tracking of lifetime, as well as immediate simulation of deformation and wear [106]. After these inventions, the aforementioned issues would become seminal in high-precision fields such as biomedical implants, aircraft components, and autonomous systems.

It is said that the next 4D and 5D printing research in biomedicine will lead to advanced-generation bioprinting technology capable of printing biologically dynamic, functional tissues and organs. 4D bioprinting has been used to create scaffolds that respond to external stimuli and can self-assemble or change shape after implantation [107]. Scientists are working on combining cell-laden hydrogels with microvascular and neural networks, thereby improving viability and function. 5D bioprinting capable of producing curved, physiologically relevant geometries could facilitate the engineering of tubular tissues (e.g., blood vessels, tracheae) and complex organ interfaces [108]. The integration of biomechanical modelling, in-situ monitoring, and multi-axis precision deposition will enable real-time printing correction, thus bringing clinical translation closer.

One of the key future points is the development of environmentally friendly, recyclable, and biodegradable materials for 4D and 5D printing in the long run. Biopolymers such as polylactic acid (PLA), polycaprolactone (PCL), and chitosan-based composites are being engineered to be eco-friendly while retaining mechanical strength [109]. By using green nanomaterials and biodegradable hydrogels with stimuli-responsive features, the environmental impact can be minimised, especially for transient biomedical implants, food packaging, and disposable sensors. The implementation of circular manufacturing principles, which enable the dismantling and recreation of printed objects, will be a major trend in future sustainable manufacturing [110].

It's still a big hurdle to scale up 4D and 5D printing for industrial applications. Next-generation manufacturing systems would be hybrids that combine additive and subtractive processes, enabling multi-material deposition and surface finishing from a single platform. 5D printing could be highly instrumental in the aerospace and automotive sectors for the production of stress-optimised, lightweight parts with reduced waste and increased structural strength. 4D printing technology can be used to create self-healing coatings, morphing aerodynamics, and control surfaces that respond to and adjust to real-time flight or driving conditions. By means of integration with robotic arms and 5-axis CNC systems, the fabrication of multi-directional curved structures would be possible with the reduced requirement of supports and post-processing. It is forecasted that future machines can come with automated calibration, closed-loop control, and distributed production networks, which can be made possible by IoT connectivity and blockchain-based traceability [111].

The next scientific inquiry in 5D printing would concentrate on functionally graded materials (FGMs) and multi-material architectures that imitate biological systems such as bone or nacre. By enabling changes in composition and orientation even along curved axes, one can create structures with local mechanical and functional features [112]. Case study on 4D printing for solar PV and thermal systems. It can create concentrators that change shape to focus sunlight more effectively at different times of the day or year, maximising energy capture. For instance, a concentrator could transition from a parabolic shape to a hyperbolic shape based on the sun's position, improving optical efficiency by over 25% compared to fixed concentrators [81]. Shape memory polymers (SMPs) can be 4D-printed into solar trackers that automatically adjust their position to follow the sun's movement, thereby maximising light absorption. Thus, the smart solar materials can be programmed to return to a pre-defined shape when exposed to a specific stimulus, such as heat. Further, by integrating in-situ sensors and multi-material printing heads, constructions would not only be smart but also inherently functional, thus bridging gaps in structural and electronic performance. Utilising smart materials, multi-axis production, and digital intelligence, 4D and 5D printing would be the core of Industry 5.0, which is expected to be characterised by human-machine collaboration, sustainability, and personalisation. The coming era presents to us systems that are fully adaptable, multifunctional, and autonomous, and which indistinctly separate structure, function, and intelligence.

15.9 CHALLENGES AND OPPORTUNITIES

One of the major obstacles in 4D printing is the scarcity of smart, stimuli-responsive materials that exhibit predictable, reproducible, and reversible shape transformations. Shape-memory polymers, hydrogels, and stimuli-responsive composites often exhibit low mechanical strength, limited durability, slow response times, and deterioration under cyclic loading. Material selection for 5D printing is further limited by the need for materials that can withstand multi-axis deposition and complex loading conditions, especially for structural and load-bearing applications [113]. 4D printing involves precise control over material composition, printing conditions, and activation methods (thermal, electrical, magnetic, or chemical), making the process complex and sensitive. In 5D printing, synchronised control of several rotational and translational axes increases system complexity, calibration needs, and error sensitivity. Toolpath planning for curved-layer deposition remains computationally demanding and requires standardised procedures [114].

The design of 4D-printed structures requires complex multiphysics modelling to predict time-dependent deformation, which is still in its early stages. The accurate simulation of stimulus-response behaviour, residual stressors, and long-term performance remains a significant challenge. In 5D printing, traditional CAD tools are insufficient for modelling curved-layer geometries and multi-axis deposition routes, restricting design flexibility. Both 4D and 5D printing systems require significant initial investment, specialised hardware, and professional operation. The lack of mass-production capabilities, sluggish printing rates, and post-processing requirements all impede large-scale industrial use. These technologies are currently more suited to prototyping and high-value applications than low-cost manufacturing.

4D printing offers significant opportunities in biomedical engineering, including self-deploying stents, smart implants, drug-delivery systems, and tissue scaffolds that adapt to physiological conditions. 5D printing enables the creation of anatomically complex, high-strength implants through curved layer deposition, enhancing load distribution and mechanical integrity. The combination of 4D printing and robotics enables the creation of soft actuators, adaptive grippers, and self-assembling robotic systems. 5D printing increases design flexibility for complicated components, eliminating the requirement for assembly and support structures and thereby increasing manufacturing efficiency. Both 4D and 5D printing promote sustainability by reducing material waste, allowing for multifunctional components, and prolonging service life through adaptive and self-responsive behaviour. The use of bio-based, recyclable smart materials improves their environmental performance.

Despite existing obstacles, advances in smart materials, multi-axis printing technology, computational design tools, and process automation are projected to propel the progress of 4D and 5D printing. As new standards, cost-effective systems, and scalable manufacturing solutions emerge, these technologies have the potential to transform next-generation adaptive, intelligent, and high-performance production.

15.10 CONCLUSION

The 4D and 5D printing are advances in additive manufacturing that leverage smart materials, multi-axis motion control, and digital intelligence to create adaptive, high-performance structures. 4D printing essentially introduces time as the fourth dimension to the functionality of the printed object, using stimuli-responsive materials that can morph or self-heal when exposed to specific environmental triggers. Certainly, the 4D printing process's ability to create self-transforming structures that adapt to environmental conditions offers promising advances for solar photovoltaic (PV) and thermal systems. In contrast, 5D printing enhances the mechanical strength, geometric freedom, and durability of printed parts by enabling curved-layer fabrication in five degrees of freedom. Thus, these innovations are revolutionising the vast industries spectrum, like biomedical engineering, which is inventing the dynamic tissue scaffolds and responsive implants, and aerospace, automotive, food, and manufacturing, which are going for sophisticated, lightweight and multifunctional parts.

The next step in evolution would include multi-material integration, AI-driven process control, sustainable bio-based materials, and digital twin systems that can make production more precise, intelligent, and sustainable. As research advances towards integrating 4D responsiveness with 5D geometrical flexibility, the line between structure and function could blur, leading to self-evolving, adaptive, and sustainable products that can define the Industry 5.0 era.

ACKNOWLEDGEMENTS

AI tools were used to refine language, rephrase, and enhance artwork in the chapter.

REFERENCES

1. Li, Z., & Loh, X. J. (2017). Four-dimensional (4D) printing: Applying soft adaptive materials to additive manufacturing. *Journal of Molecular and Engineering Materials*, 5(2), 1740003.
2. Ryan, K. R., Down, M. P., & Banks, C. E. (2021). Future of additive manufacturing: Overview of 4D and 3D printed smart and advanced materials and their applications. *Chemical Engineering Journal*, 403, 126162.
3. Reddy, P. R., & Devi, P. A. (2018). Review on the advancements of additive manufacturing-4D and 5D printing. *International Journal of Mechanical and Production Engineering Research and Development*, 8(4), 397–402.

4. Georgantzinos, S. K., Giannopoulos, G. I., & Bakalis, P. A. (2021). Additive manufacturing for effective smart structures: The idea of 6D printing. *Journal of Composites Science*, 5(5), 119.
5. Ding, A., Tang, F., & Alsberg, E. (2025). 4D printing: A comprehensive review of technologies, materials, stimuli, design, and emerging applications. *Chemical Reviews*, 125(7), 3663–3771.
6. Parhi, R., & Garg, A. (2024). Recent advances in 4D printing: A review of current smart materials, technologies, and drug delivery systems. *Current Pharmaceutical Design*, 31(15), 1180–1204.
7. Patdiya, J., & Kandasubramanian, B. (2021). Progress in 4D printing of stimuli-responsive materials. *Polymer-Plastics Technology and Materials*, 60(17), 1845–1883.
8. Aldawood, F. K. (2023). A comprehensive review of 4D printing: State of the arts, opportunities, and challenges. *Actuators*, 12(3).
9. Lai, J., & Wang, M. (2023). Developments of additive manufacturing and 5D printing in tissue engineering. *Journal of Materials Research*, 38(21), 4692–4725.
10. Vasishta, B. V., Preetham, T. H., Akhil, U. V., & Radhika, N. (2025). An overview of the evolution of additive manufacturing from 3d to 5D printing in biomedical applications. *Progress in Additive Manufacturing*, 10(5), 3287–3303.
11. Ghazal, A. F., Zhang, M., Mujumdar, A. S., & Ghamry, M. (2023). Progress in 4D/5D/6D printing of foods: Applications and R&D opportunities. *Critical Reviews in Food Science and Nutrition*, 63(25), 7399–7422.
12. Apsite, I., Salehi, S., & Ionov, L. (2021). Materials for smart soft actuator systems. *Chemical Reviews*, 122(1), 1349–1415.
13. Zhang, Z., Wang, Y., Wang, Q., & Shang, L. (2022). Smart film actuators for biomedical applications. *Small*, 18(36), 2105116.
14. Sood, K., Monika, & Sood, S. K. (2025). Mapping the future: A scientometric review of 3D/4D printing research trends. *Progress in Additive Manufacturing*, 10(5), 3305–3319.
15. Parida, S. K. (2025). Processing techniques and challenges of 3D-, 4D-, and 5D-printed polymer composites. In S. Moharana, B. B. Sahu, S. K. Satpathy, S. Chakroborty (Eds.), *Polymer Nanocomposites for 3D, 4D and 5D Printing: Fundamental to Applications* (pp. 87–108). Singapore: Springer Nature Singapore.
16. Tamay, D. G., Dursun Usal, T., Alagoz, A. S., Yucel, D., Hasirci, N., & Hasirci, V. (2019). 3D and 4D printing of polymers for tissue engineering applications. *Frontiers in Bioengineering and Biotechnology*, 7, 164.
17. Pei, E., & Loh, G. H. (2018). Technological considerations for 4D printing: An overview. *Progress in Additive Manufacturing*, 3(1), 95–107.
18. Sydney Gladman, A., Matsumoto, E. A., Nuzzo, R. G., Mahadevan, L., & Lewis, J. A. (2016). Biomimetic 4D printing. *Nature Materials*, 15(4), 413–418.
19. Zhang, H., Huang, S., Sheng, J., Fan, L., Zhou, J., Shan, M., . . . & Lu, J. (2022). 4D printing of Ag nanowire-embedded shape memory composites with stable and controllable electrical responsivity: Implications for flexible actuators. *ACS Applied Nano Materials*, 5(5), 6221–6231.
20. Cui, H., Miao, S., Esworthy, T., et al. (2019). A novel near-infrared light responsive 4D printed nanoarchitecture with dynamically and remotely controllable transformation. *Nano Research*, 12, 1381–1388. https://doi.org/10.1007/s12274-019-2340-9
21. Lai, J., & Wang, M. (2023). Developments of additive manufacturing and 5D printing in tissue engineering. *Journal of Materials Research*, 38(21), 4692–4725.
22. Han, X., Saiding, Q., Cai, X., Xiao, Y., Wang, P., Cai, Z., . . . & Cui, W. (2023). Intelligent vascularized 3D/4D/5D/6D-printed tissue scaffolds. *Nano-Micro Letters*, 15(1), 239.
23. Zhou, F., Hong, Y., Liang, R., Zhang, X., Liao, Y., Jiang, D., Zhang, J., Sheng, Z., Xie, C., Peng, Z., Zhuang, X., Bunpetch, V., Zou, Y., Huang, W., Zhang, Q., Alakpa, E. V., Zhang, S., & Ouyang, H. (2020). Rapid printing of bio-inspired 3D tissue constructs for skin regeneration. *Biomaterials*, 258, 120287.
24. Lai, J., Li, J., & Wang, M. (2020). 3D Printed porous tissue engineering scaffolds with the self-folding ability and controlled release of growth factor. MRS Communications, 10(4), 579.
25. Haleem, A., & Javaid, M. (2019). Future applications of five-dimensional printing in dentistry. *Current Medicine Research and Practice*, 9(2), 85–86.

26. Lai, J., & Wang, M. (2023). Developments of additive manufacturing and 5D printing in tissue engineering. *Journal of Materials Research*, 38, 4692–4725. https://doi.org/10.1557/s43578-023-01193-5
27. Ghazal, A. F., Zhang, M., Mujumdar, A. S., & Ghamry, M. (2022). Progress in 4D/5D/6D printing of foods: Applications and R&D opportunities. *Critical Reviews in Food Science and Nutrition*, 63(25), 7399–7422. https://doi.org/10.1080/10408398.2022.2045896
28. Reddy, P. R., and Devi, P. A. (2018). Review of the advancements to additive manufacturing-4D and 5D printing. *International Journal of Mechanical and Production Engineering Research and Development*, 8(4), 397–402.
29. Molitch-Hou, M. (2016). *3D Printing+Five Axes = 5D Printing*? Accessed October 14, 2021. https://www.engineering.com/story/3d-printing-five-axes-5d-printing
30. Sathe, G. (2017). *Meet Ethereal Machine's Halo, a '5D Printer' That's Just Won the Best of Innovation CES 2018 Award.* Accessed November 27, 2021. https://gadgets.ndtv.com/others/features/meet-ethereal-machines-halo-a-5d-printer-thats-just-won-the-best-of-innovation-ces-2018-award-1779765
31. Williams, A. (2021). *3D Printing in Five Axes Makes the World Flat.* Accessed January 12, 2021. https://hackaday.com/2021/01/12/3d-printing-in-five-axes-makes-the-world-flat/
32. Wang, Y., Cui, H., Esworthy, T., Mei, D., Wang, Y., & Zhang, L. G. (2022). Emerging 4D printing strategies for next-generation tissue regeneration and medical devices. *Advanced Materials*, 34(20), 2109198.
33. Boydston, B., Cao, A., Nelson, R. J., Ono, A., Saha, J. J., & Schwartz, C. J. (2018). Thrasher additive manufacturing with stimuli-responsive materials. *Journal of Materials Chemistry A*, 6(42), 20621–20645.
34. Bajpai, A., Baigent, S., Raghav, C., Ó. Brádaigh, C. M., Koutsos, V., & Radacsi, N. (2020). 4D printing: Materials, technologies, and future applications in the biomedical field. *Sustainability*, 12(24), 10628.
35. Wan, X., Luo, L., Liu, Y., & Leng, J. (2020). Direct ink writing based 4D printing of materials and their applications. *Advanced Science*, 7(16), 2001000.
36. Peng, X., Xiao, K., Devin, J. R., Yaoqing, W., Craig, M. H., Chunliang, L., & Jerry Qi, H. (2021). Integrating digital light processing with direct ink writing for hybrid 3D printing of functional structures and devices. *Additive Manufacturing*, 40, 101911.
37. McLellan, K., Sun, Y. C., & Naguib, H. E. (2022). A review of 4D printing: Materials, structures, and designs towards the printing of biomedical wearable devices. *Bioprinting*, 27, e00217.
38. Khorsandi, D., Fahimipour, A., Abasian, P., Saber, S. S., Seyedi, M., Ghanavati, S., Ahmad, A., DeStephanis, A. A., Taghavinezhaddilami, F., A. Leonova, & Mohammadinejad, R. (2021). 3D and 4D printing in dentistry and maxillofacial surgery: Printing techniques, materials, and applications. *Acta Biomaterialia*, 122, 26–49.
39. https://www.trendhunter.com/trends/5d-printing (Son Erişim Tarihi: 15/05/2023).
40. Sha, S. H., & Patil, P. (2020). Review on 4D and 5D printing technology composites view project additive manufacturing view project review on 4D and 5D printing technology. *International Research Journal of Engineering and Technology*, 07, 744–751.
41. Amaya-Rivas, J. L., Perero, B. S., Helguero, C. G., Hurel, J. L., Peralta, J. M., Flores, F. A., & Alvarado, J. D. (2024). Future trends of additive manufacturing in medical applications: An overview. *Heliyon*, 10(5).
42. Raviv, D., Zhao, W., McKnelly, C., Papadopoulou, A., Kadambi, A., Shi, B., . . . & Tibbits, S. (2014). Active printed materials for complex self-evolving deformations. *Scientific Reports*, 4(1), 7422.
43. Reddy, P. R., & Devi, P. A. (2018). Review on the advancements of additive manufacturing-4D and 5D printing. *International Journal of Mechanical and Production Engineering Research and Development*, 8(4), 397–402.
44. Hamidi, M. N., Abdullah, J., Shuib, R. K., Aziz, I., & Namazi, H. (2024). 4D printing of polylactic acid (PLA)/thermoplastic polyurethane (TPU) shape memory polymer–a review. *Engineering Research Express*, 6(1), 012402.
45. Subash, A., & Kandasubramanian, B. (2020). 4D printing of shape memory polymers. *European Polymer Journal*, 134, 109771.

46. Yan, S., Zhang, F., Luo, L., Wang, L., Liu, Y., & Leng, J. (2023). Shape memory polymer composites: 4D printing, smart structures, and applications. *Research*, 6, 0234.
47. Khalid, M. Y., Arif, Z. U., Noroozi, R., Zolfagharian, A., & Bodaghi, M. (2022). 4D printing of shape memory polymer composites: A review on fabrication techniques, applications, and future perspectives. *Journal of Manufacturing Processes*, 81, 759–797.
48. Narayana, S., Gowda, B. J., Hani, U., Ahmed, M. G., Asiri, Z. A., & Paul, K. (2025). Smart poly (N-isopropylacrylamide)-based hydrogels: A tour D'horizon of biomedical applications. *Gels*, 11(3), 207.
49. Choi, H., Choi, W. S., & Jeong, J. O. (2024). A review of advanced hydrogel applications for tissue engineering and drug delivery systems as biomaterials. *Gels*, 10(11), 693.
50. Shoukat, H., Buksh, K., Noreen, S., Pervaiz, F., & Maqbool, I. (2021). Hydrogels as potential drug-delivery systems: Network design and applications. *Therapeutic Delivery*, 12(5), 375–396.
51. Sun, D., Zhang, J., Li, H., Shi, Z., Meng, Q., Liu, S., . . . & Liu, X. (2021). Toward application of liquid crystalline elastomer for smart robotics: State of the art and challenges. *Polymers*, 13(11), 1889.
52. Ince, J. C., Duffy, A. R., & Salim, N. V. (2024). Silver coated multifunctional liquid crystalline elastomer polymeric composites as electro-responsive and piezo-resistive artificial muscles. *Macromolecular Rapid Communications*, 45(20), 2400370.
53. Wu, S., Hu, W., Ze, Q., Sitti, M., & Zhao, R. (2020). Multifunctional magnetic soft composites: A review. *Multifunctional Materials*, 3(4), 042003.
54. Lai, J., & Wang, M. (2023). Developments of additive manufacturing and 5D printing in tissue engineering. *Journal of Materials Research*, 38(21), 4692–4725.
55. Li, K., Chen, W., Gong, N., Pu, H., Luo, J., Zhang, D. Z., & Murr, L. E. (2023). A critical review on wire-arc directed energy deposition of high-performance steels. *Journal of Materials Research and Technology*, 24, 9369–9412.
56. Lai, J., & Wang, M. (2023). Developments of additive manufacturing and 5D printing in tissue engineering. *Journal of Materials Research*, 38(21), 4692–4725.
57. Momeni, F., Mehdi Hassani, N. S., Liu, X., & Ni, J. (2017). A review of 4D printing. *Materials & Design*, 122, 42–79. https://doi.org/10.1016/j.matdes.2017.02.068.
58. Gladman, A. S., Matsumoto, E. A., Nuzzo, R. G., Mahadevan, L., & Lewis, J. A. (2016). Biomimetic 4D printing. *Science Advances*, 2(9), e1601014. https://doi.org/10.1126/sciadv.1601014
59. Kotikian, A., Truby, R. L., Boley, J. W., White, T. J., & Lewis, J. A. (2019). 3D printing of liquid crystal elastomeric actuators with spatially programed nematic order. *Advanced Materials*, 31(25), 1901591. https://doi.org/10.1002/adma.201901591
60. Zhao, Q., Qi, H. J., & Xie, T. (2017). Recent progress in shape memory polymer: New behavior, enabling materials, and mechanistic understanding. *Progress in Polymer Science*, 49–50, 79–120. https://doi.org/10.1016/j.progpolymsci.2015.04.001
61. Kim, Y., Yuk, H., Zhao, R., Chester, S. A., & Zhao, X. (2018). Printing ferromagnetic domains for untethered fast-transforming soft materials. *Nature Materials*, 17(10), 896–904. https://doi.org/10.1038/s41563-018-0139-8
62. Luo, B., Zhu, Z., Xu, X., & Bian, C. (2022). 4D-printed low-voltage electroactive polymers modeling and fabrication. In *Smart Materials in Additive Manufacturing* (pp. 107–150). Elsevier.
63. Choi, J., Goh, G. D., & An, J. (2020). Review on 5-axis additive manufacturing: Materials, processes, and applications. *Additive Manufacturing*, 32, 101051. https://doi.org/10.1016/j.addma.2020.101051
64. Goh, G. D., Dikshit, V., Nagalingam, A. P., & Yeong, W. Y. (2021). 5-axis additive manufacturing: Technologies and future directions. *Composites Part B: Engineering*, 224, 109192. https://doi.org/10.1016/j.compositesb.2021.109192
65. Wang, H., Zhang, C., & Li, X. (2021). Functionally graded materials in additive manufacturing: A review. *Additive Manufacturing*, 48, 102398. https://doi.org/10.1016/j.addma.2021.102398
66. Chakraborty, D., Tripathi, A., & Dey, A. (2021). Mechanical characterization and modeling of 5D printed structures. *Rapid Prototyping Journal*, 27(8), 1724–1740. https://doi.org/10.1108/RPJ-12-2020-0282
67. Kumar, S., Singh, R., & Ahuja, I. P. S. (2020). Multi-axis additive manufacturing: A review on process, materials, and challenges. *Materials & Design*, 196, 109094. https://doi.org/10.1016/j.matdes.2020.109094

68. Li, Y., Zhao, H., & Zhang, X. (2023). Mechanical and structural optimization in 5D printed metallic and composite structures. *Additive Manufacturing*, 61, 103367. https://doi.org/10.1016/j.addma.2023.103367
69. Kumar, P. R. A. M. O. D., Tech, M., Roy, S. U. B. A. R. N. A., Hegde, H. A. R. S. H. A., Bharti, S., & Kumar, M. (2019). 4D and 5D printing: Healthcare's new edge. *3D Print Technol Nanomed*, 143, 143–163.
70. Ghazal, A. F., Zhang, M., Mujumdar, A. S., & Ghamry, M. (2023). Progress in 4D/5D/6D printing of foods: Applications and R&D opportunities. *Critical Reviews in Food Science and Nutrition*, 63(25), 7399–7422.
71. Wan, X., Xiao, Z., Tian, Y., Chen, M., Liu, F., Wang, D., . . . & Zhou, K. (2024). Recent advances in 4D printing of advanced materials and structures for functional applications. *Advanced Materials*, 36(34), 2312263.
72. Yarali, E., Mirzaali, M. J., Ghalayaniesfahani, A., Accardo, A., Diaz-Payno, P. J., & Zadpoor, A. A. (2024). 4D printing for biomedical applications. *Advanced Materials*, 36(31), 2402301.
73. Choi, J., Goh, G. D., & An, J. (2020). Review on 5-axis additive manufacturing: Materials, processes, and applications. *Additive Manufacturing*, 32, 101051.
74. Lv, X., Zhang, Y., & Zhang, Y. (2022). Review of multi-axis additive manufacturing and 5D printing technologies. *Materials Today Communications*, 33, 104635.
75. Mahmood, A., Akram, T., Shenggui, C., & Chen, H. (2023). Revolutionizing manufacturing: A review of 4D printing materials, stimuli, and cutting-edge applications. *Composites Part B: Engineering*, 266, 110952.
76. Kotikian, A., Truby, R. L., Boley, J. W., White, T. J., & Lewis, J. A. (2019). 3D printing of liquid crystal elastomeric actuators. *Advanced Materials*, 31(25), 1901591.
77. Goh, G. D., Dikshit, V., Nagalingam, A. P., & Yeong, W. Y. (2021). 5-axis additive manufacturing: Technologies and future directions. *Composites Part B: Engineering*, 224, 109192.
78. Zhang, J., Xu, C., & Zhou, X. (2021). Application of 4D printing technology in food design and manufacturing. *Trends in Food Science & Technology*, 111, 351–366.
79. Kouzani, A. Z., Adams, S., & Singh, B. (2023). Advances in multi-axis additive manufacturing for food applications. *Food Engineering Reviews*, 15, 112–128.
80. Bodaghi, M., Damanpack, A. R., & Liao, W. H. (2019). Self-expanding structures by 4D printing. *Smart Materials and Structures*, 28(4), 045019.
81. Kumar, S., Singh, R., & Ahuja, I. P. S. (2020). Multi-axis additive manufacturing: A review on process, materials, and challenges. *Materials & Design*, 196, 109094.
82. Momeni, F., & Ni, J. (2018). Nature-inspired smart solar concentrators by 4D printing. *Renewable Energy*, 122, 35–44.
83. Craveiro, F., Duarte, J. P., Bártolo, H., & Bártolo, P. (2019, September). Moving forward to 3D/4D printed building facades. In *International Conference of Progress in Digital and Physical Manufacturing* (pp. 277–282). Cham: Springer International Publishing.
84. Arif, Z. U., Khalid, M. Y., Zolfagharian, A., & Bodaghi, M. (2022). 4D bioprinting of smart polymers for biomedical applications: Recent progress, challenges, and future perspectives. *Reactive and Functional Polymers*, 179, 105374.
85. Han, S., Xiong, F., Qin, M., Shen, Z., Han, H., Jin, Y., . . . & Zou, R. (2024). Polyethylene glycol/polypyrrole aerogel shape-stabilized phase change material for solar-thermal energy storage and thermoelectric power generation. *Solar Energy Materials and Solar Cells*, 268, 112745.
86. Zhakeyev, A., Wang, L., Zhang, W., Shu, H., Wang, H., & Xuan, J. (2017). Additive manufacturing: Unlocking the evolution of energy materials. *Advanced Science*, 4, 700187, https://doi.org/10.1002/advs.374.
87. Parida, B., Iniyan, S., & Goic, R. (2011). A review of solar photovoltaic technologies. *Renewable and Sustainable Energy Reviews*, 15(3), 1625–1636.
88. Tian, Y., & Zhao, C. Y. (2013). A review of solar collectors and thermal energy storage in solar thermal applications, *Applied Energy*, 104, 538–553.
89. Madala, S., & Boehm, R. F. (2016). A review of nonimaging solar concentrators for stationary and passive tracking applications. *Renewable and Sustainable Energy Reviews*, 71.

90. Tytov, S., Al-Modaf, F., Su, S., & El-Atab, N. (2022). 4D-printed shape memory polymer based solar tracker. In *Proceedings of the IEEE 49th Photovoltaics Specialists Conference (PVSC)*, Philadelphia, PA, 2022, pp. 1248–1251. https://doi.org/10.1109/PVSC48317.2022.9938697.
91. Mahmood, T., Akram, C., Shenggui, C., & Chen, H. (2023). Revolutionizing manufacturing: A review of 4D printing materials, stimuli, and cutting-edge applications. *Composites Part B: Engineering*, 266, 110952. https://doi.org/10.1016/j.compositesb.2023.110952.
92. Wang, Z., Zhang, Y., Niu, Y., Chen, X., & Song, J. (2023). Solar-radiation-dependent anisotropic thermal management device with net zero energy from 4D printing shape memory polymer-based composites. *Materials*, 16(10), 3805. https://doi.org/10.3390/ma16103805.
93. Drymiskianaki, Z., Viskadourakis, Z., & Kenanakis, G. (2024). Hybrid microwave/solar energy harvesting system using 3D-printed metasurfaces. *Materials*, 17(23), 5969. https://doi.org/10.3390/ma17235969.
94. Qi, Z. (2020). 3D printing of metal-based materials for renewable energy applications. *Nano Research*, 14. https://doi.org/10.1007/s12274-020-3230-x.
95. Zhakeyev, P., Wang, L., Zhang, W., Shu, H., Wang, H., & Xuan, J. (2017). Additive manufacturing: Unlocking the evolution of energy materials. *Advanced Science*, 4, 700187, https://doi.org/10.1002/advs.374.
96. Jäger-Waldau, A. (2023). Snapshot of photovoltaics—February 2023. *Renewable and Sustainable Energy Reviews*, 173, 113058.
97. Tibbits, S. (2014). 4D printing: Multi-material shape change. *Architectural Design*, 84(1), 116–121.
98. International Renewable Energy Agency (IRENA). *Renewable Power Generation Costs in 2022*. Abu Dhabi: IRENA, 2023.
99. Henning, H. M., & Palzer, A. (2020). A comprehensive model for the economies of solar energy in different regions. *Solar Energy*, 204, 108–120.
100. Ding, A., Tang, F., & Alsberg, E. (2025). 4D printing: A comprehensive review of technologies, materials, stimuli, design, and emerging applications. *Chemical Reviews*, 125(7), 3663–3771.
101. Khan, R. K. (2025). 4D printing with liquid crystal elastomers: Materials, methods, and emerging applications. *ChemPlusChem*, 90(8), e202500049.
102. Li, S., Gao, M., Shi, W., Sun, X., Zhou, Y., Liu, L., . . . & Gu, Z. (2025). Engineering stimuli-responsive shape-morphing through high-resolution 4D printing. *Responsive Materials*, e70025.
103. Rautela, A., Kumar, A., & Pandey, R. K. (2023). AI-enhanced 4D printing: Toward intelligent material systems. *Additive Manufacturing Letters*, 6, 101028.
104. Yazdi, M. (2024). Digital twins and virtual prototyping for industrial systems. In *Advances in Computational Mathematics for Industrial System Reliability and Maintainability* (pp. 155–168). Cham: Springer Nature Switzerland.
105. Ghosh, S., Chaudhuri, S., Roy, P., & Lahiri, D. (2023). 4D Printing in biomedical engineering: A state-of-the-art review of technologies, biomaterials, and application. *Regenerative Engineering and Translational Medicine*, 9(3), 339–365.
106. Li, Y., Zhao, H., & Zhang, X. (2023). Mechanical and structural optimization in 5D printed metallic and composite structures. *Additive Manufacturing*, 61, 103367.
107. Sarkar, O., Mourya, Y., Kavya, K. L., Mutthuraj, D., Rao, P. V., & Basalingappa, K. M. (2025). 3D printing and 4D printing: Sustainable manufacturing techniques for green biomaterials. In *Sustainable Green Biomaterials as Drug Delivery Systems* (pp. 103–130). Cham: Springer Nature Switzerland.
108. Thakur, A., Anadebe, V. C., Dagdag, O., Sharma, D., & Om, H. (2025). Fundamentals and recent applications of 3D, 4D, and 5D printing of polymer and its nanocomposites: Challenges and its future prospects. *Polymer Nanocomposites for 3D, 4D and 5D Printing: Fundamental to Applications* (pp. 345–392). Engineering Materials. Springer, Singapore. https://doi.org/10.1007/978-981-96-4214-4_14
109. Wang, H., Zhang, C., & Li, X. (2023). Integration of AI and digital twin in multi-axis additive manufacturing: A review. *Journal of Manufacturing Systems*, 68, 32–51.

110. Hussain, S. A., Charoo, M. S., & Haq, M. I. U. (2025). 3D printing of functionally graded materials: Overcoming challenges and expanding applications. In *Multi-material Additive Manufacturing* (pp. 67–97). Elsevier.
111. Ahlers, D., Schmolzi, T., Junca, G., Zhang, J., & Wasserfall, F. (2025). Calibration and compensation of 5-axis 3D-printers for printed electronics. *Additive Manufacturing Letters*, 12, 100265.
112. Fayaz, A., Velu, R., Jayakumar, A., & Mallikarjuna, B. (2026). Additive manufacturing of flow field plates for polymer electrolyte membrane fuel cells: Scopes, challenges, and future prospects towards sustainable mobility. *Journal of Power Sources*, 666, 239070.
113. Rane, N. (2023). 3D, 4D, and 5D printing in architecture, engineering, and construction (AEC) industry: Applications, challenges, and future scope. *Engineering, and Construction (AEC) Industry: Applications, Challenges, and Future Scope (September 28, 2023)*. https://ssrn.com/abstract=4609912 or http://dx.doi.org/10.2139/ssrn.4609912
114. Ding, A., Tang, F., & Alsberg, E. (2025). 4D printing: A comprehensive review of technologies, materials, stimuli, design, and emerging applications. *Chemical Reviews*, 125(7), 3663–3771.

Chapter 16

Case Studies Based on the Latest Developments and Requirements in ML/AI-Based AM

Syed Quadir Moinuddin, Monaf Hodhod, Vadavalasa Harsha Vardhan, Suresh Goka, Dhanunjay Kumar Ammisetti, and Mohammad Abdul Razack

16.1 INTRODUCTION

The design and development of equipment and components capable of performing tasks that normally require human intelligence, such as decision-making, problem-solving, and language comprehension, is the broad field of artificial intelligence (AI). A form of artificial intelligence called machine learning (ML) enables systems to learn from data, spot trends, and make judgements with minimal human assistance. Machine learning is an AI technique that enables analysis and decision-making without the need for specifically designed software. Recent years have seen an increase in academic fields, including medical diagnostics, smart manufacturing, autonomous driving, object recognition, natural language analysis, and material properties [1]. The relationship between AI and ML is shown in **Figure 16.1** (**a**). The first version's existing systems or entirely new ones may function better due to the modifications [2]. Manufacturing techniques have advanced significantly to meet the need for creating complex components. Additive manufacturing (AM), a cutting-edge technology, enables the efficient and precise fabrication of complex, custom-tailored components. It is used in cutting-edge industries, including healthcare, automotive, and aerospace, and it is beginning to attract interest in a wide range of other fields. ML offers process efficiency, defect identification, quality assurance, and predictive modelling of mechanical characteristics, has emerged as a potent tool for resolving issues in AM. The emphasis is on hybrid data-collection methods that yield strong predictive models by integrating machine learning with physics-aware features. The research also assesses several machine learning methods that are used to characterise AM processes, optimise process parameters, and forecast mechanical characteristics. The measurements show ground-breaking advancements in ML-powered solutions, providing increased accuracy, convenience, and reliability in AM, including real-time process monitoring, automated parameter adaptation, and fault mitigation. However, there is a lack of data, computational difficulties, and a disconnect between ML research and practical applications. These issues must be resolved if ML in AM is to reach its full potential. It concludes with the identification of interesting research avenues, such as creating an interdisciplinary research team, standardising data enhancement, and creating new, sophisticated machine-learning algorithms [2].

An ML exercise known as "monitored learning" builds a function using tagged training data. A collection of the training occurrences is called training data. Every example in supervised learning has an input vector and a valued output, commonly known as the supervisory signal. A supervised learning system constructs an inferred function using training data to map fresh instances [3, 4]. Semi-supervised, reinforcement, supervised, and unsupervised learning are the typical classifications for machine learning algorithms. **Figure 16.1** (**b**) shows the categories in machine learning.

DOI: 10.1201/9781003590774-16

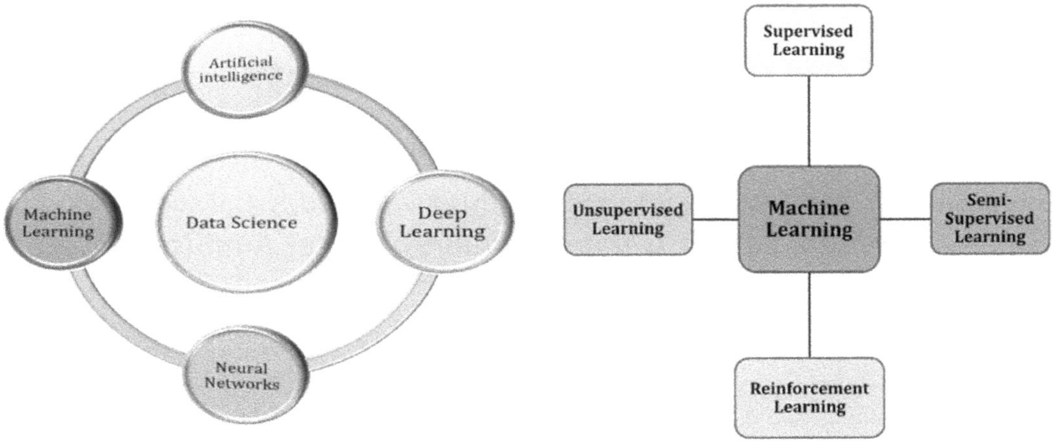

Figure 16.1 Background of AI and ML: (a) Relation of AI and ML, (b) Classification of ML.

By offering design evaluation, process optimisation [5], and production control innovation, ML revolutionises AM, as this article explains. The study examines case studies of ML applications to AM, as well as recent research. Thus, this chapter provides a thorough analysis of case studies that illustrate the technology's developments, challenges, and practical applications. These case studies highlight important innovations that have advanced the technology and provide insights into its progress. They also examine the main challenges faced by researchers and business experts, as well as the creative solutions used to overcome them. Additionally, the technology's useful applications in other fields are investigated, highlighting its significance and room for expansion.

The chapter ends by providing practical strategies for developing the technology going forward through case studies. This entails spotting new trends, suggesting enhancements in light of existing constraints, and identifying opportunities for multidisciplinary integration [6]. This chapter offers a path to greater innovation by addressing technical and practical concerns, ensuring ongoing advancement in the manufacturing sector.

16.2 LITERATURE BACKGROUND

Despite advancements, process reliability and consistent component quality remain obstacles in the AM industry. The AM process concurrently shapes a part's shape and material properties, which are the main sources of this problem. A multi-stage process involving intricate interactions between design, material, and process creates AM components. The five fundamental phases are testing and validation, postprocessing, construction, process design, and process planning. Each of these steps must be meticulously followed to create a certified part. Current efforts have focused on thorough monitoring and inspection to reduce AM part variability. Advances in data collection methods, sensor fusion, and sensor technology have led to a notable increase in AM data, which encompasses several aspects of big data.

The extremely complex and comprehensive datasets demand sophisticated computational and analytical management techniques. There are exciting opportunities for closing this gap thanks to recent advances in machine learning. By identifying implicit (previously unknown) knowledge and uncovering connections in large production datasets, machine learning techniques can transform enormous volumes of data into useful and illuminating insights [7].

This work examines the unique role that machine learning plays in addressing AM-related problems. This chapter analyses how ML methods could improve several stages of the AM lifecycle by methodically reviewing current knowledge. It focuses on how ML enhances the performance and dependability of AM processes by enabling real-time parameter changes, predictive analytics, and improved process monitoring.

Similarly, the increasing need for complex components has led to advancements in fabrication procedures, particularly in multi-material printing, which enables the creation of multiple material characteristics within a single construct to improve functionality and performance [8]. Innovations in powder bed fusion and direct energy deposition also help produce complex geometries with minimal waste and the highest level of precision [9]. Furthermore, manufacturers may reduce weight and maximise structural efficiency to satisfy the needs of the automotive and aerospace industries by using sophisticated modelling tools and AI-generated design variations [10]. These advancements have the potential to alter the production environment, as well as to improve the ability to evaluate and optimise AM processes.

Machine learning enhances traditional manufacturing by optimising processes, improving quality control, and reducing waste through predictive analytics. Additive manufacturing accelerates design, enhances material utilisation, and ensures precise control over complex production workflows. Kumar et al. [11] identified that ML has revolutionised traditional manufacturing by enhancing efficiency, reducing costs, and improving product quality. The integration of ML with cyber-physical systems, IoT, and AI has enabled the development of smart factories, characterised by faster, more flexible, and efficient processes. However, challenges persist, such as high costs and the need for specialised expertise. The chapter reviews various ML techniques applied in AM, emphasising their potential to address design, process, and production control challenges while highlighting areas for further research. Simultaneously, researchers also focused on additive manufacturing processes, such as stereolithography (SLA), fused deposition modelling (FDM), and selective laser sintering (SLS), to build objects layer by layer. These technologies enable complex geometries, customisation, and material efficiency across aerospace, healthcare, and automotive industries.

According to Wang et al. [12], AM creates 3D objects layer by layer, offering benefits such as intricate part fabrication, reduced lead time, and cost savings. Despite its advantages, AM adoption is limited by high design barriers, material constraints, processing defects, and inconsistent product quality. The authors explore ML's role in addressing these challenges by optimising process parameters, monitoring defects, and improving quality control in design for additive manufacturing (DfAM). Wang et al. [13] proposed a closed-loop control system that integrates vision-based techniques and neural networks for real-time inspection and stabilisation, demonstrating improved jetting behaviour and stability. Liquid Metal Jet Printing (LMJP) is an innovative AM technique that utilises magneto-hydrodynamic properties for high-speed, cost-effective metal printing. The process faces challenges related to stability and quality.

Researchers emphasise that predictive maintenance leverages machine learning and IoT to minimise downtime and extend equipment lifespan. They also highlight its role in enhancing quality assurance by enabling real-time defect detection and process optimisation. Also, predictive maintenance methodologies leveraging ML are vital for minimising downtime and costs. Susto et al. [14] introduced a framework to generate "health factors" and predict operational risks. The system offers trade-offs in performance and cost-based decision-making by integrating classification modules with varying prediction horizons. Its application to semiconductor manufacturing validates its effectiveness. Wu et al. [15] underscored RF's potential for precise maintenance strategies in smart manufacturing. Tool wear prediction using ML is explored, comparing Random Forest (RF), artificial neural networks (ANNs), and support vector regression (SVR). RF outperformed other methods, achieving high prediction accuracy with a coefficient of determination (R^2) of 0.992. Authors also assert that machine learning in additive manufacturing design enables automated optimisation of geometries, improving performance and material efficiency.

It also enables data-driven insights to predict and mitigate defects, ensuring more reliable, cost-effective production. Lynch et al. [16] introduced a Bayesian optimisation framework for tuning parameters in topology optimisation, reducing reliance on manual iterations. Demonstrated through a cantilevered beam problem, the approach improves efficiency and scalability, making it applicable to various complex optimisation scenarios. Lei et al. [17] presented an ML-driven framework for real-time topology optimisation, utilising SVR and k-nearest neighbours (KNN). The approach streamlines data dimensions and design parameters, facilitating an intuitive understanding of optimised structures. Numerical examples validate its effectiveness. Meanwhile, Yao et al. [18] proposed a hybrid ML algorithm combining hierarchical clustering and supervised classification via support vector machines (SVMs) for AM design-feature recommendations. Applied to R/C car components, the system assists inexperienced designers in exploring and selecting feasible design solutions, enhancing computational design methodologies. Some researchers highlight that advanced defect detection in manufacturing, powered by AI and sensors, ensures early identification of flaws, reducing waste and rework. Real-time process monitoring enables adaptive control, improving consistency and overall product quality.

Scime et al. [19] focused on in-situ defect detection in Laser Powder Bed Fusion (L-PBF) using ML and computer vision. Unsupervised ML identifies melt-pool morphologies, while supervised methods classify defects such as keyholing and balling. This enhances quality assurance in critical industries through real-time defect monitoring. Okaro et al. [20] used a semi-supervised ML approach to address quality uncertainties in L-PBF, classifying builds as faulty or acceptable using photodiode data and tensile strength tests. The method reduces certification costs and improves efficiency across sectors such as aerospace and medical. Douard et al. [21] applied ML to Electron Beam Melting (EBM), focusing on optimising support structures and part quality.

On the other hand, testing various algorithms reveals that the number of supports is a key parameter. Despite limited data, ML demonstrates its utility in identifying critical factors influencing quality. Researchers emphasise that machine learning revolutionises material design by predicting optimal compositions for enhanced properties. Additionally, it aids in microstructure analysis, uncovering correlations between processing conditions and material performance for tailored applications. Gu et al. [22, 23] explored ML and biomimicry in the design of hierarchical materials with superior mechanical properties. The ML models optimise microstructural patterns, validated through AM and tensile testing, demonstrating improvements in material toughness and strength. The approaches expand design possibilities in fields like tissue engineering. Zhang et al. [24] employed a neuro-fuzzy ML model to predict high-cycle fatigue life in L-PBF stainless steel. By analysing datasets of varying process conditions and stresses, the model achieves high accuracy. Integrating literature data enhances prediction reliability, underscoring ML's role in fatigue property assessment. Bulgarevich et al. [25] applied Random Forest algorithms to classify steel microstructures from microscopy images, achieving high accuracy in segmenting structures like ferrite and martensite. The automated process enhances quality control and accelerates the development of new materials. Gola et al. [26] emphasised the importance of objective and reproducible microstructure classification for modern steels. Using SVMs and correlative microscopy, the study identifies key parameters that influence classification success, thereby aiding steel development and quality management.

Researchers underscore the importance of robust security frameworks in additive manufacturing to protect intellectual property and prevent cyberattacks. They also highlight the importance of anomaly detection systems for identifying irregularities in real time, ensuring the integrity and reliability of AM processes. AM systems' vulnerability to malicious attacks can result in internal defects. Wu et al. [27] used ML algorithms such as Random Forests and k-Nearest Neighbours to detect such anomalies, achieving high accuracy in defect identification. Future work will explore additional materials and security measures. Khanzadeh et al. [28] applied unsupervised ML techniques like Self-Organising Maps (SOM) to quantify geometric deviations in AM parts. By reducing dataset dimensions, the study links process conditions to geometric accuracy, facilitating precision improvements in AM. ML applications in AM face challenges such as limited training datasets, data variability, and high computational costs. Zhang et al. [24] and Scime et al. [19] emphasised the need for standardised datasets and further exploration of ML's capabilities in optimising materials, designs, and defect detection.

In this way, applications for machine learning in 3D printing are numerous and include process optimisation, in situ monitoring, and 3D printing design. ML has proven to be an effective tool for cybersecurity, real-time anomaly identification, design feature recommendation, and data-driven numerical simulation, as summarised in **Figure 16.2**. Particularly when working with high-dimensional data, machine learning has proven to perform better than traditional optimisation techniques like second-order polynomial regression. The most popular and effective machine learning method for process optimisation is ANN. Up to 98% accuracy can be attained with just a three-layer ANN. Because a convolutional neural network (CNN) can capture spatial characteristics, it is more effective than an ANN for 2D images and 3D models. CNN has thus been used for anomaly detection through in situ monitoring, feature recognition, and feature suggestion in 3D-printed item design.

Thus, AI/ML has been applied to additive manufacturing technology across various aspects. Their growth and development are discussed in this chapter through case studies explored next.

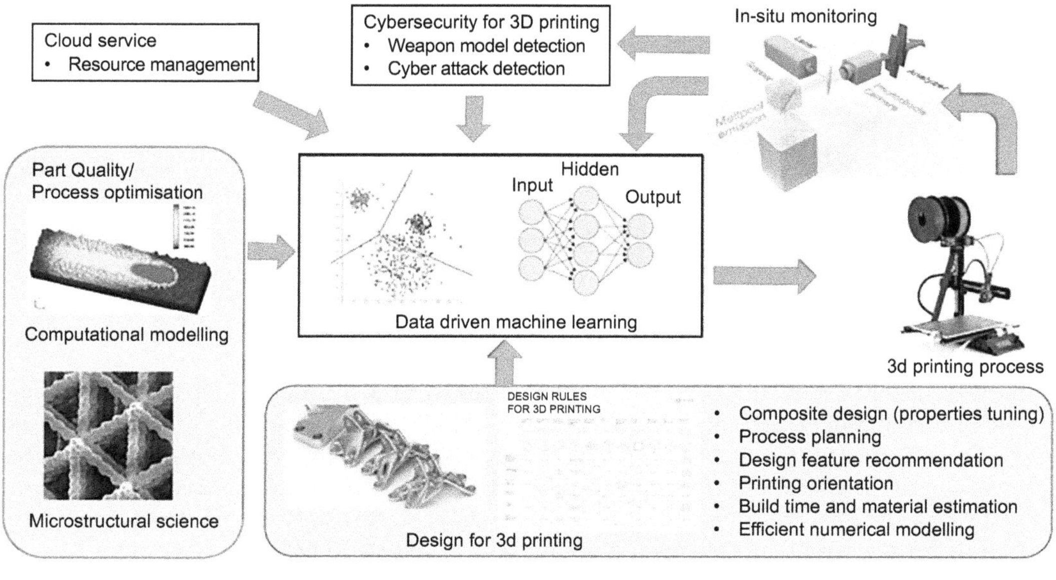

Figure 16.2 Summary of AI in additive manufacturing [29].

16.3 CASE STUDIES

16.3.1 Evaluation and Enhancement of Bead Geometry Prediction

Among the challenges faced by wire arc additive manufacturing (WAAM) are large residual stresses and insufficient deposition accuracy [30–32]. Researchers have modelled the connection between process parameters and bead shape to overcome these obstacles using Machine Learning approaches [33].

The goal of the project is to evaluate and enhance the transferability of machine learning models for predicting bead geometry. To assess and improve model transferability, it first gathers data from multiple public sources, verifies the data for physical coherence, and then conducts four investigations. In the first study, ML models are trained and tested on each dataset separately; in the second, only fully transferable input features are used; and in the third, models are trained on all but one dataset to evaluate transferability to the remaining datasets. Physics-informed ML (PIML) is incorporated in the fourth study by incorporating mass conservation principles into the loss function.

Nevertheless, current machine learning models for predicting bead shape are often overfit to the specific research testbed that generated the training data and are not transferable to different WAAM configurations.

a. Integrating data from several published sources to produce a more comprehensive and transferable machine learning model, this study seeks to increase the transferability of bead geometry prediction models.
b. Evaluating how well various ML models, including those informed by physics, transfer across different WAAM process parameters and configurations. The authors' approach was as mentioned next.

Following a physical coherence check to ensure data quality, the authors gathered data from various published sources on bead geometry prediction and conducted four studies to evaluate and enhance the transferability of the models:

a. Training and testing machine learning models (ANN and Ridge regression) on the full feature space of each dataset separately;
b. Only the completely transferable input characteristics (wire feed speed and travel speed) are used for model testing and training across datasets.
c. Testing transferability to the remaining dataset, including the impact of superordinate characteristics such as wire diameter and material parameters, after training the models on all but one dataset.
d. Using physics-informed machine learning by adding mass conservation into the loss function, the models were trained on all but one dataset.

Results obtained with this case study are as follows:

- With a mean absolute percentage error (MAPE) of 5.3% and R^2 of 0.69, the Ridge regression model fared better than the ANN model when trained on individual datasets.
- The transferability performance was inconsistent when only the transferable features of wire feed speed and travel speed were included; some models performed poorly on foreign datasets.
- Adding the wire diameter-related parameter "Provided Material" and combining data from other datasets enhanced the models' transferability.

- Comparing PIML training to ordinary ML models, the transferability was not considerably enhanced by including mass conservation in the loss function.
- When evaluated on a separate experimental dataset, the top model showed good transferability across several WAAM settings, achieving a MAPE of about 10%.

16.3.2 Validation of the Proposed AI-Based Integrated Framework for Lean and Smart Manufacturing

In the rail industry, AI technology for predictive maintenance was also presented in this research to introduce students and manufacturing staff to new technologies and how they apply to production. Specifically, the presentation focuses on how AI can inform preventative maintenance by estimating a component's remaining useful life (RUL). Wheel-bearing temperature history data that was trained in the MATLAB 2020b environment was used to assess the prediction model's performance [34]. A machine's or component's expected operating time before failing is estimated by the RUL. The RUL of a component may therefore be used to help staff plan maintenance, maximise operating effectiveness, and prevent unscheduled maintenance or downtime [35].

This research provides information on the wheel bearing's RUL. The deterioration model for the railcar high-speed bearing and the supplied threshold data served as the foundation for the RUL forecast. The outcomes of the predictive maintenance of a wheel bearing's RUL are shown in **Figure 16.3**. For monitoring, the anticipated RUL value was calculated and shown. The RUL gradually decreases, and the confidence limits on the anticipated RUL narrow whenever a slope is identified in the plot. The data enables time-series models to be fitted to condition indicators derived from sensor temperature data over time.

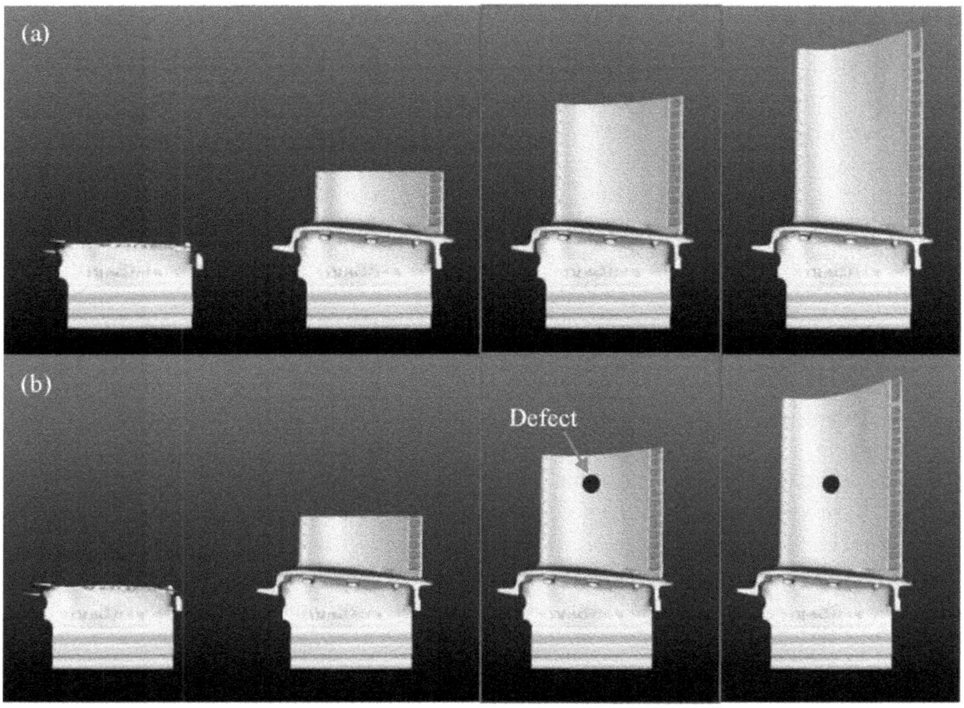

Figure 16.3 The RUL of the wheel bearing [33].

The degradation model forecasts when the condition indicator will surpass the threshold to estimate the RUL. The bearing's current state data and an exponential deterioration model were used to forecast the RUL, with a 95% confidence interval of 450–600 days. **Figure 16.3** displays the predictive model, which provides the slope detection instant, the confidence bound, and the anticipated RUL. According to the results, the wheel bearing's RUL is set at 500 hours for the ensuing 40 days. This suggests that the bearing will only survive 500 hours before failing to function for the next 40 days of operation.

The results of this investigation are consistent with those of Daniyan et al. [36], who documented a comparable predictive maintenance task utilising temperature-time graphs for wheel bearings. The findings showed that another method for estimating the RUL of a part, like a wheel bearing, is to use a temperature-time plot. Out of the four bearings that were examined, the fourth one shows a healthy condition, while the first three exhibit indications of approaching failure. Their RUL was established to prevent catastrophic failure.

16.3.3 Defect Detection

The intricacy of AM components and the limited reproducibility of AM procedures make the geometrical quality assessment for AM extremely difficult. The primary characteristics used by current 3D defect identification methods are the point-to-point distance differences between the made and designed items. This work included a new curvature feature called the discrete mean curvature measure in the training data for ML algorithms to collect macro-level information that goes beyond distances [37, 38].

Defects, including those in AM components, may be found using ML methods. SVM, a supervised machine learning method, was proposed for detecting defects in AM components in real time using RGB data, without providing an accuracy assessment. In their earlier work [38], we built machine learning models using 3D point cloud data based only on distance characteristics. They think that curvature is a crucial characteristic for determining the continuity of a mesh's neighbouring vertices. For things that should be flat, such as pipes or ceramic tiles. **Figure 16.4** represents the phases in additive-manufactured turbine blades with a proper blade and a defective blade with a hemispherical hole in the body of the blade. Golkar et al. employed one-dimensional continuous curvature as the only characteristic to detect a flatness fault in the literature. They did not apply any sophisticated analytical methods to more complex forms. Instead of utilising curvature solely for 3D purposes, Li et al. used it as an additional element in this study. The discrete curvature measure has been used as a feature in machine learning methods because 3D meshes lack a continuous surface.

Figure 16.5 illustrates the schematic proposed framework based on a synthetic dataset and machine learning. The framework is a flowchart of automated defect detection using an ML technique. The three primary components of the defect detection framework are (1) data generation/collection, (2) data preparation, and (3) model analysis. Each section is explored in greater detail.

The data collection process includes scanning experimental builds and generating fake data. After that, both datasets undergo data preprocessing. The basic mesh was a barrel mesh obtained from the internet and scaled to fit a cubic box with a 50 mm edge length. Synthetic data were generated for training and validation to ensure adequate data for machine learning. The AM original planned mesh and other faulty meshes (such as tiny spherical, cuboidal, or polyhedral flaws) may be combined or subtracted to generate groups of synthetic 3D meshes with defects. Due to their practicality and simplicity, hemispherical flaws were chosen in this investigation. Such elementary faults can be considered composites of more complex defects. A hemispherical defect's radius determines its size. Due to the prevalence of errors of this magnitude in AM practice, the faulty radii in this research varied from 1.5 to 2 mm.

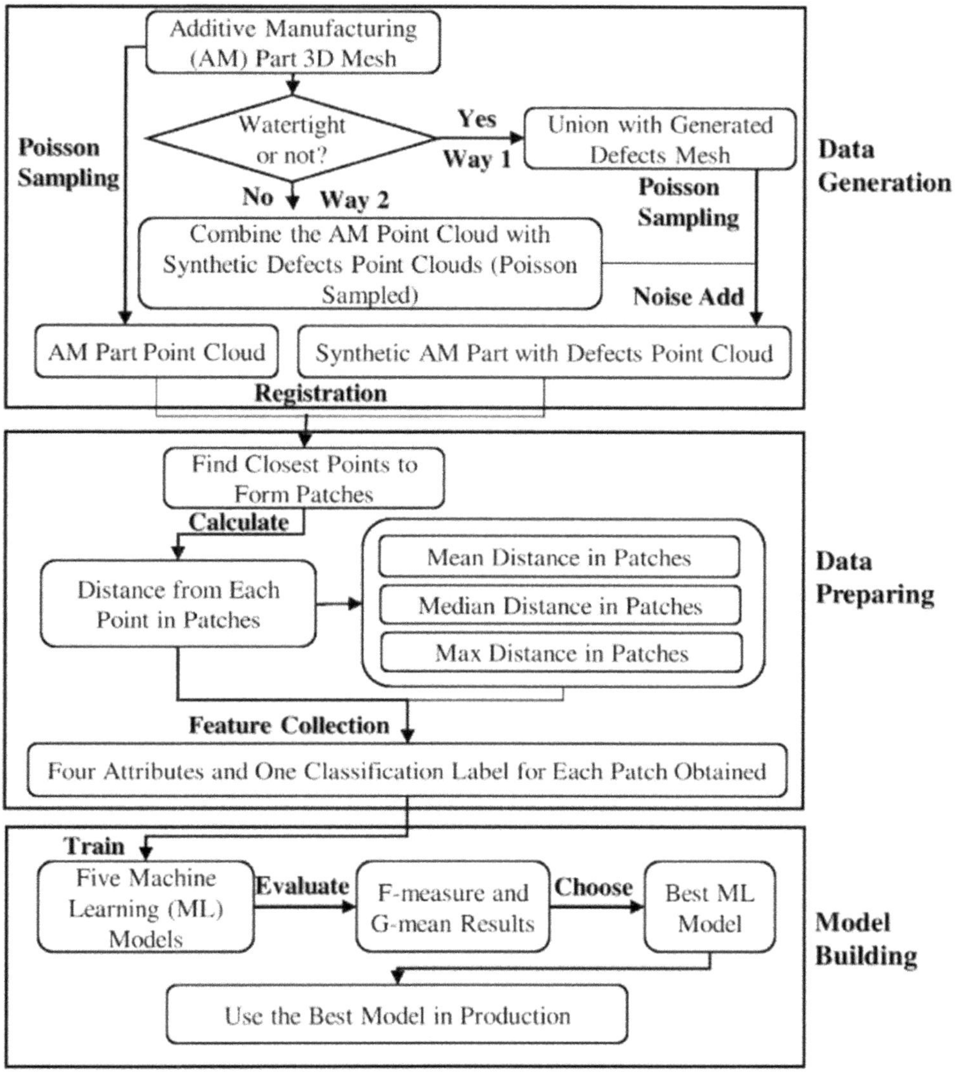

Figure 16.4 Phases in additive-manufactured turbine blade: (a) Proper blade, (b) Blade with a hemispherical defect [37].

The data pre-processing step in **Figure 16.6** receives input data comprising both scanned and synthetically generated meshes. Synthetic/experimental faulty meshes are aligned with well-constructed models using Open3D's Iterative Closest Point (ICP) Registration at Cornell University. For every point in the faulty mesh, a point-to-point distance is calculated, which is the separation between the nearest point pairs in the perfect and defective vertices. Next, neighbourhood information is captured using the patch idea, first introduced in ML models [39]. In essence, a patch is a collection of points that are closest to the chosen point. Three characteristics are taken from each patch: the mean, median, and maximum of all distances.

Figure 16.7 (**a**) illustrates the procedure for creating a defective barrel mesh. After several attempts, it was discovered that curvature is significantly influenced by edge flaws and defect overlap.

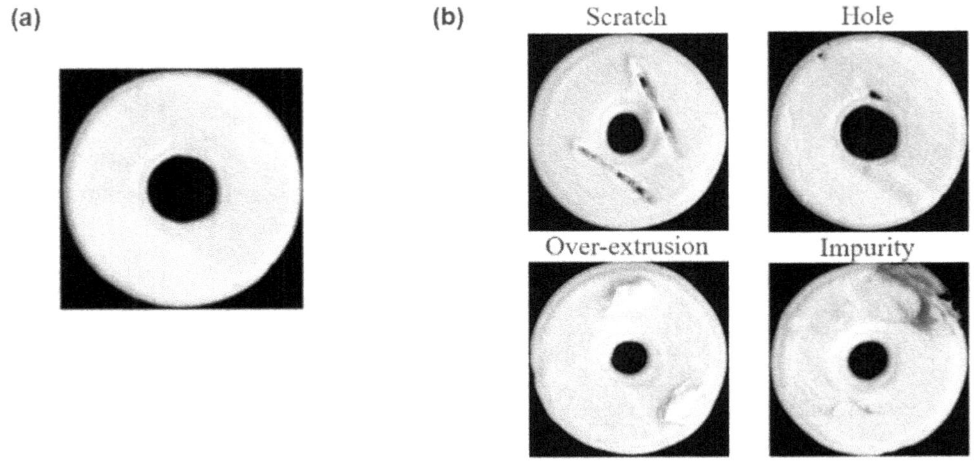

Figure 16.5 Schematic proposed framework based on a synthetic dataset and machine learning [37].

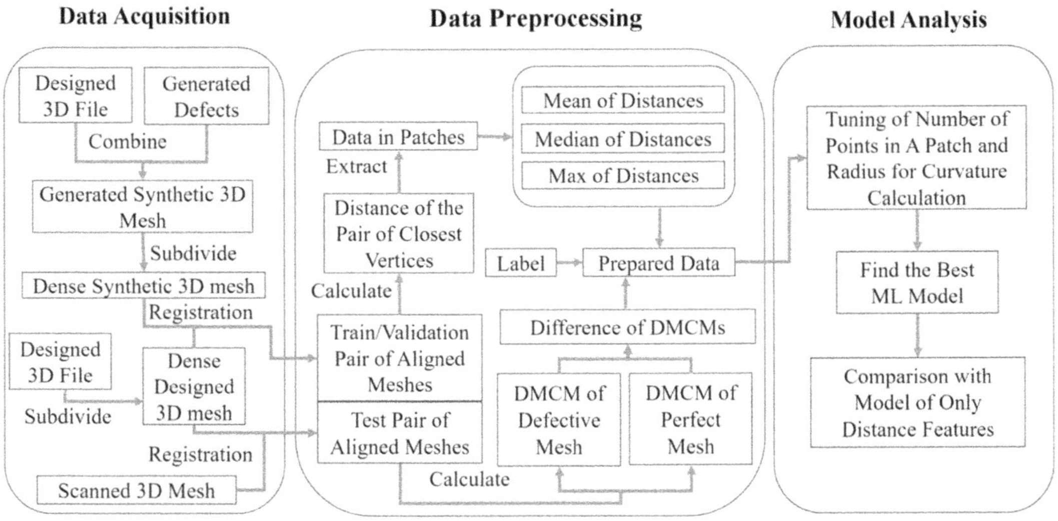

Figure 16.6 Machine learning-aided flow chart of automated defect detection [36].

First, two discrete curvature measures were examined in this study: the discrete Gaussian curvature measure (DGCM) and the discrete mean curvature measure (DMCM). Because of its stability on flat surfaces, DMCM is a better option, as demonstrated by experimental results on a flat sample. This study showed that detection accuracy increased when DMCM was included as an additional feature alongside earlier distance characteristics. For DMCM, the curvature radius was adjusted for varying fault sizes. Random Forest outperformed the other five machine learning methods, including Bagging of Trees, Gradient Boosting, Random Forest, Linear Support Vector Machine, and K-Nearest Neighbours.

DMCM is obtained by finding the vertex in the perfect mesh closest to each vertex in the defective mesh. After calculating the DMCMs at these two vertices, the difference between them is utilised as a feature in machine learning. A threshold is set to show the DMCM

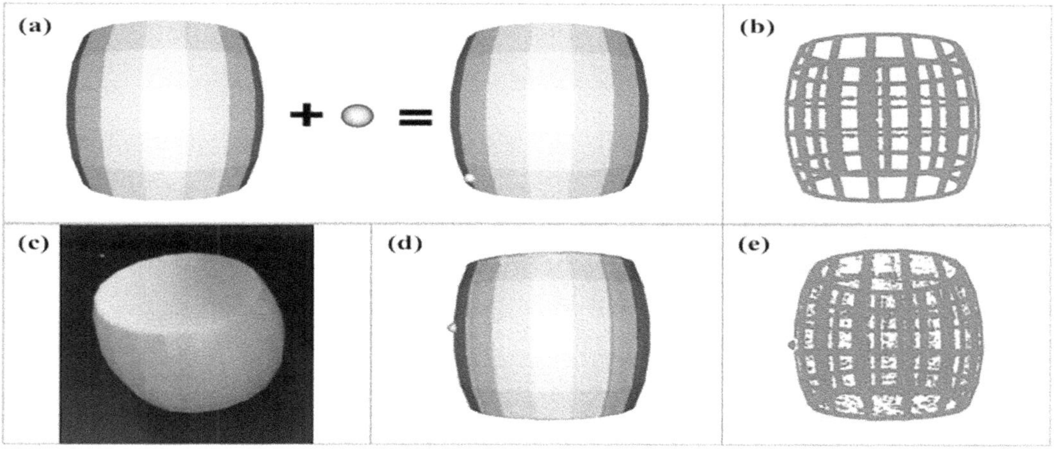

Figure 16.7 Representation of synthetic defective barrel: (a) Synthetic mesh model, (b) DMCM display, (c) Experimental defective sample, (d) Canned mesh of experimental sample, (c) DMCM display of experimental scanned data [36].

values on the barrel meshes, and values above the threshold are indicated at their calculation positions. The target mesh's condition is shown in **Figure 16.7** (**b**), whereas the source mesh's is shown in **Figure 16.7** (**e**). The last characteristic of the data used as input in the machine learning models will be the difference.

Thus, this research suggests an ML-based method that trains on synthetic data to identify geometric flaws in AM components. Time and money can be saved significantly by using synthetic data instead of printing and scanning multiple physical components for every design during training. Using the novel "patch" idea to incorporate data from neighbouring spots, the suggested approach improves detection accuracy over the conventional statistical method. To achieve in-situ detection in real-time AM production, the trained ML models may be called and executed rapidly, as they can be trained using synthetic data and saved in a ".sav" file. The authors suggest adding a feature that assesses local curvatures to machine learning models to discover geometric faults in AM.

16.3.4 Counteracting Thermal Distortion for Quality Control

Variation in production quality between machines within the same process, or even from construction to production, is a major obstacle to AM quality assurance. It may lead to differences in material characteristics, density, process stability, and dimensional accuracy. Therefore, much research is being conducted to use ML algorithms to assess the quality of AM parts. Three techniques for removing geometric flaws are using quality metrics, changing the primary Computer-Aided Design (CAD), and rescaling entire components. A multi-layer perceptron (MLP) or a CNN can predict the scaling ratio and adjust the overall component size before manufacture. Using ML techniques, shape-dependent geometric alterations caused by heat stress can be modelled, enabling the CAD model to receive the appropriate geometric corrections. MLP was used to correct geometric distortion [41] to reduce the thermal effects caused by SLM. To anticipate the deformed regions and modify the existing CAD data [42], a finite element model (FEM) was generated, as shown in **Figure 16.8**. In FDM printing, Noriega et al. used a similar technique, substituting test results for simulated data. SOM may connect different geometric aberrations for particular processing parameters to provide process control. This approach can reduce the amount of cloud data needed to assess

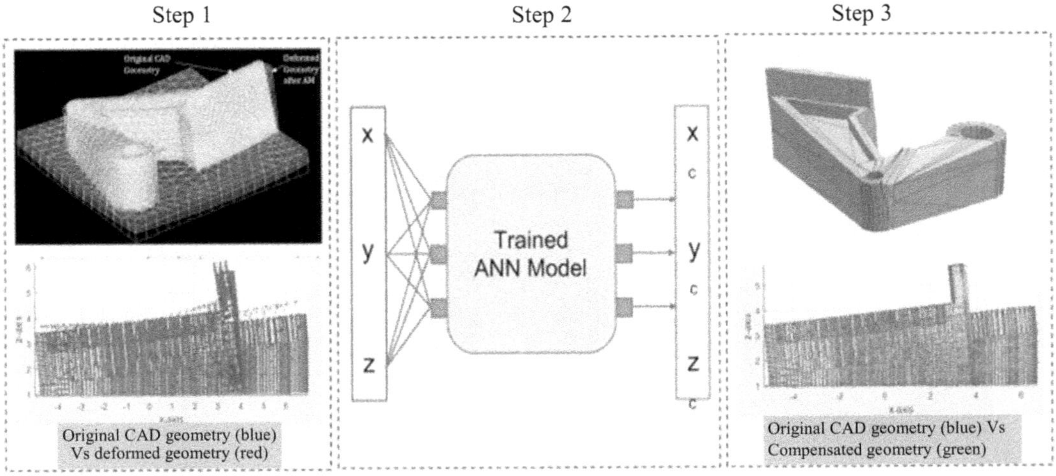

Figure 16.8 ML methodology: thermal distortion in SLM process [39, 40].

the dimensional correctness of additively manufactured components when combined with a laser scanner, compared with several existing supervised machine learning techniques.

16.4 CONCLUSION

This chapter examines case studies that highlight significant developments, challenges, and practical applications of artificial intelligence and machine learning. As it tackles challenges such as data scarcity, bias, and interpretability, it showcases significant advances in model design, training techniques, and computational efficiency. The chapter also examines the creative ways in which industry experts and researchers have addressed these problems.

It also examines AI/ML applications across a range of industries, including natural language processing, healthcare, finance, and autonomous systems, showcasing their revolutionary impact and promising future. The chapter concludes with ways to develop AI/ML, emphasising multidisciplinary approaches, addressing ethical issues, and enhancing model robustness. It offers a road map for ongoing progress in the sector by fusing technical insights with pragmatic concerns. The case studies provide valuable information about the advancement of the technology and highlight significant developments that have contributed to its growth. They also examine the primary obstacles that researchers and business professionals must overcome, along with the innovative approaches used to address them. Furthermore, the technology's practical uses in various domains are examined, emphasising its importance and potential for growth.

ACKNOWLEDGEMENT

This work was supported by the Deanship of Scientific Research, Vice Presidency for Graduate Studies and Scientific Research, King Faisal University, Saudi Arabia [Grant No. KFU262370].

REFERENCES

[1] T. Wuest, D. Weimer, C. Irgens, and K.-D. Thoben, "Machine Learning in Manufacturing: Advantages, Challenges, and Applications," *Prod. Manuf. Res.*, vol. 4, no. 1, pp. 23–45, 2016, https://doi.org/10.1080/21693277.2016.1192517.

[2] S. Inayathullah and R. Buddala, "Review of Machine Learning Applications in Additive Manufacturing," *Results Eng.*, vol. 25, no. November 2024, p. 103676, 2025, https://doi.org/10.1016/j.rineng.2024.103676.

[3] S. Goka, G. S. Narayana, G. Divya Jyothi, H. S. Shaik, and S. Q. Moinuddin, "AI and ML in Welding Technologies," in *Automation in Welding Industry*, Wiley, 2024, pp. 73–90, https://doi.org/10.1002/9781394172948.ch5.

[4] C. M. Thakar, R. J. M. Ventayen, E. H. Ramirez-Asis, J. E. Vilchez-Carcamo, M. E. Maguiña-Palma, and A. Thommandru, "An Investigation on Impact of Machine Learning in Additive Manufacturing," 2023, pp. 307–313, https://doi.org/10.1007/978-981-19-0108-9_32.

[5] S. Goka, M. Mallaiah, and M. J. Davidson, "Optimization of Bead Morphology for Cold Metal Transfer Wire Arc Additive Manufacturing of AZ31 Magnesium Alloy Wires," 2024, pp. 129–137, https://doi.org/10.1007/978-3-031-50240-8_27.

[6] S. Goka, S. Q. Moinuddin, M. Cheepu, and A. K. Dewangan, "Welding Practices in Industry 5.0," in *Automation in Welding Industry*, Wiley, 2024, pp. 263–279, https://doi.org/10.1002/9781394172948.ch15.

[7] R. Nascimento, I. Martins, T. A. Dutra, and L. Moreira, "Computer Vision Based Quality Control for Additive Manufacturing Parts," *Int. J. Adv. Manuf. Technol.*, vol. 124, no. 10, pp. 3241–3256, 2023, https://doi.org/10.1007/s00170-022-10683-5.

[8] A. García-Collado, J. M. Blanco, M. K. Gupta, and R. Dorado-Vicente, "Advances in Polymers Based Multi-Material Additive-Manufacturing Techniques: State-of-Art Review on Properties and Applications," *Addit. Manuf.*, vol. 50, p. 102577, 2022, https://doi.org/10.1016/j.addma.2021.102577.

[9] J. Liu, J. Ye, D. Silva Izquierdo, A. Vinel, N. Shamsaei, and S. Shao, "A Review of Machine Learning Techniques for Process and Performance Optimization in Laser Beam Powder Bed Fusion Additive Manufacturing," *J. Intell. Manuf.*, vol. 34, no. 8, pp. 3249–3275, 2023, https://doi.org/10.1007/s10845-022-02012-0.

[10] Z. Wang et al., "Data-Driven Modeling of Process, Structure and Property in Additive Manufacturing: A Review and Future Directions," *J. Manuf. Process.*, vol. 77, no. March, pp. 13–31, 2022, https://doi.org/10.1016/j.jmapro.2022.02.053.

[11] S. Kumar et al., "Machine Learning Techniques in Additive Manufacturing: A State of the Art Review on Design, Processes and Production Control," *J. Intell. Manuf.*, vol. 34, no. 1, pp. 21–55, 2023, https://doi.org/10.1007/s10845-022-02029-5.

[12] C. Wang, X. P. Tan, S. B. Tor, and C. S. Lim, "Machine Learning in Additive Manufacturing: State-of-the-Art and Perspectives," *Addit. Manuf.*, vol. 36, no. August, p. 101538, 2020, https://doi.org/10.1016/j.addma.2020.101538.

[13] T. Wang, T. H. Kwok, C. Zhou, and S. Vader, "In-Situ Droplet Inspection and Closed-Loop Control System Using Machine Learning for Liquid Metal Jet Printing," *J. Manuf. Syst.*, vol. 47, no. May, pp. 83–92, 2018, https://doi.org/10.1016/j.jmsy.2018.04.003.

[14] G. A. Susto, A. Schirru, S. Pampuri, S. McLoone, and A. Beghi, "Machine Learning for Predictive Maintenance: A Multiple Classifier Approach," *IEEE Trans. Ind. Informatics*, vol. 11, no. 3, pp. 812–820, 2015, https://doi.org/10.1109/TII.2014.2349359.

[15] D. Wu, C. Jennings, J. Terpenny, R. X. Gao, and S. Kumara, "A Comparative Study on Machine Learning Algorithms for Smart Manufacturing: Tool Wear Prediction Using Random Forests," *J. Manuf. Sci. Eng. Trans. ASME*, vol. 139, no. 7, 2017, https://doi.org/10.1115/1.4036350.

[16] M. E. Lynch, S. Sarkar, and K. Maute, "Machine Learning to Aid Tuning of Numerical Parameters in Topology Optimization," *J. Mech. Des.*, vol. 141, no. 11, 2019, https://doi.org/10.1115/1.4044228.

[17] X. Lei, C. Liu, Z. Du, W. Zhang, and X. Guo, "Machine Learning-Driven Real-Time Topology Optimization Under Moving Morphable Component-Based Framework," *J. Appl. Mech. Trans. ASME*, vol. 86, no. 1, 2019, https://doi.org/10.1115/1.4041319.

[18] X. Yao, S. K. Moon, and G. Bi, "A Hybrid Machine Learning Approach for Additive Manufacturing Design Feature Recommendation," *Rapid Prototyping Journal*, vol. 23, no. 6, 2017, https://doi.org/10.1108/RPJ-03-2016-0041.

[19] L. Scime and J. Beuth, "Using Machine Learning to Identify In-Situ Melt Pool Signatures Indicative of Flaw Formation in a Laser Powder Bed Fusion Additive Manufacturing Process," *Addit. Manuf.*, vol. 25, no. September 2018, pp. 151–165, 2019, https://doi.org/10.1016/j.addma.2018.11.010.

[20] I. A. Okaro, S. Jayasinghe, C. Sutcliffe, K. Black, P. Paoletti, and P. L. Green, "Automatic Fault Detection for Laser Powder-Bed Fusion Using Semi-Supervised Machine Learning," *Addit. Manuf.*, vol. 27, no. December 2018, pp. 42–53, 2019, https://doi.org/10.1016/j.addma.2019.01.006.

[21] A. Douard, C. Grandvallet, F. Pourroy, and F. Vignat, "An Example of Machine Learning Applied in Additive Manufacturing," *IEEE Int. Conf. Ind. Eng. Eng. Manag.*, vol. 2019-December, pp. 1746–1750, 2018, https://doi.org/10.1109/IEEM.2018.8607275.

[22] G. X. Gu, C. T. Chen, D. J. Richmond, and M. J. Buehler, "Bioinspired Hierarchical Composite Design Using Machine Learning: Simulation, Additive Manufacturing, and Experiment," *Mater. Horizons*, vol. 5, no. 5, pp. 939–945, 2018, https://doi.org/10.1039/c8mh00653a.

[23] G. X. Gu, C. T. Chen, and M. J. Buehler, "De Novo Composite Design Based on Machine Learning Algorithm," *Extrem. Mech. Lett.*, vol. 18, pp. 19–28, 2018, https://doi.org/10.1016/j.eml.2017.10.001.

[24] M. Zhang et al., "High Cycle Fatigue Life Prediction of Laser Additive Manufactured Stainless Steel: A Machine Learning Approach," *Int. J. Fatigue*, vol. 128, no. May, p. 105194, 2019, https://doi.org/10.1016/j.ijfatigue.2019.105194.

[25] D. S. Bulgarevich, S. Tsukamoto, T. Kasuya, M. Demura, and M. Watanabe, "Pattern Recognition with Machine Learning on Optical Microscopy Images of Typical Metallurgical Microstructures," *Sci. Rep.*, vol. 8, no. 1, pp. 3–9, 2018, https://doi.org/10.1038/s41598-018-20438-6.

[26] J. Gola et al., "Objective Microstructure Classification by Support Vector Machine (SVM) Using a Combination of Morphological Parameters and Textural Features for Low Carbon Steels," *Comput. Mater. Sci.*, vol. 160, no. October 2018, pp. 186–196, 2019, https://doi.org/10.1016/j.commatsci.2019.01.006.

[27] M. Wu et al., "Detecting Attacks in Cybermanufacturing Systems: Additive Manufacturing Example," *MATEC Web Conf.*, vol. 108, pp. 8–11, 2017, https://doi.org/10.1051/matecconf/201710806005.

[28] M. Khanzadeh, P. Rao, R. Jafari-Marandi, B. K. Smith, M. A. Tschopp, and L. Bian, "Quantifying Geometric Accuracy with Unsupervised Machine Learning: Using Self-Organizing Map on Fused Filament Fabrication Additive Manufacturing Parts," *J. Manuf. Sci. Eng. Trans. ASME*, vol. 140, no. 3, 2018, https://doi.org/10.1115/1.4038598.

[29] G. D. Goh, S. L. Sing, and W. Y. Yeong, "A Review on Machine Learning in 3D Printing: Applications, Potential, and Challenges," *Artificial Intelligence Review*, vol. 54, no. 1, 2021, https://doi.org/10.1007/s10462-020-09876-9.

[30] S. Q. Moinuddin, M. M. Cheepu, and V. C. Kantumuchu, *Gas Metal Arc Welding: Metal Joining and Additive Manufacturing*. CRC Press, 2026.

[31] S. Goka, M. Manjaiah, M. J. Davidson et al., "Review: State-of-art for wire arc additive manufacturing of metallic alloys—insights into materials, methods, and future prospects," *J. Mater. Sci.,* vol. 61, pp. 23409–23439, 2026, https://doi.org/10.1007/s10853-026-13199-5

[32] Moinuddin Syed Quadir, Patel Md Saad, Goka Suresh, Khan Mohammad Faseeulla, Jribi Skander, and Alsharif Asim. "A Review on Microstructure and Magnetic Property Control in Additively Manufactured Magnetic Materials," *Advanced Engineering Materials*, p. e202502792, 2026, https://doi.org/10.1002/adem.202502792

[33] M. Gihr, A. Rashid, and S. N. Melkote, "Transferability of Bead Geometry Prediction Model in Wire Arc Additive Manufacturing," *Procedia CIRP*, vol. 126, pp. 615–620, 2024, https://doi.org/10.1016/j.procir.2024.08.249.

[34] F. Ale, I. Daniyan, O. Aderoba, and A. A. Adediran, "An Artificial Intelligence-based Integrated Framework for Lean and Smart Manufacturing: A Case Study of the Rail Industry," *Int. Conf.*

Sci. Eng. Bus. Driv. Sustain. Dev. Goals, SEB4SDG 2024, pp. 1–10, 2024, https://doi.org/10.1109/SEB4SDG60871.2024.10629752.

[35] X. Han, Z. Wang, M. Xie, Y. He, Y. Li, and W. Wang, "Remaining Useful Life Prediction and Predictive Maintenance Strategies for Multi-State Manufacturing Systems Considering Functional Dependence," *Reliab. Eng. Syst. Saf.*, vol. 210, p. 107560, June 2021, https://doi.org/10.1016/j.ress.2021.107560.

[36] I. Daniyan, K. Mpofu, M. Oyesola, B. Ramatsetse, and A. Adeodu, "Artificial Intelligence for Predictive Maintenance in the Railcar Learning Factories," *Procedia Manuf.*, vol. 45, pp. 13–18, 2020, https://doi.org/10.1016/j.promfg.2020.04.032.

[37] R. Li, M. Jin, Z. Pei, and D. Wang, "Geometrical Defect Detection on Additive Manufacturing Parts with Curvature Feature and Machine Learning," *Int. J. Adv. Manuf. Technol.*, vol. 120, no. 5–6, pp. 3719–3729, 2022, https://doi.org/10.1007/s00170-022-08973-z.

[38] S. Q. Moinuddin, S. S. Hameed, A. K. Dewangan, K. R. Kumar, and A. S. Kumari, "A study on weld defects classification in gas metal arc welding process using machine learning techniques," *Materials. Today: Proceedings*, vol. 43, pp. 623–628, 2021, https://doi.org/10.1016/j.matpr.2020.12.159

[39] R. Li, M. Jin, and V. C. Paquit, "Geometrical Defect Detection for Additive Manufacturing with Machine Learning Models," *Mater. Des.*, vol. 206, p. 109726, 2021, https://doi.org/10.1016/j.matdes.2021.109726.

[40] S. Chowdhury, K. Mhapsekar, and S. Anand, "Part Build Orientation Optimization and Neural Network-Based Geometry Compensation for Additive Manufacturing Process," *J. Manuf. Sci. Eng.*, vol. 140, no. 3, 2018, https://doi.org/10.1115/1.4038293.

[41] C. Wang, X. P. Tan, S. B. Tor, and C. S. Lim, "Machine Learning in Additive Manufacturing: State-of-the-Art and Perspectives," *Addit. Manuf.*, vol. 36, no. January, p. 101538, 2020, https://doi.org/10.1016/j.addma.2020.101538.

[42] S. Goka, S. Q. Moinuddin, A. K. Dewangan, S. H. Saheb, and B. Madhavi, "Augmented Reality in Computer-Aided Design (CAD)," in *Metaverse and Immersive Technologies*, Wiley, 2023, pp. 217–234, https://doi.org/10.1002/9781394177165.ch8.

Index

U

V

W

For Product Safety Concerns and Information please contact our EU representative GPSR@taylorandfrancis.com
Taylor & Francis Verlag GmbH, Kaufingerstraße 24, 80331 München, Germany

www.ingramcontent.com/pod-product-compliance
Ingram Content Group UK Ltd.
Pitfield, Milton Keynes, MK11 3LW, UK
UKHW050927191026
476747UK00005B/180
* 9 7 8 1 0 3 2 9 6 8 0 9 4 *